Area under the Standard Normal Curve to the Right of z (continued)

z	0.00	0.01	0.02	0.03	0.04	0.05	0.06	0.07	0.08	0.09
0.0	.5000	.4960	.4920	.4880	.4840	.4801	.4761	.4721	.4681	.4641
0.1	.4602	.4562	.4522	.4483	.4443	.4404	.4364	.4325	.4286	.4247
0.2	.4207	.4168	.4129	.4090	.4052	.4013	.3974	.3936	.3897	.3859
0.3	.3821	.3783	.3745	.3707	.3669	.3632	.3594	.3557	.3520	.3483
0.4	.3446	.3409	.3372	.3336	.3300	.3264	.3228	.3192	.3156	.3121
0.5	.3085	.3050	.3015	.2981	.2946	.2912	.2877	.2843	.2810	.2776
0.6	.2743	.2709	.2676	.2643	.2611	.2578	.2546	.2514	.2483	.2451
0.7	.2420	.2389	.2358	.2327	.2297	.2266	.2236	.2206	.2177	.2148
0.8	.2119	.2090	.2061	.2033	.2005	.1977	.1949	.1922	.1894	.1867
0.9	.1841	.1814	.1788	.1762	.1736	.1711	.1685	.1660	.1635	.1611
1.0	.1587	.1562	.1539	.1515	.1492	.1469	.1446	.1423	.1401	.1379
1.1	.1357	.1335	.1314	.1292	.1271	.1251	.1230	.1210	.1190	.1170
1.2	.1151	.1131	.1112	.1093	.1075	.1056	.1038	.1020	.1003	.0985
1.3	.0968	.0951	.0934	.0918	.0901	.0885	.0869	.0853	.0838	.0823
1.4	.0808	.0793	.0778	.0764	.0749	.0735	.0721	.0708	.0694	.0681
1.5	.0668	.0655	.0643	.0630	.0618	.0606	.0594	.0582	.0571	.0559
1.6	.0548	.0537	.0526	.0516	.0505	.0495	.0485	.0475	.0465	.0455
1.7	.0446	.0436	.0427	.0418	.0409	.0401	.0392	.0384	.0375	.0367
1.8	.0359	.0351	.0344	.0336	.0329	.0322	.0314	.0307	.0301	.0294
1.9	.0287	.0281	.0274	.0268	.0262	.0256	.0250	.0244	.0239	.0233
2.0	.0228	.0222	.0217	.0212	.0207	.0202	.0197	.0192	.0188	.0183
2.1	.0179	.0174	.0170	.0166	.0162	.0158	.0154	.0150	.0146	.0143
2.2	.0139	.0136	.0132	.0129	.0125	.0122	.0119	.0116	.0113	.0110
2.3	.0107	.0104	.0102	.0099	.0096	.0094	.0091	.0089	.0087	.0084
2.4	.0082	.0080	.0078	.0075	.0073	.0071	.0069	.0068	.0066	.0064
2.5	.0062	.0060	.0059	.0057	.0055	.0054	.0052	.0051	.0049	.0048
2.6	.0047	.0045	.0044	.0043	.0041	.0040	.0039	.0038	.0037	.0036
2.7	.0035	.0034	.0033	.0032	.0031	.0030	.0029	.0028	.0027	.0026
2.8	.0026	.0025	.0024	.0023	.0023	.0022	.0021	.0021	.0020	.0019
2.9	.0019	.0018	.0018	.0017	.0016	.0016	.0015	.0015	.0014	.0014
3.0	.0013	.0013	.0013	.0012	.0012	.0011	.0011	.0011	.0010	.0010
3.1	.0010	.0009	.0009	.0009	.0008	.0008	.0008	.0008	.0007	.0007
3.2	.0007	.0007	.0006	.0006	.0006	.0006	.0006	.0005	.0005	.0005
3.3	.0005	.0005	.0005	.0004	.0004	.0004	.0004	.0004	.0004	.0003
3.4	.0003	.0003	.0003	.0003	.0003	.0003	.0003	.0003	.0003	.0002
3.5	.0002	.0002	.0002	.0002	.0002	.0002	.0002	.0002	.0002	.0002

STATISTICS
Learning in the Presence of Variation

STATISTICS
Learning in the Presence of Variation

Robert L. Wardrop

Statistics Department
University of Wisconsin–Madison

WCB Wm. C. Brown Publishers
Dubuque, Iowa • Melbourne, Australia • Oxford, England

Book Team

Editor *Paula-Christy Heighton*
Developmental Editor *Daryl Bruflodt*
Publishing Services Coordinator/Design *Barbara J. Hodgson*

Wm. C. Brown Publishers
A Division of Wm. C. Brown Communications, Inc.

Vice President and General Manager *Beverly Kolz*
Vice President, Publisher *Earl McPeek*
Vice President, Director of Sales and Marketing *Virginia S. Moffat*
Vice President, Director of Production *Colleen A. Yonda*
National Sales Manager *Douglas J. DiNardo*
Marketing Manager *Julie Joyce Keck*
Advertising Manager *Janelle Keeffer*
Production Editorial Manager *Renée Menne*
Publishing Services Manager *Karen J. Slaght*
Permissions/Records Manager *Connie Allendorf*

Wm. C. Brown Communications, Inc.

President and Chief Executive Officer *G. Franklin Lewis*
Corporate Senior Vice President, President of WCB Manufacturing *Roger Meyer*
Corporate Senior Vice President and Chief Financial Officer *Robert Chesterman*

Production/design by Publication Services

Copyright © 1995 by Wm. C. Brown Communications, Inc. All rights reserved

A Times Mirror Company

Library of Congress Catalog Card Number: 93–74864

ISBN 0–697–21593–8

No part of this publication may be reproduced, stored in a retrieval system, or transmitted, in any form or by any means, electronic, mechanical, photocopying, recording, or otherwise, without the prior written permission of the publisher.

Printed in the United States of America by Wm. C. Brown Communications, Inc., 2460 Kerper Boulevard, Dubuque, IA 52001

10 9 8 7 6 5 4 3 2 1

To Roger Wardrop, Helen Wardrop, and the memory of William Harold Wardrop.

CONTENTS

Preface — xv

I DICHOTOMOUS RESPONSES — 1

1 The Completely Randomized Design — 2

1.1 Introduction 3
   1.1.1 Assigning Subjects at Random 7
   1.1.2 Two Additional Examples 11
   1.1.3 Observational Studies 12

1.2 Data Presentation and Summary 13
   1.2.1 Medical Studies 18
   1.2.2 Additional Notation 22

1.3 A Sequence of Trials 28
   1.3.1 Why Randomize a Sequence of Trials? 31
   1.3.2 Observational Studies 32

References 36

2 Hypothesis Testing — 38

2.1 Fisher's Test: Getting Started 39

2.2 Step 1: The Hypotheses 42

2.3 Step 2: The Test Statistic and Its Sampling Distribution 49
2.3.1 The Test Statistic 50
2.3.2 Probability 50

2.4 Working with Probability 58
2.4.1 The Meaning of Probability 58
2.4.2 The Rules of Probability 60

2.5 Steps 3 and 4 65
2.5.1 Step 3: The Rule of Evidence 65
2.5.2 Step 4: The P-value 68
2.5.3 Interpretation of the P-value 70
2.5.4 Miscellaneous Remarks 72

2.6 Beyond Brute Force 74

References 77

3 Approximating a Sampling Distribution   78

3.1 The Probability Histogram 79
3.2 The Simulation Experiment 83
3.3 Measures of Center and Spread 88
3.4 The Standard Normal Curve 97
3.5 A Comparison of Methods 105
3.6 Computing the Mean and Standard Deviation* 108
3.6.1 The Mean 108
3.6.2 The Standard Deviation 110

4 The Randomized Block Design   114

4.1 Data Presentation and Summary 116
4.1.1 Medical Studies 116
4.1.2 Combining Similar Studies 119
4.1.3 The Collapsed Table 121

4.2 The Mantel-Haenszel Test 123
4.2.1 Step 1: The Hypotheses 126
4.2.2 Step 2: The Test Statistic and Its Sampling Distribution 127
4.2.3 Step 3: The Rule of Evidence 128
4.2.4 Step 4: Computing the P-value 129
4.3 The Randomized Pairs Design 134
References 143

5 One-Population Models 144
5.1 Surveys 146
5.2 Probability 151
5.2.1 Dichotomous Boxes 151
5.2.2 The Multiplication Rule 153
5.2.3 The Binomial Distribution 157
5.3 Bernoulli Trials 167
5.4 Some Practical Considerations 178
5.4.1 Problems with Sampling Human Populations 178
5.4.2 Birthday Coincidences and Random Samples 180
References 182

6 Inference for a Population 184
6.1 Introduction 185
6.2 Confidence Intervals 187
6.2.1 Derivation of the 95 Percent Confidence Interval 189
6.2.2 Other Confidence Levels 191
6.3 More on Confidence Intervals 195
6.3.1 Interpretation of Confidence Intervals 195
6.3.2 The Half-Width 198
6.4 Hypothesis Testing 203
6.4.1 General Case 203
6.4.2 The Special Case of $p = 0.5$ 211

6.5 Predictions* 215
- 6.5.1 The Case in Which p Is Known 215
- 6.5.2 The Case in Which p Is Unknown 217

6.6 How Many Analyses?* 220
- 6.6.1 Hypothesis Testing, Independent Analyses 221
- 6.6.2 Hypothesis Testing, Dependent Analyses 223
- 6.6.3 Confidence Intervals 224

References 226

7 Comparing Two Populations — 228

7.1 The Four Types of Studies 229
- 7.1.1 Sampling 232
- 7.1.2 Data Presentation and Summary 233

7.2 Inference 235
- 7.2.1 Derivation of the Confidence Interval Formula* 238
- 7.2.2 Hypothesis Testing 239

7.3 Interpretation of Studies 245
- 7.3.1 Practical Importance 247
- 7.3.2 Statistical Power* 248
- 7.3.3 The Assumption of Independent Random Samples 251

References 255

8 Two Dichotomous Responses — 256

8.1 Introduction 257
8.2 Conditional Probability and Screening Tests 259
8.3 Estimating Probabilities and Conditional Probabilities .. 269
8.4 Hypothesis Testing and Confidence Intervals 274
- 8.4.1 Hypothesis Testing 274
- 8.4.2 Confidence Intervals 276

8.5 Comparing p_A with p_B 281

References 287

9 Adjusting for a Factor — 288

9.1 The Hazards of Collapsing Tables — 289
9.1.1 Simpson's Paradox — 289
9.1.2 Standardized Rates — 297

9.2 Hypothesis Testing — 300
9.2.1 Free Throws — 304
9.2.2 Elder Abuse — 304

References — 311

II MULTICATEGORY RESPONSES — 312

10 One Multicategory Response — 314

10.1 Data Presentation and Summary — 316

10.2 The Population Model and Estimation — 322
10.2.1 Multinomial Trials — 324
10.2.2 Estimation — 325
10.2.3 The Chi-squared Curves — 328

10.3 The Chi-squared Goodness of Fit Test — 334
10.3.1 Summary — 339
10.3.2 Canada Geese — 342
10.3.3 The Daily Game — 345
10.3.4 Degrees of Freedom — 346

References — 348

11 Tests of Homogeneity and Independence — 350

11.1 Controlled Studies — 352

11.2 Two or More Populations — 359

11.3 One Population, Two Responses — 362

11.4 The 2×2 Table Revisited* — 366

References — 375

xii • Contents

III NUMERICAL RESPONSES — 376

12 Describing One Numerical Response — 378

12.1 Introduction — 379

12.2 Plots and Histograms — 380
- 12.2.1 The Dot Plot — 381
- 12.2.2 Histograms — 385
- 12.2.3 Stem and Leaf Plots — 392
- 12.2.4 Unequal Class Intervals — 393

12.3 Measures of Center — 399

12.4 Measures of Spread — 414
- 12.4.1 The Box Plot — 416
- 12.4.2 The Sample Standard Deviation — 418
- 12.4.3 The Effect of Outliers — 425

12.5 Assorted Topics* — 430
- 12.5.1 Granular Distributions — 430
- 12.5.2 Combining Samples — 430
- 12.5.3 Change of Units — 432
- 12.5.4 Shortcut Formulas — 434

Reference — 436

13 Correlation and Regression — 438

13.1 Introduction — 439

13.2 The Scatterplot and the Correlation Coefficient — 441
- 13.2.1 The Scatterplot — 445
- 13.2.2 The Correlation Coefficient — 450

13.3 The Regression Line — 465
- 13.3.1 Comparing Two Prediction Lines — 465
- 13.3.2 The Least Squares Line — 468
- 13.3.3 The Regression Effect — 473
- 13.3.4 The Coefficient of Determination — 475

13.4	Residuals, Outliers, and Isolated Cases	481
13.4.1	Residuals and Outliers	481
13.4.2	The Residual Plot	484
13.4.3	Isolated Cases	485
13.5	Data from Two Sources*	494
	References	498

14 Time Series — 500

14.1	The Time Series Plot	502
14.2	Autocorrelation and Smoothing	507
14.2.1	Autocorrelation	507
14.2.2	Smoothing a Time Series	511
	References	518

15 Inference for One Numerical Population — 520

15.1	Introduction	521
15.1.1	Estimation of a Population	526
15.1.2	The Family of Normal Curves	530
15.1.3	Two Final Cautions	532
15.2	Inference for the Population Median	533
15.2.1	Confidence Interval for $n = 2$	533
15.2.2	Confidence Interval for Arbitrary n	535
15.2.3	Robustness of the Confidence Interval	538
15.2.4	Hypothesis Test for the Median	540
15.2.5	Robustness of the Test	541
15.3	Inference for the Population Mean	544
15.3.1	Confidence Interval for μ	547
15.3.2	Hypothesis Testing for μ	550
15.3.3	Robustness	555
15.4	Inference for Time Series Data*	564

15.5 Predictions* 569
   15.5.1 Predictions for a Normal Population 570
   15.5.2 Predictions for an Arbitrary Unknown PDF 573

References 576

16 Numerical Data from Two Sources   578

16.1 Comparison of Means for a CRD 580

16.2 Independent Random Samples 587
   16.2.1 Case 1: Normal Populations with Equal Variances 588
   16.2.2 Case 2: Normal Populations 592
   16.2.3 Case 3: Large Sample Approximation 594
   16.2.4 Comparison of Methods 597

16.3 The Randomized Pairs Design 602

16.4 A Population Model for Paired Data 607

References 614

Solutions to Odd-Numbered Exercises   615

Appendix   647

Index   657

PREFACE

Unarguably, there is a standard format for introductory statistics texts. The standard approach begins with descriptive statistics, followed by probability theory, including distributions and sampling theory, and finally proceeds to inference, but inference only for random samples from populations. To be sure, books differ in the amount of attention paid to these three parts, but every introductory text I have seen follows this standard format. Even a cursory examination of the table of contents reveals that this text does not resemble the standard format. Most of this preface is devoted to explaining how and why this book is different.

The primary difference between this text and others is the organization of the material; the secondary difference is the choice of topics. Regarding the choice of topics, this book covers the standard material, but much less emphasis is placed on probability and sampling theory, and much greater emphasis is placed on experimental design and simulation.

Regarding the organization of material, the standard approach separates topics by type of method. First the student learns a variety of ways to describe data. Next, the student learns the mathematical foundations of statistical inference for random samples. Finally, the student learns statistical inference for random samples. A particular method, such as the confidence interval for the difference of two means, is introduced with great attention paid to the mathematical assumptions underlying its derivation. Data are then trotted out to illustrate the use of the method. This organizational structure gives the student the very strong message that methods and their mathematical assumptions are much more important than the scientific issues that motivated the study that generated the data.

By contrast, this book separates topics by the type of data and then, within each type of data, by the type of study. I have sorted topics in this way because it helps me achieve my fundamental goal:

> To enable my students to discover that statistics can be an important tool in daily life.

I will explain how I achieve this goal by dramatizing the experiences of some students of mine. These stories will provide you with a rough outline of the organization of material in this text.

Chapter 1 introduces the completely randomized design (CRD) with two treatments and a dichotomous response. I challenge each student to find an issue *that interests him or her* and that can be investigated with a CRD. I then tell the students to investigate their issues.

Therese Nyswonger was interested in marital infidelity. More precisely, she wanted to know whether the gender of the cheater would influence a person's decision on whether to tell the wronged spouse. Therese decided it would be bad science to give each subject both versions of her question (the husband as the cheater and the wife as the cheater) and settled on a CRD. Therese's subjects were 20 female coworkers. Therese found that 70 percent of the subjects who were told the husband was cheating would tell the wife, compared with only 40 percent who would tell if the wife was cheating. Clearly, the version read did not *determine* the response, but there was evidence that the version read *influenced* the response. The meaning of the evidence, however, was not clear.

In Chapter 2 Therese learned of the Skeptic's Argument, which states that the pattern in her data is completely due to chance. More precisely, this argument says that the version read is totally irrelevant. It is merely the case, the Skeptic claims, that 11 of Therese's friends were "tellers," and nine were not. Confronted with either version, the tellers would tell and the others would not. Thus, the pattern in Therese's data, 70 percent versus 40 percent, is only the result of the chance assignment of seven tellers to the cheating husband version. Debating the Skeptic is the Advocate, who acknowledges that the Skeptic could be correct, but argues that the pattern is unlikely to be due to chance. The Skeptic and Advocate agree on the central role of chance, so Chapter 2 proceeds to give chance a careful definition, as well as enough of a development to address Therese's curiosity. This leads to hypothesis testing (Fisher's test) which yields the P-value, the number that quantifies the debate between the Skeptic and the Advocate. For Therese's data, the P-value supports the Skeptic. The difference between 70 and 40 percent is large, but it is obtained from too few subjects to be convincing.

Therese learned that computing the exact P-value can be very tedious, so Chapter 3 presents two ways to obtain an approximation, either through simulation or by using the standard normal curve. First, a computer simulation experiment is presented, and it is shown that 10,000 runs yield an excellent approximation to the sampling distribution of the test statistic. This is a pedagogically important topic because

- A simulation experiment is very intuitive; no fancy mathematical arguments were needed to convince Therese that looking at some assignments is a reasonable approximation to looking at all of them.
- Simulation is a powerful tool for studying robustness (Chapters 10 and 15), and power (Chapter 7).

If the probability histograms of the sampling distribution of the test statistic for Fisher's test are drawn for each of several studies, their most striking difference is in the amount of spread. If these test statistics are standardized, the resulting probability histograms are similar to each other and to the standard normal curve. Thus, a P-value can be approximated by using the standard normal curve. This approach has two noteworthy features. First, the standard normal curve is introduced as an approximation device without any reference to the abstract notion of a continuous random variable. Second, the student learns that the standard deviation is a reasonable way to measure spread because it works: standardizing by dividing by the standard deviation yields probability histograms that are similar.

Faced with the Chapter 1 challenge, Clyde Gaines opted for a study of his basketball shooting prowess. Clyde performed a CRD of his three-point basket shooting ability in a practice session by comparing 50 shots from in front of the basket to 50 shots from the left corner. Using the same techniques that Therese learned in Chapters 2 and 3, Clyde concluded that on the day of his experiment, there was no significant difference between the two locations.

In Chapter 5 Clyde learned about Bernoulli trials, which led to the questions, Were his shots from in front Bernoulli trials? Were his shots from the left corner Bernoulli trials? Chapter 5 also provided Clyde with some techniques for looking at his data to informally answer these questions. On the assumption of Bernoulli trials, Clyde learned in Chapter 6 how to obtain point and interval estimates of his probability of success from each location. He also learned how to use the results of his experimentation to compute point and interval predictions for the results of future experimentation.

Chapter 7 introduces the student to comparative observational studies and the two population model for a CRD. In this chapter the students learn from studies performed by others. None of the comparative studies in this text shows a deterministic relation between population and response. A good scientist will consider which other factors might influence the response and hence contribute to the lack of determinism. For a controlled study, there is reason to believe that the process of randomizing has achieved good balance in the distribution to treatments of these other factors. In fact, for a controlled study the P-value can be interpreted, in a certain sense, as the probability of a misleading assignment. For an observational study, however, there is no basis for this belief in approximate balance. As is demonstrated in Chapter 9, if a researcher controls for one or more factors, this lack of balance can result in a reversal of the pattern in the data (Simpson's paradox). It is important that users of statistics understand this basic weakness of observational studies.

Part II of this book, Chapters 10 and 11, provide a brief generalization of the material in Part I to multicategory responses.

In Part III, Chapters 12 through 16, the methods of Part I are extended to numerical responses. Like the first chapter, Chapter 12 begins by challenging the student to investigate an issue of interest. Sara Lamers performed an 80-trial CRD in which each trial consisted of hitting a golf ball with a 3-wood or a 3-iron and the response was the distance the ball traveled. Sara learned from her study by plotting the data and computing various summary measures. But Sara also found that her learning involved more than creating plots and computing numbers; she had to interpret the plots and numbers, and she had to make judgments. For example, the measures of center of the distributions suggested that she was better (hit the ball farther) with the wood. The various measures of spread were similar for the two distributions, and both distributions were skewed to the left, yielding no reason to question the superiority of the wood suggested by the measures of center. The wood, however, had more small outliers than the iron. Thus, Sara learned that if there is a nearby water hazard, it might pay to reach for the iron instead of the wood.

In Chapter 16 Sara learned how to perform a hypothesis test to compare her performance with the two clubs with and without the assumption of independent random samples from populations. In addition, on the assumption that the trials with a club were a random sample from a population, Sara learned in Chapter 15 how to obtain point and interval estimates of the mean and median of the club's population, and point and interval predictions of the distance of a future shot with the club.

Phil Coan and Jean Schoeni performed a randomized pairs design to compare the time required to complete two similar sterilization tasks in a laboratory. In Chapter 16 they learned how to analyze their data, with and without the assumption of a random sample from a population of differences.

Students who have access to a computer and statistical software can perform a project that utilizes the techniques of regression and correlation (Chapter 13). Susan Robords enjoyed camping and obtained data to investigate the accuracy of using the rate of cricket chirping to predict temperature.

I hope that these dramatizations have been helpful in conveying the approach to statistics used in this book. In addition to my fundamental goal, stated above, there are several features of this book that I consider to be important. They are described below.

STUDENT PROJECTS

The narrative and exercises contain descriptions of and data from approximately 80 projects conducted by my students, including the ones described above. I strongly advocate such active learning in statisics courses. Unfortunately, many teachers do not have the time to grade written reports. Teachers who class-tested this book without assigning projects have noted that their students enjoy reading about other student's projects. As an intermediate strategy, a teacher could reduce the required grading by assigning group projects or by requiring only a cursory verbal analysis.

There is a movement in higher education in the United States to teach quantitative reasoning and communication skills across the curriculum. Projects in a statisics course are an ideal way to achieve both objectives simultaneously.

EARLY INTRODUCTION TO INFERENCE

I believe it is important to challenge students early in a course. An early introduction to hypothesis testing certainly has that effect! In addition, the most highly motivated students enter an introductory course with an interest in learning inference, and they are typically frustrated by the standard format's delaying of this topic to the last third of the course. Finally, I believe that learning, especially the learning of difficult concepts, is enhanced by repetition. Inference appears early and throughout this book, providing the desired repetition.

EMPHASIS ON EXPERIMENTAL DESIGN

My emphasis on experimental design is a natural consequence of my emphasis on statistics as a tool for scientists. (I use the term *scientist* in the broad sense of anyone who wants to understand his or her environment.) Once an issue has been chosen for study, a researcher must decide how to collect information, that is, how to design an experiment. Experimental design encourages a researcher to think about sources of variation in the data and provides tools, randomization and blocking, to deal with that variation. By contrast, only teaching students how to compare two populations discourages thinking about variation, because identifying other sources of variation will lead to defining additional populations, something the student is not equipped to handle.

It is important for educated citizens to understand the difference between controlled and observational studies. The standard format, in which every inferential problem is stated in terms of random samples from populations, is poor at illuminating this distinction.

PRESENTING TOPICS BY TYPE OF RESPONSE

The quickest, and most unfair, analysis of this book is that nine chapters is too much coverage for dichotomous responses. The first nine chapters of this book introduce many of the fundamental ideas and methods of statistics *illustrated with* a dichotomous response. These ideas and methods are equally relevant for a study of multicategory and numerical responses. With the solid background of the first nine chapters, the material in Part III, on numerical responses, can be covered quite quickly. But I have only argued that my organization is not as bad as it might seem; I also must convince you that it is good.

For a dichotomous response, data presentation and summary are straightforward. This simplicity enables the first nine chapters of the book to focus on important statistical ideas without undue attention to issues that are basically mathematical or arithmetic. But there is another important point. For a dichotomous response a population is characterized completely by a single number, p. As a result, studying the population and studying p are equivalent. For a numerical response, however, the population need not be characterized by a single number. Thus data presentation and summary are not straightforward but require considerable judgment. I believe students are better equipped to make those judgments when they have the knowledge of what statistics can do that is provided in Part I of this book.

More importantly, inference for a numerical response is very complicated. To be sure, if a person is told that the data are a random sample from a normal population, inference is simple. But in real life nobody tells a researcher what the population is. Consider, for example, a researcher who wants to study a population. The first question is whether knowledge of the population mean or median, as opposed to the entire population, would be sufficiently insightful to the scientific issue that motivated the study. I am not arguing that the mean and median are not widely useful; but they are not universally useful. (Note that this issue does not arise for a dichotomous response.) Suppose the researcher decides that knowledge of the mean would be fruitful. The researcher will probably use a confidence interval based on the t-distribution, but the validity of the answer depends on a number of possible complicating contingencies, of which some require a careful examination of the data, some require the researcher to be knowledgeable in the subject area, and some may remain poorly understood.

My point is not that you must (or even should) overload an introductory course with all possible concerns about each set of data. I do believe, however, that it is wrong to present inference merely by saying, "If you have a random sample from a normal population, use the following formula." My point is that it is immensely helpful for students to have a strong understanding of hypothesis testing and confidence intervals before they are introduced to the difficulties inherent in numerical responses.

Finally, my decision to sort material by type of response is related to two features stated earlier. An early introduction to inference would not be possible if I first covered ways to present and summarize all types of responses. Without an early introduction to inference, either student projects would need to be delayed or their inference component would have to be removed. Both these modifications are unacceptable to me: the former because it is important to get students involved in the course early, and the latter because projects without inference are viewed as unsatisfactory and incomplete by students.

INTERDEPENDENCE OF CHAPTERS

The core material of this text is presented in Chapters 1 through 3, 5 through 7, 12, 13, 15, and 16. With the starred (optional) sections omitted, these 10 chapters would provide about the correct amount of material for a one-quarter course. In my one-semester course I have covered the entire book, but I generally prefer to omit one or two chapters.

The core chapters should be covered in numerical order, except that Chapter 13 can be delayed until after Chapter 15 or Chapter 16. The noncore chapters are independent of each other, except that Chapter 9 refers to Chapter 4, and Chapter 11 refers to Chapter 10. Also, not surprisingly, noncore chapters refer to earlier core chapters. Occasionally, a core chapter contains a brief reference to material covered in an earlier noncore chapter. To my knowledge, this has never resulted in any confusion for my students.

ACKNOWLEDGMENTS

In July 1990 I signed a contract with the C.V. Mosby Company to write this book. My thanks to Ed Murphy and J. P. Lenney for their support and encouragement. Thanks also to J. P. for having the wisdom to assign John Murdzek to be my developmental editor. John's careful attention to detail improved the manuscript immensely.

Late in 1992 I came to William C. Brown Publishers. I am impressed by and thankful for the enthusiasm and care that everyone at Wm. C. Brown has brought to this inherited project. Special thanks go to Jane Parrigin, Paula-Christy Heighton, Marilyn Sulzer, Barb Hodgson, Julie Keck, and Daryl Bruflodt. I also want to thank the people at Publication Services, especially my production coordinator, Greg Martel; my copy editor, Joe Vittitow; and my art coordinator, Michele Runyon.

Special thanks to Bruce Craig for his accuracy check of the solutions to the exercises, Sam Rosenthal and especially Paula Waite for their help compiling the index, and Tracy Bearman for her help with permissions.

Earlier versions of this book were class-tested by Bob Miller, Xiaodong (James) Zheng, Michael Kosorok, and Gary Schroeder, at the University of Wisconsin-Madison; Ron Wasserstein, at Washburn University; and David Madigan, at the University of Washington in Seattle. The following persons served as reviewers of this text:

Robert Woodle,
Jamestown College

Winson Taam,
Oakland University

Chris Franklin,
University of Georgia

Patricia Buchanan,
Penn State University

Ronald L. Wasserstein,
Washburn University

Eugene Enneking,
Portland State University

David Madigan,
University of Washington

Paul J. Campbell,
Beloit College

Philip Steitz,
Beloit College

William Notz,
Ohio State University

Donald Ramirez,
University of Virginia

M. Lawrence Clevenson,
California State University

Smiley W. Cheng,
University of Manitoba

Robert K. Smidt,
California Polytechnic State University

Alan Agresti,
University of Florida

David R. Lund,
University of Wisconsin

Shu-ping Hodgson,
Central Michigan University

Peter Matthews,
University of Maryland

Richard J. Rossi,
Montana State University

William Gratzer,
Iona College

Dean Fearn,
California State University

Martin Buntinas,
Loyola University

Vasant B. Waikar,
Miami University

James E. Holstein,
University of Missouri

Gerry Hobbs,
West Virginia University

Mike Jacroux,
Washington State University

Douglas Zahn,
Florida State University

I owe all of these persons a great debt of gratitude. Thank you very much.

Since 1974 I have had the privilege of being "a small fish in the big pond" of the University of Wisconsin-Madison. I want to thank the university, especially former Dean Donald Crawford, for granting me a sabbatical to work on this book. Thanks also to Dean Judy Craig for her many years of effort to improve undergraduate education at Wisconsin, and for the support and encouragement she has given to me.

I have learned much from my faculty colleagues in the Department of Statistics. Those who had a big impact on my development as an educator include Brian Joiner, Martin Tanner, Sue Leurgans, George Box, Bob Miller, Richard Johnson, and the late Bill Hunter. Bob Miller deserves special thanks for his many years of unflagging enthusiasm, encouragement, and friendship. Thanks to Doug Bates and Murray Clayton for answering my numerous questions about computers. I have also had the good fortune of getting to know George Cobb, of Mount Holyoke College, whose innovative ideas on statistical education have had a big impact on me. As I hope this preface has made clear, I also have learned a great deal from my students.

Finally, I wish to thank my many friends and relatives who have supported me in this effort these past few years, including my mother, Helen Wardrop; my sister, Geri Pearce; my son, Roger Wardrop; and my long-time special friend, Joy Ransom. And thanks for the best cure I found for writer's block—a walk in the woods with my canine companion Casey.

ACCOMPANIED BY:

- Instructor's solution manual: Contains solutions to even-numbered exercises and comments on pedagagy for each chapter.
- Student's solution manual: Contains more detailed solutions to odd-numbered exercises (The end of the text presents solutions to odd-numbered exercises.) It is available for student purchase.
- *WCB Computerized Testing Program:* Provides you with an easy-to-use computerized testing and grade management program. No programming experience is required to generate tests randomly, by objective, by section, or by selecting specific test items. In addition, test items can be edited and new test items can be added. Also included with the *WCB Computerized Testing Program* is an on-line testing option which allows students to take tests on the computer. Tests can then be graded and the scores forwarded to the grade-keeping portion of the program.
- The *Test Item File:* A printed version of the computerized testing program that allows you to examine all of the prepared test items and choose test items based on chapter, section, or objective.

TO THE STUDENT

Every field has its own vocabulary. For statisticians, words or expressions such as "at random," "treatments," and "response" have precise technical meanings that may, but usually do not, agree with everyday usage. Such words and expressions are referred to as technical terms. In this text, the first appearance of a technical term is in **boldface** to alert you that a definition is forthcoming. (Occasionally, the first appearance of a technical term is in an overview of coming material; usually in these situations, the term is not boldfaced until it is defined.)

The phrase, "It can be shown that..." frequently appears in this text. Do not try to verify these statements! Typically, the omitted material is of a more advanced mathematical nature or is extremely tedious. Please do not think that I am "talking down to you" or am contemptuous of your mathematical skills. One of a teacher's most important responsibilities is the decision of *what* to teach, since students cannot learn everything! In this book I have decided to teach you statistics, not math. (There are many others better equipped than I to teach you math.)

Especially important concepts, definitions, and formulas are color-coded as key extracts. I hope they prove helpful.

My basic philosophy is that statistical concepts and methods gain value through their ability to help us learn about our world. Thus, the core of this book consists of the studies and their data sets that are used to introduce the statistical concepts and methods. Many of these studies were performed by professional researchers, and I want to thank them for their generosity in allowing me to share their work with you. Over 80 of the studies in this book were performed by my students as class projects. I want to thank my students for teaching me so much about my field and about how to be a more effective teacher. Finally, the text contains many sets of data that I collected myself. I will consider this book a success if it inspires you to collect some data to investigate an issue of interest to you.

Robert L. Wardrop
University of Wisconsin–Madison

STATISTICS
Learning in the Presence of Variation

Part I

DICHOTOMOUS RESPONSES

Chapter 1   **The Completely Randomized Design**
Chapter 2   **Hypothesis Testing**
Chapter 3   **Approximating a Sampling Distribution**
Chapter 4   **The Randomized Block Design**
Chapter 5   **One-population Models**
Chapter 6   **Inference for a Population**
Chapter 7   **Comparing Two Populations**
Chapter 8   **Two Dichotomous Responses**
Chapter 9   **Adjusting for a Factor**

The Completely Randomized Design

1.1 INTRODUCTION
1.2 DATA PRESENTATION AND SUMMARY
1.3 A SEQUENCE OF TRIALS

Chapter 1

1.1 INTRODUCTION

The principles, concepts, and methods of statistics have great value for persons who want to learn about their physical, biological, or social world. More precisely, statistics is concerned with learning that results from the collection and interpretation of information—more commonly called **data**—by researchers. Data must be collected in a valid manner, one that minimizes the likelihood that the data present a distorted view of the world. In addition, it is desirable to collect data in an efficient manner that minimizes the time and resources required. Concerns of validity and efficiency are the foci of **experimental design.** Data interpretation begins with **descriptive statistics,** which is the presentation and summary of data. Much can be learned from a clever and creative—or even a straightforward—presentation and summary of a set of data. There also exists a body of formal ways to learn from data, which fall under the heading of **inferential statistics.** Methods of inferential statistics are valid only for certain types of experimental design; details are given throughout the book.

Statistics often is viewed as a difficult subject. A big reason for this perception is that learning from "the collection and interpretation of information" does not receive much attention in the American educational system. All students have the experiences of learning by reading, listening, watching, and repetition. Many students, primarily but not exclusively those in the mathematical sciences, also learn through the applications of the rules of logic to a set of assumptions. The natural sciences come the closest to training students to learn from data, but unfortunately the introductory courses often present the world as deterministic. For example, my three years of chemistry consisted largely of learning the following type of fact and demonstrating it (in the negative sense) in a laboratory:

> There are 6.02×10^{23} molecules to a mole. Any experimental deviation from this value is due to bad experimental technique, measurement error, or contamination of materials.

Many important scientific problems, however, are not deterministic. For example, not everyone who smokes cigarettes develops lung cancer, and among those who do, the progression of the disease varies. This book aims to help you become a better learner in a nondeterministic setting—that is, in the presence of variation. Although it is difficult and challenging, this new way of learning has its rewards.

Because mathematics has played a central role in the development of statistics, one can easily believe that statistics is concerned with deducing the logical consequences (formulas and methods) of assumptions. It can be counterproductive, however, to adopt this point of view. Faced with a scientific question, the student can become stuck searching for the assumptions that will yield the "correct" solution. Instead, this book attempts to focus always on the scientific issues of interest. Formulas and methods gain importance not by being the optimal solution to some mathematical question, but by helping people better understand the world.

This book does not attempt to describe or outline the entire scientific process; instead, it begins at the point where one has selected an **issue** to investigate. For example, the exposition of this chapter presents studies concerned with the following issues:

- Does the wording of a question influence peoples' answers?
- Are women equally honest with their male and female friends?
- Does the promise of a sentence reduction influence prisoners' decisions to participate in a community service volunteer program?
- Is cyclosporine an effective therapy for chronic Crohn's disease?
- Is it easier for a basketball player to make a three-point basket from in front of the basket or from the left corner?
- Is a soccer goalie better at diving to her left or her right to make a save?
- Does a dancer perform pirouettes better to her left or right?

Remember that a particular study does not necessarily provide a definitive resolution to an issue and is not the only way to investigate the issue.

The first three chapters of this book present a simple case of one of the most useful experimental designs available to researchers—the completely randomized design. This first chapter defines the design and indicates how to present and summarize the data it yields. Chapters 2 and 3 present statistical inference for a completely randomized design. Chapter 2 introduces hypothesis testing and presents the exact analysis. Frequently for statistical inference the exact analysis is not practical because of computational difficulties. Chapter 3 introduces two computationally simple methods of approximate analysis. In total, the first three chapters present many of the important concepts and methods that appear in this book. If they seem difficult at first, do not despair; the ideas are reinforced through repetition in the remainder of the book.

The following example describes a study that was conducted to investigate the first issue listed above:

Does the wording of a question influence people's answers?

This example is used to illustrate many of the important ideas introduced in this chapter.

EXAMPLE 1.1: The Colloquium Study

The first component of an experimental design is the selection of the particular **subjects** to be included in the study. Subjects are the objects that are manipulated, questioned, or observed to obtain data. Obviously, the subjects for this example must be people, since the issue deals with the answers given by people. Studies usually do not include all conceivable subjects, so the experimental design specifies which subjects are included. The subjects for the current example were 28 persons who attended a particular Statistics Department colloquium at the University of Wisconsin-Madison.

The second and third components of an experimental design are the **treatments** and the **response**. It is frequently easiest to define the treatments and the response simultaneously. Below are two versions of a question; words that differ between one version and the other are in *italics* to facilitate a comparison of the versions.

- Threatened by a superior enemy force, the general faces a dilemma. His intelligence officers say his soldiers will be caught in an ambush in which 600 of them will die unless he leads them to safety by one of two available routes. If he takes the first route, *200 soldiers will be saved*. If he takes the second, there is a one-third chance that *600 soldiers will be saved*, and a two-thirds chance that *none will be saved*. Which route should he take?

- Threatened by a superior enemy force, the general faces a dilemma. His intelligence officers say his soldiers will be caught in an ambush in which 600 of them will die unless he leads them to safety by one of two available routes. If he takes the first route, *400 soldiers will die*. If he takes the second, there is a one-third chance that *no soldiers will die*, and a two-thirds chance that *600 soldiers will die*. Which route should he take?

These questions are referred to as the first and second versions of the General's Dilemma, respectively. They were created by the psychologists Kahneman and Tversky [1; Kevin McKean/© 1985 Discover Magazine. Reprinted with permission]. These versions are different wordings of exactly the same dilemma, with the first version written in terms of saving lives and the second written in terms of preventing deaths.

For this example, the first and second treatments are the first and second versions of the General's Dilemma, respectively. Each subject gives a response, namely a choice of the first or second route. Note that the response for the Colloquium study can take on two possible values, the first and second routes. A response that can take on two possible values is called a **dichotomous response**.

Briefly stated, the broad purpose of a study is to investigate whether the treatment received influences the subjects' responses. In the current example, this translates into investigating whether the version read influences the route selected.

There is an obvious way to determine whether the treatment influences the response: simply give a subject both treatments and compare the responses. This obvious method has a fatal flaw. A subject who is given both versions of the General's Dilemma will notice that they are equivalent and should therefore give the same response to each. In general terms, exposure to either treatment might change a subject in a way that would influence the subject's response to the other treatment. In short, giving each subject both treatments will destroy any chance of learning anything useful about the underlying scientific issue. Thus each subject is given only one treatment. (Chapter 4 presents the randomized pairs design, which is appropriate for those situations in which subjects can be given both treatments without invalidating the study.)

The fourth component of an experimental design is the method of assigning subjects to treatments. In the Colloquium study the researcher decided to assign 14 subjects to each treatment, and the assignment was made **at random,** also called **by randomization**. You have undoubtedly heard the expression "at random"; it is important to realize that this expression has a very precise technical meaning for researchers. This meaning is explained in the following subsection. ▲

To summarize, the four components of an experimental design are

1. The selection of the subjects to be included in the study
2. The specification of the treatments to be compared
3. The specification of the response to be obtained from each subject
4. The specification of the method by which subjects are assigned to treatments

Whenever subjects are assigned to treatments by randomization, the resulting design is called a **completely randomized design,** abbreviated CRD.

The Colloquium study is a CRD with two treatments and a dichotomous response. A CRD can have any number of treatments, but except in Chapter 11, this book restricts attention to two treatments. More importantly, the response need not be dichotomous—it can be multicategory or a number. Multicategory and numerical responses are studied in Parts II and III of this book, respectively.

Dichotomous responses are presented first because—in contrast to numerical data—presentation, summary, and inference are straightforward and—relative to multicategory data—notation and the interpretation of findings are simple. (The bases for these rather vague claims are made apparent in Parts II and III.) The purpose of Part I of the book is to introduce you to nearly all of the important concepts and techniques of statistics illustrated with dichotomous responses. Parts II and III extend then familiar ideas to multicategory and numerical responses.

1.1.1 Assigning Subjects at Random

This subsection consists of three parts:

- A mechanical description of how to assign subjects to treatments at random
- A brief discussion of why assigning subjects to treatments at random is desirable
- Some equivalent ways to assign subjects to treatments at random

These parts are illustrated by the Colloquium Study.

A Mechanical Description The technique is given first for the Colloquium study, and then a general algorithm is presented. Recall that the researcher decided to assign 14 subjects to each treatment. When an equal number of subjects is assigned to each treatment, the design is called **balanced**. The assignment by randomization was achieved by the following five steps:

1. The subjects were assigned the numbers $1, 2, \ldots, 28$ in an arbitrary manner.
2. The researcher obtained a collection of 28 cards, labeled $1, 2, \ldots, 28$ but otherwise indistinguishable.
3. The cards were placed in a box and mixed thoroughly.
4. Without looking, the researcher selected 14 cards from the box.
5. The numbers on the cards selected were

    1, 2, 6, 7, 9, 10, 12, 14, 15, 21, 22, 24, 27, and 28.

    Subjects corresponding to these numbers were given the first version of the General's Dilemma, and the remaining subjects—those corresponding to numbers 3, 4, 5, 8, 11, 13, 16, 17, 18, 19, 20, 23, 25, and 26—were given the second version.

The outcome of a randomization is two lists of numbers that indicate which subjects are to be assigned to each treatment. Each possible outcome of the process of randomizing is called an **assignment**. For example, the above assignment can be presented as

Version 1: 1, 2, 6, 7, 9, 10, 12, 14, 15, 21, 22, 24, 27, and 28.
Version 2: 3, 4, 5, 8, 11, 13, 16, 17, 18, 19, 20, 23, 25, and 26.

In Chapter 2 all possible assignments that could result from the process of randomization are considered. (It can be shown that there are 40,116,600 possible assignments of subjects to versions in the Colloquium study.)

The general method extends the above technique. First, determine the number of subjects in the study and denote this number by n. Next, decide how many subjects are to be assigned to the first and second treatments and denote these numbers by n_1 and n_2, respectively. (The symbol n_1 is read "en sub one.") Note that the sum of n_1 and n_2 equals the total number of subjects, n. Finally, complete the following five-step algorithm.

1. Assign the numbers $1, 2, \ldots, n$ in any manner to the subjects.
2. Obtain n cards that are labeled $1, 2, \ldots, n$ but are otherwise indistinguishable.
3. Place the cards in a box and mix them thoroughly.
4. Without looking, select n_1 cards from the box.
5. Write down the n_1 numbers obtained from the cards just selected. The subjects corresponding to these numbers receive the first treatment; the remaining subjects receive the second treatment.

Why Randomization Is Desirable Consider the two choices for the response in the Colloquium study. A general who selects the first route will have to accept the knowledge that he or she did not take the risk that might have saved additional lives. By contrast, a general who selects the second route will eventually either enjoy the satisfaction of saving an additional 400 lives or endure the remorse of being responsible for 200 additional deaths. This discussion suggests labeling the first route *risk averse* and the second route *risk seeking*.

The data from the Colloquium study are presented in the next section; not surprisingly, the data indicate that the relationship between version and response is not deterministic. In other words, the variation in responses among subjects cannot be attributed entirely to the version of the dilemma read. Let us embark on the mental exercise of listing a number of conceivable **factors** other than the version read that could influence a subject's response. One collection of factors is

> age, gender, religion, military experience (personal, family, friends), education, and income.

Another collection of factors is

> personality risk type (averse or seeking), degree of extroversion, intelligence, empathy, and satisfaction with life.

These lists are presented for illustration and are not meant to be exhaustive. In addition, there is no claim that any of these factors actually does influence the response—just that they might.

There is an important distinction between the above lists. The first list contains factors whose values are fairly easy to determine. For example, it takes little time to determine a subject's age, gender, or religion (provided the subject is willing to cooperate!). By contrast, the factors in the second list can be useful concepts, but the determination of their values can be impossible or, at best, difficult and imperfect.

There are ways to formally incorporate into the design and analysis any factors whose values can be obtained; this topic is discussed in Chapter 4. Formally incorporating one factor into an analysis usually is not a problem, but any attempt to incorporate many factors requires either a huge number of subjects or severe mathematical assumptions that might not be realistic. Thus, formally incorporating factors into an analysis cannot be viewed as a panacea.

One way to justify randomization is to criticize alternative methods of assigning subjects to treatments. (Another way is given in Example 1.4 later in this chapter.) Below are two other possible ways to assign subjects to treatments in the Colloquium study and criticisms of each. Remember that the goal is to obtain data that do not distort reality. The exposition below concentrates on a single factor, personality risk type, for simplicity. A similar argument can be applied to other factors or to a combination of factors. In order to illustrate the ideas, we assume that

- The version read has no effect on the response.
- The response is completely determined by the subject's personality risk type according to the following natural relationship: all risk-averse personality types select the risk-averse route (route one), and all risk-seeking personality types select the risk-seeking route (route two).
- The personality type of each subject can be determined correctly.

1. Given that the above conditions are true, an unethical researcher could deliberately assign all of the risk-averse subjects to the first version of the General's Dilemma and all risk-seeking subjects to the second version (perhaps destroying the balance). This assignment would result in a great distortion of reality: even though the version read has no effect on the response, the data would show that all subjects shown the first version selected the first route and all subjects shown the second version selected the second route. Even if the response is not *completely* determined by personality type, and even if personality type cannot be determined *perfectly*, an unethical researcher might still be able to seriously distort the data's picture of reality.

2. Assign the first 14 subjects to arrive at the room to the first treatment. This could distort reality if the reason for arriving early is related to a subject's risk status. For example, perhaps risk-averse subjects tend to arrive earlier than risk-seeking subjects; this assignment would then result in a disproportionate share of risk-averse subjects on the first treatment.

In summary, alternatives to randomization have the weakness that they might deliberately or inadvertently introduce a distortion of reality into the data. Of course, an unfortunate randomization can lead to a distortion (perhaps, by chance, all of the risk-averse subjects are assigned to the same treatment), but an advantage of assigning subjects by randomization is that the chance of obtaining an unfortunate randomization can be quantified precisely. This number can be used to help analyze the data. This notion of quantification is addressed in Chapter 2.

It can be a nuisance to assign subjects to treatments at random. Thus, a researcher faces a great temptation to assign subjects to treatments haphazardly and to pretend that the assignment was by randomization. (Statisticians use the word *haphazard* to describe any method of assigning subjects to treatments that does not follow any formal protocol yet does not satisfy the definition of randomization.) This temptation must be resisted. A number of exercises at the end of this section investigate some consequences of failing to randomize.

Equivalent Ways to Randomize It is not always convenient for a researcher to use a box and cards to randomize. One alternative is to use (part of) a deck of playing cards. The technique is illustrated for the Colloquium study. First, assign numbers to the subjects as described earlier. Next, make the following identifications:

> Ace of clubs = 1, two of clubs = 2, ..., king of clubs = 13, ace of diamonds = 14, two of diamonds = 15, ..., king of diamonds = 26, ace of hearts = 27, and two of hearts = 28.

Shuffle these 28 cards thoroughly (a theoretical result suggests that at least seven mixings are required) and choose the top 14 cards. The numbers assigned to the chosen cards indicate which subjects receive the first treatment, as before.

Another valid way to randomize is illustrated with the following dramatization. Laurian decided to study the General's Dilemma using her 100 students as subjects. She decided to have a balanced CRD—50 students on each version. At the entrance to the classroom Laurian placed a table with the following items on it (see Figure 1.1):

- A stack of 100 well-shuffled cards, lying face down. The faces of 50 of the cards were red and the remaining 50 were black. Next to the stack was a sign that read, "Take top card from deck."

- A stack of 50 copies of the first version of the General's Dilemma next to a sign that read, "If your card is black, take one of these."

- A stack of 50 copies of the second version of the General's Dilemma next to a sign that read, "If your card is red, take one of these."

- A box with a sign next to it that read, "Place card here."

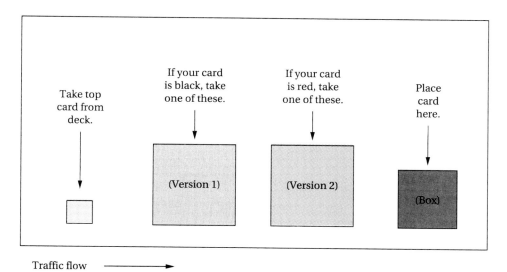

FIGURE 1.1 *An alternative valid way to assign subjects to versions by randomization.*

1.1.2 Two Additional Examples

EXAMPLE 1.2: *The Infidelity Study*

Therese Nyswonger studied 20 of her adult female friends. The women were divided into two treatment groups of size 10 by randomization. Women assigned to the first treatment group read the following question:

> You are friends with a married couple. You are equally fond of the man and the woman. You discover that the *husband* is having an affair. The *wife* suspects that something is going on. *She* asks you if you know anything about *her husband's* having an affair. Do you tell?

Women assigned to the second treatment group read the following question:

> You are friends with a married couple. You are equally fond of the man and the woman. You discover that the *wife* is having an affair. The *husband* suspects that something is going on. *He* asks you if you know anything about *his wife's* having an affair. Do you tell?

Each subject chose for her response either yes or no. ▲

EXAMPLE 1.3: *The Prisoner Study*

Ruth Parson's subjects were 50 male inmates at a minimum-security federal prison camp. All of the men were first-time nonviolent criminal offenders serving two or more years of prison time. The men were divided into two treatment groups of 25 each by randomization. Men assigned to the first treatment group were given the following question:

> The prison is beginning a program in which inmates have the opportunity to volunteer for community service with developmentally disabled adults. *Inmates who volunteer will receive a sentence reduction.* Would you participate?

Men assigned to the second treatment group were given the following question:

> The prison is beginning a program in which inmates have the opportunity to volunteer for community service with developmentally disabled adults. Would you participate?

Each subject responded either yes or no. ▲

1.1.3 Observational Studies

Consider the following study. I select 50 acquaintances, 25 men and 25 women, and give each person the first question in the Infidelity study. The goal of my study is to compare the responses of the male and female subjects. This is a perfectly reasonable study, but note that it is not a CRD because I cannot assign subjects (acquaintances) to treatments (gender) at random. Instead, I observe the gender of each subject. Such studies are called observational studies and are considered in Chapter 7. Similar remarks apply if, instead of comparing men and women, I compare the following, to name a few of many possibilities:

- Married men to single men
- Women aged 20 through 29 to women aged 40 through 49
- Married women to divorced women
- Married men who live in New York to married men who live in California

EXERCISES 1.1

1. Name a discipline in the social sciences that interests you. Give an example of an issue (as defined in the text) of interest to you that is related to that discipline and that can be investigated with a CRD.

2. Name a discipline in the natural sciences that interests you. Give an example of an issue (as defined in the text) of interest to you that is related to that discipline and that can be investigated with a CRD.

3. Name a discipline in the social sciences that interests you. Give an example of an issue (as defined in the text) of interest to you that is related to that discipline and that can be investigated with an observational study.

4. Name a discipline in the natural sciences that interests you. Give an example of an issue (as defined in the text) of interest to you that is related to that discipline and that can be investigated with an observational study.

5. A researcher decides to conduct a balanced CRD with two treatments on ten subjects. Assume that the

subjects are labeled 1, 2, 3, ..., 10. Help the researcher by selecting five subjects at random to receive the first treatment. Which assignment do you obtain? Repeat this exercise two more times. Do you obtain three different assignments?

6. A researcher decides to conduct an unbalanced CRD with $n_1 = 5$ and $n_2 = 3$. Assume that the subjects are labeled 1, 2, 3, ..., 8. Help the researcher by selecting five subjects at random to receive the first treatment. Which assignment do you obtain? Repeat this exercise two more times. Do you obtain three different assignments?

7. Refer to the alternative method of randomizing illustrated in Figure 1.1. Suppose that the deck contains 12 cards, six black (B) and six red (R). Suppose that the cards appear in the deck in the order, from top to bottom,

    B, B, R, R, R, B, R, R, B, B, B, and R.

    If the subjects select cards in the order 1, 2, 3, ..., 12 (that is, subject 1 selects first, followed by 2, then 3, and so on), which subjects will read version 1?

8. Refer to the alternative method of randomizing illustrated in Figure 1.1. Suppose that the deck contains 14 cards, eight black (B) and six red (R). Suppose that the cards appear in the deck in the order, from top to bottom,

    B, B, R, R, B, R, B, R, R, B, B, B, B and R.

    If the subjects select cards in the order 1, 2, 3, ..., 14 (that is, subject 1 selects first, followed by subject 2, then subject 3, and so on), which subjects will read version 1?

9. A researcher has 20 lab rats to use as subjects in a study to compare two diets. The rats are all in the same cage. The researcher reaches into the cage and selects 10 rats to be assigned to the first treatment; the remaining rats are assigned to the second treatment. Comment.

10. Students in a particular statistics class must also register for a discussion. Attendance at discussion is not mandatory. A researcher compares students who attended at least 75 percent of the discussion meetings to those who did not and finds that the latter group received better grades in the course. The researcher concludes that the discussion was worthless. Comment.

11. Professor X has a reputation of being very good at teaching statistics to students who have a weak math background. By contrast, Professor Y has a reputation of being very poor at teaching statistics to students who have a weak math background. At the end of the semester, all students are given a common final exam, and the students in Professor Y's course score better than the students in Professor X's course. Comment.

12. List three factors, other than the question read, which could influence a person's response to the Infidelity study.

1.2 DATA PRESENTATION AND SUMMARY

Recall that in the Colloquium study 14 subjects were assigned to each version of the General's Dilemma.

- In response to the first version, 7 subjects selected the first route and 7 subjects selected the second route.
- In response to the second version, only 1 subject selected the first route and the other 13 subjects selected the second route.

Such data are presented in a 2 × 2 **contingency table** of observed counts, as in Table 1.1. The first 2 in the term 2 × 2 counts the number of rows in the

TABLE 1.1 2 × 2 Contingency Table of Observed Counts for the Colloquium Study

Version Read	Route Selected		
	Risk Averse (1)	Risk Seeking (2)	Total
1. Saving lives	7	7	14
2. Preventing deaths	1	13	14
Total	8	20	28

table—the number of treatments. The second 2 counts the number of columns—the number of categories of the response. The row and column labeled "Total" are not counted in this notation.

The 2 × 2 contingency table presents the data from a CRD; the next goal is to summarize the data with the **table of row proportions.** The table of row proportions can be obtained easily from the contingency table of observed counts. The method is illustrated for the Colloquium study data in Table 1.1. Each number in the first row of the table, 7, 7, and 14, is divided by the total for that row, 14. Similarly, each number in the second row of the table, 1, 13, and 14, is divided by the total for that row, 14. The resulting proportions are presented in Table 1.2, which is called the table of row proportions for the Colloquium study. This table shows, for example, that 50 percent of the subjects on the first version selected the first route compared with only 7 percent of the subjects on the second version, a difference of 43 **percentage points**. The table of row proportions should present the number of subjects (**sample size**) for each row.

The row proportions also can be presented in a picture. Figure 1.2 is a **bar chart** comparison of the proportions of subjects in the Colloquium study

TABLE 1.2 Row Proportions for the Colloquium Study

Version Read	Route Selected			Sample Size
	Risk Averse (1)	Risk Seeking (2)	Total	
1. Saving lives	0.50	0.50	1.00	14
2. Preventing deaths	0.07	0.93	1.00	14

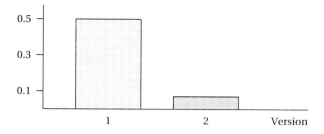

FIGURE 1.2 Bar chart of the proportion selecting the risk-averse route (route 1) for the Colloquium study.

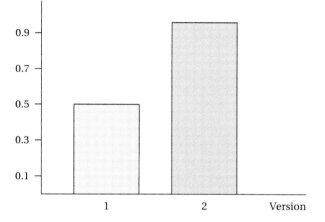

FIGURE 1.3 Bar chart of the proportion selecting the risk-seeking route (route 2) for the Colloquium study.

who selected the first route for each of the two versions. One also can construct a bar chart to compare the proportions of subjects who selected the second route, as presented in Figure 1.3.

There are two natural ways to compare the lengths of the bars in a bar chart:

1. Subtract the length of the shorter bar from the length of the longer bar.
2. Divide the length of the longer bar by the length of the shorter bar.

With the first method, subtraction, Figures 1.2 and 1.3 are in agreement; the difference between the two versions is 43 percentage points. By contrast, with the second method, division, the two figures tell different stories. Figure 1.2 shows that the proportion who selected the first route on treatment 1 (7/14) is 7 times larger than that proportion on treatment 2 (1/14). By contrast, Figure 1.3 shows that the proportion who selected the second route on treatment 2 (13/14) is 1.86 times larger than that proportion on treatment 1 (7/14). Neither of these answers, 7 or 1.86, is better. Do not discard the ratio as a meaningful way to compare proportions because of this disagreement. Rather, care is required in the interpretation of such ratios. One good analysis strategy is always to compute

TABLE 1.3 2 × 2 Contingency Table of Observed Counts for the Infidelity Study

Version Read	Response Yes	No	Total
1. Tell on husband?	7	3	10
2. Tell on wife?	4	6	10
Total	11	9	20

and report both ratios. In this text, proportions will be compared primarily by taking differences.

It is important to summarize the patterns in the data verbally. For example, here is a summary of the conclusions that can be drawn from the row proportions of the Colloquium study:

> When the dilemma is posed in terms of saving lives, subjects are equally divided between the risk-averse and risk-seeking solutions. When the dilemma is posed in terms of preventing deaths, the respondents exhibit an overwhelming preference for the risk-seeking solution.

Tables 1.3 through 1.6 present the tables of observed counts and row proportions for the Infidelity and Prisoner studies. For the Infidelity study,

> Seventy percent of the women say they would tell on a cheating husband, compared with only 40 percent who say they would tell on a cheating wife.

For the Prisoner study,

> When informed that they would receive a sentence reduction, 72 percent of the inmates say they would volunteer for community service. By contrast, when sentence reduction is not mentioned, 92 percent of the inmates say they would volunteer. This pattern is surprising!

TABLE 1.4 Row Proportions for the Infidelity Study

Version Read	Response Yes	No	Total	Sample Size
1. Tell on husband?	0.7	0.3	1.0	10
2. Tell on wife?	0.4	0.6	1.0	10

1.2 Data Presentation and Summary · 17

TABLE 1.5 2 × 2 Contingency Table of Observed Counts for the Prisoner Study

	Volunteer?		
Version Read	Yes	No	Total
1. Sentence reduction	18	7	25
2. No sentence reduction	23	2	25
Total	41	9	50

TABLE 1.6 Row Proportions for the Prisoner Study

	Volunteer?			Sample
Version Read	Yes	No	Total	Size
1. Sentence reduction	0.72	0.28	1.00	25
2. No sentence reduction	0.92	0.08	1.00	25

TABLE 1.7 Notation for a 2 × 2 Table in Format 1

	Response		
Treatment	S	F	Total
1	a	b	n_1
2	c	d	n_2
Total	m_1	m_2	n

It is convenient to have a standard notation for any 2 × 2 contingency table. This notation is given in Table 1.7. Note the following features of the standard notation:

- The possible responses are labeled S, for success, and F, for failure. Although there is no moral judgment attached to these terms, it is occasionally convenient to refer to a success as preferred to a failure. Of course, in some studies the assignment of the labels *success* and *failure* can prove offensive to certain individuals. As a small consolation, the language is improving—the labels *fertile-sterile* and *defective-nondefective* were popular in the past.

TABLE 1.8 Verbal Description of Notation for a 2 × 2 Table in Format 1

Symbol	Meaning
n	Total number of subjects in the study
n_1	Total number of subjects on treatment 1
n_2	Total number of subjects on treatment 2
m_1	Total number of subjects that respond S
m_2	Total number of subjects that respond F
a	Number of subjects on treatment 1 that respond S
b	Number of subjects on treatment 1 that respond F
c	Number of subjects on treatment 2 that respond S
d	Number of subjects on treatment 2 that respond F

The presentation of the data from the earlier studies in Tables 1.1, 1.3, and 1.5 reflect my decision to label as successes the first route, yes, and yes, respectively.

- Nine letters represent the numbers appearing in the table; their meanings are given in Table 1.8.
- Table 1.7 is called a Format 1 table. In addition to being studied in the first three chapters, Format 1 tables arise in different settings in Chapters 7 and 8. Chapters 4 and 8 consider the related Format 2 contingency table.

For the Colloquium study, you can verify from Table 1.1 that

$a = 7, b = 7, n_1 = 14, c = 1, d = 13, n_2 = 14, m_1 = 8, m_2 = 20$, and $n = 28$.

The following terminology is useful:

- The values a, b, c, and d are called the **cell counts**, **cell frequencies**, or **observed frequencies**.
- The values n_1, n_2, and n are called the **row totals** because they equal the sum of the other entries in their rows. Similarly, m_1, m_2, and n are called the **column totals**.
- The values n_1, n_2, m_1, m_2, and n are called the **marginal totals.**
- The value n is called the **grand total.**

1.2.1 Medical Studies

This section uses medical studies to introduce the important ideas of control group, placebo, blindness, and follow-up. In addition, medical studies are a good vehicle for further exploration of the importance of randomization.

A team of Scandinavian researchers was interested in finding a therapy to prevent the development of the acquired immunodeficiency syndrome (AIDS) in persons who test positive for the human immunodeficiency virus (HIV) antibody [2]. They believed that the drug inosine pranobex (IP) would be an effective therapy and decided to perform a study to determine whether their belief was justified. In the United States the drug zidovudine (AZT, taken from its former name, azidothymidine) has been studied extensively. In theory AZT is potentially effective because of its virus-attacking properties. By contrast, IP is potentially effective because of its ability to strengthen the immune system. The researchers had 59 persons available who met the criteria for inclusion in the study, namely, they were willing to participate in the study, they were HIV antibody positive with a CD4+ cell count less than 200 cells per cubic millimeter of blood, and they did not have AIDS according to the 1985 definition.

It is important to consider carefully what is meant by saying that a new therapy is effective. At any time, medical problems fall into two broad categories: those for which there is no currently accepted therapy and those for which there is a currently accepted therapy. (There could be more than one currently accepted therapy, but I want to keep the exposition simple.) In the former case a new therapy is considered effective if it is better than no therapy, whereas in the latter case a new therapy is considered effective if it is better than the currently accepted therapy. Thus, studies to determine the effectiveness of a new therapy should include two treatments—the new therapy and the standard (or no) therapy. Subjects assigned to the standard are said to form the **control group**. What is considered better is, of course, a difficult determination since in addition to the primary endpoint, *AIDS* or *not AIDS* for the current example, physicians are interested in the side effects of therapy and other quality of life issues. Whether or not there is a currently accepted therapy is not always a straightforward question. One might argue that IP should have been compared with AZT, but the researchers decided to compare it with no therapy.

In the study of IP, randomization was used to divide the 59 subjects into a group of 28 subjects who were assigned to the first treatment and a group of 31 subjects who were assigned to the second treatment. Persons on the first treatment were given IP, and those on the second treatment were given a pill that looked, felt, smelled, and tasted like IP, but contained no active drug. This pill is an example of what is called a **placebo**. Although the subjects knew that some people received IP and others the placebo, they did not know their own therapy. When subjects are ignorant of the treatment they receive, they are said to be **blind** to the treatment. The reason for this elaborate procedure—using a placebo and leaving the subjects blind to the treatment—is simple. A large body of evidence indicates that receiving medical attention of any kind, even administration of an inert drug, improves the physical condition of many people. This phenomenon is referred to as the **placebo effect.** By making the two treatments appear identical, the researchers isolated the effect of IP from the effect of the placebo.

Not only were the subjects blind to the treatment, but the attending physicians were too. This precaution was adopted for two reasons. First, a physician who knows which therapy is received might inadvertently convey this information to a subject. Second, the response of interest, AIDS or not AIDS, is potentially a somewhat subjective diagnosis; it is more valid if the physician making the assessment is ignorant of the therapy received. When both the subjects and the person(s) determining the responses are ignorant of the assignment of subjects to treatments, the study is said to be **double-blind.**

The earlier discussion of the value of randomization focused on potential weaknesses of other methods of assigning subjects to treatments. The following example highlights some positive features of randomization.

EXAMPLE 1.4: Chronic Crohn's Disease

Crohn's disease is a chronic inflammatory bowel disease. This example presents the findings of a double-blind CRD on 71 persons who were resistant to or intolerant of the standard therapy [3]. Because there was not a beneficial standard therapy for these subjects, the second treatment was a placebo. The first treatment was cyclosporine therapy. Thirty-seven subjects were assigned to the first treatment by randomization. The therapies lasted for three months, after which each subject was evaluated. Subjects were classified as improved (success) or not (failure). Table 1.9 presents the data from the study, and Table 1.10 presents the row proportions. The proportion of improved subjects was 27 percentage points higher on cyclosporine than on the placebo.

TABLE 1.9 Results of the Chronic Chrohn's Disease Study

Treatment	Clinical Result		Total
	Improved	Not Improved	
Cyclosporine	22	15	37
Placebo	11	23	34
Total	33	38	71

TABLE 1.10 Row Proportions for the Chronic Crohn's Disease Study

Treatment	Clinical Result		Total	Sample Size
	Improved	Not Improved		
Cyclosporine	0.59	0.41	1.00	37
Placebo	0.32	0.68	1.00	34

TABLE 1.11 Clinical Features of 71 Subjects with Crohn's Disease according to Study Group

Feature	Cyclosporine* ($n_1 = 37$)	Placebo* ($n_2 = 34$)
Gender		
Female	24(65)	22(65)
Previous therapy		
Corticosteriods	36(97)	34(100)
Sulfasalazine or 5-ASA	29(78)	27(79)
Immunosuppressive drugs	5(14)	5(15)
Surgical resection	19(51)	19(56)
Disease site		
Small bowel	10(27)	13(38)
Colon	7(19)	8(24)
Both preceding sites	20(54)	13(38)
Complications		
Anal lesions	7(19)	7(21)
Fistulas at other sites	0	1(3)

*The tabulated values are the numbers (percentages) of subjects having the given features.

The researchers felt that a number of other factors besides treatment could influence a subject's response. Some of these factors are listed in the first column of Table 1.11 An inspection of this table indicates that assigning subjects by randomization resulted in a very equitable distribution of possibly important factors to treatments. As examples of this, note that of the 10 features listed,

- Sixty-five percent of the subjects on each treatment were female.
- For the six features other than disease site and gender, the difference in proportions assigned to each treatment is 5 percentage points or less.
- The first treatment had a 16-percentage-point-higher proportion of subjects with dual disease sites. Thus, cyclosporine had a disproportionately high share of persons with the more serious—dual disease site—condition. Whether this makes a success (improvement) easier or more difficult for cyclosporine is not clear. (Certainly, it is ethically preferable for the persons who are more seriously ill to receive cyclosporine.)

Let us return again to the possibility of an unethical researcher, introduced on p. 9. In the current setting, an unethical researcher might assign the subjects with a better chance for success to cyclosporine and those with a poorer chance for success to the placebo. This assignment would distort the findings of the study in favor of cyclosporine. Alternatively, a compassionate researcher might reverse this assignment in order to give the subjects with a poorer chance for success access to the active drug. The compassionate researcher would distort the findings of the study in favor of the placebo. ▲

EXAMPLE 1.5: Chronic Crohn's Disease, Continued

The study of the previous example can also illustrate the importance of **follow-up** in a study. The therapies, cyclosporine and placebo, were ended after three months. Three months later the subjects were examined again; the results are presented in Table 1.12, and the row proportions are in Table 1.13. At follow-up cyclosporine still had a big advantage over the placebo, but neither treatment fared as well as it had at the end of therapy. These data illustrate that the beneficial effect of a placebo (or, alas, an active drug) can be transient. ▲

1.2.2 Additional Notation

It is convenient to define standard notation for the table of row proportions, given in Table 1.14. The symbol $\hat{p}_1$ is read "pea hat sub one" or "pea sub one hat." This table includes the previously defined n_1 and n_2 for the number of subjects on the first and second treatments, respectively. This book contains a great many p's and q's; some with hats, some without; some with subscripts, some without. The following rules might help you keep track of their meanings.

- The symbols p and q always represent proportions; that is, they are numbers between zero and one. A p always corresponds to success and a q to failure.
- The subscript on a p or a q denotes the treatment considered.
- The "hat," or circumflex, over the p or q serves as a reminder that the values are computed from data. Beginning in Chapter 5, another class of

TABLE 1.12 Results of the Follow-up to the Chronic Crohn's Disease Study

Treatment	Clinical Result		Total
	Improved	Not Improved	
Cyclosporine	14	23	37
Placebo	5	29	34
Total	19	52	71

TABLE 1.13 Row Proportions for the Follow-up to the Chronic Crohn's Disease Study

Treatment	Clinical Result		Total	Sample Size
	Improved	Not Improved		
Cyclosporine	0.38	0.62	1.00	37
Placebo	0.15	0.85	1.00	34

TABLE 1.14 Notation for the Table of Row Proportions

Treatment	Response		Total	Sample Size
	S	F		
1	$\hat{p}_1$	$\hat{q}_1$	1	n_1
2	$\hat{p}_2$	$\hat{q}_2$	1	n_2

proportions is introduced, distinguished from sample proportions by the absence of circumflexes.

- If their subscripts are identical or both absent, and if they both have or both lack circumflexes, a p and a q sum to one. This is illustrated in Table 1.14 by the equations

$$\hat{p}_1 + \hat{q}_1 = 1 \quad \text{and} \quad \hat{p}_2 + \hat{q}_2 = 1$$

You should verify the following identifications for the row proportions of the Prisoner study, Table 1.6:

$$\hat{p}_1 = 0.72, \quad \hat{q}_1 = 0.28, \quad \hat{p}_2 = 0.92, \quad \text{and} \quad \hat{q}_2 = 0.08.$$

EXERCISES 1.2

1. Refer to the Infidelity study, presented in this section. Draw a bar chart of the proportions of subjects who responded yes. Draw a bar chart of the proportions of subjects who responded no. Compare these two pictures.

2. Refer to the Prisoner study, presented in this section. Draw a bar chart of the proportions of subjects who responded yes. Draw a bar chart of the proportions of subjects who responded no. Compare these two pictures.

3. Becca Pryse modified the Infidelity study to obtain a study more relevant to her life. She conducted a balanced CRD on 50 of her sorority sisters. Here are the two versions of her question:

    - Version 1: You are the friend of a couple, a sorority sister and her boyfriend. You are equally fond of them. You are at a party and see *him* hitting on another *woman* and see the two leave together. Your *sorority sister* comes to you and tells you *she* suspects *her boyfriend* is cheating on *her*. Do you tell *her* what you have witnessed?

    - Version 2: You are the friend of a couple, a sorority sister and her boyfriend. You are equally fond of them. You are at a party and see *her* hitting on another *man* and see the two leave together. Your *sorority sister's boyfriend* comes to you and tells you *he* suspects *his girlfriend* is cheating on *him*. Do you tell *him* what you have witnessed?

    Becca obtained a total of 35 successes—a response of yes—with 20 of the successes occurring on the first treatment.

    (a) Present the data in a contingency table.
    (b) Construct the table of row proportions.
    (c) Write one or two sentences that summarize the results of the study.

4. One weakness of the General's Dilemma is that few people are generals! Thus, the dilemma can be viewed as artificial. Mike Teff, a Madison fire fighter, modified the General's Dilemma to obtain the Fire Fighter's Dilemma. Below are the two versions of his question:

   - Version 1: The interior sector officer at the Central Storage Warehouse fire faces a dilemma. The incident commander radios that the roof is about to collapse and that all 120 fire fighters inside will be killed unless they are led to safety.
     The chief informs them there are two ways to get out. If they take the first route, *40 fire fighters will be saved*. If they take the second route, there is a one-third chance that *120 fire fighters will be saved*, and a two-thirds chance that *none will be saved*. Which route should they take?
   - Version 2: The interior sector officer at the Central Storage Warehouse fire faces a dilemma. The incident commander radios that the roof is about to collapse and that all 120 fire fighters inside will be killed unless they are led to safety.
     The chief informs them there are two ways to get out. If they take the first route, *80 fire fighters will die*. If they take the second route, there is a one-third chance that *no fire fighters will die*, and a two-thirds chance that *120 fire fighters will die*. Which route should they take?

   The subjects were 50 Madison fire fighters. Mike performed a balanced CRD; 10 persons who read the first version and 4 who read the second version selected the first route (success); all others selected the second route (failure).

   (a) Present the data in a contingency table.
   (b) Construct the table of row proportions.
   (c) Write one or two sentences that summarize the results of the study.

5. Inspired by the General's Dilemma, Julie Fries, an oncology nurse, created the following Patient's Dilemma. Her subjects were 10 people between the ages of 26 and 83. She divided her subjects into two groups of five by randomization. Subjects assigned to the first treatment read the following passage:

   You are 65 years old. Your physician has informed you that you have a fatal disease. You have two choices:

   (a) Let the disease take its course—in which case you will die a very peaceful, painless death within six months.
   (b) Take a prescribed treatment for the disease. For 50 percent of the patients who take this treatment, life is lengthened by three to five years. This treatment is extremely painful for all who take it.

   Subjects assigned to the second treatment read a passage identical to the above except that the first sentence read, "You are 35 years old." The response is the choice made by the subject, with the first choice labeled a success. There were four successes on the first treatment compared with only one on the second treatment.

   (a) Present the data in a contingency table.
   (b) Construct the table of row proportions.
   (c) Write one or two sentences that summarize the results of the study.

6. Refer to the previous exercise. Theresa Stubblefield performed the Patient's Dilemma study on 56 college students. She used a balanced CRD and obtained 15 successes on the first treatment and 20 successes on the second. Compare her results to Julie Fries's findings. Comment.

7. Andre Rosay, the manager of an ice cream shop, wanted to determine whether the word *homemade* influenced his customers' purchases. His subjects were customers who asked for an ice cream but did not specify the type of cone. Subjects assigned to the first treatment were asked

   Would you like a plain cone or a homemade waffle cone?

   By contrast, subjects assigned to the second treatment were asked

   Would you like a plain cone or a waffle cone?

   A request for a waffle cone was denoted a success. Andre conducted a balanced CRD on 50 subjects and obtained a total of 25 successes. Seventeen of the successes were from persons assigned to the first treatment.

(a) Present the data in a contingency table.
(b) Construct the table of row proportions.
(c) Write one or two sentences that summarize the results of the study.

8. A balanced CRD was performed on 50 medical personnel who know CPR. Lynn Posick's two versions of her question read as follows:

- Version 1: You are standing at a bus stop on University Avenue. You notice a crowd forming around a person who is lying on the ground. No one knows anything about this person or what has happened. The person is unconscious and not breathing. Someone runs to call 911. Would you start CPR and mouth-to-mouth resuscitation?

- Version 2: You are at a distant relative's wedding. You notice a crowd forming around a member of the wedding party who is lying on the floor. No one knows anything about this person or what has happened. The person is unconscious and not breathing. Someone runs to call 911. Would you start CPR and mouth-to-mouth resuscitation?

An answer of yes was labeled a success and an answer of no or "I don't know" was labeled a failure. Lynn obtained 11 successes with the first version and 19 with the second version.

(a) Present the data in a contingency table.
(b) Construct the table of row proportions.
(c) Write one or two sentences that summarize the results of the study.

9. Anita Hecht, a social worker with the University of Wisconsin Counseling Service, performed a balanced CRD on 50 college students. The two versions of her question were

- Version 1: If you had a friend (who was a University of Wisconsin student) who you believed was experiencing depression (for example, severe and ongoing fatigue, loss of interest in things, or suicidal thoughts), would you advise them to seek help at the University Counseling Service?

- Version 2: If you were experiencing depression (for example, severe and ongoing fatigue, loss of interest in things, or suicidal thoughts), would you seek help at the University Counseling Service?

An answer of yes was labeled a success. Anita obtained a total of 26 successes, 18 with the first version and 8 with the second.

(a) Present the data in a contingency table.
(b) Construct the table of row proportions.
(c) Write one or two sentences that summarize the results of the study.

10. In May 1993, for their class project Dave Charvat and John Woodford investigated whether the wording of a question would influence a person's support for President Clinton's proposed broad-based energy tax, also called the Btu-Energy Bill. The first version of their question was

President Clinton has proposed a broad-based energy tax (Btu-Energy Bill) that would tax coal, oil, gas, nuclear, and hydroelectric power. According to the government, a family of four with an annual income of $40,000 would annually pay an additional $120 directly or $320 total, when the effects on all goods and services are included. The following government estimates indicate some specific increases:

| Gasoline | 7.5 cents per gallon | (5%) |
| Residential electricity | $2.44 per month | (3%) |

The legislation has been proposed to:

(a) **Help reduce the deficit**, *boosting long-term U.S. investment and productivity*

(b) **Improve the environment** *through reduced growth of fossil fuel consumption and modest incentives favoring cleaner fuels*

(c) **Enhance national security** *and the U.S. trade balance by reducing oil imports*

(d) **Increase energy efficiency** *for long-run competitive advantage.*

Do you support this legislation?

The second version was the same as the first version except that the words in italics above were deleted. The possible responses were yes, a success, and no.

Dave and John performed a balanced CRD on 50 persons and obtained 20 successes with the first version and 8 successes with the second version.

(a) Present the data in a contingency table.

(b) Construct the table of row proportions.

(c) Write one or two sentences that summarize the results of the study.

11. Lisa Harris performed a balanced CRD on 30 persons enrolled in the African-American studies course Race, Class, and Social Conflict at the University of Wisconsin-Madison. The two versions of her question were

- Version 1: Afro-centric schools are *designed primarily for African-American males and virtually exclude African-American females. They are* based on the commitment of producing strong socially conscious role models to help uplift the African-American community. Do you think this is a good program?

- Version 2: Afro-centric schools are based on the commitment of producing strong socially conscious role models to help uplift the African-American community. Do you think this is a good program?

The possible responses were yes, a success, and no. Lisa obtained 13 successes with the second version, but only 3 successes with the first version.

(a) Present the data in a contingency table.

(b) Construct the table of row proportions.

(c) Write one or two sentences that summarize the results of the study.

12. Andrea Litt, Michelle Huh, and Bernie Menachery performed a balanced CRD on 50 women they recruited from the terrace at the Memorial Union at the University of Wisconsin-Madison. The two versions of the "Are You Shallow?" study were

- Version 1: You are at the Terrace. An *attractive* male approaches you and invites you to join him for a drink. Do you accept?

- Version 2: You are at the Terrace. A male approaches you and invites you to join him for a drink. Do you accept?

The possible responses were yes, a success, and no. The study yielded 15 successes with the first version and 10 successes with the second version.

(a) Present the data in a contingency table.

(b) Construct the table of row proportions.

(c) Write one or two sentences that summarize the results of the study.

13. In the introduction to their class project, Tracey Deutsch and Natasha Larimer wrote

Historians have occasionally been known to debate the worth of including data tables in text. One group argues that tables of numbers in a history text are a turn-off. According to them, tables make a reader think that the text is inaccessible and that the theory must be driven by incomprehensible mathematical formulas.... These academics suggest incorporating more important numbers within the text, and if necessary, placing tables in an appendix in the back of the book. The opposing group suggests that tables within the text are handy references to have while reading. They like knowing exactly how numbers referred to in the text were derived, and just what "a majority" or "more likely" means numerically. As budding historians ourselves, we wanted to know whether tables of data in the text made a difference when making an argument.

Tracey and Natasha carried out a balanced CRD study using 44 subjects they found at the Wisconsin Historical Society Library. Their first treatment was a photocopy of a page, complete with tables, from a history text. The second treatment was the same page with the tables and references to them removed. On the page, the author makes the point that women over 40 years of age were in greater danger during the Salem outbreak (of witch-hunting). Subjects were asked whether they found the author's argument convincing. With the first treatment there were 14 successes (responses of *yes*), and with the second treatment there were 8 successes.

(a) Present the data in a contingency table.

(b) Construct the table of row proportions.

(c) Write one or two sentences that summarize the results of the study.

14. Chronic hepatitis B is a common and often progressive liver disorder. A team of researchers decided to test the effectiveness of treatment with interferon [4]. Eighty-seven subjects with chronic hepatitis were assigned by randomization to either

   - Treatment 1: Prednisone for six weeks followed by 5 million units of recombinant interferon alfa-2b daily for 16 weeks.
   - Treatment 2: An untreated control group.

   A response of success (loss of both hepatitis B viral DNA and hepatitis B antigen from serum) was given by 16 of 44 patients on treatment 1 and by 3 of 43 patients on treatment 2.

   (a) Present the data in a contingency table.
   (b) Construct the table of row proportions.
   (c) Neither the subjects nor supervising physicians were blind to the treatment received. Briefly explain how this might have influenced the results.
   (d) Write one or two sentences that summarize the results of the study.

15. It is known that if a woman suffers from pregnancy-induced hypertension, her newborn baby's chance of being ill or dying is increased. A study was conducted on 65 pregnant women who, according to a screening process, were at high risk of developing hypertension [5]. Thirty-four women were selected by randomization to receive the first treatment, a low dose of aspirin daily during the third trimester of pregnancy. The other 31 women received a placebo. A success was defined as not developing pregnancy-induced hypertension. Thirty of the women on the first treatment obtained successes compared with 20 on the second treatment.

   (a) Present the data in a contingency table.
   (b) Construct the table of row proportions.
   (c) The report said that the study was double-blind. Explain what this means.
   (d) Write one or two sentences that summarize the results of the study.

16. For women with Stage I or II breast cancer (tumor size 4 cm or less), lumpectomy followed by irradiation is an alternative to the standard treatment of total mastectomy. A CRD was conducted on 1219 women; 629 were assigned to the first treatment, lumpectomy and irradiation, and the remaining 590 were assigned to the second treatment, total mastectomy [6]. At follow-up, 412 women on the first treatment and 373 on the second treatment were alive with no evidence of disease (success). (The length of follow-up ranged from 47 to 141 months and was comparable for the two treatment groups.)

   (a) Present the data in a contingency table.
   (b) Construct the table of row percentages.
   (c) Explain why the study could not be blind.
   (d) Write one or two sentences that summarize the results of the study.

17. Lithium therapy is useful in preventing or reducing manic or depressive episodes in patients with bipolar disorder. Unfortunately, the therapy has some undesirable side effects. Ninety-four subjects with bipolar disorder participated in a study to determine whether the standard dose could be reduced [7]. The study was a balanced, double-blind CRD. Of 47 patients who received the standard dose of lithium (second treatment), 6 had relapses (failures) during the study, compared with 18 of the 47 assigned to the low dose treatment group (first treatment).

   (a) Present the data in a contingency table.
   (b) Construct the table of row proportions.
   (c) Briefly discuss why blinding is important in this study.
   (d) Write one or two sentences that summarize the results of the study.

18. Conduct a balanced CRD on 50 subjects. (You may use one of the studies described in the text, create your own research question, or take a question from another source.)

   (a) Describe your study; in particular, describe your subjects, treatments, and responses. How did you randomize?
   (b) Report your findings using the means (tables, figures, and numerical summaries) of this section.
   (c) Write one or two sentences describing what you present in part (b).
   (d) Did anything surprising happen when you conducted your study? If you were to repeat this exercise, would you do anything differently?

   Save a copy of your results for future use.

1.3 A SEQUENCE OF TRIALS

So far, all examples have had people for subjects. As this book unfolds, you will discover that almost anything imaginable can be a subject. Fortunately, there is an extremely useful dichotomous classification of the kinds of subjects possible in the studies considered in this book. All of the earlier studies have subjects that are **distinct individuals**; the other possibility is for a study to have subjects that constitute a **sequence of trials**. The next three studies are examples of trials as subjects.

EXAMPLE 1.6: Three-Point Basket Study

In high school, college, and professional basketball, a team is awarded three points, instead of the usual two, for a successful shot from beyond the three-point line. At all levels of competition, the distance from the line to the basket is shorter from either corner of the court than it is from in front of the basket. This reflects the conventional wisdom that a shot from in front of the basket is easier to make than an equidistant shot from either corner of the court. Right-handed shooter Clyde Gaines attempted 50 three-point jump shots from in front of the basket and 50 from the left corner. Clyde feared that more than 100 shots would lead to leg fatigue and a subsequent adverse effect on his performance. Clyde's study can be viewed as a CRD with the following identifications:

1. The subjects are the 100 shots, numbered consecutively 1, 2, ..., 100.
2. The treatments are shooting from in front of the basket (first treatment) and shooting from the left corner (second treatment).
3. Each shot yields a basket (success) or a miss (failure).
4. Randomization can be achieved by numbering the shots 1, 2, ..., 100, matching the order in which they will be taken, and then selecting 50 of these numbers at random using the method described earlier. The selected numbers indicate which shots are assigned to the first treatment, shooting from in front of the basket. Clyde, however, performed his randomization by the alternative method illustrated in Figure 1.1 on p. 11; namely, he took a well-shuffled deck of 50 black and 50 red cards to the gym. Before each shot he looked at the top card—if it was black, he shot from in front of the basket, and if it was red, he shot from the left corner. This alternative method of randomization is easier to implement in the field.

TABLE 1.15 Results of the Three-Point Basket Study

	Result		
Location	Basket	Miss	Total
In front	21	29	50
Left corner	20	30	50
Total	41	59	100

TABLE 1.16 Row Proportions for the Three-Point Basket Study

	Result			Sample
Location	Basket	Miss	Total	Size
In front	0.42	0.58	1.00	50
Left corner	0.40	0.60	1.00	50

The data for Clyde's study are presented in Table 1.15, and the row proportions are given in Table 1.16. For the trials of this study, Clyde shot 2 percentage points better from in front of the basket than from the left corner. ▲

EXAMPLE 1.7: Soccer Goalie Study

Maureen Sullivan wanted to investigate whether a soccer goalie was better at stopping shots to her right or to her left. The trials were 50 shots from the penalty spot (see Figure 1.4). The first treatment was a shot to the goalkeeper's right

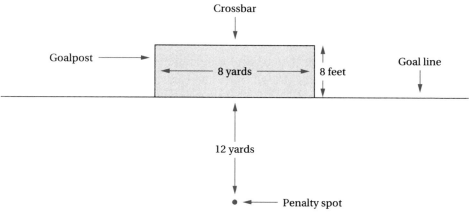

FIGURE 1.4 Soccer goal and penalty spot.

TABLE 1.17 Results of the Soccer Goalie Study

	Result		
Location	Stop	Goal	Total
Goalie's right	15	10	25
Goalie's left	7	18	25
Total	22	28	50

TABLE 1.18 Row Proportions for the Soccer Goalie Study

	Result			Sample
Location	Stop	Goal	Total	Size
Goalie's right	0.60	0.40	1.00	25
Goalie's left	0.28	0.72	1.00	25

and the second treatment was a shot to the goalkeeper's left. The response was either a stop (success) or a goal (failure). (In soccer a goal is scored if the entire ball passes over the goal line between the goalposts and under the crossbar; see Figure 1.4.) The trials were assigned to treatments by randomization. The data from the study are presented in Table 1.17, and the row proportions are given in Table 1.18. For the trials of this study, the soccer goalkeeper was much better at stopping shots to her right than shots to her left. ▲

EXAMPLE 1.8: The Ballerina Study

Julie Gleich was interested in seeing if she was better at performing a pirouette to her left or her right. Her trials were 50 pirouettes. The first treatment was spinning left, and the second was spinning right. A trial was a success if she completed three or more spins and was a failure otherwise. The trials were assigned to treatments by randomization. The data from Julie's study are presented in Table 1.19, and the row proportions are given in Table 1.20. For the trials of this study, Julie was better at performing a pirouette to her right than to her left. ▲

Each of the three examples of this section has trials for subjects. The trials are all made on a single individual—the basketball player, the soccer goalkeeper, or the dancer. Occasionally, it is convenient to refer to this individual as the **metasubject**. As mentioned at the beginning of this section, this book only

TABLE 1.19 Results of the Ballerina Study

	Result		
Direction	Success	Failure	Total
To the left	16	9	25
To the right	22	3	25
Total	38	12	50

TABLE 1.20 Row Proportions for the Ballerina Study

	Result			
Direction	Success	Failure	Total	Sample Size
To the left	0.64	0.36	1.00	25
To the right	0.88	0.12	1.00	25

considers studies with subjects that are either distinct individuals or trials. More complicated studies can be obtained by combining these types of studies, for example, a sequence of trials observed on a number of different metasubjects.

1.3.1 Why Randomize a Sequence of Trials?

In earlier studies in this chapter the subjects were distinct individuals—people, in fact. As argued earlier, because people are so different, assigning subjects to treatments by randomization is sensible and useful. By contrast, one might ask, "One pirouette trial is the same as any other. Why should I randomize?" The biggest concern in a sequence of trials—and the primary reason researchers randomize—is the possibility that a **time trend** is an important factor. It is easiest to define a time trend by giving examples.

- During the course of data collection it is possible that successes become more common, regardless of treatment, because the metasubject is improving from practice.

- During the course of data collection it is possible that successes become less common, regardless of treatment, because the metasubject is becoming bored or fatigued.

- A serial relationship between trials is possible. For example, perhaps a successful pirouette fatigues the dancer and hence tends to be followed by a failure.

1.3.2 Observational Studies

My neighborhood park has an outdoor basketball court with an electric light for night shooting. In my opinion the lighting is poor, and I decide to conduct a study to investigate whether my ability to shoot is influenced by the lighting. My subjects are 100 shots from a fixed spot on the basketball court. The treatments are shooting during bright daylight and shooting at night when the light is turned on. The response, naturally, is a success if a shot goes through the basket and a failure otherwise. Suppose that on a particular date I attempt 50 shots during the day and after dark attempt 50 more shots. Note that this is not a CRD because I do not assign shots to treatments by randomization; my design forced all of the daytime shots to precede all of the evening shots. My design is an observational study and can be analyzed using the methods presented in Chapter 7.

It is instructive to consider two ways to modify my design so that it becomes a CRD. First, I could perform the study over a period of 100 days. Beginning with a well-shuffled deck of 50 red and 50 black cards, the morning of the first day of the study I select the top card. If the card is red, that day's shot will be taken during daylight hours, and if the card is black, that day's shot will be taken after dark when the light is turned on. The card is set aside, and the next day the (new) top card determines that day's treatment as described above. This procedure is repeated until 100 shots have been obtained. You should convince yourself that this is a CRD. Although it is a CRD, this design seems absurd: it is a waste of my time to walk to the park to obtain the result of a single shot.

The second modification combines characteristics of the original design and the first modification. As in the first modification, I collect data on different days, but only on ten different days. I randomly select five days for shooting during daylight and five days for shooting after dark. As in the original design, I attempt several shots during each session, for example 20 shots at each session. If I take my subjects to be individual shots, this is not a CRD, but if I take a subject to be a session of 20 shots, the 10 subjects are assigned at random to treatments and the design is a CRD. It is important to note, however, that if I take a subject to be a session of 20 shots, a dichotomous response is no longer reasonable. The natural response is the number of shots made out of 20. This response is called a numerical response because it is a number. Methods for analyzing numerical responses are developed in Part III of this book.

EXERCISES 1.3

1. Name a hobby, game, or pastime that you enjoy. Give an example of an issue of interest to you that is related to your choice and that can be investigated with a CRD.

2. Name a hobby, game, or pastime that you enjoy. Give an example of an issue of interest to you that is related to your choice and that can be investigated with an observational study.

3. A researcher wants to investigate the effect of caffeine on shooting free throws. A basketball player attempts 25 free throws while caffeine free and then drinks three cups of coffee in 10 minutes. After drinking the coffee, the player attempts 25 more free throws. If each shot is a subject, is this a CRD?

4. A student in my Statistics 301 class conducted a study to test his ability at serving a tennis ball. Here is his description of the experimental protocol:

   > I will serve 25 to the add court and 25 to the deuce court. I will start serving from the deuce court and will rotate sides after each serve until reaching 25 attempts from each side.

   Comment.

5. Pam Crawford believed her son to be ambidextrous. To investigate this issue she performed the following experiment. He threw a ball at a target from a fixed distance 50 times. He made 25 throws with each hand with the trials properly randomized. Hitting the target was deemed a success. He obtained 12 successes with his right hand (first treatment) and 10 with his left hand (second treatment).

   (a) Present the data in a contingency table.
   (b) Construct the table of row proportions.
   (c) Write one or two sentences that summarize the results of the study.

6. Ken M., an avid tennis player, conducted a balanced CRD with 50 serves as his trials. Out of 25 serves into the advantage court (first treatment), 23 were successes, compared with just 8 of 25 serves into the deuce court (second treatment).

   (a) Present the data in a contingency table.
   (b) Construct the table of row proportions.
   (c) Write one or two sentences that summarize the results of the study.

7. Loretta Moore performed a CRD with her husband serving as the metasubject. He shot 25 basketball bank shots from a distance of 15 feet from a 45-degree angle to the left of the basket—the first treatment—and shot 25 bank shots from a distance of 15 feet from a 90-degree angle (in front of the basket). He obtained 20 successes with the first treatment and 8 with the second treatment.

   (a) Construct the contingency table for these results.
   (b) Construct the table of row proportions.
   (c) Write one or two sentences that summarize the results of the study.

8. Wayne Bell, the athletic director at Oregon High School in Oregon, Wisconsin, gave the following explanation for his selection of a statistics project:

   > The Oregon High School gym was recently refitted with new light fixtures designed to save a substantial amount of electricity while providing additional illumination. Once installed, the lighting seemed not to be dispersed evenly and seemed especially uneven behind the two main backboards. There was also increased glare at the east end of the gym.

   Wayne conducted a balanced CRD on 20 free throws. The first treatment was shooting at the east-end basket, and the second treatment was shooting at the west-end basket. He obtained six successes at the east end of the gym and eight successes at the west end of the gym.

   (a) Construct the contingency table for these results.
   (b) Construct the table of row proportions.
   (c) Write one or two sentences that summarize the results of the study.

9. Raul Mares and his roommate enjoy watching the "Bozo Show" on television. The climax of the show is the Grand Prize Game. One day, Raul's roommate remarked that the game is so easy he could do it with his eyes closed. Raul challenged him to be a metasubject in a study. Each trial was an attempt to throw a table-tennis ball into a bucket on the floor from a distance of 11 feet. In the first treatment the metasubject's eyes were open and uncovered, and in the second treatment the metasubject was blindfolded. Raul performed a balanced CRD on 50 trials; his roommate obtained 12 successes (the ball in the bucket) with his eyes open and only 4 when blindfolded.

   (a) Construct the contingency table for these results.
   (b) Construct the table of row proportions.
   (c) Write one or two sentences that summarize the results of the study.

10. Mike Derr, an avid golfer, was interested in whether he putts better on a right-to-left or on a left-to-right slope. He investigated this issue by conducting a balanced CRD on 50 putts, each from a distance of 11 feet. On the first treatment, right to left, he had five successes, and on the second treatment he had six successes.
    (a) Construct the contingency table for these results.
    (b) Construct the table of row proportions.
    (c) Write one or two sentences that summarize the results of the study.

11. Kimberly J. performed a balanced CRD on 50 tosses of horseshoes. A ringer was labeled a success and any other outcome a failure. Her first treatment was tossing from a distance of 30 feet and the second treatment was tossing from 40 feet. She obtained a total of seven ringers, but only one from the longer distance.
    (a) Construct the contingency table for these results.
    (b) Construct the table of row proportions.
    (c) Write one or two sentences that summarize the results of the study.

12. Kelly Kuss performed a balanced CRD on 50 tosses of lawn darts. Hitting a target was labeled a success and missing it a failure. Her first treatment was tossing with her left hand and the second treatment was tossing with her right hand. She obtained a total of 18 successes, 5 with her left hand and 13 with her right hand.
    (a) Construct the contingency table for these results.
    (b) Construct the table of row proportions.
    (c) Write one or two sentences that summarize the results of the study.

13. Franchell McClendon conducted a balanced CRD on 50 free throws. The first treatment was shooting free throws with a women's basketball, and the second treatment was shooting with a men's basketball. (The men's ball is larger.) Franchell made 17 free throws with the women's ball and 21 with the men's ball.
    (a) Construct the contingency table for these results.
    (b) Construct the table of row proportions.
    (c) Write one or two sentences that summarize the results of the study.

14. Thomas Lothrop wanted to determine the difference in difficulty between 4-foot and 8-foot putts in golf. He performed a balanced CRD on 50 putts. The first treatment was putting from 4 feet on a level surface, and the second treatment was putting from 8 feet on a level surface. Thomas made 28 putts, 10 from the longer distance and 18 from the shorter.
    (a) Construct the contingency table for these results.
    (b) Construct the table of row proportions.
    (c) Write one or two sentences that summarize the results of the study.

15. Kimberly Keller Sippola conducted a balanced CRD on 50 dart throws. The first treatment was throwing right-handed from a distance of 9 feet, and the second treatment was throwing left-handed from the same distance. If the dart stuck on the board anywhere within the second ring, the throw was labeled a success. Kimberly obtained a total of 28 successes, but only 8 with her left hand.
    (a) Construct the contingency table for these results.
    (b) Construct the table of row proportions.
    (c) Write one or two sentences that summarize the results of the study.

16. Robert Miller conducted a balanced CRD on 100 shots of a rifle. The first treatment was shooting from the prone position, and the second treatment was shooting from the kneeling position. The distance between Robert and the target was 50 feet, and he used a 22-caliber model 52 Winchester rifle. A shot was labeled a success if it hit a prespecified region of a target. Robert had 42 successes from the prone position and 25 from the kneeling position.
    (a) Construct the contingency table for these results.
    (b) Construct the table of row proportions.
    (c) Write one or two sentences that summarize the results of the study.

17. Paula Waite, Heather Wamsley, and Mary Yard wrote the following in their class project, "Is Pearl Ambidextrous?":

In the training of a horse, it is important to increase the horse's flexibility and strength. Just as humans are often stronger when using one hand or foot than when using the other (handedness), so too do horses exhibit handedness. However, horses will benefit greatly in use and health if they become ambidextrous through training. This project will determine whether a seven-year-old mare named Pearl, owned by Mary Yard, has approached ambidexterity through Mary's work with her.

The first treatment was canter depart on the left lead, and the second treatment was canter depart on the right lead. (The precise definition of a canter is not needed to complete this exercise.) A canter could be executed successfully or not. A balanced CRD on 50 trials was performed, and it yielded 22 successes on the first treatment and 20 successes on the second treatment.

(a) Construct the contingency table for these results.
(b) Construct the table of row proportions.
(c) Write one or two sentences that summarize the results of the study.

18. In their class project, Susan Brehm-Stecher, Becky Byom, and Boris Feldman wrote the following about Susan:

> Her father is "nuts" about roasted peanuts and has eaten them during all leisure activities for as long as Susan can remember. He has always held the strong opinion that shaking the nut in his hand before tossing it up into the air to be caught in his mouth improves his aim.

Susan enlisted her husband to serve as the metasubject for a study of her father's theory. A balanced CRD was performed with 100 trials. In treatment 1 Susan's husband shook the peanut in his hand three times before tossing it into the air. In treatment 2 he tossed the peanut into the air without shaking it. A success occurred if the peanut was caught, and a failure otherwise. Susan's husband obtained a total of 76 successes, with 41 coming on the first treatment.

(a) Construct the contingency table for these results.
(b) Construct the table of row proportions.

(c) Write one or two sentences that summarize the results of the study.

19. An archer's arrow can have plastic or feather vanes to help stabilize the arrow during flight. In their class project, Teresa Chervenka and Lan Nguyen write, "Most archers agree that feather vanes fly better than plastic vanes in dry weather." Theresa and Lan performed a balanced CRD of 100 shots to investigate which stabilizer is better in wet weather. Their first treatment was shooting an arrow with wet feathers, and the second treatment was shooting an arrow with wet plastic vanes. Teresa's boyfriend, John, served as the metasubject. If John hit a 4-inch bull's-eye from 20 yards, the trial was a success; otherwise, it was a failure. John obtained only 16 failures, with all but 2 of them being obtained with the wet feathers.

(a) Construct the contingency table for these results.
(b) Construct the table of row proportions.
(c) Write one or two sentences that summarize the results of the study.

20. Michael Kaminski performed a balanced CRD on 100 trials to study his ability at golf. Treatment 1 was putting from 40 feet while wearing a golf glove, and treatment 2 was the same putt without the glove. If Michael sank the ball in two or fewer putts, the trial was a success; otherwise, it was a failure. Michael obtained a total of 54 successes, 29 of which were with the glove removed.

(a) Construct the contingency table for these results.
(b) Construct the table of row proportions.
(c) Write one or two sentences that summarize the results of the study.

21. Perform a CRD for a sequence of 100 trials of your choice. Remember, there must be two ways to perform each trial (the two treatments), and the response must be dichotomous.

(a) Present your data in a 2×2 contingency table.
(b) Create the table of row proportions and draw a bar chart comparing the proportions of successes on each treatment.
(c) Did anything surprising happen when you conducted your study? If you were to repeat this exercise, would you do anything differently?

Save a copy of your results for future use.

REFERENCES

1. K. McKean, "Decisions, Decisions," *Discover*, June 1985, pp. 22–31. Kevin McKean/© 1985 Discover Magazine. Used with permission.
2. C. Pedersen et al., "The Efficacy of Inosine Pranobex in Preventing the Acquired Immunodeficiency Syndrome in Patients with Human Immunodeficiency Virus Infection," *New England Journal of Medicine*, Vol. 322, June 21, 1990, pp. 1757–1763.
3. J. Brynskov et. al., "A Placebo Controlled, Double-Blind, Randomized Trial of Cyclosporine Therapy in Active Chronic Crohn's Disease," *New England Journal of Medicine*, Vol. 321, September 28, 1989, pp. 845–850.
4. R. Perrillo et. al., "A Randomized, Controlled Trial of Interferon Alfa-2b Alone and after Prednisone Withdrawal for the Treatment of Chronic Hepatitis B," *New England Journal of Medicine*, Vol. 323, August 2, 1990, pp. 295–301.
5. E. Schiff et al., "The Use of Aspirin to Prevent Pregnancy-Induced Hypertension and Lower the Ratio of Thromboxane A_2 to Prostacyclin in Relatively High Risk Pregnancies," *New England Journal of Medicine*, Vol. 321, August 10, 1989, pp. 351–356.
6. B. Fisher et al., "Eight-Year Results of a Randomized Clinical Trial Comparing Total Mastectomy and Lumpectomy with or without Irradiation in the Treatment of Breast Cancer," *New England Journal of Medicine*, Vol. 320, March 30, 1989, pp. 822–828.
7. A. J. Gelenberg et al., "Comparison of Standard and Low Serum Levels of Lithium for Maintenance Treatment of Bipolar Disorder," *New England Journal of Medicine*, Vol. 321, November 30, 1989, pp. 1489–1493.

Hypothesis Testing

2.1 FISHER'S TEST: GETTING STARTED
2.2 STEP 1: THE HYPOTHESES
2.3 STEP 2: THE TEST STATISTIC AND ITS SAMPLING DISTRIBUTION
2.4 WORKING WITH PROBABILITY
2.5 STEPS 3 AND 4
2.6 BEYOND BRUTE FORCE

Chapter 2

2.1 FISHER'S TEST: GETTING STARTED

For ease of exposition, in this chapter the two possible values a dichotomous response can take on are referred to as a success and a failure. In addition, a success is considered to be better than or preferred to a failure, and the treatment with the higher proportion of successes in the study is said to have performed better.

The following sentence appeared in Chapter 1, on p. 6:

Briefly stated, the broad purpose of a study is to investigate whether the treatment received influences the subjects' responses.

Consider the Colloquium study. Of the persons who read the first version, 50 percent (7 of 14) yielded successes, but only 7 percent (1 of 14) of the persons who read the second version yielded successes. Thus, one might be tempted to say that the above stated purpose of the study has been achieved: namely, the treatment does influence the response, and the first treatment is better than the second. This answer, however, is usually considered unsatisfactory because of the following argument, which will be referred to as the **Skeptic's Argument.**

Perhaps the version read has no effect on the response for any of the subjects. In other words, perhaps 8 of the subjects would have given successes and the remaining 20 subjects would have given failures regardless of which version they read. The first version did better in the actual study only because it had the good fortune of being assigned to 7 of the 8 subjects who were bound to yield successes.

Note the following features of this argument.

- The Skeptic's Argument cannot be dismissed out of hand. The argument is possibly correct.
- This argument can be extended to any of the studies considered in Chapter 1; that is, the argument can be extended to any study in which subjects are given only one treatment.

The Skeptic's Argument can be viewed as one-half of a debate. The other half of the debate is the **Advocate's Argument:**

> Although I have to admit that the Skeptic's Argument is possibly correct, it strains credibility. If, in fact, there were exactly 8 subjects who would have given successes (and exactly 20 subjects who would have given failures) regardless of the version read, it is extremely unlikely that 7 out of these 8 would have been assigned to the first version. I do not believe that such an extremely unlikely event occurred; thus, I disagree with the Skeptic's Argument and find convincing the evidence in the data of a difference between versions.

Hypothesis testing is a technique used by statisticians and researchers to quantify the debate between the Skeptic and the Advocate. The idea is that once the debate is quantified, it is easier for people to decide which argument is more convincing. The hypothesis test for the studies of Chapter 1—a CRD with two treatments and a dichotomous response—is called **Fisher's test**, named for the famous scientist and statistician Sir Ronald A. Fisher.

It is absolutely critical that you realize that hypothesis testing is concerned with "that which cannot be seen." A hint of this fact appears in the Skeptic's Argument: the Skeptic speaks of subjects that give the same response regardless of the treatment received. This claim cannot be checked absolutely because it is impossible to give a subject both treatments (as discussed in Chapter 1, on p. 6). Although it is impossible to refute or confirm the Skeptic's Argument with certainty, the data do provide *evidence* that helps the researcher decide which argument to believe, the Skeptic's or the Advocate's.

Consider a particular subject, Mary, in the Colloquium study. Before the study was performed, it was possible that Mary would be assigned to either treatment 1 or treatment 2, and it was conceivable that Mary would yield a success or a failure as her response. The four possible combinations of treatment and response are presented in Table 2.1. Reading from this table, a type A subject is one who would respond with a success to either treatment; a type B subject would give a success to treatment 1 but a failure to treatment 2; a type C subject would give a success to treatment 2 but a failure to treatment 1; and a type D subject would respond with a failure to either treatment. Note that subjects of type A or D are not influenced by the treatment they receive; they give the same response to each treatment. By contrast, subjects of type B or C are influenced

TABLE 2.1 *The Four Types of Subjects in a CRD*

	Response to	
Type	Treatment 1	Treatment 2
A	S	S
B	S	F
C	F	S
D	F	F

by the treatment received. In particular, treatment 1 is better for type B subjects (remember that a success is the preferred response), and treatment 2 is better for type C subjects.

Unfortunately, the results of an experiment do not identify the type of a subject. Suppose that, for example, the previously mentioned Colloquium study subject, Mary, actually received treatment 1 and yielded a success. From Table 2.1, one can see that either type A or type B, but not type C or type D, subjects yield a success when assigned to treatment 1. Hence, Mary is either a type A or type B subject, in agreement with the entry in the first row of Table 2.2. You should use Table 2.1 to verify the remaining entries in Table 2.2.

It is useful to define the following notation. For a CRD, let

- n_A equal the number of subjects of type A
- n_B equal the number of subjects of type B
- n_C equal the number of subjects of type C
- n_D equal the number of subjects of type D

These numbers are unknown and will always remain unknown. They are defined for conceptual use only.

Note that if all subjects were given treatment 1, the study would yield $n_A + n_B$ successes (refer to Table 2.1). If, on the other hand, all subjects were given treatment 2, the study would yield $n_A + n_C$ successes. Thus, if both treatments could be given to each subject, treatment 1 would yield more successes than

TABLE 2.2 *Relationship between Experimental Result and Subject Type in a CRD*

Treatment	Response	Possible Types
1	S	A or B
1	F	C or D
2	S	A or C
2	F	B or D

treatment 2—and, hence, be the better treatment—if, and only if,

$$n_A + n_B > n_A + n_C$$

or, cancelling the n_A's,

$$n_B > n_C.$$

In fact, there are four possibilities of particular interest to a researcher:

- $n_B > n_C$. In words, treatment 1 is superior to treatment 2.
- $n_B < n_C$. In words, treatment 1 is inferior to treatment 2.
- $n_B = n_C = 0$. All subjects are type A or D and, hence, no subject is influenced by the treatment received. For each subject, treatments 1 and 2 have the same effect on the response.
- $n_B = n_C > 0$. Some subjects respond differently to the two treatments, but neither treatment is better overall.

A researcher would be very happy to know which of these four possibilities is correct for the study at hand. Hypothesis testing helps the researcher decide between these possibilities.

Every hypothesis test is divided into four steps.

- Step 1: The hypotheses
- Step 2: The test statistic and its sampling distribution
- Step 3: The rule of evidence
- Step 4: The P-value

The first step is explained in the next section.

2.2 STEP 1: THE HYPOTHESES

This presentation begins with some notation and comments that apply to all hypothesis-testing problems and then moves to a consideration of Fisher's test.

There are two hypotheses, called the **null hypothesis** and the **alternative hypothesis.** The null hypothesis is denoted by the symbol H_0, which is read "H sub zero" or "H naught," and the alternative hypothesis is denoted by H_1, read "H sub one." These hypotheses are nonoverlapping conjectures about the phenomenon being studied.

The null hypothesis for Fisher's test states that for every subject the two treatments have the same effect on the response. In the terminology of the previous section, the null hypothesis states that each subject is either type A or type D.

Fisher's test allows the researcher some discretion in choosing the alternative. More precisely, the researcher has three choices for the alternative hypothesis:

1. $n_B > n_C$. This is called the first alternative and is denoted by $>$.
2. $n_B < n_C$. This is called the second alternative and is denoted by $<$.
3. $n_B > n_C$ or $n_B < n_C$. This is called the third alternative and is denoted by $\neq$.

The first alternative states that the first treatment is superior to the second treatment, the second alternative states that the first treatment is inferior to the second treatment, and the third alternative states that the first treatment is either superior or inferior to the second treatment.

In a hypothesis test, the researcher begins by assuming that the null hypothesis is true. After collecting data, the researcher follows the steps outlined in this chapter to obtain a number called the **P-value**. The P-value measures how strongly the data support the selected alternative. If the P-value is less than or equal to 0.05, the researcher *rejects the null hypothesis* in favor of the selected alternative. If, however, the P-value is larger than 0.05, then the researcher *is unable to reject the null hypothesis*. (The rationale behind this procedure and the use of the phrases "rejects" and "is unable to reject" are made clear later in this chapter.)

The hypotheses for Fisher's test reveal another important general feature of hypothesis testing:

> The choice of the null and alternative hypotheses need not exhaust all possibilities.

For example, combining the null hypothesis with the first alternative does not cover the possibility that the first treatment is inferior to the second treatment.

The most controversial aspect of hypothesis testing is the choice of the alternative hypothesis. The reason the choice is controversial is that for a given set of experimental results (a given set of data) the three different choices for the alternative will yield at least two, and usually three, different P-values. This is important because the eventual conclusions of a study depend heavily on the P-value. Thus, it is useful to consider some guidelines for choosing the alternative hypothesis.

The null hypothesis states that there is no difference between treatments. Clearly, if the null hypothesis is false, there are two possibilities: the first treatment is either superior or inferior to the second treatment. The choice of the first or second alternative ($>$ or $<$) implies that the researcher is interested in only one of these possibilities, and the choice of the third alternative ($\neq$) means that the researcher is interested in either possibility. This may sound strange—why wouldn't a researcher be interested in both possibilities? This question is easiest to answer by considering several examples.

Consider the study of chronic Crohn's disease, introduced in Chapter 1 on p. 20. The first treatment was cyclosporine therapy, and the second treatment was a placebo. The researcher in the study chose the first alternative—that cyclosporine is superior to the placebo. A researcher investigating cyclosporine therapy is not interested in the second alternative, that cyclosporine is inferior to the placebo. Two explanations for the researcher's lack of interest in the second alternative are given below.

First, consider the researcher's next action at the conclusion of the study. If the researcher concludes that cyclosporine is superior to the placebo, he or she will advocate its use as a therapy for chronic Crohn's disease. (Or, more conservatively, the researcher might schedule more testing to determine if cyclosporine is beneficial for the public at large. Remember that the subjects in the study have not been assumed to be representative of the larger collection of disease sufferers.) On the other hand, if the researcher concludes that the treatments are identical or that cyclosporine is inferior to the placebo, the next action will be to discard cyclosporine as a potential therapy. In other words, from a practical point of view the second alternative, that cyclosporine is inferior to the placebo, is not of interest because knowledge that it is true would lead to the same action as knowledge that the null hypothesis is true. The appropriate alternative for the Chronic Crohn's Disease study is the only alternative of interest, the first alternative.

Second, it is helpful to consider the science of the study. Crohn's disease is an inflammation of the intestines, and cyclosporine is an immunosuppressive drug. In theory, cyclosporine should, if anything, be beneficial. In other words, the researcher might consider it inconceivable that cyclosporine could be inferior to the inert placebo. Thus, the researcher would not be interested in the inconceivable second alternative. A word of caution is necessary. Do not be hasty to label a possibility as inconceivable; experience indicates that occasionally what has been labeled inconceivable actually is true. (Humorous examples of this frequently occur in the movie *The Princess Bride*.)

It is fair to say that the researcher in the Chronic Crohn's Disease study hopes to conclude that cyclosporine is superior to the placebo. (One does not become a famous researcher, let alone help people, by proposing worthless therapies!) Thus, some people like to say that the choice of the alternative hypothesis reflects what the researcher is hoping to prove. This is a useful paradigm, but it does have its exceptions, as illustrated below.

Many medical studies, such as the one just discussed, compare an active therapy with an inert placebo. For such studies the natural alternative is that the active therapy is superior to the placebo. Other medical studies, however, compare a new therapy to an accepted active therapy. In many studies of this type the alternative would be that the new therapy is superior to the existing therapy. (Apply the first argument given above for the Chronic Crohn's Disease study along with the idea that in the case of a tie, preference should be given to an existing therapy over a new therapy.) There is, however, an exception to this approach if the new therapy is proposed because it has fewer or less severe side effects.

Exercise 17 on p. 27 introduced a study of lithium therapy for persons with bipolar disorder. The new therapy was a reduced dose of lithium, and the standard therapy was the usual dose of lithium. The new therapy, treatment 1, was of interest because lithium has some undesirable side effects and it is generally believed that reducing the dose would reduce their number or severity. The two values that the formal response in the study could take on were no relapse of manic or depressive episodes (a success) and relapse (a failure). There are, of course, three possibilities:

- Perhaps the null hypothesis is true: the low and high doses have identical effects on the response.
- Perhaps the first alternative is true: the low dose is better at preventing relapses than the high dose.
- Perhaps the second alternative is true: the low dose is worse at preventing relapses than the high dose.

Consider the two types of arguments given earlier for cyclosporine therapy.

First, if the researcher concludes that the null hypothesis or the first alternative is true, the recommendation will be to give patients the low dose of lithium because of the assumed benefit of fewer or less severe side effects. If, however, the researcher concludes that the second alternative is true, the recommendation will be to give patients the usual dose of lithium because having a higher rate of successes is deemed more important than reducing the side effects. (If that were not the case, there would have been no reason to do the study—simply give everyone the lower dose.) Thus, analysis based on what should happen next leads to the second alternative, that the new therapy is *inferior* to the existing therapy.

Second, considering the science of the problem, it is inconceivable that a lower dose of lithium is more effective than the regular dose at preventing relapses. Thus, the first alternative is simply not of interest.

The lithium study illustrates another important point. Unlike the investigator of Crohn's disease, the lithium researcher hopes to conclude that the null hypothesis is true. (Why?) This is the first counterexample to the paradigm that the researcher wants to conclude that the alternative is true. (Technical note: A hypothesis test begins by assuming that the null hypothesis is true and rejects that belief only if the data are sufficiently strong in their support of the alternative. As will be seen repeatedly in this text, for any study with a small number of subjects the data must have a very strong pattern in order to allow rejection of the null hypothesis. Thus, an unethical researcher can conclude that the null hypothesis is true simply by conducting a very small study! An ethical researcher who hopes to conclude that the null hypothesis is correct should be cautious when interpreting the results of a study with a small number of subjects. The technique of estimation, introduced in Chapter 6, can be viewed as a more useful analytical technique than hypothesis testing for small studies.)

Next, consider the Three-Point Basket study introduced in Chapter 1 on p. 28. The null hypothesis states that the location of the shot—in front of the basket or in the left corner—has no effect on the outcome. The first alternative states that Clyde Gaines was better from in front of the basket, and the second alternative states that he was better from the left corner. Unlike the alternatives in the medical studies, neither of these possibilities has the same practical effect as the null hypothesis. For example, if Clyde wants to use the results of the study as a guide for future competition, then if he concludes that the null hypothesis is true, he will be indifferent to the location of his future shots. If, however, he concludes that there is a difference, he will want to know what the difference is. Moreover, neither of the possible alternatives is inconceivable. With interest in both the first and second alternatives, the researcher should choose the third alternative.

The choice of the alternative hypothesis may hinge on information available only to the researcher. For example, in the Ballerina study, introduced on p. 30, the metasubject, Julie Gleich, considered the first alternative—that she was better at performing pirouettes to her left than to her right—to be inconceivable. Thus, she chose the second alternative.

Sometimes what seems inconceivable before collecting data looks very plausible afterwards. For example, consider the Prisoner study, introduced on p. 11. Before collecting data, Ruth Parsons thought that it was inconceivable that promising a sentence reduction would reduce the likelihood of volunteering (I agreed with her). The actual data, however, gave strong evidence that the inconceivable was true: 72 percent of the inmates promised a sentence reduction said they would volunteer, compared with 92 percent of the inmates not promised a reduction! This example suggests that a researcher should not be too hasty to label either alternative inconceivable.

Finally, it is extremely important that the researcher selects the alternative hypothesis before collecting data. As the following example suggests, it is easy to be fooled by hindsight.

EXAMPLE 2.1: *Hindsight Study*

Motivated by a presentation in a social psychology class, Tracy Morovits decided to investigate whether people exhibit 20/20 hindsight when interpreting the findings of research studies. (If this statement is unclear, it will become clear shortly.) Tracy conducted a balanced CRD with 50 subjects. Subjects assigned to the first treatment were told

> Social psychology research has shown that the old proverb "Birds of a feather flock together" is true.

Subjects assigned to the second treatment were told

> Social psychology research has shown that the old proverb "Opposites attract" is true.

2.2 Step 1: The Hypotheses · 47

TABLE 2.3 Results of the Hindsight Study

Reported "Truth"	No (Not Surprised)	Yes (Surprised)	Total
"Birds of a feather...."	21	4	25
"Opposites attract."	18	7	25
Total	39	11	50

TABLE 2.4 Row proportions for the Hindsight Study

Reported "Truth"	No (Not Surprised)	Yes (Surprised)	Total	Sample Size
"Birds of a feather...."	0.84	0.16	1.00	25
"Opposites attract."	0.72	0.28	1.00	25

Note that these proverbs can be viewed as contradictory. Each subject was asked whether or not the finding is surprising. The response "No, not surprising" was defined to be a success, and the response "Yes, surprising" was a failure. The tables of observed frequencies and row proportions for Tracy's data are given in Table 2.3 and Table 2.4, respectively.

The key feature of Tracy's data is that no matter which of the contradictory proverbs was represented as proven true, an overwhelming majority of the subjects were not surprised. Tracy's data suggest that a researcher who obtains evidence that the first treatment is better (that is, $\hat{p}_1 > \hat{p}_2$) might decide, "Oh yes, that is what I expected to find, so I should use the first alternative," but if that same researcher with the same study had obtained evidence that the first treatment was inferior ($\hat{p}_1 < \hat{p}_2$), the decision would have been, "Oh yes, that is what I expected to find, so I should use the second alternative." In reality, this researcher clearly is interested in either alternative, and by our dictum should use the third alternative. ▲

Here is a technical issue that can be read and then, if desired, forgotten. You may have noticed that a fourth alternative is missing from the earlier list, namely that $n_B = n_C > 0$. Note that the null hypothesis is false because some subjects respond differently to the two treatments, but neither treatment is better overall. It can be shown that a CRD cannot provide any information to distinguish this newest alternative from the null hypothesis. In fact, this newest possibility is excluded from the null hypothesis largely for technical mathematical reasons. For the goals of this book, it is acceptable to view this fourth alternative as having the same practical impact as the null hypothesis.

EXERCISES FOR SECTIONS 2.1 AND 2.2

1. Refer to the data from the Chronic Crohn's Disease study in Table 1.9 on p. 20. What would the Skeptic say about these data? What would the Advocate say?

2. Refer to the data from the Ballerina study in Table 1.19 on p. 31. What would the Skeptic say about these data? What would the Advocate say?

3. A balanced CRD is run on 20 subjects, and all subjects yield failures. This information allows you to compute the value of one of the numbers n_A, n_B, n_C, and n_D. Which number is it, and what is its value?

4. A balanced CRD is run on 20 subjects, and all subjects yield successes. This information allows you to compute the value of one of the numbers n_A, n_B, n_C, and n_D. Which number is it, and what is its value?

5. A balanced CRD is run on 20 subjects. All subjects on treatment 1 yield successes and all subjects on treatment 2 yield failures. This information allows you to compute the value of one of the numbers n_A, n_B, n_C, and n_D. Which number is it, and what is its value?

6. A balanced CRD is run on 20 subjects. All subjects on treatment 1 yield failures and all subjects on treatment 2 yield successes. This information allows you to compute the value of one of the numbers n_A, n_B, n_C, and n_D. Which number is it, and what is its value?

7. Refer to the study of sorority sisters introduced in Exercise 3 on p. 23. Becca believed it was inconceivable that women would be more likely to tell on a sister than to tell on a sister's boyfriend. Which alternative should she have used?

8. Refer to the study of fire fighters introduced in Exercise 4 on p. 24. If Mike was interested in any difference between versions of his question, which alternative should he have used?

9. Refer to the study of ice cream cones introduced in Exercise 7 on p. 24. The shop makes a bigger profit from selling waffle cones, so Andre was interested in finding any difference between the two questions. Which alternative should be used?

10. Refer to the study of giving CPR and mouth-to-mouth resuscitation to a stranger introduced in Exercise 8 on p. 25. Lynn believed that the only conceivable alternative was that her subjects would be more likely to give CPR at the wedding. Which alternative should she have used?

11. Refer to the study of the University Counseling Service introduced in Exercise 9 on p. 25. If Anita was interested in finding any difference between the two questions, which alternative should she have used?

12. Refer to the study of the wording of the description of the Btu-Energy Bill introduced in Exercise 10 on p. 25. Dave and John believed that the only conceivable alternative was that giving persons more information would increase the chance of a success. Which alternative should they have used?

13. Refer to the study of Afro-centric schools introduced in Exercise 11 on p. 26. Lisa believed that the only conceivable alternative was that not mentioning the exclusion of girls would increase the chance of a success. Which alternative should she have used?

14. Refer to the "Are You Shallow?" study introduced in Exercise 12 on p. 26. The researchers believed that the only conceivable alternative was that including the word *attractive* would increase the chance of a success. Which alternative should they have used?

15. Refer to the study of tables in history text introduced in Exercise 13 on p. 26. The researchers were interested in finding any difference between the two treatments. Which alternative should they have used?

16. Refer to the study of chronic hepatitis introduced in Exercise 14 on p. 27. Which alternative should be used?

17. Refer to the study of pregnancy-induced hypertension introduced in Exercise 15 on p. 27. Which alternative should be used?

18. Refer to the study of two treatments for breast cancer introduced in Exercise 16 on p. 27. Which alternative should be used? (Hint: Refer to the lithium study in the text. What are some side effects of the two treatments of this exercise?)

19. Refer to the study of the ambidexterity of Pam Crawford's son introduced in Exercise 5 on p. 33. She chose the third alternative. Briefly explain why that is a reasonable choice.

20. Refer to the study of tennis serves introduced in Exercise 6 on p. 33. Before performing the study, Ken thought it was inconceivable that he would serve better into the deuce court. Which alternative should be used?

21. Refer to the study of basketball bank shots introduced in Exercise 7 on p. 33. If Loretta believed it was inconceivable that her husband would shoot better from the front, which alternative should she have used?

22. Carefully read Wayne Bell's motivation for the free-throw study at his high school given in Exercise 8 on p. 33. Which alternative should be used?

23. Refer to the study of Bozo's *Grand Prize Game* introduced in Exercise 9 on p. 33. Choose the alternative and justify your choice.

24. Refer to the study of putting introduced in Exercise 10 on p. 34. Before performing the study, Mike had no idea which type of break (right-to-left or left-to-right) was more difficult for him. Which alternative should be used?

25. Refer to the study of horseshoes introduced in Exercise 11 on p. 34. Choose the alternative and justify your choice.

26. Refer to the study of throwing lawn darts introduced in Exercise 12 on p. 34.
    (a) If it is known that Kelly always plays with her right hand, which alternative should be used?
    (b) If it is known that Kelly always plays with her left hand, which alternative should be used?
    (c) If it is known only that Kelly is ambidextrous, which alternative should be used?

27. Refer to the study to compare men's and women's basketballs introduced in Exercise 13 on p. 34. If Francell was interested in finding any difference between balls, which alternative Should have been used?

28. Refer to the study of putting from two distances introduced in Exercise 14 on p. 34. Choose the alternative and justify your choice.

29. Refer to the study of dart throwing introduced in Exercise 15 on p. 34. Given that Kim always throws darts with her right hand, choose the alternative and justify your choice.

30. Refer to the study of shooting a gun introduced in Exercise 16 on p. 34. If Robert was interested in finding any difference between positions, which alternative should he have used?

31. Refer to the study "Is Pearl Ambidextrous?" introduced in Exercise 17 on p. 34. If Mary discovered that Pearl was weaker at either the left or right canter, then she would use that information to direct future training of Pearl. Which alternative should she have used?

32. Refer to the study of tossing peanuts into the air introduced in Exercise 18 on p. 35. If Susan's father had been the metasubject, which alternative would she have chosen? Explain your answer.

33. Refer to the study of the archer who compared wet feathers and wet plastic vanes introduced in Exercise 19 on p. 35. The researchers chose the third alternative. Write a brief justification of their choice.

34. Refer to the study of long putts introduced in Exercise 20 on p. 35. The researcher chose the third alternative. Write a brief justification of his choice.

35. Refer to your balanced CRD on 100 trials introduced in Exercise 21 on p. 35. Which alternative should you use? Explain your answer.

36. Refer to your balanced CRD on 50 subjects introduced in Exercise 18 on p. 27. Which alternative should you use? Explain your answer.

2.3 STEP 2: THE TEST STATISTIC AND ITS SAMPLING DISTRIBUTION

As mentioned above, the P-value measures how strongly the data support the alternative hypothesis. For Fisher's test, the data are elements of the 2 × 2 contingency table of observed counts. Some of the information in the data, however, is irrelevant for the purpose of comparing hypotheses. For example, the values n, n_1, and n_2 simply reflect the experimenter's decision on how many subjects to include in the study and how to divide them between the treatments. They give no insight into the relative merits of the treatments. The numbers m_1 and m_2 give information about how common successes and failures are, but provide no information on which treatment is better.

2.3.1 The Test Statistic

The **test statistic** is the number that summarizes the information in the data that is relevant to the problem of deciding between the hypotheses. It is sometimes referred to informally as the evidence in the data. For Fisher's test, and all other tests in this book, the test statistic is the same regardless of the alternative selected by the researcher. A mathematical proof that a particular formula is the best, or even a reasonable, choice for the test statistic can be a formidable task. In this text each test statistic is given without a mathematical proof of its merit. Fortunately, in all cases the test statistic is intuitively reasonable.

The hypotheses are concerned with comparing how the treatments would perform if both could be given to every subject. Since these performances cannot be observed, it is natural to use as evidence a comparison of the treatments based on the actual data, or, in other words, a comparison of the sample proportions of successes on each treatment. Thus,

> The test statistic for Fisher's test is the sample proportion of successes on the first treatment minus the sample proportion of successes on the second treatment.

The observed value of the test statistic for any particular study is denoted by x, and using the notation of Chapter 1, $x = \hat{p}_1 - \hat{p}_2$. For the Colloquium study, for example, reading from Table 1.2 (p. 14) the value of the test statistic is

$$x = \hat{p}_1 - \hat{p}_2 = 0.50 - 0.07 = 0.43.$$

The value of the test statistic reflects the evidence in the data: the proportion of successes with the first version is 43 percentage points higher than the proportion of successes with the second version. The researcher selected the first alternative for the Colloquium study. The P-value measures how strongly this observed value, 0.43, supports the alternative. To proceed, one must study **probability,** the language of the P-value.

2.3.2 Probability

You are no doubt familiar with the word *probability*. All that is needed now is a relatively gentle introduction to the aspects of probability that are required to perform Fisher's test. Probability is studied more extensively later in the book; the lessons learned now will make the later material easier to understand and use.

The fundamental notion behind probability is the idea of a chance mechanism. A **chance mechanism** is any phenomenon that leads to an **outcome** whose value cannot be predicted, with certainty, in advance. The chance mechanism of interest in this chapter is the assignment of subjects to treatments by randomization. An outcome is any particular assignment that *could* result from the process of randomizing. The first step in studying a chance mechanism is to list its possible outcomes, as illustrated in the following example.

EXAMPLE 2.2: The Class Study

A researcher decided to perform a balanced CRD of the General's Dilemma on four students in a class, Ken, Monice, Pam, and Sarah. Table 2.5 lists the six possible assignments of subjects to treatments. It is convenient, for ease of reference, to number the assignments in an arbitrary manner, as given in the first column of the table. ▲

Any specified collection of outcomes is called an **event**. The event that consists of all possible outcomes is called the **sample space** and is denoted by S, which is read "script ess." The sample space for the Class study consists of the six assignments presented in Table 2.5. With the exception of the sample space S, events are usually denoted by capital letters near the beginning of the alphabet. An event can be described either verbally or by listing its outcomes. Below are three events for the Class study, each presented both ways.

- $A = \{$Ken is assigned to the first version.$\} = \{1, 2, 3\}$.
- $B = \{$Ken and Monice read the same version.$\} = \{1, 6\}$.
- $C = \{$Ken and Monice are assigned to the first version.$\} = \{1\}$.

If an outcome satisfies the conditions of an event, it is said to be **in** or **belong to** the event. For example, assignments 1, 2, and 3 belong to event A, assignments 1 and 6 belong to event B, and assignment 1 is in event C. After the chance mechanism yields its outcome, an event is said to have **occurred** if the observed outcome belongs to the event. For example, suppose that the actual randomization for

TABLE 2.5 Possible Assignments of Subjects to Treatments for the Class Study

	Subjects Assigned	
Assignment Number	Version 1	Version 2
1	Ken, Monice	Pam, Sarah
2	Ken, Pam	Monice, Sarah
3	Ken, Sarah	Monice, Pam
4	Monice, Sarah	Ken, Pam
5	Monice, Pam	Ken, Sarah
6	Pam, Sarah	Ken, Monice

the Class study yielded assignment 2; then event A occurred, but events B and C did not occur. Similarly, if the randomization yielded assignment 1, all three of the events occurred; if it yielded assignment 4 or 5, none of these three events occurred.

As noted above, before the outcome of the chance mechanism is observed, there is no way to predict, with certainty, what it will be. Clearly, however, some events are more likely to occur than others. For example, event A is more likely to occur than event C because A **contains** C; that is, A includes the only outcome in C, plus other outcomes. Similarly, B is more likely to occur than C. To be really useful, however, this idea of "more likely" must be made more precise and must be extended to comparisons of events that do not satisfy containment relationships. For example, is A more likely than B? If so, how much more likely? Probability theory provides the answers to these and other questions.

There are two main questions in probability theory:

1. How should probabilities be assigned to the outcomes of a chance mechanism?
2. Once these probabilities are assigned, what rules must they obey?

The first question is basically scientific and the second, mathematical. It turns out that the simplest way to assign probabilities to outcomes is appropriate for Fisher's test.

The condition of the **equally likely case** is that any possible outcome is just as likely to occur as any other. Quite often a researcher is willing to assume that the chance mechanism under consideration satisfies this condition. For example, one might be willing to assume that the two possible outcomes of tossing a coin, heads and tails, are equally likely to occur or that the six possible outcomes of casting a die are equally likely to occur. More to the point of this chapter, if the cards used to randomize are indistinguishable and thoroughly mixed and the researcher does not look when selecting cards, it seems reasonable to assume that the possible assignments of subjects to treatments are equally likely. For example, the six assignments listed in Table 2.5 are equally likely to occur.

Probabilities for the equally likely case are assigned by the equation

$$P(\text{any particular outcome}) = \frac{1}{\text{The number of outcomes in } S}.$$

For the Class study this rule gives

$$P(\text{any particular assignment}) = \tfrac{1}{6}.$$

For the equally likely case, the probability of an event is computed by

$$P(\text{an event}) = \frac{\text{The number of outcomes in the event}}{\text{The number of outcomes in } S}.$$

For the three events defined earlier for the Class study,

$$P(A) = \tfrac{3}{6}, \qquad P(B) = \tfrac{2}{6}, \quad \text{and} \quad P(C) = \tfrac{1}{6}.$$

In order to be useful to a researcher, probabilities must be extended from assignments to data. This is achieved by remembering that the researcher begins the hypothesis test by assuming that the null hypothesis is true. As shown below, this assumption provides the necessary bridge between assignments and data. It is important to remember that all probability computations for data in this and the following two chapters assume that the null hypothesis is true.

It is now necessary to reveal some information about the results of the Class study.

Ken and Monice yielded successes, and Pam and Sarah yielded failures.

Assuming that the null hypothesis is true, each subject is either type A or type D, where these types are as defined in Table 2.1 (p. 41). Ken and Monice gave successes, so they must be type A; similarly, Pam and Sarah are type D. Thus, according to the null hypothesis, Ken and Monice would have given successes after reading either version, and Pam and Sarah would have given failures after reading either version. The information on responses and the assumption that the null hypothesis is true enable one to compute the data that would have been obtained with any of the six possible assignments. Refer to Table 2.5. Assignment 1 has Ken and Monice read version 1 and Pam and Sarah read version 2. Thus, this assignment would give two successes and no failures with the first version, and no successes and two failures with the second version. These results are presented in Table 2.6.

By a similar line of reasoning, assignment 2 would result in one success and one failure with each version. This result is presented in Table 2.7. Continuing this process, assignments 3, 4, and 5 also would each give one success and one failure with each version. Finally, assignment 6 would give no successes and two failures with the first version and two successes and no failures with the second version. This result is presented in Table 2.8.

For Table 2.6 it is easy to verify that

$$\hat{p}_1 = 1 \quad \text{and} \quad \hat{p}_2 = 0.$$

Thus, if assignment 1 was selected, the value of the test statistic would be

$$x = \hat{p}_1 - \hat{p}_2 = 1 - 0 = 1.$$

TABLE 2.6 *Class Study Data That Would Result from Assignment 1*

| | Route Selected | | |
Version	First	Second	Total
1	2	0	2
2	0	2	2
Total	2	2	4

TABLE 2.7 Class Study Data That Would Result from Assignments 2, 3, 4, or 5

	Route Selected		
Version	First	Second	Total
1	1	1	2
2	1	1	2
Total	2	2	4

TABLE 2.8 Class Study Data That Would Result from Assignment 6

	Route Selected		
Version	First	Second	Total
1	0	2	2
2	2	0	2
Total	2	2	4

Similarly, if assignment 2, 3, 4, or 5 was selected, then
$$\hat{p}_1 = 0.5, \quad \hat{p}_2 = 0.5,$$
and the value of the test statistic would be
$$x = \hat{p}_1 - \hat{p}_2 = 0.5 - 0.5 = 0.$$
Finally, if assignment 6 was selected, then
$$\hat{p}_1 = 0, \quad \hat{p}_2 = 1,$$
and the value of the test statistic would be
$$x = \hat{p}_1 - \hat{p}_2 = 0 - 1 = -1.$$

The earlier definition of probability can be applied to the above observations to give the following equations:

$$P(\text{The test statistic} = 1) = \frac{\text{The number of assignments that give } x = 1}{6}$$
$$= \frac{1}{6}.$$

$$P(\text{The test statistic} = 0) = \frac{\text{The number of assignments that give } x = 0}{6}$$
$$= \frac{4}{6}.$$

$$P(\text{The test statistic} = -1) = \frac{\text{The number of assignments that give } x = -1}{6}$$

$$= \frac{1}{6}.$$

It is tedious to write "The test statistic" in a probability statement. Usually, this phrase is replaced by the symbol X, transforming these three equations to

$$P(X = 1) = \tfrac{1}{6}, \quad P(X = 0) = \tfrac{4}{6}, \quad \text{and} \quad P(X = -1) = \tfrac{1}{6}. \tag{2.1}$$

The three equations in Display 2.1 are referred to jointly as the (null) sampling distribution of the test statistic X. The word *null* is suppressed unless one wants to draw explicit attention to the fact that the sampling distribution is computed under the assumption that the null hypothesis is true. These equations often are presented in a table, as in Table 2.9. Each equation above is represented by a row in the table. For example, the first row is read, "When x equals -1, then $P(X = x) = \tfrac{1}{6}$"; that is, $P(X = -1) = \tfrac{1}{6}$. The standard practice in such tables is to list the values x in order from smallest to largest. It is sometimes convenient to think of a sampling distribution as providing two types of information, namely, the possible values of the test statistic and their corresponding probabilities.

A sampling distribution also is called a **probability distribution** since it indicates how the total probability, 1, is distributed among the possible values of the test statistic.

Many of my students have found the distinction between X and x to be confusing and statements such as $P(X = x)$ to be borderline nonsensical. As a result, I will try to give further insight into this notation.

Refer to Figure 2.1 as you read this paragraph. The probabilistic aspects of Fisher's test (as opposed to the researcher's selection of the alternative hypothesis, the professional statistician's justification of the choice of the test statistic, and the rule of evidence, discussed in Section 2.5.1) begin with the chance mechanism of randomization. This chance mechanism yields an outcome—an assignment of subjects to treatments. The researcher then uses this assignment and collects data. The result of the data collection is the 2×2 contingency table.

TABLE 2.9 The (Null) Sampling Distribution of the Test Statistic X for the Class Study

x	$P(X = x)$
-1	$\tfrac{1}{6}$
0	$\tfrac{4}{6}$
1	$\tfrac{1}{6}$
Total	1

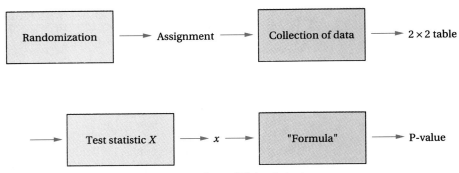

FIGURE 2.1 Schematic representation of Fisher's test.

The test statistic X can be thought of as a machine that has as its input the 2×2 table and as its output the number x, which is obtained from the rule

$$x = \hat{p}_1 - \hat{p}_2.$$

Thus, x always represents a number, whereas X represents the rule the researcher uses to obtain the number x from the data. As a crude analogy, think of the 2×2 table as a piece of fruit or vegetable, X as a juicer that the piece of fruit or vegetable is placed in, and x as the juice that comes out of X. Clearly, the juice is different from the juicer, just as x is different from X.

Another source of potential confusion is the expression $P(X = x)$. First, note that in any application of this expression, the symbol x will be replaced by a number. For example, for the Class study, the analyst might be interested in $P(X = -1)$, $P(X = 0)$, or $P(X = 1)$. Second, remember that a probability is always associated with a chance mechanism. For Fisher's test, the chance mechanism is the assignment of subjects to treatments by randomization. The meaning of the equation $P(X = 1) = \frac{1}{6}$, for example, is explained in the following sentence.

> Before the chance mechanism is operated, the probability is one-sixth that the outcome of the chance mechanism will result in a computed value of 1 for the test statistic.

Once one becomes familiar with working with probabilities, it is too tedious to use the previous sentence; instead, we say $P(X = 1) = \frac{1}{6}$.

A **random variable** is any rule (function) that assigns a number to each outcome of a chance mechanism. Thus, the test statistic X is a special case of a random variable. When appropriate, results of the next two chapters are stated in terms of random variables so they can be used later in the book. The only interesting applications of the results in the next two chapters, however, are to random variables X that are also the test statistic for Fisher's test.

A great deal of effort has been expended above to find the sampling distribution of the test statistic for a small study with only six possible assignments of subjects to treatments. Fortunately, the above derivation presents the essential steps in finding the null sampling distribution of the test statistic for a study of any size; namely,

1. List all the possible assignments that can arise from the randomization process.
2. Peek at the results to learn which subjects give a success for the response. Assuming that the null hypothesis is true, these are exactly the subjects who would give successes under any assignment of subjects to treatments.
3. For each possible assignment use the information from the previous item to determine the 2×2 table it would yield.
4. For each of the 2×2 tables just obtained, compute the value of the test statistic, x.
5. The probability of any particular value x is simply the number of assignments that would yield x divided by the total number of possible assignments.

It can be shown that there are 184,756 possible assignments of subjects to treatments for the 20 persons participating in the Infidelity study. To find the sampling distribution of its test statistic, the statistician follows the five steps listed above, but uses the following two facts to greatly reduce the tedium:

- Knowing exactly which subjects give successes and which give failures is not necessary; it suffices to know the total number of successes and failures in the study, that is, m_1 and m_2. As discussed earlier, knowing m_1 and m_2 gives no information about which treatment performed better in the study.
- Listing all assignments is not necessary. The statistician only needs to count the total number of possible assignments and count the number of assignments that lead to each different value x of the test statistic. Mathematicians have developed clever ways to obtain the necessary counts.

Verifying the validity and mastering the mechanics of these facts are beyond the scope of this book; sampling distributions are simply presented as they are needed.

Before proceeding to steps 3 and 4 of Fisher's test, the next section of the book digresses into mathematics to consider some useful properties of probability.

EXERCISES 2.3

1. A balanced CRD will be performed on six subjects. Let A, B, C, D, E, and F denote the subjects.
   (a) List all twenty possible assignments of subjects to treatments.
   (b) What is the probability of any particular assignment?
   (c) Compute the probability that A and B are assigned to the same treatment.

(d) Compute the probability that A and B are assigned to different treatments.

2. A CRD with $n_1 = 3$ and $n_2 = 2$ will be performed on five subjects. Let A, B, C, D, and E denote the subjects.

   (a) List all ten possible assignments of subjects to treatments.
   (b) What is the probability of any particular assignment?
   (c) Compute the probability that A and B are assigned to the same treatment.
   (d) Compute the probability that A and B are assigned to the first treatment.
   (e) Compute the probability that A and B are assigned to different treatments.

3. Refer to Exercise 1. You determine that subjects A, B, and C gave successes and subjects D, E, and F gave failures. Find the null sampling distribution of the test statistic X.

4. Refer to Exercise 2. You are aware that subjects A, B, and C gave successes and subjects D and E gave failures. Find the null sampling distribution of the test statistic X. In the narrative to Chapter 3 this study is referred to as the Memorization study.

5. Repeat Exercise 3 assuming that subjects A and B gave successes and subjects C, D, E, and F gave failures.

6. Repeat Exercise 4 assuming that subjects A and B gave successes and subjects C, D, and E gave failures.

2.4 WORKING WITH PROBABILITY

2.4.1 The Meaning of Probability

Let us return to the possible assignments of subjects to treatments for the Class study. The sample space for the chance mechanism is

$$S = \{1, 2, 3, 4, 5, 6\},$$

and the six outcomes in the sample space are equally likely. The events

$$A = \{1, 2, 3\}, \quad B = \{1, 6\}, \quad \text{and} \quad C = \{1\}$$

were defined and their probabilities were shown to be

$$P(A) = \tfrac{3}{6}, \quad P(B) = \tfrac{2}{6}, \quad \text{and} \quad P(C) = \tfrac{1}{6}.$$

This section begins by addressing the specific question,

What does it mean to say that the probability of an event equals some number?

One answer to this question is given by the long-run relative frequency interpretation of probability, which is illustrated and then stated below.

Some chance mechanisms can be observed over and over again, but others can be observed only once. For example, the chance mechanism of tossing a coin and observing heads or tails can be repeated any number of times, but observing which baseball team wins the World Series in a given year cannot. The long-run relative frequency interpretation of probability is applicable only to chance mechanisms that can be observed repeatedly under identical conditions.

TABLE 2.10 Results of 60,000 Generated Assignments of Subjects to Treatments for the Class Study

Assignment	Frequency	Relative Frequency	Probability	Difference
1	10,123	0.1687	0.1667	0.0020
2	9,996	0.1666	0.1667	−0.0001
3	9,976	0.1663	0.1667	−0.0004
4	9,989	0.1665	0.1667	−0.0002
5	9,925	0.1654	0.1667	−0.0013
6	9,991	0.1665	0.1667	−0.0002
Total	60,000	1.0000		1.0002

The chance mechanism of assigning subjects to treatments by randomization can be observed repeatedly under identical conditions; after obtaining a particular assignment, the researcher simply returns the selected cards to the box, remixes the cards, and starts again. Given enough time, an especially industrious researcher could generate any number of observed assignments. Such tedious work, however, is best left to an electronic computer. The statistical software package Minitab was used to generate the outcomes of 60,000 randomizations for the Class study. The results are given in Table 2.10. The first column of this table lists the six possible outcomes (assignments) of the chance mechanism. The second column presents the observed frequencies of each assignment; for example, 10,123 generated randomizations yielded the first assignment, 9,996 yielded the second assignment, and so on. The entries in the third column are obtained by dividing each entry in the second column by 60,000, the total number of generated outcomes. These numbers are the relative frequencies of the six assignments. The fourth column presents the probability of each outcome, $\frac{1}{6}$, written in decimal form. The similarity of the values in the third and fourth columns is striking; to facilitate a comparison, the differences of these numbers (third column minus fourth column) are given in the fifth column. For each assignment the absolute value of the difference between its relative frequency and its probability is at most 0.0020.

The number of times the event $A = \{1, 2, 3\}$ occurred in the generated outcomes is obtained by adding together the number of times each of its members—1, 2, or 3—occurred:

$$10{,}123 + 9{,}996 + 9{,}976 = 30{,}095.$$

Thus, the relative frequency of the occurrence of A is $30{,}095/60{,}000 = 0.5016$, which is very close to $P(A) = 0.5$. Similarly, the event $B = \{1, 6\}$ occurred $10{,}123 + 9{,}991 = 20{,}114$ times, giving a relative frequency of $20{,}114/60{,}000 = 0.3352$. This relative frequency is very close to $P(B) = 0.3333$.

The above computations illustrate the following mathematical result, called the **long-run relative frequency interpretation of probability:**

Suppose a chance mechanism can be repeated a large number of times under identical conditions. The relative frequency of occurrence of any event is approximately equal to its probability.

This result gives one answer to the question posed at the beginning of this section, What does it mean to say that the probability of an event equals some number? If the chance mechanism can be repeated a large number of times under identical conditions, knowing the probability of an event enables one to predict fairly accurately what proportion of the observed outcomes of the chance mechanism will yield an occurrence of the event. For example, suppose an event of interest has probability equal to 0.40; in a large number of observations of the chance mechanism, the proportion of observations in which the event occurs will be approximately 0.40.

The preceding paragraph is deliberately imprecise. What constitutes "identical conditions" and what it means to "predict fairly accurately" are considered carefully in Chapters 5 and 6.

2.4.2 The Rules of Probability

The previous section stated that the two main questions in probability theory are

1. How should probabilities be assigned to the outcomes of a particular chance mechanism?
2. Once assigned, what rules must probabilities obey?

Further, the simplest way to assign probabilities to outcomes, the equally likely case, was defined and investigated. Other methods of assignment are presented in the exercises and narrative of this book. Now, however, attention is turned to the second question.

Every assignment of probabilities must obey three rules. Learning these rules serves two purposes. First, the rules can help reveal bogus assignments of probabilities. Second, the rules and their logical consequences can be used to change difficult computational problems into simple ones.

- The first rule of probability is that the probability of the sample space must equal 1:

$$P(S) = 1. \qquad (2.2)$$

- The second rule of probability is that for any event A, the probability of A must be between 0 and 1, inclusive:

$$0 \leq P(A) \leq 1. \qquad (2.3)$$

Let A and B be two events. These events can be used to create two new events. First, AB denotes the event that consists of all outcomes that are in both A and B. Second, $(A \text{ or } B)$ denotes the event that consists of all outcomes that are in A or B or both. The event with no outcomes is called the **empty event,** or **impossible** event. Two events *A* and *B* are called **disjoint** if they have no outcomes in common: *A* and *B* are disjoint if *AB* equals the empty event.

- The third rule of probability is called the addition rule:
$$P(A \text{ or } B) = P(A) + P(B), \tag{2.4}$$
provided that A and B are disjoint.

In words, the third rule states that the probability that (*A* or *B*) occurs is equal to the sum of the probabilities of the individual events. It is crucial that the events be disjoint; otherwise, the right side of Equation 2.4 will double count some outcomes (see the exercises). If you find Venn diagrams to be useful, you may want to examine the left diagram in Figure 2.2.

Those who enjoy mathematics can verify that the assignment of probabilities to events for the equally likely case satisfies these three rules.

The addition rule can be extended to any number of events, provided all pairs of events are disjoint. For example, if *A* and *B*, *A* and *C*, and *B* and *C* are disjoint,

$$P(A \text{ or } B \text{ or } C) = P(A) + P(B) + P(C).$$

The use of the third rule is illustrated for the Class study of the previous section. The sampling distribution of the test statistic *X*, given in Display 2.1, is reproduced here for convenience:

$$P(X = 1) = \tfrac{1}{6}, \quad P(X = 0) = \tfrac{4}{6}, \quad \text{and} \quad P(X = -1) = \tfrac{1}{6}.$$

Suppose a researcher wants to compute $P(X \geq 0)$. Certainly, one approach is to return to the original definition of probability and count the number of assignments that lead to a value of 0 or larger for the test statistic. Returning to the definition can be tedious; using the sampling distribution and the rules of probability is much more efficient.

Of the three rules of probability, only the third enables one to compute new probabilities from old. Before one can use this rule, the event of interest, $(X \geq 0)$, must be rewritten using the word *or*. Because the possible values of *X* are $-1, 0,$ and 1,

$$P(X \geq 0) = P(X = 0 \text{ or } X = 1).$$

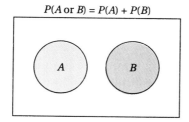
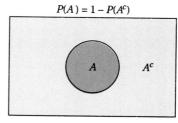

FIGURE 2.2 *Venn diagrams that illustrate the addition and complement rules.*

TABLE 2.11 *Sampling Distribution of the Test Statistic for the Infidelity Study*

x	$P(X = x)$	$P(X \leq x)$	$P(X \geq x)$
−0.9	0.0001	0.0001	1.0000
−0.7	0.0027	0.0027	0.9999
−0.5	0.0322	0.0349	0.9973
−0.3	0.1500	0.1849	0.9651
−0.1	0.3151	0.5000	0.8151
0.1	0.3151	0.8151	0.5000
0.3	0.1500	0.9651	0.1849
0.5	0.0322	0.9973	0.0349
0.7	0.0027	0.9999	0.0027
0.9	0.0001	1.0000	0.0001
Total	1.0002		

Next, note that the events $(X = 0)$ and $(X = 1)$ have no outcomes in common (each assignment gives one value of the test statistic, not two), so they are disjoint events. Thus, by the third rule of probability,

$$P(X \geq 0) = P(X = 0 \text{ or } X = 1) = P(X = 0) + P(X = 1).$$

These last two probabilities are obtained from the sampling distribution of X; thus,

$$P(X \geq 0) = P(X = 0 \text{ or } X = 1) = P(X = 0) + P(X = 1) = \tfrac{4}{6} + \tfrac{1}{6} = \tfrac{5}{6}.$$

It can be shown that the first two columns of Table 2.11 give the sampling distribution of the test statistic X for the Infidelity study. For example, the fourth row of this table yields $P(X = -0.3) = 0.1500$. The numbers in the third and fourth columns of this table are obtained from the first two columns of the table and the repeated application of the addition rule. For example,

$$P(X \geq 0.3) = P(X = 0.3 \text{ or } X = 0.5 \text{ or } X = 0.7 \text{ or } X = 0.9)$$
$$= P(X = 0.3) + P(X = 0.5) + P(X = 0.7) + P(X = 0.9)$$
$$= 0.1500 + 0.0322 + 0.0027 + 0.0001 = 0.1850,$$

which agrees, except for round-off error, with the appropriate value in the fourth column of the table.

The three rules of probability have a useful logical consequence called the complement rule. Let A be any event. The **complement** of A is the event that consists of all outcomes in the sample space that are not in A. The complement of A is denoted by A^c.

- The complement rule states that

$$P(A) = 1 - P(A^c). \tag{2.5}$$

TABLE 2.12 Two Bogus Sampling Distributions

I		II	
x	$P(X = x)$	x	$P(X = x)$
0	0.4	0	0.3
1	0.7	1	0.4
2	−0.1	2	0.2
Total	1.0	Total	0.9

If you find Venn diagrams to be useful, you may want to examine the right diagram in Figure 2.2.

Consider the Infidelity study again. Suppose the analyst wants to compute $P(X \leq 0.7)$ using only the first two columns of Table 2.11. The answer could be obtained by using the addition rule, but it would be very tedious. An examination of the first column of Table 2.11 shows that the complement of the event $(X \leq 0.7)$ is the event $(X = 0.9)$. The probability of this latter event can be obtained immediately from Table 2.11. Thus, by the complement rule,

$$P(X \leq 0.7) = 1 - P(X = 0.9) = 1 - 0.0001 = 0.9999.$$

Table 2.12 presents two bogus sampling distributions. Can you spot what is wrong with each one? Check your answers against the following.

- For bogus sampling distribution I, the second rule of probability is violated because $P(X = 2) = -0.1$, which is not a number between 0 and 1.
- For bogus sampling distribution II, the first rule of probability is violated because

$$P(S) = P(X = 0 \text{ or } X = 1 \text{ or } X = 2) = 0.3 + 0.4 + 0.2 = .0.9.$$

EXERCISES 2.4

1. Suppose that the sample space for a chance mechanism is

$$S = \{1, 2, 3, 4, 5, 6, 7, 8, 9, 10\}.$$

Suppose further that these ten possible outcomes are equally likely. Let A denote the event that the outcome is an odd number and let B denote the event that the outcome is larger than 8.

(a) List the outcomes in the event A; list the outcomes in the event B; list the outcomes in the event AB; list the outcomes in the event $(A \text{ or } B)$.
(b) Compute $P(A)$, $P(B)$, $P(AB)$, and $P(A \text{ or } B)$.
(c) The addition rule states that

$$P(A \text{ or } B) = P(A) + P(B).$$

Is the addition rule true for this example? Explain why or why not.

2. Suppose that the sample space for a chance mechanism is

$$S = \{1, 2, 3, \ldots, 20\}.$$

Suppose further that these twenty possible outcomes are equally likely. Let A denote the event that the outcome is larger than 14 and let B denote the event that the outcome is smaller than 8.

(a) List the outcomes in the event A; list the outcomes in the event B; list the outcomes in the event AB; list the outcomes in the event (A or B).

(b) Compute $P(A)$, $P(B)$, $P(AB)$, and $P(A \text{ or } B)$.

(c) The addition rule states that

$$P(A \text{ or } B) = P(A) + P(B).$$

Is the addition rule true for this example? Explain why or why not.

3. Below is the sampling distribution of a random variable:

x	$P(X = x)$
1	0.1
2	0.4
3	0.3
4	0.2
Total	1.0

(a) Compute $P(X = 2)$ and $P(X = 4)$.
(b) Use the addition rule to compute $P(X \geq 2)$ and $P(X \leq 3)$.
(c) Use the complement rule to compute $P(X \geq 2)$ and $P(X \leq 3)$.

4. Below is the sampling distribution of a random variable:

x	$P(X = x)$
1	0.6
3	0.3
6	0.1
Total	1.0

(a) Compute $P(X = 1)$ and $P(X = 3)$.
(b) Use the addition rule to compute $P(X \geq 3)$ and $P(X \leq 3)$.
(c) Use the complement rule to compute $P(X \geq 3)$ and $P(X \leq 3)$.

5. Refer to Table 2.12. Create two other bogus sampling distributions.

6. The "odds against" and the "odds in favor of" an event are alternatives to probability as a measure of uncertainty. The three measures are equivalent. For example, if the odds against an event B are 3-1 (read "three to one"), the odds in favor of B are 1-3 and the probability of B is $1/(3 + 1) = 1/4 = 0.25$. The general rule is

> If the odds against an event are r to t, then the odds in favor of the event are t to r and the probability of the event is $t/(r + t)$.

This rule can be rewritten as follows:

> If the probability of an event is $P(B)$, the odds against the event are $[1 - P(B)]$ to $P(B)$ and the odds in favor of the event are $P(B)$ to $[1 - P(B)]$.

For example, if $P(B) = 0.20$, the odds against B are 0.80 to 0.20 or, multiplying each term by five, 4 to 1. The odds in favor of B are 1 to 4.

Odds are particularly popular with sports fans. By March 18, 1987, the NCAA men's basketball championship tournament field had been reduced to 16 teams. A newspaper article gave the odds against winning the tournament for each of the 16 teams; the odds are reproduced in the following display [1]:

Team	Odds Against
Georgetown, Indiana	1-1
Alabama, Iowa	2-1
North Carolina, UNLV	3-1
DePaul	4-1
Syracuse	5-1
Florida, Providence, Wyoming	6-1
Kansas, Notre Dame, Oklahoma	8-1
LSU	10-1
Duke	25-1

Convert these odds to probabilities and find the sum of the sixteen probabilities. Comment. (Note: Odds or probabilities that reflect someone's opinion are referred to as subjective or personal. Thus, the above are subjective odds. There is nothing wrong with subjective odds, but they must obey the rules of probability.

Although the author of the article was poor at assigning odds, she did know basketball—Indiana won the tournament.)

7. Refer to the previous exercise. Find odds in a published source and determine whether they obey the first rule of probability.

2.5 STEPS 3 AND 4

This section completes the four steps of Fisher's test. Since there is only one null hypothesis, the first step consists of choosing among the three possible alternatives. The second step is the specification of the test statistic and the determination of its sampling distribution. The second step is the same for all three alternatives. Steps 3 and 4 are presented in this section; there are three versions of each step, one for each possible alternative hypothesis.

2.5.1 Step 3: The Rule of Evidence

Recall that hypothesis testing seeks to measure the strength of the evidence in the data. Step 2 defined the evidence in the data, or the test statistic, and set the foundation for measuring its strength via probability. Step 3 begins the task of deciding which event's probability should be computed.

The First Alternative (>) Recall that the first alternative states

The first treatment is superior to the second treatment.

This alternative was selected by the researchers for several of the studies introduced in Chapter 1. The Colloquium study is used to introduce the ideas and results for this case.

It can be shown that Table 2.13 presents the sampling distribution of the test statistic for the Colloquium study. Step 3 ignores probabilities and focuses on the possible values of the test statistic:

$$-\frac{4}{7}, -\frac{3}{7}, -\frac{2}{7}, -\frac{1}{7}, 0, \frac{1}{7}, \frac{2}{7}, \frac{3}{7}, \text{ and } \frac{4}{7}.$$

Select any two of these possible values; for illustration, select $\frac{3}{7}$ and $\frac{4}{7}$. Consider the question,

Which of these two possible values provides stronger evidence in support of the first alternative hypothesis?

Each value reflects that the first treatment does better in the data, but $\frac{4}{7}$ reflects a stronger thrashing of the second treatment than does $\frac{3}{7}$. So the answer is

The value $\frac{4}{7}$ provides stronger evidence than $\frac{3}{7}$ in support of the first alternative hypothesis.

TABLE 2.13 The Sampling Distribution of the Test Statistic for the Colloquium Study

x	$P(X = x)$	$P(X \leq x)$	$P(X \geq x)$
$-\frac{4}{7}$	0.0010	0.0010	1.0000
$-\frac{3}{7}$	0.0155	0.0165	0.9990
$-\frac{2}{7}$	0.0879	0.1043	0.9836
$-\frac{1}{7}$	0.2345	0.3388	0.8957
$\frac{0}{7}$	0.3224	0.6612	0.6612
$\frac{1}{7}$	0.2345	0.8957	0.3388
$\frac{2}{7}$	0.0879	0.9836	0.1043
$\frac{3}{7}$	0.0155	0.9990	0.0165
$\frac{4}{7}$	0.0010	1.0000	0.0010
Total	1.0002		

It can be shown that a simple general rule exists for comparing possible values of the test statistic; it is the following:

The *larger* the value of the test statistic, the stronger is the evidence in support of the first alternative.

Although the above rule is fairly simple, one can easily be misled by reading too much into it. Keep in mind the following points:

1. The rule of evidence specifies the *relative* strength of evidence in a value of the test statistic; it does not specify whether the evidence is *convincing*. For example, for the Colloquium study the rule of evidence states that 0 provides stronger evidence than $-\frac{1}{7}$ in support of the first alternative. It is intuitively reasonable and, as discussed below, correct (see P-values), to conclude that neither of these values provides convincing evidence in support of the first alternative.

2. The rule of evidence is concerned with evidence in support of the alternative, not evidence in support of the null hypothesis. For the Colloquium study, for example, the value $-\frac{4}{7}$ is the weakest possible evidence for the first alternative, but it is not the strongest evidence for the null hypothesis, which is that no difference exists between the treatments. Rather, $-\frac{4}{7}$ supports the possibility, which is uninteresting to the researcher, that the first treatment is inferior to the second treatment.

The Second Alternative (<) Recall that the second alternative states

The first treatment is inferior to the second treatment.

An easy modification of the above argument for the first alternative yields the following rule:

> The *smaller* the value of the test statistic, the stronger is the evidence in support of the second alternative.

The Third Alternative ($\neq$)   The third alternative states

The first treatment is either superior or inferior to the second treatment.

The researcher who conducted the Infidelity study chose the third alternative. Table 2.11 presents the sampling distribution of the test statistic. The possible values of the test statistic are

$-0.9, -0.7, -0.5, -0.3, -0.1, 0.1, 0.3, 0.5, 0.7$, and 0.9.

Consider any two of these values of the same magnitude but opposite sign; for example, 0.3 and -0.3. Clearly, these values present contrary evidence: the value 0.3 provides evidence that the first treatment is superior to the second, and the value -0.3 provides evidence that the first treatment is inferior to the second. The rule of evidence states that because these numbers are the same distance from 0, they provide the *same* strength of evidence in support of the third alternative.

The rule of evidence also must handle values that are a different distance from 0. It does, and the complete rule is

> The *further* the value of the test statistic is *from 0 in either direction*, the stronger is the evidence in support of the third alternative.

Thus, for example, for the Infidelity study -0.5 provides stronger evidence than 0.3 in support of the third alternative.

The three rules of evidence are presented graphically in Figure 2.3.

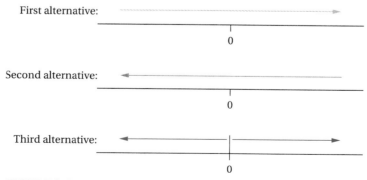

FIGURE 2.3 *The rules of evidence. The horizontal number lines denote the values of the test statistic, and the arrows point in the direction of stronger evidence in support of the alternative.*

2.5.2 Step 4: The P-value

The basic ingredients for a hypothesis test have been obtained: the alternative, the test statistic and its null sampling distribution, and the rule of evidence. These ingredients are combined to yield the P-value, the most popular single-number summary of a hypothesis test. The following definition is appropriate for any hypothesis-testing problem.

Let X denote the test statistic and let x denote its observed value. The **P-value** is equal to the (null) probability of the following event:

> The test statistic achieves a value that gives either the same evidence as x in support of the alternative hypothesis or stronger evidence than x in support of the alternative hypothesis.

The P-value also is denoted by **P**.

The above is a general definition of the P-value that is appropriate for all hypothesis-testing problems. It is important to determine what it means for Fisher's test; this is achieved by considering the rule of evidence. Recall that X denotes the test statistic and x denotes the observed value of the test statistic for the actual data.

For the first alternative, the actual evidence is x; any value of the test statistic larger than x provides stronger evidence in support of the alternative. The P-value is the probability of the obtained evidence or stronger evidence; that is,

$$\mathbf{P} = P(X \geq x) \quad \text{for the first alternative.} \tag{2.6}$$

The researcher who performed the Colloquium study selected the first alternative. The observed value of the test statistic was $x = \frac{3}{7}$. Thus, using Equation 2.6 and Table 2.13,

$$\mathbf{P} = P(X \geq x) = P(X \geq \tfrac{3}{7}) = 0.0165.$$

For the second alternative, the actual evidence is x; any value of the test statistic smaller than x provides stronger evidence in support of the alternative. The P-value is the probability of the obtained evidence or stronger evidence; that is,

$$\mathbf{P} = P(X \leq x) \quad \text{for the second alternative.} \tag{2.7}$$

For example, in the Ballerina study Julie Gleich used the second alternative, and the observed value of the test statistic was $x = -0.24$. Thus, from Equation 2.7,

$$\mathbf{P} = P(X \leq x) = P(X \leq -0.24).$$

It can be shown that this probability equals 0.0477. Thus, $\mathbf{P} = 0.0477$.

For the third alternative, the actual evidence is x; any value of the test statistic further from 0 than x, in either direction, provides stronger evidence in support of

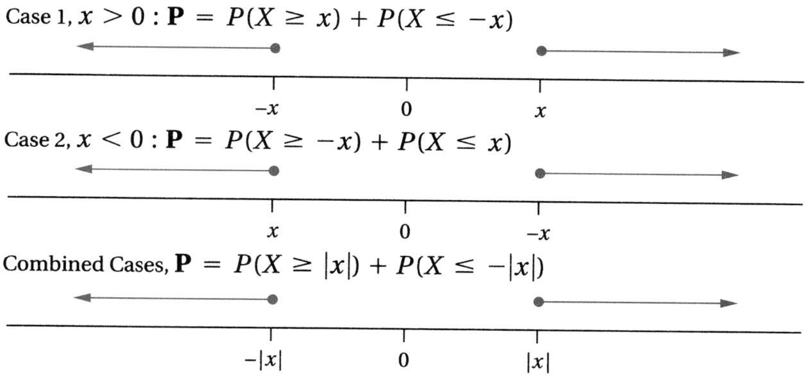

Case 1, $x > 0$: $\mathbf{P} = P(X \geq x) + P(X \leq -x)$

Case 2, $x < 0$: $\mathbf{P} = P(X \geq -x) + P(X \leq x)$

Combined Cases, $\mathbf{P} = P(X \geq |x|) + P(X \leq -|x|)$

FIGURE 2.4 *Computing the P-value for the third alternative.*

the alternative. Thus, $x = 0$ is the weakest evidence in support of the alternative, and this value of x gives a P-value of 1. If x does not equal 0, the procedure is a bit messy; you might want to refer to Figure 2.4 while perusing the remainder of this paragraph. If x is a positive number, then

$$\mathbf{P} = P(X \geq x \text{ or } X \leq -x) = P(X \geq x) + P(X \leq -x).$$

Similarly, if x is a negative number, then $-x$ is a positive number and

$$\mathbf{P} = P(X \geq -x) + P(X \leq x).$$

It is awkward to work with cases; fortunately, the above two equations can be written as one. Recall from algebra that the absolute value of the number x is denoted by $|x|$. The absolute value function has no effect on a positive number, but changes the sign of a negative number. For example,

$$|0.35| = 0.35 \quad \text{and} \quad |-0.25| = 0.25.$$

The absolute value of x can be interpreted as the distance from the point x to 0 on the number line (remember distances are always nonnegative). With the absolute value notation the above two equations for obtaining the P-value can be written as one:

$$\mathbf{P} = P(X \geq |x|) + P(X \leq -|x|) \quad \text{for the third alternative.} \tag{2.8}$$

For example, the researcher who performed the Infidelity study selected the third alternative, and the observed value of the test statistic was $x = 0.3$. Thus, from Equation 2.8 and Table 2.11,

$$\mathbf{P} = P(X \geq |x|) + P(X \leq -|x|) = P(X \geq 0.3) + P(X \leq -0.3)$$
$$= 0.1849 + 0.1849 = 0.3698.$$

2.5.3 Interpretation of the P-value

A researcher begins a study by assuming that the null hypothesis is true. The data are collected and are then summarized by the value of the test statistic. A central question of hypothesis testing is

> Is the evidence in the value of the test statistic sufficiently strong to justify discarding the assumption that the null hypothesis is true in favor of the alternative?

The P-value is used to answer this question. The P-value is a probability, so it is always a number between 0 and 1. Suppose the researcher obtains a very small P-value, for example, one in one million (0.000001). This means, literally, that the probability of obtaining evidence equal to or stronger than the evidence actually obtained is one in one million. In other words, only one-millionth of the possible assignments would yield the same or stronger evidence than what was obtained by the actual assignment. Thus, the actual evidence is very strong in its support of the alternative!

By contrast, suppose a researcher obtains **P** = 0.50. One-half of all assignments would yield the same or stronger evidence than the actual assignment. Thus, the evidence in the data is not very strong in its support of the alternative.

The above two examples, **P** = 0.000001 and **P** = 0.50, illustrate the following basic rule:

> The smaller the P-value, the stronger is the evidence in support of the alternative.

This rule can be difficult to remember because it is an inverse rule: *smaller* values of **P** provide *stronger* evidence for the alternative.

A standard terminology describes various values of **P**:

- If $P > 0.05$, the data are **not statistically significant.**
- If $0.01 < P \leq 0.05$, the data are **statistically significant.**
- If $P \leq 0.01$, the data are **highly statistically significant.**

Figure 2.5 provides a picture of this terminology.

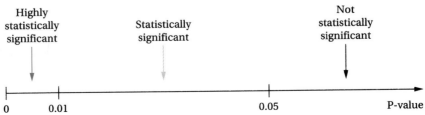

FIGURE 2.5 *Standard terminology for describing the P-value.*

TABLE 2.14 *P-values for Studies Introduced in Chapter 1*

Study	Alternative	P-value
Colloquium	First	0.0165
Chronic Crohn's Disease	First	0.0198
Soccer Goalie	First	0.0225
Chronic Crohn's Disease (at follow-up)	First	0.0256
Ballerina	Second	0.0477
Infidelity	Third	0.3698
Prisoner	First	0.9884
Three-Point Basket	Third	1.0000

The central question posed above has not been answered. For a given study, should the null hypothesis be replaced by the alternative? The standard practice among researchers is to

Reject the null hypothesis in favor of the alternative if, and only if, $P \leq 0.05$; that is, if, and only if, the data are either statistically significant or highly statistically significant.

In some studies the number 0.05 in this guideline is replaced by 0.01 or, occasionally, 0.10. Rarely are numbers below 0.01 or above 0.10 used. This guideline (using 0.05) will suffice for the purposes of this text.

Table 2.14 presents the P-values, ordered from smallest to largest, for selected studies introduced in Chapter 1. (Do not try to verify these entries!) You should note the following features:

- The Chronic Crohn's Disease study appears twice in this table; once for the data obtained at the end of the three months of treatment and once for the data obtained at follow-up, three months later.
- Henceforth in this book, P-values are rounded off to at most four digits. As a result, a P-value could be reported as 0.0000; this literally means that the P-value is smaller than 0.00005 and has been rounded to 0.0000.
- Using the standard terminology,
  - The Colloquium, Chronic Crohn's Disease (both times), Soccer Goalie, and Ballerina studies gave statistically significant data.
  - The Infidelity, Prisoner, and Three-Point Basket studies gave data that are not statistically significant.
- Applying the standard practice, there is insufficient evidence to reject the null hypothesis for the last three studies listed in the table, but sufficient evidence to reject it for the other studies. In other words, there is insufficient evidence from the subjects studied to reject the beliefs that
  - The gender of the cheater has no influence on a woman's answer.

- The promise of sentence reduction does not influence a decision to perform volunteer work.
- Shots from in front of the basket and from the left corner are equally difficult.

Moreover, there is sufficient evidence from the subjects studied to conclude that

- Posing the General's Dilemma in terms of saving lives, as compared with preventing deaths, does lead to a preference for the risk-averse route.
- Cyclosporine is superior to a placebo as a therapy for chronic Crohn's disease, both at the end of therapy and at the three-month follow-up.
- The soccer goalie and the dancer are better at moving to their right than to their left.
- It is important to remember that the conclusions of these hypothesis tests are restricted to the subjects in the studies. For example, the findings of the Soccer Goalie study do not imply that
  - All soccer goalies are better at diving to their right.
  - The metasubject is better at diving to her right in a game.
  - The metasubject would be better at diving to her right on a different practice occasion.

There are methods for generalizing findings to other subjects; they are considered beginning in Chapter 6.

2.5.4 Miscellaneous Remarks

Suppose that a report states, "The findings are not statistically significant." This is very unsatisfactory because the reader cannot tell whether the P-value is 0.0501 or 1 or something in between. Similarly, "Highly statistically significant" could mean that, among other possibilities, the P-value is one in one-hundred or one in one-billion! Thus, the P-value should always be reported to allow the reader to make a personal interpretation of the strength of the evidence.

Consider the Prisoner study again. The researcher wanted to learn whether the promise of a sentence reduction would lead to an increase in volunteerism, but the data showed the reverse pattern. This reversal is reflected in the huge P-value of 0.9884, as reported in Table 2.14. Suppose the researcher had looked at her data and decided that the second alternative was, in fact, the one that she wanted. It can be shown that for the second alternative, the P-value equals 0.0692. This huge swing in the P-value is one of the reasons the scientific community believes the alternative should be selected before looking at the data. As general advice, a unidirectional alternative (that is, the first or second alternative) should be used only if the researcher is quite certain that only one direction of effect is of interest. Otherwise, it is better to use the third alternative.

EXERCISES 2.5

1. For a CRD, two possible values of the test statistic are 0.30 and 0.50.
   (a) Which of these values provides stronger evidence in support of the first alternative?
   (b) Which of these values provides stronger evidence in support of the second alternative?
   (c) Which of these values provides stronger evidence in support of the third alternative?

2. For a CRD, two possible values of the test statistic are 0.30 and −0.50.
   (a) Which of these values provides stronger evidence in support of the first alternative?
   (b) Which of these values provides stronger evidence in support of the second alternative?
   (c) Which of these values provides stronger evidence in support of the third alternative?

3. For a CRD, two possible values of the test statistic are −0.30 and −0.50.
   (a) Which of these values provides stronger evidence in support of the first alternative?
   (b) Which of these values provides stronger evidence in support of the second alternative?
   (c) Which of these values provides stronger evidence in support of the third alternative?

4. For a CRD, two possible values of the test statistic are 0.50 and −0.50.
   (a) Which of these values provides stronger evidence in support of the first alternative?
   (b) Which of these values provides stronger evidence in support of the second alternative?
   (c) Which of these values provides stronger evidence in support of the third alternative?

5. Table 2.15 is the sampling distribution of the test statistic for a balanced CRD. Use this table to compute the following P-values.
   (a) Suppose that the researcher selected the first alternative. Find the P-value if $x = 0.24$; if $x = 0.08$.
   (b) Suppose that the researcher selected the second alternative. Find the P-value if $x = -0.32$; if $x = 0.08$.
   (c) Suppose that the researcher selected the third alternative. Find the P-value if $x = -0.48$; if $x = 0.08$.

6. Table 2.15 is the sampling distribution of the test statistic for a balanced CRD. Use this table to compute the following P-values.
   (a) Suppose that the researcher selected the first alternative. Find the P-value if $x = 0.40$; if $x = -0.16$.
   (b) Suppose that the researcher selected the second alternative. Find the P-value if $x = -0.56$; if $x = 0.00$.
   (c) Suppose that the researcher selected the third alternative. Find the P-value if $x = -0.24$; if $x = 0.48$.

7. Refer to Exercise 5. For each P-value you computed, decide if the data are not statistically significant, statistically significant, or highly statistically significant.

TABLE 2.15 Sampling Distribution for the Test Statistic for a Balanced CRD

x	$P(X = x)$	$P(X \leq x)$	$P(X \geq x)$
−0.56	0.0001	0.0001	1.0000
−0.48	0.0007	0.0008	0.9999
−0.40	0.0043	0.0051	0.9992
−0.32	0.0182	0.0232	0.9949
−0.24	0.0549	0.0782	0.9768
−0.16	0.1199	0.1981	0.9218
−0.08	0.1907	0.3888	0.8019
0.00	0.2225	0.6112	0.6112
0.08	0.1907	0.8019	0.3888
0.16	0.1199	0.9218	0.1981
0.24	0.0549	0.9768	0.0782
0.32	0.0182	0.9949	0.0232
0.40	0.0043	0.9992	0.0051
0.48	0.0007	0.9999	0.0008
0.56	0.0001	1.0000	0.0001
Total	1.0001		

TABLE 2.16 Sampling Distribution for the Test Statistic for an Unbalanced CRD

x	$P(X = x)$	$P(X \le x)$	$P(X \ge x)$
−0.55	0.0031	0.0031	1.0000
−0.40	0.0338	0.0369	0.9969
−0.25	0.1384	0.1753	0.9631
−0.10	0.2767	0.4520	0.8247
0.05	0.2980	0.7500	0.5480
0.20	0.1788	0.9288	0.2500
0.35	0.0596	0.9884	0.0712
0.50	0.0106	0.9990	0.0116
0.65	0.0009	1.0000	0.0010
0.80	0.0000	1.0000	0.0000
0.95	0.0000	1.0000	0.0000
Total	0.9999		

8. Refer to Exercise 6. For each P-value you computed, decide if the data are not statistically significant, statistically significant, or highly statistically significant.

9. Table 2.16 is the sampling distribution of the test statistic for a CRD. Use this table to compute the following P-values.

    (a) Suppose that the researcher selected the first alternative. Find the P-value if $x = 0.35$; if $x = 0.50$.

    (b) Suppose that the researcher selected the second alternative. Find the P-value if $x = -0.25$; if $x = -0.40$.

    (c) Suppose that the researcher selected the third alternative. Find the P-value if $x = 0.65$; if $x = -0.40$. [Hint: Note that $P(X \ge 0.40) = P(X \ge 0.50)$.]

10. Table 2.16 is the sampling distribution of the test statistic for a CRD. Use this table to compute the following P-values.

    (a) Suppose that the researcher selected the first alternative. Find the P-value if $x = 0.20$; if $x = 0.65$.

    (b) Suppose that the researcher selected the second alternative. Find the P-value if $x = -0.10$; if $x = -0.55$.

    (c) Suppose that the researcher selected the third alternative. Find the P-value if $x = 0.80$; if $x = -0.25$. [Hint: Note that $P(X \ge 0.25) = P(X \ge 0.35)$.]

11. Refer to Exercise 9. For each P-value you computed, decide if the data are not statistically significant, statistically significant, or highly statistically significant.

12. Refer to Exercise 10. For each P-value you computed, decide if the data are not statistically significant, statistically significant, or highly statistically significant.

2.6 BEYOND BRUTE FORCE

The most time-consuming step in conducting a hypothesis test is the determination of the sampling distribution of the test statistic. Four methods for determining the sampling distribution are discussed in this book:

1. Brute force
2. Exact mathematical solution
3. Approximate simulation solution
4. Approximate mathematical solution

The brute force method was demonstrated with the Class study; it consists of listing all the possible assignments and examining each one. It is a manageable method only for very small studies, say, three or fewer subjects on each treatment. The exact mathematical solution was used to obtain the sampling distributions presented in Table 2.11 (p. 62) and Table 2.13 (p. 66), as well as the P-values in Table 2.14. A derivation of the exact mathematical solution or even instruction on its use is beyond the scope of this text. The exact mathematical solution works well if the total number of subjects is 100 or fewer, but for larger studies the computations can become messy, even with a computer.

For a small, balanced study, Table A.1 (in the Appendix) provides values of **P** for Fisher's test that are obtained from the exact mathematical solution. To use this table, recall the standard notation for a 2 × 2 contingency table in Format 1, which was introduced in Chapter 1. It is reproduced here for convenience:

Treatment	S	F	Total
1	a	b	n_1
2	c	d	n_2
Total	m_1	m_2	n

Table A.1 can be used only if $n_1 = n_2$ and if this common value is 14 or less. (More extensive tables are available, for example, in [2].) The researcher must determine the values of R, C, and O from the data to use Table A.1. The rules for determining these values and **P** are as follows:

1. R is equal to the common value of n_1 and n_2.
2. C is equal to the smaller of m_1 and m_2.
3. For the first alternative (>),
   a. If $m_1 \leq m_2$, then $O = a$; otherwise, $O = d$.
   b. Locate the entry in the Prob. column of Table A.1 for the above values of R, C, and O. The entry equals the P-value. If the above values of R, C, and O cannot be located in the table, the P-value is greater than or equal to 0.2500.

   For the second alternative (<),
   a. If $m_1 \leq m_2$, then $O = c$; otherwise, $O = b$.
   b. Locate the entry in the Prob. column of Table A.1 for the above values of R, C, and O. The entry equals the P-value. If the above values of R, C, and O cannot be located in the table, the P-value is greater than or equal to 0.2500.

   For the third alternative (≠),
   a. If $m_1 \leq m_2$, then O equals the larger of a and c; otherwise, O equals the larger of b and d.

b. Locate the entry in the Prob. column of Table A.1 for the above values of R, C, and O. Two times the entry equals the P-value. If the above values of R, C, and O cannot be located in the table, the P-value is greater than or equal to 0.5000.

These rules are illustrated with two examples.

For the Colloquium study, reading from Table 1.1 (p. 14),

$$a = 7, b = 7, c = 1, d = 13, n_1 = 14, n_2 = 14, m_1 = 8, \text{ and } m_2 = 20.$$

Thus, for the first alternative,

1. $R = 14$, the common value of n_1 and n_2.
2. $C = 8$, the smaller of m_1 and m_2.
3. Because $m_1 \leq m_2$, $O = a$, which is 7. Entering Table A.1 at $R = 14$, $C = 8$, and $O = 7$ gives 0.0165 in the Prob. column; thus, **P** = 0.0165, which agrees with the answer found earlier (see Table 2.14).

For the Infidelity study, reading from Table 1.3 (p. 16),

$$a = 7, b = 3, c = 4, d = 6, n_1 = 10, n_2 = 10, m_1 = 11, m_2 = 9.$$

Thus, for the third alternative,

1. $R = 10$, the common value of n_1 and n_2.
2. $C = 9$, the smaller of m_1 and m_2.
3. Because $m_1 > m_2$, O equals the larger b and d, which is 6. Entering Table A.1 at $R = 10$, $C = 9$, and $O = 6$ gives 0.1849 in the Prob. column. Thus **P** = 2(0.1849) = 0.3698, which agrees with the answer obtained earlier (see Table 2.14).

The exercises provide other opportunities for the application of this rule.

The two approximation methods mentioned at the beginning of this section are the main topics of Chapter 3.

EXERCISES 2.6

1. For the following data, use Table A.1 to obtain the P-value for the first alternative; for the third alternative.

Treatment	S	F	Total
1	9	1	10
2	3	7	10
Total	12	8	20

2. For the following data, use Table A.1 to obtain the P-value for the first alternative; for the third alternative.

Treatment	S	F	Total
1	9	5	14
2	7	7	14
Total	16	12	28

3. For the following data, use Table A.1 to obtain the P-value for the second alternative; for the third alternative.

Treatment	S	F	Total
1	5	6	11
2	8	3	11
Total	13	9	22

4. For the following data, use Table A.1 to obtain the P-value for the second alternative; for the third alternative.

Treatment	S	F	Total
1	3	9	12
2	8	4	12
Total	11	13	24

5. Exercise 5 on p. 24 introduced the Patient's Dilemma. The earlier information implies that $a = d = 4$ and $b = c = 1$. Use Table A.1 to obtain the P-value for the first alternative.

6. Exercise 8 on p. 33 introduced a study of the effect of the lighting in the gymnasium at Oregon High School. The earlier information implies that $a = 6$, $b = 4$, $c = 8$, and $d = 2$. Use Table A.1 to obtain the P-value for the second alternative.

REFERENCES

1. Jackie MacMullen, "Indiana, Georgetown May Have Inside Track," *Wisconsin State Journal*, March 18, 1987, pp. 266–281.

2. W. Beyer, ed., *Handbook of Tables for Probability and Statistics*, 2d ed. (Cleveland: Chemical Rubber Company, 1974).

Approximating a Sampling Distribution

- **3.1 THE PROBABILITY HISTOGRAM**
- **3.2 THE SIMULATION EXPERIMENT**
- **3.3 MEASURES OF CENTER AND SPREAD**
- **3.4 THE STANDARD NORMAL CURVE**
- **3.5 A COMPARISON OF METHODS**
- **3.6 COMPUTING THE MEAN AND STANDARD DEVIATION***

Chapter 3

As mentioned in Chapter 2, if a completely randomized design includes a large number of subjects, the sampling distribution and the P-value can be computationally inaccessible, even with an electronic computer. In this case the researcher must settle for an approximate answer. One can approximate a sampling distribution either by using a simulation experiment, usually with the aid of a computer, or by using an appropriate mathematical formula. Both methods are introduced in this chapter. Having a visual representation of a sampling distribution makes it easier to understand the approximation methods. Thus, this chapter begins with the probability histogram, which is a picture of the sampling distribution.

3.1 THE PROBABILITY HISTOGRAM

Figure 3.1 displays the **probability histogram** of the test statistic for the Infidelity study. Two skills are developed in this section:

- How to construct a probability histogram from a sampling distribution.
- How to use a probability histogram to compute probabilities.

Construction of the probability histogram is explored first.

Table 2.11 (p. 62) presents the sampling distribution of the test statistic for the Infidelity study. If the possible values of the test statistic are written in numerical order,

$-0.9, -0.7, -0.5, -0.3, -0.1, 0.1, 0.3, 0.5, 0.7$, and 0.9,

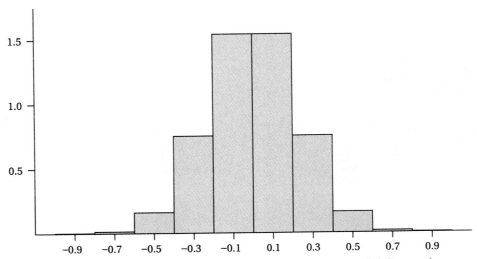

FIGURE 3.1 Probability histogram of the test statistic for the Infidelity study.

one notices that successive values differ by a constant amount, 0.2. This feature is not unique to the Infidelity study; it can be shown always to be true for the test statistic for Fisher's test. Note, however, that the constant typically differs from study to study; for example, the constant equals $\frac{1}{7}$ for the Colloquium study.

Here are the steps in constructing a probability histogram *if the successive distinct possible values of the random variable differ by a constant.* As claimed above, this method covers the test statistic for any Fisher's test, but it does not cover all possible random variables.

1. Locate each of the possible values of the random variable along a horizontal number line. Let δ denote the value of the constant difference between successive possible values (the symbol δ is the lowercase Greek letter delta).
2. Draw and label a vertical axis for reference.
3. Above each possible value of the random variable, a rectangle is drawn, centered on the value. The base of the rectangle is δ, and its height is the probability of the value divided by δ.

This algorithm can be used to obtain the probability histogram in Figure 3.1:

1. The values $-0.9, -0.7, \ldots, 0.9$ are identified on the horizontal number line. Note that $\delta = 0.2$.
2. The vertical axis has been drawn and labeled for reference.
3. Consider the rectangle centered at 0.1: from Table 2.11, on p. 62, $P(X = 0.1) = 0.3151$. Next, divide this value by $\delta = 0.2$ to obtain $(0.3151/0.2) = 1.5755$. Thus, the rectangle centered at 0.1 in Figure 3.1 has height equal to 1.5755 and base equal to 0.2, or δ. This process is

repeated for the other nine possible values of the test statistic. Note that two of the values, 0.9 and −0.9, have rectangles that are too short to be seen at the scale of the picture.

A very important feature of the probability histogram in Figure 3.1 is its **symmetry** about the point 0—the histogram to the right of 0 is a mirror image of the histogram to the left of 0. For this histogram, 0 is the **point of symmetry;** other points of symmetry are also interesting, as will be demonstrated later in the book.

It can be shown that if the researcher chooses a balanced CRD or if it turns out that the total number of successes in the study equals the total number of failures (in symbols, $m_1 = m_2$), the probability histogram for the test statistic for Fisher's test will be symmetric about the point 0.

The second skill is now developed, that of using a probability histogram to compute probabilities. Consider again the probability histogram for the test statistic for the Infidelity study and focus attention on the rectangle centered at the value 0.1. The area of this rectangle is

$$\text{Base} \times \text{Height} = 0.2 \times 1.5755 = 0.3151.$$

This last number is recognized as the probability that the test statistic takes on the value 0.1. This is no coincidence; symbolically, the area of the rectangle centered at any value is

$$\text{Base} \times \text{Height} = \delta \times \frac{\text{Probability of value}}{\delta}.$$

The δ's cancel, giving the following result:

> Every rectangle in a probability histogram has a number at which it is centered and an area. The probability that the random variable equals the former number (the center) is equal to the latter number (the area). In short, probability histograms represent probabilities by areas.

The sampling distribution is equivalent to the probability histogram in the sense that the knowledge of either allows one to construct the other. Thus, sometimes researchers abuse the language by calling the probability histogram the sampling distribution. The tabular presentation of a sampling distribution is certainly preferable to the probability histogram if one wants accurate probabilities, but the histogram is better for gaining insight into certain techniques, as illustrated repeatedly below.

Recall the following features of the Infidelity study:

- The researcher selected the third alternative ($\neq$), and the value of the test statistic was 0.3.
- The P-value equals $P(X \geq 0.3) + P(X \leq -0.3)$; with the aid of the sampling distribution presented in Table 2.11 (p. 62), the P-value can be found to equal 0.3698.

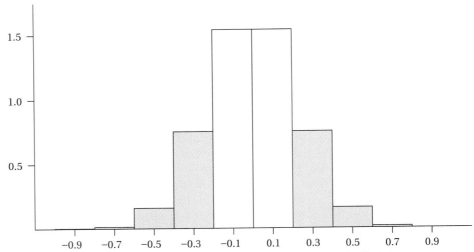

FIGURE 3.2 Probability histogram of the test statistic for the Infidelity study. The P-value equals the sum of the areas of the shaded rectangles.

Given the results of this section, one can *visualize* the P-value as the sum of the areas of the rectangles centered at

$$0.3, 0.5, 0.7, 0.9, -0.3, -0.5, -0.7, \text{ and } -0.9,$$

as illustrated in Figure 3.2.

EXERCISES 3.1

1. Draw a probability histogram for the following sampling distribution:

x	$P(X = x)$
1	0.1
2	0.2
3	0.3
4	0.4
Total	1.0

Use your picture to compute $P(X = 3)$ and compare your answer to the tabulated value.

2. Draw a probability histogram for the following sampling distribution:

x	$P(X = x)$
1	0.1
3	0.3
5	0.6
Total	1.0

Use your picture to compute $P(X = 5)$ and compare your answer to the tabulated value.

3. Draw a probability histogram for the following sampling distribution:

x	$P(X = x)$
1	0.1
2	0.2
3	0.4
4	0.2
5	0.1
Total	1.0

Use your picture to compute $P(X = 4)$ and compare your answer to the tabulated value.

4. Draw a probability histogram for the following sampling distribution:

x	$P(X = x)$
2	0.1
4	0.2
6	0.3
8	0.4
Total	1.0

Use your picture to compute $P(X = 4)$ and compare your answer to the tabulated value.

3.2 THE SIMULATION EXPERIMENT

The sampling distribution of the test statistic for Fisher's test is obtained by examining all possible assignments of subjects to treatments, as described in Chapter 2. For example, it can be shown that there are 40,116,600 different possible assignments of subjects to treatments for the 28 persons participating in the Colloquium study. Further, it can be shown, assuming that the null hypothesis is true, that 12,932,920 of these assignments give the value 0 for the test statistic. Thus, according to the definition of probability,

$$P(X = 0) = \frac{12,932,920}{40,116,600} = 0.3224,$$

as reported in Table 2.13 (p. 66). As discussed earlier, for large studies the number of possible assignments can cause computational difficulties, even if one utilizes clever mathematical arguments. The approximation method of this section is motivated by the following idea:

> Instead of examining all of the possible assignments, just look at some of them.

This raises two questions: how many assignments should be examined, and which ones?

- **How many?** The number of assignments examined is called the number of **runs**. The consequences of different numbers of runs are examined below.
- **Which ones?** The method of selecting assignments for examination was introduced in a different context in Section 2.4, Working with Probability.

The chance mechanism of assigning subjects to treatments is observed repeatedly under identical conditions; that is, after obtaining a particular assignment, the researcher returns the selected cards to the box, remixes the cards, and starts again. This process is continued until the desired number of runs has been obtained. If this method is used for selecting assignments, the process is referred to as a **simulation experiment.**

I conducted a 10,000-run simulation experiment for the Colloquium study. Figure 3.3 displays the elements that constitute a single run of my simulation experiment. While reading the items below, please refer to Figure 3.3.

- Recall that the Colloquium study had 28 subjects, who were arbitrarily assigned the numbers 1, 2, ..., 28. When the study was performed, 8 subjects yielded successes and 20 yielded failures. For ease of exposition, suppose that the subjects numbered 1, 7, 9, 12, 15, 23, 27, and 28 were the ones who yielded successes. On the assumption that the null hypothesis is true—and remember that the sampling distribution is computed on that assumption—these 8 subjects would have yielded successes and the remaining 20 subjects would have yielded failures, regardless of the assignment of subjects to treatments.

- Imagine a box containing 28 cards, one for each subject, as illustrated in Figure 3.3. On each card is a subject number and a response, S for success or F for failure.

- The simulation run begins with the selection of 14 cards at random from the box. The subjects corresponding to the cards selected are placed on the first treatment, and the remaining subjects are placed on the second treatment. Figure 3.3 shows one possible result of the randomization.

- The next step in the simulation run is to count the number of successes and failures on each treatment.

- Next, the proportions $\hat{p}_1$ and $\hat{p}_2$, and their difference, x, are computed.

To summarize, the end result of each run of the simulation experiment is a value of the test statistic.

A summary of the results of my 10,000-run simulation experiment is presented in Table 3.1. The first column of the table lists the different values of the test statistic that were obtained in the simulation experiment. The second column lists the frequencies and the third column the relative frequencies of each of these values in the simulation experiment. The fourth column lists the probabilities of the values. Note that for every value of the test statistic the difference between the relative frequency and the probability is small. This similarity of values is not surprising in view of the long-run relative frequency interpretation of probability introduced in Chapter 2.

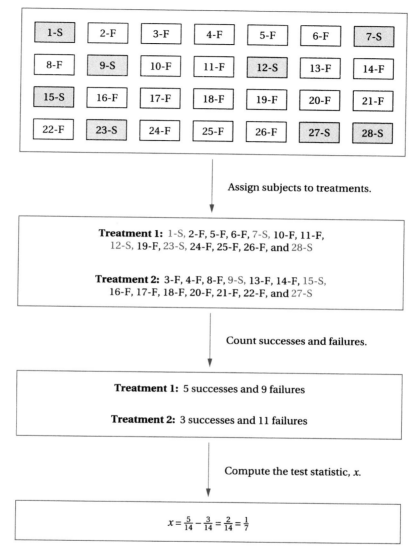

FIGURE 3.3 *A single run of the simulation experiment for the Colloquium study.*

There is a convenient way to visualize the difference between the relative frequencies obtained by a simulation experiment and the actual probabilities. The bottom picture in Figure 3.4 contains two probability histograms. The solid rectangles are the probability histogram of the sampling distribution of the test statistic for the Colloquium study, and the dashed rectangles are the

TABLE 3.1 Results of a Simulation Experiment with 10,000 Runs and Actual Probabilities for the Test Statistic for the Colloquium Study

Value	Frequency	Relative Frequency	Probability
$-\frac{4}{7}$	5	0.0005	0.0010
$-\frac{3}{7}$	143	0.0143	0.0155
$-\frac{2}{7}$	848	0.0848	0.0879
$-\frac{1}{7}$	2,354	0.2354	0.2345
0	3,267	0.3267	0.3224
$\frac{1}{7}$	2,330	0.2330	0.2345
$\frac{2}{7}$	902	0.0902	0.0879
$\frac{3}{7}$	140	0.0140	0.0155
$\frac{4}{7}$	11	0.0011	0.0010
Total	10,000	1.0000	1.0002

probability histogram that one obtains by using the relative frequencies from the simulation experiment instead of the actual probabilities. Note that at the scale of this drawing, the probability histograms are nearly indistinguishable. Since area corresponds to probability (or relative frequency for the simulation experiment results) this near congruence of the pictures implies a near agreement of the values of the probabilities and the relative frequencies, as already noted in Table 3.1. Figure 3.4 also contains the results of 100-run and 1,000-run simulation experiments for the Colloquium study. Note that the agreement between the simulation experiment relative frequencies and the probabilities looks pretty good for 1,000 runs but is noticeably inadequate for only 100 runs.

Recall that for the Colloquium study the observed value of the test statistic is $\frac{3}{7}$ and the P-value for the first alternative is

$$\mathbf{P} = P(X \geq \tfrac{3}{7}) = 0.0165,$$

from Table 2.13 (p. 66). Suppose, however, that the analyst does not know the exact sampling distribution, but only has the results of the 10,000-run simulation experiment given in Table 3.1. The analyst could approximate the P-value by adding the relative frequencies of the values $\frac{3}{7}$ and $\frac{4}{7}$:

$$\mathbf{P} \approx 0.0140 + 0.0011 = 0.0151.$$

This answer is incorrect, but it does provide a close approximation to the exact P-value.

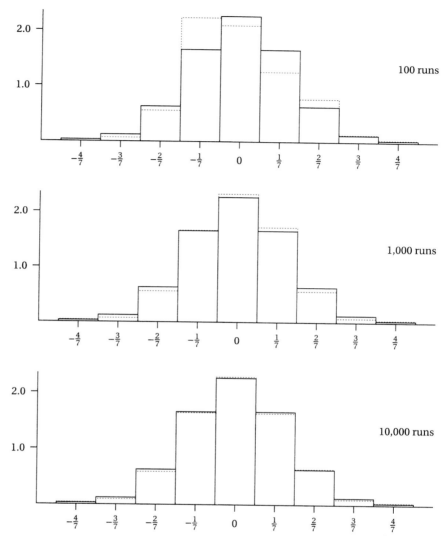

FIGURE 3.4 *Comparisons of the computer simulation experiment approximation (dashed lines) with the sampling distribution (solid lines) for the test statistic for the Colloquium study.*

EXERCISES 3.2

1. The Patient's Dilemma was introduced in Exercise 5 (p. 24). The value of the test statistic is 0.60, and the P-value for the first alternative from Table A.1 is 0.1032. A simulation experiment with 1,000 runs gives the following relative frequency distribution of values of the test statistic:

88 · Chapter 3 / Approximating a Sampling Distribution

Value	Frequency	Relative Frequency
−1.0	3	0.003
−0.6	93	0.093
−0.2	413	0.413
0.2	390	0.390
0.6	96	0.096
1.0	5	0.005
Total	1,000	1.000

What is the simulation estimate of the P-value for the first alternative? Briefly discuss how it compares with the exact value.

2. Recall that in the Infidelity study, for the third alternative

$$P = P(X \geq 0.3) + P(X \leq -0.3) = 0.1849 + 0.1849$$
$$= 0.3698.$$

A simulation experiment with 1,000 runs gives the following relative frequency distribution of values of the test statistic:

Value	Frequency	Relative Frequency
−0.9	1	0.001
−0.7	1	0.001
−0.5	41	0.041
−0.3	143	0.143
−0.1	302	0.302
0.1	317	0.317
0.3	144	0.144
0.5	42	0.042
0.7	9	0.009
Total	1,000	1.000

What is the simulation estimate of the P-value for the third alternative? Briefly discuss how it compares with the exact value.

3.3 MEASURES OF CENTER AND SPREAD

The material in this section is fairly technical. In addition, the reason for this development is not apparent until the very end of the section.

Figure 3.5 displays the probability histograms of the test statistic for Fisher's test for the Infidelity, Colloquium, Soccer Goalie, and Chronic Crohn's Disease studies. In each of these probability histograms the P-value is equal to the sum of the areas of the shaded rectangles. The scales on the axes are identical for the four histograms, so it is easy to compare the pictures. Several features stand out.

- In agreement with the statement on p. 81, the probability histograms for the Infidelity, Colloquium, and Soccer Goalie studies are symmetric about the point zero because the designs are balanced, but the probability histogram for the Chronic Crohn's Disease study is not symmetric—not only is the design not balanced, but the total number of successes does not equal the total number of failures. Note, however, that the probability histogram for the Chronic Crohn's Disease study can be viewed as being approximately symmetric.
- Because of the symmetry, it is natural to say that the probability histograms for the Infidelity, Colloquium, and Soccer Goalie studies are centered at 0.

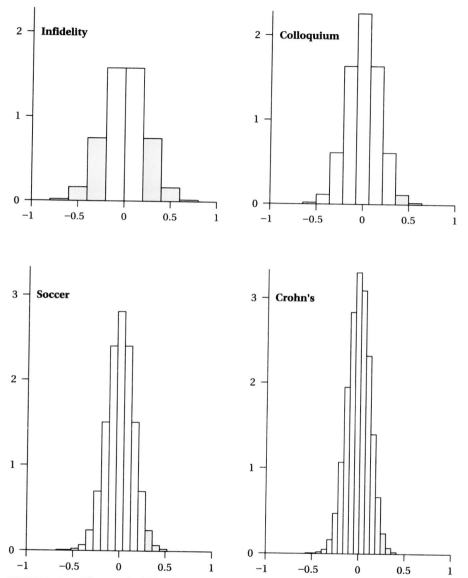

FIGURE 3.5 The probability histograms of the test statistic for Fisher's test for the Infidelity, Colloquium, Chronic Crohn's Disease, and Soccer Goalie studies. The areas of the shaded regions equal the P-values.

Even though it is not symmetric, it is natural to call 0 the center of the probability histogram for the Crohn's Disease study. (Why this is "natural" is discussed later in this chapter.)

- Each probability histogram has a dominant peak located above or nearly above 0.
- The most striking difference between the probability histograms is in their amounts of **spread**. More precisely, compare the probability histograms of the Chronic Crohn's Disease and Infidelity studies. Examining the rectangles located near the center, 0, one sees that those for the Chronic Crohn's Disease study are much taller than those for the Infidelity study. Since probabilities correspond to areas, this means that the histogram of the Chronic Crohn's Disease study has more probability close to its center than does that of the Infidelity study. Next, consider values of the test statistic far from the center, 0. Near ± 0.5, the Infidelity study rectangles are fairly tall, implying substantial probability in that vicinity. By contrast, the rectangles for the Chronic Crohn's Disease study, near ± 0.5 are not visible at the scale of the figure, indicating that there is very little probability in that vicinity. In summary, the Crohn's disease histogram has more probability close to its center and less probability far from its center than does the Infidelity histogram. In short, the sampling distribution of the Infidelity study has more spread, or is more dispersed, than the sampling distribution of the Chronic Crohn's Disease study.
- The above comparison of the histograms of the Infidelity and Chronic Crohn's Disease studies can be extended to all four probability histograms; it is clear from an examination of Figure 3.5 that the Infidelity study histogram has the most spread, followed by the Colloquium study histogram, then the Soccer Goalie study histogram, and, finally, the Chronic Crohn's Disease study histogram.

Throughout this book we will need precise notions for the center and spread of a sampling distribution. The center of a sampling distribution is measured by its **mean,** which is denoted by the symbol μ, the lowercase Greek letter mu. The spread of a sampling distribution is measured by its **standard deviation,** which is denoted by the symbol σ, the lowercase Greek letter sigma. The computation of the mean and standard deviation are presented in the last section of this chapter; for now, we will focus on how to interpret and use these measures.

The mean is also called the **expected value of the random variable X**, written as $E(X)$. The standard deviation is sometimes called the standard deviation of the random variable X, written as $SD(X)$. Closely related to the standard deviation is the variance. The variance is equal to the square of the standard deviation and is thus denoted by σ^2. The variance is also called the variance of the random variable X and is written as $Var(X)$. You need to know about the variance because some methods are easier to use if one works with the variance instead of the standard deviation.

Examine the probability histograms in Figure 3.5 again and imagine that the rectangles are made of a material of uniform mass. Consider the following physical question (the seesaw problem):

> Where should a fulcrum be placed under the rectangles so that the probability histogram will balance?

The answer to this question is called the center of gravity of the probability histogram. This question is easy to answer if the probability histogram is symmetric; namely, the center of gravity is the point of symmetry. For example, if a fulcrum is placed at 0, the probability histogram of the test statistic for the Infidelity study will balance.

The following fact often is very helpful:

> The mean of a sampling distribution is equal to the center of gravity of its probability histogram.

Thus, the mean of the sampling distribution of the test statistic is 0 for the Infidelity, Colloquium, and Soccer Goalie studies.

The following algebraic fact is very useful:

> For any Fisher's test, the mean of the sampling distribution of the test statistic is 0.

Thus, just like the other studies pictured in Figure 3.5, the mean of the sampling distribution of the test statistic for the Chronic Crohn's Disease study is 0. In other words, the center of gravity of the probability histogram for the Chronic Crohn's Disease study is equal to 0.

The mean has a simple physical interpretation—it is the center of gravity of the probability histogram. Unfortunately, a simple interpretation of the standard deviation must be delayed until Chapter 12. For now, the following properties of the standard deviation will suffice:

- The standard deviation is always greater than or equal to 0. In fact, the standard deviation equals 0 only in the uninteresting case in which the sampling distribution is such that all of the probability (that is, 1) is at a single value. In symbols, if a sampling distribution is

$$P(X = x) = 1$$

for some number x, then the standard deviation of the sampling distribution equals 0.

- Of any two sampling distributions, the sampling distribution with the greater spread has the larger standard deviation. *Remember: the greater the spread, the larger the standard deviation.*

Because of this second remark, it is immediately clear that of the sampling distributions pictured in Figure 3.5, the Infidelity study histogram has the largest standard deviation, followed by the Colloquium, Soccer Goalie, and Chronic Crohn's Disease study histograms, in that order.

Fortunately, there is a simple formula for computing the standard deviation of the test statistic for Fisher's test. Recall the symbols m_1, m_2, n_1, n_2, and n that were defined in Chapter 1 (p. 18) as part of the standard notation for a 2×2 contingency table. For the test statistic for Fisher's test,

$$\sigma = \sqrt{\frac{m_1 m_2}{n_1 n_2 (n-1)}}. \tag{3.1}$$

The use of this formula is demonstrated with the four distributions pictured in Figure 3.5:

- For the Infidelity study,

$$\sigma = \sqrt{\frac{11(9)}{10(10)(19)}} = 0.2283.$$

- For the Colloquium study,

$$\sigma = \sqrt{\frac{8(20)}{14(14)(27)}} = 0.1739.$$

- For the Soccer Goalie study,

$$\sigma = \sqrt{\frac{22(28)}{25(25)(49)}} = 0.1418.$$

- For the Chronic Crohn's Disease study,

$$\sigma = \sqrt{\frac{33(38)}{37(34)(70)}} = 0.1193.$$

Let X be a random variable, and let μ denote its mean and σ denote its standard deviation. The **standardized version** of X is the random variable Z defined by

$$Z = \frac{X - \mu}{\sigma}. \tag{3.2}$$

This operation, which creates a new random variable from an existing one, can be confusing. Remember that a random variable X is simply a rule that assigns a number to each possible outcome of a chance mechanism. The standardized version Z is also a rule: it takes the number obtained from the rule X, subtracts μ from that number, and divides the result by σ.

The observed value of the random variable Z is denoted by z, given by the formula

$$z = \frac{x - \mu}{\sigma}. \tag{3.3}$$

In words, to obtain the observed value of Z, take the observed value of X, subtract μ, and divide the result by σ.

In the remainder of the narrative of this chapter the notion of the standardized version is restricted to test statistics for Fisher's test. This simplifies matters because $\mu = 0$ and the standardized version reduces to

$$Z = \frac{X}{\sigma},$$

the test statistic divided by its standard deviation.

It was demonstrated earlier that the sampling distribution of the test statistic for the Infidelity study has $\sigma = 0.2283$. Thus, the standardized version of this test statistic is

$$Z = \frac{X}{\sigma} = \frac{X}{0.2283}.$$

It is instructive to consider the possible values of the random variables X and Z. The possible values of X are

$$-0.9,\ -0.7,\ -0.5,\ -0.3,\ -0.1,\ 0.1,\ 0.3,\ 0.5,\ 0.7,\ \text{and } 0.9.$$

Consider the smallest of these values, -0.9. If $X = -0.9$, then

$$Z = \frac{X}{0.2283} = \frac{-0.9}{0.2283} = -3.94.$$

Similarly, if $X = -0.7$, then

$$Z = \frac{X}{0.2283} = \frac{-0.7}{0.2283} = -3.07.$$

Continuing this process yields the correspondence presented in Table 3.2. Note that knowledge of the value of X is mathematically equivalent to knowledge of the value of Z. For example, knowing $X = -0.50$ implies that $Z = -2.19$; knowing $Z = 1.31$ implies $X = 0.30$. In addition, the sampling distribution of X automatically yields the sampling distribution of Z in the following way. The equation $P(X = -0.50) = 0.0322$, taken from Table 2.11 (p. 62), implies that $P(Z = -2.19)$ also equals 0.0322. By continuing this reasoning, one can obtain

TABLE 3.2 Correspondence between X and $Z = X/0.2283$ for the Infidelity Study

X	Z	X	Z
−0.90	−3.94	0.10	0.44
−0.70	−3.07	0.30	1.31
−0.50	−2.19	0.50	2.19
−0.30	−1.31	0.70	3.07
−0.10	−0.44	0.90	3.94

TABLE 3.3 The Sampling Distribution of Z for the Infidelity Study

z	P(Z = z)	z	P(Z = z)
−3.94	0.0001	0.44	0.3151
−3.07	0.0027	1.31	0.1500
−2.19	0.0322	2.19	0.0322
−1.31	0.1500	3.07	0.0027
−0.44	0.3151	3.94	0.0001

the sampling distribution of Z; it is presented in Table 3.3. Finally, note the following two facts:

- The sampling distribution of Z can be used to obtain the P-value. To see this, recall that the P-value for the Infidelity study and the third alternative is

$$P(X \geq 0.3) + P(X \leq -0.3).$$

Examining the above displays, it is seen that this sum is equal to

$$P(Z \geq 1.31) + P(Z \leq -1.31).$$

- One can draw the probability histogram of the sampling distribution of Z. Thus, the P-value can be visualized as the sum of the areas of the appropriate rectangles of the probability histogram of Z.

Figure 3.6 presents the probability histogram for Z, the standardized version of the test statistic. The P-value is represented as a sum of areas of rectangles. For now, ignore the smooth curve in the picture.

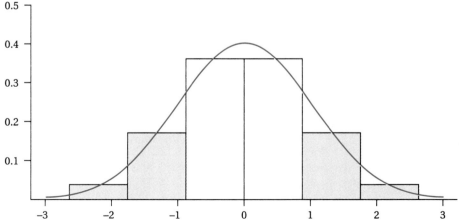

FIGURE 3.6 A comparison of the standard normal curve and the probability histogram of the standardized version of the test statistic for the Infidelity study. The sum of the areas of the shaded rectangles equals the P-value.

As demonstrated above for the Infidelity study, the test statistic can be standardized for any Fisher's test. Figures 3.7, 3.8, and 3.9 present the resulting probability histograms for the Colloquium, Soccer Goalie, and Chronic Crohn's Disease studies, respectively. (It is not important for you to be able to create these pictures; it is important that you understand that these constructions are possible.) These three figures are all drawn to the same scale used in

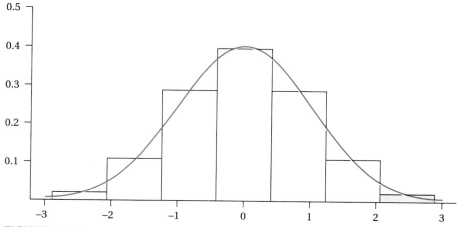

FIGURE 3.7 *A comparison of the standard normal curve and the probability histogram of the standardized version of the test statistic for the Colloquium study. The sum of the areas of the shaded rectangles equals the P-value.*

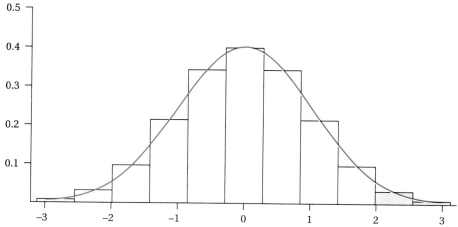

FIGURE 3.8 *A comparison of the standard normal curve and the probability histogram of the standardized version of the test statistic for the Soccer Goalie study. The sum of the areas of the shaded rectangles equals the P-value.*

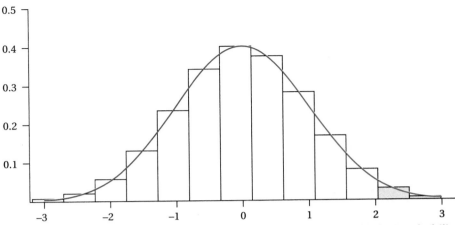

FIGURE 3.9 *A comparison of the standard normal curve and the probability histogram of the standardized version of the test statistic for the Chronic Crohn's Disease study. The sum of the areas of the shaded rectangles equals the P-value.*

Figure 3.6 so that comparisons can easily be made between pictures. For the moment, disregard the smooth curve in these figures and note that the probability histograms are very similar: they are centered at zero, they have a single peak, they are symmetric or nearly symmetric, and they have a similar amount of spread. (It can be shown algebraically that the standardized version of any random variable has mean equal to zero and standard deviation equal to one.) This similarity between these histograms suggests *approximating* areas under these histograms by areas under the smooth curve in these pictures. This smooth curve is called the standard normal curve. Fortunately, it is very easy to obtain areas under the standard normal curve. The technique is presented in the next section.

In summary, standardizing a random variable is useful because it changes dissimilar probability histograms from different studies into similar probability histograms. This similarity enables us to use one method—one approximating curve—to approximate probabilities for any of the different studies. The standard deviation is a good way to measure spread because it makes standardizing work!

EXERCISES 3.3

1. Refer to the study of sorority sisters introduced in Exercise 3 on p. 23. Use Equation 3.1 to obtain the standard deviation of the test statistic for Fisher's test.

2. Refer to the study of fire fighters introduced in Exercise 4 on p. 24. Use Equation 3.1 to obtain the standard deviation of the test statistic for Fisher's test.

3. Refer to the study of pregnancy-induced hypertension introduced in Exercise 15 on p. 27. Use Equation 3.1 to obtain the standard deviation of the test statistic for Fisher's test.

4. Refer to the study of two treatments for breast cancer introduced in Exercise 16 on p. 27. Use Equation 3.1 to obtain the standard deviation of the test statistic for Fisher's test.

5. Refer to the study of basketball bank shots introduced in Exercise 7 on p. 33. Use Equation 3.1 to obtain the standard deviation of the test statistic for Fisher's test.

6. Refer to the study of shooting a gun introduced in Exercise 16 on p. 34. Use Equation 3.1 to obtain the standard deviation of the test statistic for Fisher's test.

7. Consider the following sampling distribution for X. It can be shown that $\mu = 6$ and $\sigma = 2$.

x	$P(X = x)$
2	0.1
4	0.2
6	0.3
8	0.4
Total	1.0

(a) Write down the formula for Z, the standardized version of X.
(b) What are the four possible values of Z?
(c) Draw the probability histogram for X; for Z. Compare the pictures.

8. Consider the following sampling distribution for X. It can be shown that $\mu = 6$ and $\sigma = 4$.

x	$P(X = x)$
0	0.15
2	0.15
4	0.05
6	0.30
8	0.05
10	0.15
12	0.15
Total	1.00

(a) Write down the formula for Z, the standardized version of X.
(b) What are the seven possible values of Z?
(c) Draw the probability histogram for X; for Z. Compare the pictures.

3.4 THE STANDARD NORMAL CURVE

The standard normal curve is drawn in Figures 3.6 through 3.9; its most important features for the purposes of the developments in this book are the following:

- The height of the curve is always positive, and the total area under the curve equals 1.
- The curve is symmetric about the point 0.

As argued in the previous section, it seems reasonable to approximate areas under the probability histogram for Z by areas under the standard normal curve; thus, the first task is to learn how to compute areas under the standard normal curve.

All areas must come from the two facts given above—the total area under the curve equals 1 and the curve is symmetric about 0—and one source, Table A.2, in the Appendix. The generation of Table A.2 is a complicated mathematical problem that is beyond the scope of this book.

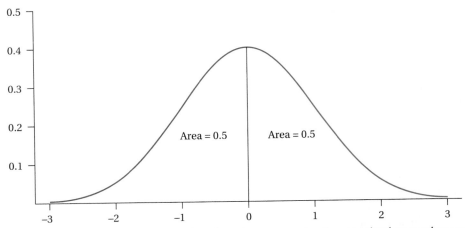

FIGURE 3.10 *Visual demonstration that the area under the standard normal curve to the right of 0 equals the area to the left of 0.*

The first result to note is that, by symmetry, the area under the standard normal curve to the left of 0 equals the area under the standard normal curve to the right of 0. Since the total area under the curve equals 1, each of these areas must equal $\frac{1}{2}$. This fact is presented pictorially in Figure 3.10.

In order to obtain other areas, Table A.2 is needed. The table gives areas under the standard normal curve to the right of the number denoted by z. For example, suppose one wanted to find the area under the standard normal curve to the right of $z = 1.26$. To find this area, enter Table A.2 at $z = 1.26$, as illustrated below with a portion of the table.

z	$\cdots$	0.06
.	$\cdots$	.
.	$\cdots$	.
.	$\cdots$	.
1.2	$\cdots$	.1038 ← Area to the right of 1.26

Thus, the area under the standard normal curve to the right of 1.26 equals 0.1038. As another example, the portion of Table A.2 given below shows that the area under the standard normal curve to the right of -0.32 is equal to 0.6255.

z	$\cdots$	0.02
.	$\cdots$	.
.	$\cdots$	.
.	$\cdots$	.
-0.3	$\cdots$	.6255 ← Area to the right of -0.32

Table A.2 can be used to verify the following statements:

- The area under the standard normal curve to the right of 1.16 equals 0.1230.
- The area under the standard normal curve to the right of −1.35 equals 0.9115.
- The area under the standard normal curve to the right of 2.93 equals 0.0017.

Table A.2 can also be used to find the area under the standard normal curve to the left of a number. Figure 3.11 demonstrates visually that the area to the left of −1.00 equals the area to the right of +1.00, because of symmetry. Thus, the area under the standard normal curve to the left of −1.00 can be obtained by looking up $z = +1.00$ in Table A.2; the answer is 0.1587. As another example, the area under the standard normal curve to the left of 2.11 equals the area under the curve to the right of −2.11, which, from Table A.2, equals 0.9826. This argument can be generalized, giving the result

The area under the standard normal curve to the left of a number z can be obtained by entering Table A.2 at the number $-z$.

Table A.2 can be used to verify the following statements:

- The area under the standard normal curve to the left of 1.16 equals 0.8770.
- The area under the standard normal curve to the left of −1.35 equals 0.0885.
- The area under the standard normal curve to the left of 2.93 equals 0.9983.

For some problems, the area under the standard normal curve between two numbers is required. For example, suppose the area under the standard normal curve between −2.00 and +1.00 is needed. Let A denote the desired area.

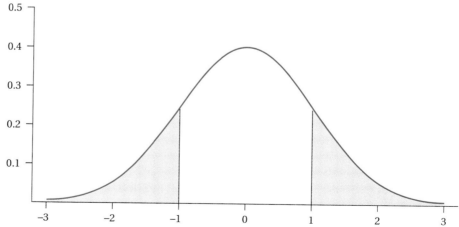

FIGURE 3.11 Visual demonstration that the area under the standard normal curve to the left of −1.00 equals the area to the right of +1.00.

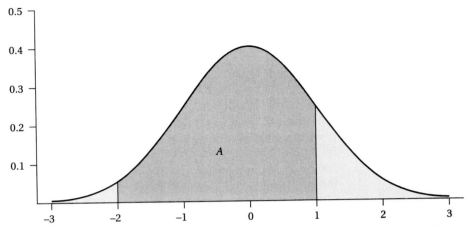

FIGURE 3.12 Visual demonstration that the area under the standard normal curve between −2.00 and +1.00 equals the area to the right of −2.00 minus the area to the right of +1.00.

Figure 3.12 shows that the area to the right of +1.00 added to A equals the area to the right of −2.00. Thus, A can be computed by first obtaining the area to the right of −2.00, which equals 0.9772, and then subtracting the area to the right of +1.00, which equals 0.1587. The result is

$$0.9772 - 0.1587 = 0.8185.$$

This argument can be generalized, yielding the following rule.

Let a and b be any numbers, with a smaller than b. The area under the standard normal curve between a and b is equal to the area to the right of a minus the area to the right of b.

This result and Table A.2 can be used to verify the following:

- The area under the standard normal curve between −0.03 and 1.62 equals
$$0.5120 - 0.0526 = 0.4594.$$
- The area under the standard normal curve between −2.78 and −1.62 equals
$$0.9973 - 0.9474 = 0.0499.$$
- The area under the standard normal curve between 0.37 and 1.52 equals
$$0.3557 - 0.0643 = 0.2914.$$
- The area under the standard normal curve between −1.96 and 1.96 equals
$$0.9750 - 0.0250 = 0.9500.$$

Table A.2 gives only approximate answers to some questions. For example, suppose the area under the standard normal curve to the right of 1.238 is needed. The number of interest, 1.238, has one digit too many to use Table A.2. The method used in this book to handle such problems is to round off this number to the nearest hundredth. Thus, the area to the right of 1.238 is approximated by the area to the right of 1.24, which equals 0.1075. Next, suppose that the area to the right of +4.00 is needed. The number +4.00 is larger than the largest entry, +3.59, in Table A.2. Clearly the area under the standard normal curve to the right of +4.00 is smaller than the area under the standard normal curve to the right of +3.59 since it is the area of a smaller region. Thus, with the help of Table A.2, the area to the right of +4.00 is smaller than 0.0002, which is a sufficiently precise statement for most statistical applications. Similarly,

- For any number z that is larger than 3.59, the area under the standard normal curve to the right of z is less than 0.0002.
- For any number z that is smaller than -3.59, the area under the standard normal curve to the right of z is greater than 0.9998.
- For any number z that is smaller than -3.59, the area under the standard normal curve to the left of z is less than 0.0002.
- For any number z that is larger than 3.59, the area under the standard normal curve to the left of z is greater than 0.9998.

Recall that for the Chronic Crohn's Disease study the observed value of the test statistic is $x = 0.27$ and the standard deviation of the sampling distribution of the test statistic equals 0.1193. Thus, the standardized version of X is

$$Z = \frac{X}{\sigma} = \frac{X}{0.1193}.$$

It was stated in Chapter 2 that the P-value equals 0.0198. Alternatively, the standard normal curve can be used to approximate the P-value as follows:

1. $\mathbf{P} = P(X \geq x) = P(X \geq 0.27)$, from Equation 2.6 (p. 68).
2. Standardize X to obtain

$$P(X \geq 0.27) = P(X/0.1193 \geq 0.27/0.1193) = P(Z \geq 2.26).$$

3. Assume that the exact sampling distribution is not available and hence that this last probability is unknown. It is equal to the area under the probability histogram of Z of all rectangles at or to the right of 2.26. The standard normal curve approximation technique replaces this unknown area with the area under the standard normal curve to the right of 2.26. From Table A.2 this latter area is found to be 0.0119.

In summary, the exact P-value is 0.0198. The standard normal curve approximation to the P-value is 0.0119. The accuracy of the approximation is discussed in a later section of this chapter.

The above method can be applied to the problem of obtaining an approximate P-value for any Fisher's test with the first alternative. In fact, a few minor modifications extend the technique to the second and third alternatives. It is not necessary, however, to recreate the above reasoning each time; just perform these two steps:

1. Compute $z = x/\sigma$, the observed value of Z. For computational purposes, the following algebraic identity is useful:

$$z = \frac{\sqrt{n-1}(ad - bc)}{\sqrt{n_1 n_2 m_1 m_2}}. \qquad (3.4)$$

2. Evaluate one of the following:
   - For the first alternative (>) the approximate P-value equals the area under the standard normal curve to the right of z.
   - For the second alternative (<) the approximate P-value equals the area under the standard normal curve to the right of $-z$.
   - For the third alternative ($\neq$) the approximate P-value equals twice the area under the standard normal curve to the right of $|z|$.

This method is illustrated for a number of studies from Chapter 1.

In the Soccer Goalie study the researcher selected the first alternative. The observed frequencies and row proportions are presented in Tables 1.17 and 1.18 (p. 30). These tables yield

$$a = 15, b = 10, c = 7, d = 18, n_1 = 25, n_2 = 25, m_1 = 22, m_2 = 28,$$
$$n = 50, \hat{p}_1 = 0.60, \hat{p}_2 = 0.28, \text{ and } x = \hat{p}_1 - \hat{p}_2 = 0.60 - 0.28 = 0.32.$$

These values give

$$z = \frac{\sqrt{n-1}(ad - bc)}{\sqrt{n_1 n_2 m_1 m_2}} = \frac{\sqrt{49}[15(18) - 10(7)]}{\sqrt{25(25)(22)(28)}} = 2.26.$$

The approximate P-value for the first alternative is the area under the standard normal curve to the right of 2.26, which is 0.0119. The exact P-value is 0.0225.

In the Ballerina study the researcher selected the second alternative. The observed frequencies and row proportions are presented in Tables 1.19 and 1.20 (p. 31). These tables yield

$$a = 16, b = 9, c = 22, d = 3, n_1 = 25, n_2 = 25, m_1 = 38, m_2 = 12,$$
$$n = 50, \hat{p}_1 = 0.64, \hat{p}_2 = 0.88, \text{ and } x = \hat{p}_1 - \hat{p}_2 = 0.64 - 0.88 = -0.24.$$

These values give

$$z = \frac{\sqrt{n-1}(ad - bc)}{\sqrt{n_1 n_2 m_1 m_2}} = \frac{\sqrt{49}[16(3) - 9(22)]}{\sqrt{25(25)(38)(12)}} = -1.97.$$

The approximate P-value for the second alternative is the area under the standard normal curve to the right of 1.97, which equals 0.0244. The exact P-value is 0.0477.

In the Three-Point Basket study the researcher selected the third alternative. The observed frequencies and row proportions are presented in Tables 1.15 and 1.16 (p. 29). These tables yield

$a = 21, b = 29, c = 20, d = 30, n_1 = 50, n_2 = 50, m_1 = 41, m_2 = 59,$
$n = 100, \hat{p}_1 = 0.42, \hat{p}_2 = 0.40,$ and $x = \hat{p}_1 - \hat{p}_2 = 0.42 - 0.40 = 0.02.$

These values give

$$z = \frac{\sqrt{n-1}(ad-bc)}{\sqrt{n_1 n_2 m_1 m_2}} = \frac{\sqrt{99}[21(30)-29(20)]}{\sqrt{50(50)(41)(59)}} = 0.20.$$

The approximate P-value for the third alternative is twice the area under the standard normal curve to the right of 0.20, which is 2(0.4207), or 0.8414. The exact P-value is 1.

EXERCISES 3.4

1. Find the area under the standard normal curve to the right of

    0.33, −2.28, −0.62, 1.22, 3.11, −0.08, and 3.71.

2. Find the area under the standard normal curve to the right of

    0.99, −2.65, −1.07, 2.81, 0.06, −3.17, and −3.82.

3. Find the area under the standard normal curve to the left of

    0.33, −2.28, −0.62, 1.22, 3.11, −0.08, and 3.71.

4. Find the area under the standard normal curve to the left of

    0.99, −2.65, −1.07, 2.81, 0.06, −3.17, and −3.82.

5. Find the area under the standard normal curve between

    (a) 1.25 and 1.97
    (b) −0.34 and 2.87
    (c) −0.67 and 0.67
    (d) −2.00 and 0

6. Find the area under the standard normal curve between

    (a) 0.09 and 2.17
    (b) −0.97 and 1.24
    (c) −1.63 and −0.19
    (d) −2.36 and 2.36

7. Refer to the study of tables in history texts introduced in Exercise 13 on p. 26. The results of the study give

    $a = 14, b = 8, c = 8, d = 14, n_1 = 22, n_2 = 22, m_1 = 22, m_2 = 22, n = 44, \hat{p}_1 = 0.636,$ and $\hat{p}_2 = 0.364.$

    (a) Compute the standard normal curve approximation to the P-value for the third alternative.
    (b) Write one or two sentences that summarize what you have learned.

8. Exercise 14 on p. 27 introduced a study of the use of interferon to treat hepatitis B. The results of the study give

    $a = 16, b = 28, c = 3, d = 40, n_1 = 44, n_2 = 43, m_1 = 19, m_2 = 68, n = 87, \hat{p}_1 = 0.364,$ and $\hat{p}_2 = 0.070.$

    (a) Compute the standard normal curve approximation to the P-value for the first alternative.
    (b) One way to summarize this study is to state, "Interferon has been shown to be a cure for hepatitis B." Criticize this summary and write one or two sentences to summarize the results.

9. Exercise 15 on p. 27 introduced a study of the use of aspirin to prevent pregnancy-induced hypertension. The results of the study give

$a = 30$, $b = 4$, $c = 20$, $d = 11$, $n_1 = 34$, $n_2 = 31$, $m_1 = 50$, $m_2 = 15$, $n = 65$, $\hat{p}_1 = 0.882$, and $\hat{p}_2 = 0.645$.

(a) Compute the standard normal curve approximation to the P-value for the third alternative.

(b) Write one or two sentences that summarize what you have learned.

10. Exercise 16 on p. 27 introduced a study that compared lumpectomy plus irradiation to total mastectomy as treatments for women suffering from Stage I or II breast cancer. The results of the study give

$a = 412$, $b = 217$, $c = 373$, $d = 217$, $n_1 = 629$, $n_2 = 590$, $m_1 = 785$, $m_2 = 434$, $n = 1{,}219$, $\hat{p}_1 = 0.655$, and $\hat{p}_2 = 0.632$.

(a) Compute the standard normal curve approximation to the P-value for the second alternative.

(b) Write one or two sentences that summarize what you have learned.

11. Exercise 17 on p. 27 introduced a study that compared two doses of lithium in the treatment of persons with bipolar disorder. The results of the study are

$a = 29$, $b = 18$, $c = 41$, $d = 6$, $n_1 = 47$, $n_2 = 47$, $m_1 = 70$, $m_2 = 24$, $n = 94$, $\hat{p}_1 = 0.617$, and $\hat{p}_2 = 0.872$.

(a) Compute the standard normal curve approximation to the P-value for the second alternative.

(b) Write one or two sentences that summarize what you have learned.

12. Refer to the study of shooting a gun introduced in Exercise 16 on p. 34. The results of the study give

$a = 42$, $b = 8$, $c = 25$, $d = 25$, $n_1 = 50$, $n_2 = 50$, $m_1 = 67$, $m_2 = 33$, $n = 100$, $\hat{p}_1 = 0.84$, and $\hat{p}_2 = 0.50$.

(a) Compute the standard normal curve approximation to the P-value for the third alternative.

(b) Write one or two sentences that summarize what you have learned.

13. Refer to the study "Is Pearl Ambidextrous?" introduced in Exercise 17 on p. 34. The results of the study give

$a = 22$, $b = 3$, $c = 20$, $d = 5$, $n_1 = 25$, $n_2 = 25$, $m_1 = 42$, $m_2 = 8$, $n = 50$, $\hat{p}_1 = 0.88$, and $\hat{p}_2 = 0.80$.

(a) Compute the standard normal curve approximation to the P-value for the third alternative.

(b) Write one or two sentences that summarize what you have learned.

14. Refer to the study of tossing peanuts into the air introduced in Exercise 18 on p. 35. The results of the study give

$a = 41$, $b = 9$, $c = 35$, $d = 15$, $n_1 = 50$, $n_2 = 50$, $m_1 = 76$, $m_2 = 24$, $n = 100$, $\hat{p}_1 = 0.82$, and $\hat{p}_2 = 0.70$.

(a) Compute the standard normal curve approximation to the P-value for the third alternative.

(b) Write one or two sentences that summarize what you have learned.

15. Refer to the study of the archer who compared wet feather vanes and wet plastic vanes introduced in Exercise 19 on p. 35. The results of the study give

$a = 36$, $b = 14$, $c = 48$, $d = 2$, $n_1 = 50$, $n_2 = 50$, $m_1 = 84$, $m_2 = 16$, $n = 100$, $\hat{p}_1 = 0.72$, and $\hat{p}_2 = 0.96$.

(a) Compute the standard normal curve approximation to the P-value for the third alternative.

(b) Write one or two sentences that summarize what you have learned.

16. Refer to the study of long putts introduced in Exercise 20 on p. 35. The results of the study give

$a = 25$, $b = 25$, $c = 29$, $d = 21$, $n_1 = 50$, $n_2 = 50$, $m_1 = 54$, $m_2 = 46$, $n = 100$, $\hat{p}_1 = 0.50$, and $\hat{p}_2 = 0.58$.

  (a) Compute the standard normal curve approximation to the P-value for the third alternative.
  (b) Write one or two sentences that summarize what you have learned.

17. Exercise 21 on p. 35 asked you to perform a CRD for a sequence of 100 trials. In Chapter 2 you were asked to specify the alternative hypothesis for your study.

  (a) Compute the standard normal curve approximation to the P-value for your alternative.
  (b) Write one or two sentences that summarize what you have learned.

18. Exercise 18 on p. 27 asked you to perform a CRD on 50 subjects. In Chapter 2 you were asked to specify the alternative hypothesis for your study.

  (a) Compute the standard normal curve approximation to the P-value for your alternative.
  (b) Write one or two sentences that summarize what you have learned.

3.5 A COMPARISON OF METHODS

The problem of computing the P-value for Fisher's test has received a great deal of attention in the current and previous chapters. A researcher may compute the exact P-value, the approximate P-value based on a simulation experiment, or the approximate P-value based on the standard normal curve. The exact solution is always preferred if it is readily available. For sufficiently large studies, however, the exact solution becomes inaccessible regardless of the researcher's mathematical skill and computer facility. The researcher must then either resort to an approximation or abandon the goal of learning about the P-value. The researcher will want to use an approximation method if the answer it yields is close to the unknown exact P-value. Unfortunately, there are no mathematical theorems that indicate exactly how much an approximate answer differs from the exact answer, but some insight can be gained by examining the performance of the two possible approximations for a number of examples for which the exact P-value is known.

For several studies introduced in Chapter 1, 10,000-run simulation experiments were performed to obtain approximate P-values. The results are presented in Table 3.4. This table also presents the exact P-values and the values obtained from the standard normal curve approximation method introduced in the previous section. (If you wish, you can practice the method of the previous section by verifying the values in the fifth column of the table.)

Consider an approximate P-value computed from a simulation experiment with 10,000 runs. The following two statements are true; they follow from the development of confidence intervals in Section 6.2.

  1. If the approximate P-value obtained from the simulation experiment is less than or equal to 0.05, you can be very confident that the absolute value of the difference between the approximate P-value and the exact P-value is at most 0.0066.

TABLE 3.4 Exact and Approximate P-values for Studies Introduced in Chapter 1

Study	Alternative	Exact P-value	Simulation Approximation	Normal Approximation
Colloquium	>	0.0165	0.0151	0.0069
Chronic Crohn's Disease	>	0.0198	0.0179	0.0119
Soccer Goalie	>	0.0225	0.0202	0.0119
Chronic Crohn's Disease (at follow-up)	>	0.0256	0.0250	0.0146
Ballerina	<	0.0477	0.0467	0.0244
Infidelity	≠	0.3698	0.3623	0.1902
Prisoner	>	0.9884	0.9874	0.9656
Three-Point Basket	≠	1.0000	1.0000	0.8414

2. If the approximate P-value obtained from the simulation experiment is greater than 0.05, you can be very confident that the absolute value of the difference between the approximate P-value and the exact P-value is at most 0.0150.

In each of these statements, "very confident" means 99.7 percent confident, in the sense described in Section 6.2. The numbers 0.0066 and 0.0150 in the above two statements are called **error bounds.**

The mathematical results on the accuracy of the standard normal curve approximation are not nearly as simple to discuss. In fact, even the statement, without proof, of the mathematical results is well beyond the level of this text. Instead, it is common to state a **general guideline** that specifies the situations in which the approximation should be used. A commonly recommended general guideline for the standard normal curve approximation to the P-value for a CRD is

Let r_1 equal the smaller of n_1 and n_2, and let r_2 equal the smaller of m_1 and m_2. The standard normal curve approximation to the exact P-value should be used if, and only if,

$$\frac{r_1 r_2}{n} \geq 5.$$

The condition in the guideline is mysterious; its origin will be discussed in a more general context in Chapter 11. A few examples will illustrate the condition. For the Colloquium study, $n_1 = 14$, $n_2 = 14$, $m_1 = 8$, $m_2 = 20$, and $n = 28$. These values yield

$$r_1 = 14, \quad r_2 = 8, \quad \text{and} \quad \frac{r_1 r_2}{n} = \frac{14(8)}{28} = 4.$$

Thus, according to the general guideline, the standard normal curve approximation should not be used.

Similarly, for the Infidelity and Three-Point Basket studies you can verify that
$$\frac{r_1 r_2}{n} = \frac{10(9)}{20} = 4.5 \quad \text{and} \quad \frac{r_1 r_2}{n} = \frac{50(41)}{100} = 20.5,$$
respectively. Thus, according to the general guideline, the standard normal curve approximation should not be used for the Infidelity study but can be used for the Three-Point Basket study.

The general guideline sometimes gives questionable directives, as can be seen in cases in which the exact P-value is known. For example, it states that the standard normal curve approximation should not be used for the Colloquium study, even though the approximate P-value arguably is adequate. This is not a serious difficulty. The Colloquium study had a small number of subjects, and its exact P-value is easy to obtain from Table A.1; thus, it is not a study for which a researcher would want to use the standard normal curve approximation. Moreover, the numerical value of the criterion, 4, is close to the cutoff, 5, for using the approximation. In fact, the value of the cutoff is somewhat arbitrary; 5 is a widely promoted value. By contrast, for the Three-Point Basket study, the value of the criterion, 20.5, is much larger than the cutoff, but the approximate P-value, 0.8414, is not very close to the exact value, 1. In fact, it can be shown that unless the sample size is very large, the standard normal curve often provides a poor approximation to P-values that are not statistically significant. (This problem can be alleviated considerably by employing a **continuity correction,** which is a modification of the standard normal curve approximation. This modification is not presented in this book.) For better or worse, this feature does not bother those researchers who are primarily interested in the accuracy of the approximate P-value only when the results are statistically significant.

One unfortunate consequence of using approximate answers is illustrated by the Colloquium study. An analyst using the standard normal curve approximation would label the data highly statistically significant ($P \leq 0.01$) when in reality they are only statistically significant ($0.01 < P \leq 0.05$). This observation speaks more to the arbitrariness of categorizing P-values than it does to the inaccuracy of the standard normal curve approximation. For instance, suppose the exact P-value for a study is 0.0501 and the approximation is 0.0499. The approximation is very accurate, but it incorrectly labels the data statistically significant.

There is a characteristic of the approximation based on a simulation experiment that some people find disconcerting. If 100 persons are asked to find the standard normal curve approximation to the P-value for the Colloquium study, then, barring arithmetic errors, all will return 0.0069 as the approximation for the true value of 0.0165. For comparison, I asked each of 100 persons to perform a 10,000-run simulation experiment. (Actually, the persons themselves were simulated.) My 100 persons returned 45 different approximate P-values, ranging from a low of 0.0132 to a high of 0.0199. Note that every one of the simulation based approximate P-values is better than the value obtained using the standard normal curve approximation in the sense of being closer to the exact P-value. Some people find it a bit strange that many different answers can be preferred to a consistent, though incorrect, answer.

EXERCISES 3.5

1. Refer to Table 3.4. For each study, determine whether the simulation approximation or the standard normal curve approximation is closer to the exact P-value.

2. Refer to Table 3.4. For each study, determine whether the absolute difference between the simulation approximation and the exact P-value is less than the appropriate error bound given in the text.

3.6 COMPUTING THE MEAN AND STANDARD DEVIATION*

The material in this section is optional. Chapter 12 will discuss the computation of the related sample mean and sample standard deviation.

3.6.1 The Mean

Table 3.5 presents three sampling distributions:

- The test statistic for Fisher's test for the Class study, which was derived in Chapter 2
- The test statistic for Fisher's test for the Memorization study, which was introduced in Exercise 4 on p. 58
- The random variable for the game of chance described below

The test statistic for the Memorization study has three possible values, $-\frac{4}{6}$, $\frac{1}{6}$, and 1. One possible measure of center is the arithmetic average of these values, which is obtained by adding the values and dividing by three:

$$\text{Arithmetic average of possible values} = \frac{-\frac{4}{6} + \frac{1}{6} + 1}{3} = \frac{1}{6}.$$

Note that the arithmetic average can also be written as

$$\tfrac{1}{3}\left(-\tfrac{4}{6}\right) + \tfrac{1}{3}\left(\tfrac{1}{6}\right) + \tfrac{1}{3}(1).$$

This expression is called the **weighted average** of the values $-\frac{4}{6}$, $\frac{1}{6}$, and 1, using the weights $\frac{1}{3}$, $\frac{1}{3}$, and $\frac{1}{3}$. In fact, when all the weights are equal, the weighted average is called the unweighted average, or, as above, the arithmetic average.

The arithmetic average of the possible values turns out not to be a useful measure of center because it ignores the **relative likelihood** of the different values. For example, in the Memorization study the value $\frac{1}{6}$ is six times as likely as the value 1, but this difference in relative likelihood is ignored by the arithmetic average.

TABLE 3.5 Computation of the Mean for Three Sampling Distributions

Class Study			Memorization Study			Game of Chance		
x	$P(X=x)$	$xP(X=x)$	x	$P(X=x)$	$xP(X=x)$	x	$P(X=x)$	$xP(X=x)$
-1	$\frac{1}{6}$	$-\frac{1}{6}$	$-\frac{4}{6}$	$\frac{3}{10}$	$-\frac{12}{60}$	1	0.5	0.5
0	$\frac{4}{6}$	0	$\frac{1}{6}$	$\frac{6}{10}$	$\frac{6}{60}$	2	0.4	0.8
1	$\frac{1}{6}$	$\frac{1}{6}$	1	$\frac{1}{10}$	$\frac{6}{60}$	3	0.1	0.3
Total	1	$\mu = 0$		1	$\mu = 0$		1	$\mu = 1.6$

By contrast, the mean of the sampling distribution acknowledges the different relative likelihoods of the possible values by computing the weighted average of the possible values using the probabilities as weights. In particular, for the Memorization study the mean is equal to

$$\tfrac{3}{10}\left(-\tfrac{4}{6}\right) + \tfrac{6}{10}\left(\tfrac{1}{6}\right) + \tfrac{1}{10}(1) = 0.$$

In this sum, the numbers in parentheses are the possible values of the test statistic and the coefficients are their respective probabilities.

There is another way to motivate the formula for the mean, which also illustrates why the mean is sometimes called the expected value of the random variable X.

Suppose you are planning to play a game of chance a large number of times and that your reward, in dollars, for each play will be a realization of the random variable X, whose sampling distribution is given in Table 3.5 under "Game of Chance." The mean, or the expected value of X, is the answer to the question,

How much should you expect to win, on average, per play of this game?

This question is answered by appealing to the long-run relative frequency interpretation of probability that was introduced and discussed on p. 60.

On each play, you will win either 1, 2, or 3 dollars because these are the only possible values of the random variable X. Because the probability of the value 1 is 0.5, you expect to win 1 dollar on about 50 percent of your plays. Similarly, you expect to win 2 dollars on about 40 percent of your plays, and you expect to win 3 dollars on about 10 percent of your plays. Thus, you expect to win

$$.5(1) + .4(2) + .1(3) = 1.6 \text{ dollars per play.}$$

This last sum is recognized as the weighted average of the possible values of X with weights equal to the respective probabilities of the values.

Table 3.5 demonstrates an easy way to compute the mean of a sampling distribution. Simply create a third column in the table of the sampling distribution.

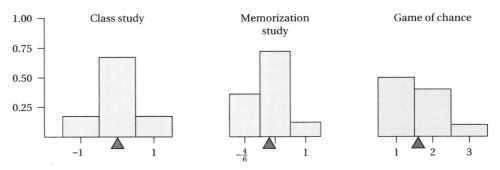

FIGURE 3.13 *The center of gravity interpretation of the mean.*

This column is labeled $xP(X = x)$, and its entries equal the product of the entries in the x column (the possible values of X) and the entries in the $P(X = x)$ column (the probabilities, or weights). You should verify the computations presented in Table 3.5.

Figure 3.13 presents the probability histograms of the three sampling distributions in Table 3.5 with a fulcrum drawn at the mean, which, as you will recall, equals the center of gravity of the probability histogram.

3.6.2 The Standard Deviation

Given a sampling distribution, first compute the value of its variance. Once the variance is known, the standard deviation is obtained by computing the square root of the variance.

Table 3.6 shows how to compute the variances of the three sampling distributions given in Table 3.5. The first two columns of each table are the sampling distribution, and the third column is used to compute the mean. The remaining columns correspond to the following principles of what a measure of spread should be:

1. Any measure of spread in the values of a random variable should take into account the relative likelihoods of the different values. As seen below, this principle leads to the computation of a weighted average with weights given by the probabilities.
2. If only one value of the test statistic is possible, the measure of spread should equal 0; otherwise, the measure of spread should be a positive number.
3. Of any two distributions, the distribution with the greater dispersion should be the distribution with the larger measure of spread.
4. The spread in a set of values should be measured by examining how far each value is from the center of the sampling distribution. The idea is that if the values are tightly clustered, they are all near the center, but if one or more values are far from the others, then one or more values are far

TABLE 3.6 Computation of the Variance for Three Sampling Distributions

x	$P(X = x)$	$xP(X = x)$	$x - \mu$	$(x - \mu)^2$	$(x - \mu)^2 P(X = x)$
		Class Study			
-1	$\frac{1}{6}$	$-\frac{1}{6}$	-1	1	$\frac{1}{6}$
0	$\frac{4}{6}$	0	0	0	0
1	$\frac{1}{6}$	$\frac{1}{6}$	1	1	$\frac{1}{6}$
Total		$\mu = 0$			$\sigma^2 = \frac{2}{6}$
		Memorization Study			
$-\frac{4}{6}$	$\frac{3}{10}$	$-\frac{12}{60}$	$-\frac{4}{6}$	$\frac{16}{36}$	$\frac{48}{360}$
$\frac{1}{6}$	$\frac{6}{10}$	$\frac{6}{60}$	$\frac{1}{6}$	$\frac{1}{36}$	$\frac{6}{360}$
$\frac{6}{6}$	$\frac{1}{10}$	$\frac{6}{60}$	$\frac{6}{6}$	$\frac{36}{36}$	$\frac{36}{360}$
Total		$\mu = 0$			$\sigma^2 = \frac{90}{360}$
		Game of Chance			
1	0.5	0.5	-0.6	0.36	0.180
2	0.4	0.8	0.4	0.16	0.064
3	0.1	0.3	1.4	1.96	0.196
Total		$\mu = 1.6$			$\sigma^2 = 0.440$

from the center. This principle is achieved by computing $(x - \mu)$ for each possible value, x, of the random variable. The quantity $(x - \mu)$ is called the **deviation** (from the mean) of the value x. The values of the deviations appear in the fourth column of Table 3.6. Note that if the mean equals 0 (as it does for the test statistic for Fisher's test), the deviation is $x - 0 = x$; that is, the value and the deviation are the same number.

5. A deviation of, say, -2, reflects the same amount of spread as a deviation of $+2$. Thus, any summary of the deviations should treat -2 and $+2$ as interchangeable. More precisely, the summary is a function of the *absolute values* of the deviations.

6. In view of item 3 in this list, the further a deviation is from 0, in either direction, the more it should contribute to the measure of spread. For example, $+5$ should contribute more to the measure of spread than does -2.

7. There are many ways to achieve the goals of the previous two items in this list. A method that turns out to be very useful in statistics is to compute the squared deviations, $(x - \mu)^2$. Squaring the deviations satisfies item 5 because, for example, $(-2)^2 = 4$ and $2^2 = 4$, and it satisfies item 6 because $5^2 = 25$ is larger than $(-2)^2 = 4$. The squared deviations appear in the fifth column of Table 3.6.

8. The variance is defined to be the weighted average of the squared deviations, with weights given by the corresponding probabilities. The products of the weights and squared deviations appear in the final column of Table 3.6. The sum of these entries equals the variance.

There are many reasons statisticians prefer the standard deviation over the variance as a measure of spread. One of these reasons is presented now.

Consider the sampling distribution for the game of chance. The unit of the random variable X is dollars. Thus, deviations are also measured in dollars, but squared deviations and the variance (being a weighted average of squared deviations) are measured in *squared dollars*. The standard deviation is the square root of the variance and, thus, its unit is dollars, the same unit as that of the random variable X.

EXERCISES 3.6

1. Compute the mean, variance, and standard deviation of the following sampling distribution:

x	$P(X = x)$
1	0.1
2	0.2
3	0.3
4	0.4
Total	1.0

2. Compute the mean, variance, and standard deviation of the following sampling distribution:

x	$P(X = x)$
1	0.1
3	0.3
5	0.6
Total	1.0

3. Compute the mean, variance, and standard deviation of the following sampling distribution:

x	$P(X = x)$
1	0.1
2	0.2
3	0.4
4	0.2
5	0.1
Total	1.0

4. Compute the mean, variance, and standard deviation of the following sampling distribution:

x	$P(X = x)$
2	0.1
4	0.2
6	0.3
8	0.4
Total	1.0

The Randomized Block Design

4.1 DATA PRESENTATION AND SUMMARY
4.2 THE MANTEL–HAENSZEL TEST
4.3 THE RANDOMIZED PAIRS DESIGN

Chapter 4

The material in this chapter is optional; the uninterested reader may proceed to Chapter 5.

Recall that the completely randomized design was introduced in Chapter 1 as a method for comparing two treatments on a collection of subjects. A key component of a CRD is the random assignment of subjects to treatments. Sometimes a researcher divides the subjects into two or more groups, called **blocks**, before assigning them to treatments. There are many motivations for dividing subjects into blocks; the examples and exercises of this chapter provide some insight into the reasons.

A **randomized block design**, abbreviated **RBD**, is simply a completely randomized design on each block of subjects. A randomized block design offers two possible strategies for data summary and analysis:

- Handle each block separately, using the methods and ideas introduced in Chapters 1 through 3.
- Combine the information across the blocks to obtain an overall comparison of the treatments.

These strategies should not be viewed as mutually exclusive. Quite often, in fact, applying both strategies leads to a better insight into the scientific issue motivating the study than either strategy alone. Not surprisingly, this chapter emphasizes the second strategy, since you are familiar with the mechanics of the first strategy.

4.1 DATA PRESENTATION AND SUMMARY

This section discusses data presentation and summary for two types of studies in which blocks are used.

4.1.1 Medical Studies

You may wish to review the material presented in Section 1.2.1 of Chapter 1.

In nearly every interesting study the relation between treatment and response is not deterministic. (This is true for all types of studies, not just medical ones.) For example, in the Chronic Crohn's Disease study of Chapter 1, some subjects assigned to cyclosporine therapy improved, and some did not; some subjects assigned to the placebo therapy improved, and some did not. This is not surprising, because the subjects are people, and people are complex biological and psychological organisms. There are a myriad of factors that measure differences between subjects, in addition to the treatment received, that *could* influence the response. The values of one or more of these factors could be used to divide the subjects into blocks. For example, a very popular factor in medical studies is the subject's prognosis at the time of entry into a study. Of course, prognosis can be difficult to measure, so researchers often settle for a clinical or laboratory measure that is related to prognosis and easy to determine. These ideas are made more precise through a lengthy examination of the AIDS study introduced in Section 1.2.1. All salient features are presented below. See [1] and [2] for further details.

A team of Scandinavian researchers wanted to investigate the effectiveness of the drug inosine pranobex (IP) as a treatment for persons who have tested positive for the human immunodeficiency virus (HIV) antibody. The 826 subjects in this study all tested positive for the HIV antibody on both the enzyme immunoassay and the immunoblot assay; subjects with acquired immunodeficiency syndrome (AIDS) according to the 1985 definition of the Centers for Disease Control were excluded from the study. HIV attacks the human immune system; thus, a measure of the strength of each subject's immune system at entry into the study is a natural gauge for prognosis. The standard measure used in HIV research is the subject's CD4+ cell count, also called the T-cell count. The normal range for this count in uninfected adults is from 900 to 1200 (all cell counts given here are cells per cubic millimeter of blood); smaller values of the count reflect a weakened immune system. The researchers divided subjects into blocks based on cell count at study entry according to the criterion given in Table 4.1.

Subjects in block 1 had the highest and subjects in block 3 had the lowest T-cell counts at the beginning of the study. Thus, subjects in block 1 had the best and subjects in block 3 had the worst prognoses at the beginning of the study. There were 427 subjects in block 1, 340 in block 2, and 59 in block 3. Within each block the patients were randomly assigned to receive either IP or a placebo three times daily for 24 weeks. The response was measured at the end of the 24-week

TABLE 4.1 Definition of Blocks for the AIDS-IP Study

Block	CD4+ Cell Count (cells/mm³)
1	>500
2	200–500
3	<200

treatment. Development of AIDS at any time during the 24 weeks was defined to be a failure, and not developing AIDS was a success.

The study was double blind, so neither the subjects nor the physicians knew which subjects were receiving IP and which were receiving the placebo. This required some care, however. Quoting from the paper,

> Inosine pranobex has a distinctive bitter taste. In order to improve blinding, a small amount of quinine hydrochloride was added to each placebo tablet. The placebo tablets were thus indistinguishable in appearance from the inosine pranobex tablets and similar in taste and smell.

A code that linked patient to treatment was not broken until after the end of the study, when all the data had been collected.

Table 4.2 presents the observed counts and row proportions for each block. Figure 4.1 presents bar charts of the proportion who developed AIDS (failures) in each block. The juxtaposition of these bar charts facilitates the comparison of treatments within and between blocks. For example,

- Within each block, IP performed better than the placebo.
- For IP, moving from subjects with the best prognosis to those with the worst: the proportion of failures is 0 in block 1; it grows slightly, to 0.6 percent, in block 2; and it makes a substantial jump, to 3.6 percent, in block 3.
- The across-block pattern for the placebo is similar: the proportion of failures grows from 1.9 percent in block 1 to 4 percent in block 2, and jumps to 19.4 percent in block 3.

These comparisons illustrate one of the reasons for creating blocks, namely, to gain more insight into the phenomenon being studied.

Figure 4.2 is an alternative to adjacent bar charts as a visual representation of the data; it is called an **interaction graph**. In fact, the comparisons within and between blocks are arguably easier to make with the interaction graph than with the bar charts.

The remainder of this section digresses to a notion introduced in Chapter 1. In a discussion of chronic Crohn's disease (p. 9) the hypothetical unethical and compassionate researchers were introduced. The unethical researcher, in an attempt to exaggerate the value of the new therapy, would assign subjects with the best prognoses to the new therapy and subjects with the worst prognoses

TABLE 4.2 Results of AIDS-IP Study for Each Block

	Observed Counts				Row Proportions		
Treatment	S*	F	Total	Treatment	S*	F	Total
Block 1. CD4+ Cell Count Greater than 500							
IP	217	0	217	IP	1.000	0.000	1.000
Placebo	206	4	210	Placebo	0.981	0.019	1.000
Total	423	4	427				
Block 2. CD4+ Cell Count between 200 and 500							
IP	163	1	164	IP	0.994	0.006	1.000
Placebo	169	7	176	Placebo	0.960	0.040	1.000
Total	332	8	340				
Block 3. CD4+ Cell Count Less than 200							
IP	27	1	28	IP	0.964	0.036	1.000
Placebo	25	6	31	Placebo	0.806	0.194	1.000
Total	52	7	59				

*A success is defined as not developing AIDS.

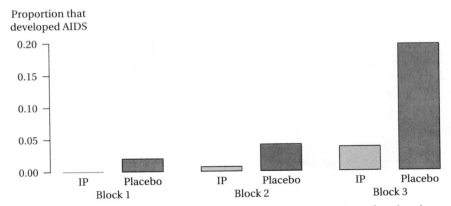

FIGURE 4.1 Bar charts of the proportion that developed AIDS for the three blocks. CD4+ cell counts are above 500 in block 1, between 200 and 500 in block 2, and below 200 in block 3.

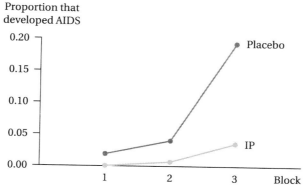

FIGURE 4.2 Interaction graph of the proportion that developed AIDS by block.

to the placebo. By contrast, the compassionate researcher would reverse the assignments in an attempt to help those most in need.

What would have happened if the AIDS-IP study in this section had been conducted by the unethical researcher? This question cannot be answered with certainty, but there is a simple way to gain some useful insight. Assume that the unethical researcher had assigned all subjects with a CD4+ count below 200 to the placebo, assigned all subjects with a cell count above 500 to IP, and dropped from the study all subjects with a cell count between 200 and 500. Again, there is no way to tell what would have happened, but the values in Table 4.2 *suggest* that none of the IP group and 19.4 percent of the placebo group would have been failures. This difference exaggerates the observed effectiveness of IP.

Now consider what would have happened if the AIDS-IP study of this section had been conducted by the compassionate researcher. Assume that the compassionate researcher had assigned all subjects with a CD4+ count below 200 to IP, assigned all subjects with a cell count above 500 to the placebo, and dropped from the study all subjects with a cell count between 200 and 500. Again, there is no way to tell what would have happened, but the values in Table 4.2 *suggest* that 3.6 percent of the subjects on IP and 1.9 percent of the subjects on placebo would have been failures. In words, the placebo would have performed better than IP—a serious contradiction of the actual study results.

4.1.2 Combining Similar Studies

Eight students from my statistics class performed the Infidelity study for their class projects. Each student conducted a balanced completely randomized design on 50 subjects. These eight studies can be viewed as one randomized block design with eight blocks, with each block being a different study. Figure 4.3 presents the bar charts of the proportions of subjects who responded yes (tell on the cheater) for each of the studies. Recall that the husband is the cheater in the first version and the wife is the cheater in the second version. The blocks are labeled

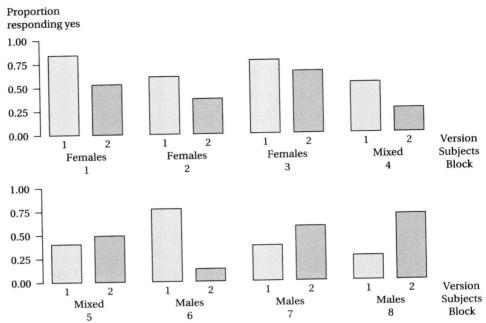

FIGURE 4.3 Bar charts of proportions responding yes for eight Infidelity studies. In version 1 the husband is cheating; in version 2 the wife is cheating.

1, 2, ..., 8, for ease of reference. Three of the studies, blocks 1 through 3, used only female subjects; three studies, blocks 6 through 8, used only male subjects; and the remaining two studies had both female and male subjects in unknown proportions. A wealth of information is in this figure, including the following:

- The proportion who would tell on a cheating husband ranges from a low of 0.24, in block 8, to a high of 0.84, in block 1. (Such precise numbers cannot be easily obtained from the figure; they are taken from the complete data set. The patterns reflected by the numbers, however, are easy to see in the figure.) Similarly, the proportion who would tell on a cheating wife varies considerably, from a low of 0.12, in block 6, to a high of 0.68, in block 8.
- In five of the blocks (1 through 4, and 6), the proportion who would tell on the husband exceeds the proportion who would tell on the wife. This difference agrees with the original findings reported in Chapter 1 (Table 1.3, p. 16) for a much smaller study of female subjects. Three blocks, however, show the reverse pattern. The value of $\hat{p}_1 - \hat{p}_2$ ranges from a low of -0.44, in block 8, to a high of 0.64, in block 6.
- The studies on female subjects show some consistency of findings. The studies with male subjects have wildly different results. The two studies with subjects of both genders have different findings, but interpretation of the difference is difficult without the precise proportions of genders.

- Although it is not noted in the figure, the researchers for blocks 6 and 8 were male, and the other researchers were female. The two male researchers certainly found wildly different patterns! Further, in all studies but block 4, subjects were coworkers or friends of the researcher. The block 4 researcher used a haphazard scheme of calling strangers to obtain her data.

The topic of this section, grouping similar studies, is appropriate for multicenter medical studies if each center has its own set of subjects and its own randomization.

4.1.3 The Collapsed Table

In the AIDS-IP study the value of $\hat{p}_1 - \hat{p}_2$ was positive in all blocks. By contrast, in the eight Infidelity studies, the sign of this value varied (five positives and three negatives). The pattern in the AIDS-IP study data is an example of **consistency of evidence across blocks,** whereas the eight Infidelity studies illustrate **inconsistency of evidence across blocks.** Do not interpret these definitions too rigidly; it is better to think of consistency as being measured on a continuum, like time or distance, rather than as a dichotomy. In particular, if the number of subjects in some or all blocks is small or the real difference between the effects of the treatments is small, some inconsistency should be expected simply because of chance variation. This remark is especially appropriate if the number of blocks is large.

It is not uncommon for researchers to collapse the data from several blocks into one 2×2 table. For example, Table 4.3 presents the **collapsed table** of observed frequencies and its row proportions for the AIDS-IP study. Each cell count in the collapsed table is obtained by summing the corresponding cell counts from the tables for each block. For example, the entry $a = 407$ in the collapsed table is obtained by summing the a values from the block tables in Table 4.2, namely, $217 + 163 + 27 = 407$. Similarly, $d = 17$ in the collapsed table is obtained by summing the d's in the three block tables, $4 + 7 + 6 = 17$.

The row proportions in the collapsed table indicate that $\frac{1}{2}$ of 1 percent of the subjects on IP developed AIDS during the study, compared with about 4 percent of the subjects on the placebo. Note that the pattern in the collapsed table is nearly identical to the pattern in block 2. The main advantage of collapsing the data is that several tables (three tables in the AIDS-IP study) are replaced by one, leading to a substantial simplification. The main disadvantage of collapsing the

TABLE 4.3 The Collapsed Table and Its Row Proportions for the AIDS-IP Study

	Observed Counts				Row Proportions		
Treatment	S	F	Total	Treatment	S	F	Total
IP	407	2	409	IP	0.995	0.005	1.000
Placebo	400	17	417	Placebo	0.959	0.041	1.000
Total	807	19	826				

TABLE 4.4 The Collapsed Table and Its Row Proportions for the Eight Infidelity Studies

	Observed Counts			Row Proportions		
Version Read	Yes	No	Total	Yes	No	Total
1. Tell on husband?	112	88	200	0.56	0.44	1.00
2. Tell on wife?	90	110	200	0.45	0.55	1.00
Total	202	198	400			

data is that the collapsed table can be misleading. For example, the collapsed table row proportions are a good summary of the experiences of the block 2 subjects, but they are too pessimistic for block 1 subjects and too optimistic for block 3 subjects. (Note: Chapter 9 gives an alternative to collapsing, called standardized rates, that can be used as a descriptive summary of the relation across blocks.)

Table 4.4 contains the collapsed table for the eight Infidelity studies and its row proportions. The collapsed table does not show the tremendous variation between blocks or the different patterns for male and female subjects.

1. Below are the results of a hypothetical RBD:

	Block 1		
Treatment	S	F	Total
1	55	45	100
2	45	55	100
Total	100	100	200

	Block 2		
Treatment	S	F	Total
1	75	25	100
2	65	35	100
Total	140	60	200

Compute the table of row proportions for each block and draw the interaction graph. Construct the collapsed table and its table of row proportions. Comment.

2. Below are the results of a hypothetical RBD:

	Block 1		
Treatment	S	F	Total
1	70	30	100
2	30	70	100
Total	100	100	200

	Block 2		
Treatment	S	F	Total
1	30	70	100
2	70	30	100
Total	100	100	200

Compute the table of row proportions for each block and draw the interaction graph. Construct the collapsed table and its table of row proportions. Comment.

3. Five students conducted a study of the General's Dilemma for their class project. The first student conducted a balanced CRD on 52 subjects and obtained 11 successes with the first version and 8 successes with the second version. The second student conducted a balanced CRD on 16 subjects and obtained 5 successes with the first version and 2 successes with the second version. The third and fourth students obtained identical results; each conducted a balanced CRD on 10 subjects and obtained 3 successes with the first version and 1 success with the second version. The fifth student conducted a balanced CRD on 10 subjects and obtained 3 successes with the first version and 2 successes with the second version. Treat each student researcher's subjects as a block.

   (a) Construct the tables of observed frequencies and row proportions for each block.

   (b) Construct the interaction graph.

   (c) Construct the collapsed table and its table of row proportions.

   (d) Briefly describe what you have learned.

4. Consider only the first three blocks in Figure 4.3. These blocks correspond to the three performances of the Infidelity study using only female subjects. Recall that each block is a balanced CRD with 50 subjects. The values of a and c are 21 and 13 in block 1; 15 and 9 in block 2; and 19 and 16 in block 3.

   (a) Construct the tables of observed frequencies and row proportions for each block.

   (b) Construct the interaction graph.

   (c) Construct the collapsed table and its table of row proportions.

   (d) Briefly describe what you have learned.

4.2 THE MANTEL-HAENSZEL TEST

A randomized block design begins by assigning subjects to blocks. Next, within each block a completely randomized design is performed. One obvious approach to analyzing an RBD is to analyze each block separately using Fisher's test. This approach is illustrated below for the AIDS-IP study.

In each block a Fisher's test was performed with the first alternative—that IP is superior to the placebo. The computational details will not be given, but Table 4.5 presents the exact P-values for these three tests. In each block the P-value is very close to 0.05.

It is also possible to use the standard normal curve to obtain approximate P-values for the tests. Recall from Chapter 3 that an approximate P-value is

TABLE 4.5 Exact P-values for the First Alternative for Fisher's Test on Each Block of the AIDS-IP Study

Block	P
1	0.0576
2	0.0418
3	0.0681

TABLE 4.6 Approximate P-values for the First Alternative for Fisher's Test on Each Block of the AIDS-IP Study

Block	z	Approximate P
1	2.04	0.0207
2	2.04	0.0207
3	1.86	0.0314

obtained by the following two steps:

1. Compute the observed value of the standardized version of the test statistic,

$$z = \frac{\sqrt{n-1}(ad - bc)}{\sqrt{n_1 n_2 m_1 m_2}}$$

2. Approximate the P-value (for the first alternative) by the area under the standard normal curve to the right of z.

Table 4.6 presents the values of z and the approximate P-values for the AIDS-IP Study. The technique of computing these numbers is familiar, and the details are not given. If interested, you are encouraged to verify the table's entries. Note that there is poor agreement between the exact and the approximate P-values. This is not surprising, because the condition of the general guideline for using the standard normal curve approximation is not met for any of these tests.

The results of the individual block analyses are frustrating. Each analysis gives strong, but not overwhelming, support to the notion that IP is superior to the placebo. The strongest possible evidence in support of the alternative that IP is superior to the placebo is obtained when all failures occur on the placebo. This is precisely the case in block 1. Unfortunately, with only four failures, the null probability that all failures would occur on the placebo, 0.0576, is substantial. (A researcher, of course, would never hope for many failures, but if the number of failures is small, as in block 1, a difference between the treatments cannot be demonstrated statistically.) Clearly, a method is needed for combining the evidence from the different blocks into one overall test. The **Mantel-Haenszel (MH) test** is such a method. The MH test is very similar to Fisher's test, and Fisher's test is used in its derivation.

To develop the MH test, it is convenient to reexamine Fisher's test. Recall that the test statistic X for Fisher's test has observed value

$$x = \hat{p}_1 - \hat{p}_2.$$

Now X is not the only random variable associated with Fisher's test. Recall that a denotes the number of successes on the first treatment. Let A denote the random variable whose observed value is a. It can be shown that a simple algebraic relationship exists between the random variables A and X:

$$A = \frac{n_1 n_2}{n} X + \frac{n_1 m_1}{n}. \tag{4.1}$$

For example, Chapter 2 (p. 51) introduced the Class study, which had

$$n_1 = n_2 = m_1 = m_2 = 2 \quad \text{and} \quad n = 4.$$

Substituting these values into Equation 4.1 gives

$$A = \frac{2(2)}{4} X + \frac{2(2)}{4} = X + 1,$$

a fact that appeared, but was not highlighted, in Chapter 2. It was demonstrated in Section 3.3 that working with X is equivalent to working with its standardized version, Z. By the same type of argument, working with X is equivalent to working with A. In other words, Fisher's test can be written in terms of the random variable A instead of the random variable X. It can be shown that the null sampling distribution of A has the following properties:

$$E(A) = \frac{n_1 m_1}{n} \tag{4.2}$$

and

$$\text{Var}(A) = \frac{n_1 n_2 m_1 m_2}{n^2 (n-1)}. \tag{4.3}$$

Note that the expression for $E(A)$ is the second term in Equation 4.1. Here is a final feature to note from Equation 4.1 about the relationship between A and X:

> The relationship between A and X is direct; that is, if either value increases, so does the other. In particular, X is larger than 0 if, and only if, A is larger than $E(A)$.

Consider block 1 in the AIDS-IP study. The observed value of the random variable A is $a = 217$. Equations 4.2 and 4.3 yield

$$E(A) = \frac{n_1 m_1}{n} = \frac{217(423)}{427} = 214.97$$

and

$$\text{Var}(A) = \frac{n_1 n_2 m_1 m_2}{n^2(n-1)} = \frac{217(210)(423)(4)}{(427)^2 (426)} = 0.9927.$$

Thus, the standardized value of $a = 217$ is given by

$$\frac{a - E(A)}{\sqrt{\text{Var}(A)}} = \frac{217 - 214.97}{\sqrt{0.9927}} = 2.04.$$

Note that this value coincides with the standardized value of x that was presented in Table 4.6. This agreement is no accident: the standardized values of x and a always agree. The value z needed for the standard normal curve approximation to the P-value can be computed with the equation

$$z = \frac{a - E(A)}{\sqrt{\text{Var}(A)}}$$

or with the more familiar

$$z = \frac{\sqrt{n-1}(ad - bc)}{\sqrt{n_1 n_2 m_1 m_2}}.$$

4.2.1 Step 1: The Hypotheses

Recall that for Fisher's test the null hypothesis states that no subject would respond differently to the two treatments. This is also the null hypothesis for the MH test.

Fisher's test has three choices of alternative. Using the terminology of Chapter 2, they are

1. The first treatment is superior to the second treatment.
2. The first treatment is inferior to the second treatment.
3. The first treatment is either superior or inferior to the second treatment.

For the MH test, the three choices of alternative are

1. The first treatment is superior to the second treatment in every block.
2. The first treatment is inferior to the second treatment in every block.
3. The first treatment is either superior to the second treatment in every block or inferior to the second treatment in every block.

There are two important features to note about these hypotheses:

- The null hypothesis for the MH test says *nothing* about comparing different blocks. In particular, it does not imply that the first treatment is equally effective in all blocks. Quite the contrary, one often forms blocks, as in the AIDS-IP study, with the goal of making subjects in different blocks dissimilar.
- Each choice for the alternative hypothesis specifies a consistent pattern in the relationship between treatment and response across blocks; that is, the first treatment must be consistently superior or inferior to the second. As demonstrated below in a number of examples and exercises, the MH test is not useful for investigating inconsistent patterns.

4.2.2 Step 2: The Test Statistic and Its Sampling Distribution

Each block of an RBD has its own random variable A with an observed value a. It is convenient to present these observed values in a table, as illustrated in the first two columns of Table 4.7 for the AIDS-IP study. Next, use Equations 4.2 and 4.3 to compute the values of $E(A)$ and $\text{Var}(A)$ and place the results in the third and fourth columns of the table. It is easy to analyze each individual block; simply compute

$$z = \frac{a - E(A)}{\sqrt{\text{Var}(A)}}$$

for each block and place these values in the fifth column of the table. The standard normal curve can now be used to obtain an approximate P-value, as given in the sixth column of Table 4.7.

Define the random variable U by the equation

$$U = \sum A,$$

where the sum is taken over all of the blocks. The observed value of the random variable U is denoted by u and is computed by the equation

$$u = \sum a.$$

For the AIDS-IP study this value is obtained by summing the entries in the second column of Table 4.7,

$$u = \sum a = 217 + 163 + 27 = 407.$$

The exact null sampling distribution of U is difficult to obtain, and only a mathematical approximation is presented in this chapter. It can be shown that

$$E(U) = \sum E(A) \quad \text{and} \quad \text{Var}(U) = \sum \text{Var}(A),$$

TABLE 4.7 Calculations for the MH Test for the AIDS-IP Study

Block	a	$E(A)$	$\text{Var}(A)$	z	P*
1	217	214.97	0.993	2.04	0.0207
2	163	160.14	1.956	2.04	0.0207
3	27	24.68	1.565	1.85	0.0322
MH test:	$u = 407$	$E(U) = 399.79$	$\text{Var}(U) = 4.514$	3.40	0.0003

*Entries for the P-value are for the first alternative and were obtained with the standard normal curve approximation.

where again the sums are taken over all blocks. Thus, if one sums the three entries in the third column of Table 4.7, the result is $E(U)$, which is presented in the bottom row of that column. Similarly, the sum of the three entries in the fourth column is $\text{Var}(U)$, also presented in the bottom row.

Finally, let Z denote the standardized version of U,

$$Z = \frac{U - E(U)}{\sqrt{\text{Var}(U)}}.$$

It can be shown that the null sampling distribution of Z can be approximated well by the standard normal curve.

The observed value z can be computed from the equation

$$z = \frac{u - E(U)}{\sqrt{\text{Var}(U)}}. \tag{4.4}$$

For the AIDS-IP study, for example, z can be computed easily from the information in the bottom row of Table 4.7:

$$z = \frac{407 - 399.79}{\sqrt{4.514}} = 3.40.$$

4.2.3 Step 3: The Rule of Evidence

The First Alternative (>) Recall that the first alternative states

Treatment 1 is superior to treatment 2 in every block.

As argued in Chapter 2, in any particular block the larger the value of x, the stronger is the evidence in support of the first alternative. As noted in step 2, the larger the value of x, the larger is the value of a. Thus, in terms of a the rule of evidence is

In any particular block, the larger the value of A, the stronger is the evidence that the first treatment is superior to the second treatment in that block.

The evidence across blocks is obtained by summing the individual values of a to obtain u. The rule of evidence for the first alternative for the MH test is

The larger the value of U, the stronger is the evidence in support of the first alternative.

The Second Alternative (<) Recall that the second alternative states

Treatment 1 is inferior to treatment 2 in every block.

A simple modification of the argument for the first alternative gives

The smaller the value of U, the stronger is the evidence in support of the second alternative.

The Third Alternative ($\neq$)   Recall that the third alternative states

> Treatment 1 is either superior to treatment 2 in every block or inferior to treatment 2 in every block.

It can be shown that the rule of evidence is

> The further the value of U from its mean $E(U)$, in either direction, the stronger is the evidence in support of the third alternative.

4.2.4 Step 4: Computing the P-value

Recall that the P-value is equal to the probability of obtaining a value of the test statistic that gives the same or stronger evidence (in support of the alternative) than what was obtained in the actual study. (The smaller the numerical value of **P**, the stronger is the evidence in support of the alternative.) Let z denote the standardized value of u. A simple modification of the argument given in Chapter 2 indicates the following:

- For the first alternative, $\mathbf{P} = P(Z \geq z)$.
- For the second alternative, $\mathbf{P} = P(Z \leq z)$.
- For the third alternative, $\mathbf{P} = P(Z \geq |z|) + P(Z \leq -|z|)$.

Therefore, the standard normal curve approximation of the P-value is given by the following:

- For the first alternative ($>$) the approximate P-value is the area under the standard normal curve to the right of z.
- For the second alternative ($<$) the approximate P-value is the area under the standard normal curve to the right of $-z$.
- For the third alternative ($\neq$) the approximate P-value is twice the area under the standard normal curve to the right of $|z|$.

Recall that $z = 3.40$ is the observed value of the MH test statistic for the AIDS-IP study. Thus, for the first alternative the standard normal curve approximation to the P-value is the area under the standard normal curve to the right of 3.40, which equals 0.0003, from Table A.2. The MH test is a big improvement over the analysis of one block at a time given earlier; the combined evidence of the entire study is overwhelmingly in support of the first alternative, that IP is superior to the placebo. It can be shown that the exact P-value for the MH test for the AIDS-IP study is 0.0004. Interestingly, the standard normal curve approximation is very accurate for the MH test even though it is inaccurate for each of the Fisher's tests.

TABLE 4.8 *Calculations for the MH Test for the Eight Infidelity Studies*

Block	a	E(A)	Var(A)	z	P*
1	21	17.0	2.776	2.40	0.0164
2	15	12.0	3.184	1.68	0.0930
3	19	17.5	2.679	0.92	0.3576
4	13	9.5	3.005	2.02	0.0434
5	10	11.0	3.143	−0.56	0.5754
6	19	11.0	3.143	4.51	<0.0004
7	9	11.5	3.168	−1.40	0.1616
8	6	11.5	3.168	−3.09	0.0020
MH test:	u = 112	E(U) = 101.0	Var(U) = 24.266	2.23	0.0258

*Entries for the P-value are for the third alternative and were obtained with the standard normal curve approximation.

The MH test can be performed for the eight Infidelity studies; Table 4.8 presents the necessary numbers. (These numbers are taken from the students' projects; they are not easy to obtain from Figure 4.3.) The P-values in this table are computed for the third alternative using the standard normal curve approximation.

Before you use the MH test, it is prudent to examine the individual block analyses. The individual Fisher's tests yield highly statistically significant results in blocks 6 and 8. In block 6, the evidence indicates that the subjects prefer to tell on a husband, but in block 8 the evidence indicates that the subjects prefer to tell on a wife. Blocks 1 and 4 yield statistically significant evidence that subjects prefer to tell on husbands. Blocks 2 and 7 provide evidence in opposite directions, each of which is close to being statistically significant. Finally, blocks 3 and 5 provide evidence in opposite directions, but in each block the evidence is weak.

The entries in the bottom row of Table 4.8 indicate that the MH test is statistically significant. In fact, because $u = 112$ is larger than $E(U) = 101$, the direction of the evidence is that subjects prefer to tell on the husband *in every block*. This conclusion is misleading! It obliterates, by combining data, one of the most interesting features in the studies, namely, the large variation in the magnitude and direction of evidence from block to block. This example illustrates the need for care in applying the MH test. The MH test looks for a consistent pattern of evidence in every block and is therefore not appropriate when such a pattern is clearly contradicted by the data.

In summary, for an RBD a researcher should perform Fisher's test on each block before performing the MH test because something interesting might be learned from the individual block analyses and because by comparing the direction and magnitude of the evidence in each block, the researcher might decide that the MH test is inappropriate.

EXERCISES 4.2

1. Refer to Exercise 1 on p. 122.
   (a) Conduct Fisher's test within each block with the first alternative. (Use the standard normal curve to obtain approximate P-values.)
   (b) Conduct the MH test with the first alternative.
   (c) Write a few sentences comparing the results from the two hypothesis tests.

2. Refer to Exercise 2 on p. 122.
   (a) Conduct Fisher's test within each block with the third alternative. (Use the standard normal curve to obtain approximate P-values.)
   (b) Conduct the MH test with the third alternative.
   (c) Write a few sentences comparing the results from the two hypothesis tests.

3. Refer to Exercise 3 on p. 123.
   (a) Conduct Fisher's test within each block with the first alternative. (If possible, use Table A.1 to obtain the exact P-values; otherwise use the standard normal curve approximation.)
   (b) Conduct the MH test with the first alternative.
   (c) Write a few sentences comparing the results from the two hypothesis tests.

4. Refer to Exercise 4 on p. 123.
   (a) Conduct Fisher's test within each block with the first alternative.
   (b) Conduct the MH test with the first alternative.
   (c) Write a few sentences comparing the results from the two hypothesis tests.

Table 4.9 is taken from a double-blind study of the effectiveness of zidovudine (formerly AZT) in preventing the development of AIDS or advanced AIDS-related complex (ARC) among subjects who have tested positive for the HIV antibody [3]. The design was a randomized block design with subjects divided into two blocks according to CD4+ cell count at the beginning of the study. Use the information in Table 4.9 to complete Exercises 5 through 18.

Be careful. Several characteristics of this problem can easily lead to confusion. First, Table 4.9 does not look like the tables in this book. Second, there are two choices for the dichotomous response. One choice has the categories "AIDS" (a failure) and "not AIDS" (a success); the other choice has categories "AIDS or ARC" (a failure) and "neither AIDS nor ARC" (a success). Read each exercise carefully to determine which response is appropriate. Third, there were three treatments in the actual study

TABLE 4.9 *Frequency of Clinical Progression of Disease to AIDS or Advanced AIDS-Related Complex*

CD4+ Cell Count at Entry (cells/mm^3)	Characteristic	Study Group		
		Placebo	500 mg Zidovudine	1,500 mg Zidovudine
< 200	No. of subjects	56	55	51
	AIDS	12	7	3
	AIDS or ARC	13	10	6
200–499	No. of subjects	372	398	406
	AIDS	21	4	11
	AIDS or ARC	25	7	13

instead of the usual two. The treatments were a placebo, 500 mg of zidovudine, and 1,500 mg of zidovudine. Some exercises ask you to compare the two doses of zidovudine, and others ask for a comparison of the placebo with either dose of zidovudine or with a combination of the doses.

5. Use the dichotomous response that defines "not AIDS" a success and "AIDS" a failure. Use the information in Table 4.9 to complete the following table:

	CD4+ Cell Count					
	< 200			200–499		
Treatment	S	F	Total	S	F	Total
1,500 mg AZT	48	3	51	395	11	406
500 mg AZT		7	55			
Placebo		12	56			
Total		22	162			

6. Use the dichotomous response that defines "neither AIDS nor ARC" a success and "AIDS or ARC" a failure. Use the information in Table 4.9 to complete the following table:

	CD4+ Cell Count					
	< 200			200–499		
Treatment	S	F	Total	S	F	Total
1,500 mg AZT	45	6	51	393	13	406
500 mg AZT		10	55			
Placebo		13	56			
Total		29	162			

7. Construct a table of row proportions for each block in Exercise 5.

8. Construct a table of row proportions for each block in the Exercise 6.

9. Use the results of Exercise 7 to draw an interaction graph of the proportion of failures on each treatment by block. Briefly describe what it reveals.

10. Use the results of Exercise 8 to draw an interaction graph of the proportion of failures on each treatment by block. Briefly describe what it reveals.

11. Use the results from Exercise 5 to perform an MH test that compares the low and high doses of zidovudine. Use the third alternative. Briefly describe what the test reveals.

12. Use the results from Exercise 6 to perform an MH test that compares the low and high doses of zidovudine. Use the third alternative. Briefly describe what the test reveals.

13. Use the results from Exercise 5 to perform an MH test that compares the placebo to the low dose of zidovudine. Use the alternative that states that zidovudine increases the chance of success. Briefly describe what the test reveals.

14. Use the results from Exercise 6 to perform an MH test that compares the placebo to the low dose of zidovudine. Use the alternative that states that zidovudine increases the chance of success. Briefly describe what the test reveals.

15. Use the results from Exercise 5 to perform an MH test that compares the placebo to the high dose of zidovudine. Use the alternative that states that zidovudine increases the chance of success. Briefly describe what the test reveals.

16. Use the results from Exercise 6 to perform an MH test that compares the placebo to the high dose of zidovudine. Use the alternative that states that zidovudine increases the chance of success. Briefly describe what the test reveals.

17. Use the results from Exercise 5 to perform an MH test that compares the subjects on placebo with all subjects on zidovudine (that is, collapse the subjects on low dose with those on high dose to form one sample). Use the alternative that states that zidovudine increases the chance of success. Briefly describe what the test reveals.

18. Use the results from Exercise 6 to perform an MH test that compares the subjects on placebo with all subjects on zidovudine (that is, collapse the subjects on low dose with those on high dose to form one sample). Use the alternative that states that zidovudine increases the chance of success. Briefly describe what the test reveals.

Table 4.10 presents some results from a double-blind study of the effectiveness of AZT (zidovudine) in the treatment of patients with AIDS or AIDS-related complex (ARC) [4]. Exercises 19 and 21 ask you to analyze other data from this study, and Exercises 20 and 22 refer to the data in this table. Each subject was classified into one of two blocks based on CD4+ count at the beginning of the study.

19. Overall, ignoring block and diagnosis, 1 of 145 subjects assigned to AZT and 19 of 137 assigned to the placebo died.

    (a) Present this collapsed data in a 2 × 2 table.
    (b) Construct the table of row proportions.
    (c) Pretend that the randomization was performed on the entire collection of 282 subjects instead of within blocks. Conduct a hypothesis test of the equivalence of treatment versus the alternative that AZT is better than the placebo at preventing death.

20. Refer to Table 4.10. A positive reaction to at least one skin antigen during the study was considered to be a sign that the subject's immune system was

TABLE 4.10 *Frequency of Skin-Test Conversions during the Study*

	No. Positive/No. Tested
CD4+ ≤ 100	
AZT	16/80
Placebo	3/76
101 < CD4+ < 500	
AZT	21/49
Placebo	8/41

becoming stronger. Complete the following table; note that a positive skin test is defined to be a success.

	CD4+ Cell Count					
	≤ 100			101–499		
Treatment	S	F	Total	S	F	Total
AZT	16	64	80	21	28	49
Placebo						
Total						

21. Refer to Exercise 19. Overall, only 20 of 282 patients died, so the researchers were interested in other measures in addition to mortality to compare the treatments. Twenty-four of the subjects in the AZT group and 45 in the placebo group developed at least one opportunistic infection during the study. These 24 + 45 = 69 subjects included the 20 who died during the study.

    (a) Present this collapsed data in a 2 × 2 table.
    (b) Construct the table of row proportions.
    (c) Pretend that the randomization was performed on the entire collection of 282 subjects instead of individual blocks. Conduct a hypothesis test of the equivalence of treatment versus the alternative that AZT is better than the placebo at preventing opportunistic infection.

22. Refer to Exercise 20.

    (a) Construct a table of row proportions for each block.
    (b) Draw an interaction graph of the proportion of successes on each treatment by block. Briefly describe what it reveals.
    (c) Perform an MH test that compares the placebo to AZT with the alternative that AZT increases the chance of a positive skin test. Briefly describe what the test reveals.

4.3 THE RANDOMIZED PAIRS DESIGN

Consider a randomized block design with the following two characteristics:

- There are exactly two subjects in each block.
- In each block one subject is assigned to each treatment.

Such an RBD is called a randomized pairs design (RPD); the name reflects the pairing of subjects into blocks of size two.

An RPD is often used in place of a CRD when a sequence of trials is studied. The first two trials are paired, as are the third and fourth, and so on. As a design for a sequence of trials, an RPD has two potential advantages over a CRD:

- The RPD guards against a **familiarity effect,** because no treatment will ever be used more than twice in succession (the last trial of one pair followed by the first trial of the next pair).
- The RPD guards against a time trend; for a CRD it is possible, by chance, for one treatment to have an excess of trials early or late in the study.

If neither a familiarity effect nor a time trend actually exists, the CRD has some advantages over an RPD, although the latter is still valid. Using an RPD instead of a CRD is analogous to purchasing insurance. If it is not needed, it is a waste of money; if needed, however, it can be very valuable.

These ideas are illustrated with a class project conducted by Craig Bender, referred to as the Catcher study. The motivation for the study, taken from Craig's report, was as follows:

> My 12-year-old son plays catcher on a Little League team, and he needs to practice his throws to second base. I have not been able to persuade him to step in front of home plate to make the throw. I decided to test the accuracy of his throws to second by having him throw 25 times from in front of home plate and 25 times from the catcher's position behind the plate. The purpose of getting him to throw in front of the plate is to decrease the distance of the throw, to get his body into a better throwing position, and to create a natural follow-through.

Treatment 1 was throwing from in front of the plate, and treatment 2 was throwing from behind the plate. The subjects were the 50 throws, and the metasubject was Craig's son. A throw was a success if the second baseman could reach the thrown ball while remaining in contact with the base, and a failure otherwise. Craig was concerned with the possibility of a familiarity effect. With several consecutive throws from the same location, the catcher might have used the immediate feedback of observing his result to adjust his effort. In itself, this is not bad, but Craig wanted the study to resemble a game situation as closely as possible.

Throws to second base occur irregularly during the course of a baseball game, ruling out the possibility of improving through immediate feedback.

Craig also was concerned with the possibility of a time trend; his son might have grown tired or become bored during the study, and his performance might have deteriorated. Alternatively, he might have improved from practice. For these reasons Craig selected an RPD for his study.

The Catcher study had 25 pairs of subjects (throws), and a separate randomization had to be performed for each pair. Craig used a box containing two indistinguishable tickets, with one ticket marked "Front" and the other marked "Behind." In order to randomize the first pair of subjects, the tickets were mixed and one was selected at random. If the ticket marked "Front" was selected, the first subject (throw) was with treatment 1 (in front of the plate) and the second subject was with treatment 2. If the ticket marked "Behind" was selected, the order was reversed and the first throw was made from behind the plate. The selected ticket was replaced in the box, and the two tickets were mixed again. The above procedure was repeated for the second pair of subjects, and then the third, and so on until all 25 pairs had been randomized. The results of the Catcher study are given in Table 4.11.

In the AIDS-IP and the eight Infidelity studies a 2 × 2 table was constructed for each block of subjects. Although this idea could be applied to the Catcher study, it would result in the construction of 25 tables. Twenty-five tables are too many to examine visually! The Catcher study is not unusual: large numbers of pairs of subjects in an RPD are typical. Fortunately, there exists a simple alternate method of presenting the data from an RPD. Because the block sizes are so small—two subjects per block with one subject assigned to each treatment—

TABLE 4.11 Raw Data for the Catcher Study

Pair	1	2	3	4	5	6	7	8	9
Trial	1 2	3 4	5 6	7 8	9 10	11 12	13 14	15 16	17 18
Treatment*	2 1	1 2	2 1	1 2	1 2	2 1	1 2	2 1	1 2
Response	S S	F F	F S	S S	S S	F F	S F	S F	S S

Pair	10	11	12	13	14	15	16	17
Trial	19 20	21 22	23 24	25 26	27 28	29 30	31 32	33 34
Treatment*	1 2	1 2	1 2	2 1	2 1	1 2	2 1	2 1
Response	F S	S F	F F	S S	F S	S F	F S	F S

Pair	18	19	20	21	22	23	24	25
Trial	35 36	37 38	39 40	41 42	43 44	45 46	47 48	49 50
Treatment*	1 2	2 1	2 1	1 2	2 1	2 1	1 2	1 2
Response	F S	S S	F S	S F	F F	F S	F S	S F

*Treatment 1 is throwing from in front of the plate, and treatment 2 is throwing from behind the plate.

TABLE 4.12 The Four Possible Outcomes for a Block in an RPD

A		S	F	Total
	Treatment 1	1	0	1
	Treatment 2	1	0	1
	Total	2	0	2

B		S	F	Total
	Treatment 1	1	0	1
	Treatment 2	0	1	1
	Total	1	1	2

C		S	F	Total
	Treatment 1	0	1	1
	Treatment 2	1	0	1
	Total	1	1	2

D		S	F	Total
	Treatment 1	0	1	1
	Treatment 2	0	1	1
	Total	0	2	2

there are only four tables that could possibly occur in a block. They are presented in Table 4.12 and are labeled A, B, C, and D, for convenience. Table A corresponds to a success with each treatment, table B corresponds to a success with treatment 1 only, table C denotes a success with treatment 2 only, and table D represents a failure with each treatment.

For the Catcher study, six pairs (numbers 1, 4, 5, 9, 13, and 19) give table A, eleven pairs (3, 7, 11, 14, 15, 16, 17, 20, 21, 23, and 25) give table B, four pairs (8, 10, 18, and 24) give table C, and four pairs (2, 6, 12, and 22) give table D.

Let

a = The number of pairs that yield table A.

b = The number of pairs that yield table B.

c = The number of pairs that yield table C.

d = The number of pairs that yield table D.

For the Catcher study, $a = 6$, $b = 11$, $c = 4$, and $d = 4$.

A 2×2 table in Format 2 is a convenient way to present these values. The general form of the 2×2 table in Format 2 is given in Table 4.13, the definitions of its symbols are in Table 4.14, and the specific Format 2 table for the Catcher

4.3 The Randomized Pairs Design

TABLE 4.13 2 × 2 Table in Format 2

Treatment 1	Treatment 2		Total
	S	F	
S	a	b	n_1
F	c	d	n_2
Total	m_1	m_2	n

study is given in Table 4.15. Note that the Format 2 table uses the same symbols $(a, b, \ldots)$ as the Format 1 table and that the symbols appear at the same locations in the table. The difference between the formats is in the labels of the rows and columns. For a Format 2 table the row proportions are not of interest; instead, the analyst computes the following summaries:

1. The proportion of successes with treatment 1 is denoted by $\hat{p}_1$ and is computed by the formula $\hat{p}_1 = n_1/n$.
2. The proportion of failures with treatment 1 is denoted by $\hat{q}_1$ and is computed by the formula $\hat{q}_1 = n_2/n$. Alternatively, $\hat{q}_1$ can be computed by using the equation $\hat{q}_1 = 1 - \hat{p}_1$.
3. The proportion of successes with treatment 2 is denoted by $\hat{p}_2$ and is computed by the formula $\hat{p}_2 = m_1/n$.

TABLE 4.14 Meaning of the Symbols in a 2 × 2 Table in Format 2

Symbol	Meaning
n	Total number of pairs in the study; also equals the total number of subjects on each treatment
n_1	Total number of successes with treatment 1
n_2	Total number of failures with treatment 1
m_1	Total number of successes with treatment 2
m_2	Total number of failures with treatment 2
a	Number of pairs that yield two successes
b	Number of pairs that yield a success with treatment 1 and a failure with treatment 2
c	Number of pairs that yield a failure with treatment 1 and a success with treatment 2
d	Number of pairs that yield two failures

TABLE 4.15 2 × 2 Table for the Catcher Study

Front	Behind S	F	Total
S	6	11	17
F	4	4	8
Total	10	15	25

4. The proportion of failures with treatment 2 is denoted by $\hat{q}_2$ and is computed by the formula $\hat{q}_2 = m_2/n$. Alternatively, $\hat{q}_2$ can be computed by using the equation $\hat{q}_2 = 1 - \hat{p}_2$.

This notation is both familiar and confusing. It is familiar in that the p's and q's, the hats, and the subscripts are used in exactly the way introduced in Chapter 1. It is confusing because you must remember that the formula (for example, for $\hat{p}_1$) depends on the format of the table.

The difference of the proportions of successes with the two treatments is

$$\hat{p}_1 - \hat{p}_2 = \frac{n_1}{n} - \frac{m_1}{n} = \frac{n_1 - m_1}{n}. \tag{4.5}$$

Note that

$$n_1 - m_1 = (a + b) - (a + c) = b - c.$$

Thus, Equation 4.5 can be written as

$$\hat{p}_1 - \hat{p}_2 = \frac{b - c}{n}. \tag{4.6}$$

For the Catcher study the success rates with the two treatments are

$$\hat{p}_1 = \frac{n_1}{n} = \frac{17}{25} = 0.68 \quad \text{and} \quad \hat{p}_2 = \frac{m_1}{n} = \frac{10}{25} = 0.40.$$

The difference in success rates is $\hat{p}_1 - \hat{p}_2 = 0.68 - 0.40 = 0.28$, which also can be computed with Equation 4.6,

$$\hat{p}_1 - \hat{p}_2 = \frac{b - c}{n} = \frac{11 - 4}{25} = 0.28.$$

One can also create the collapsed table for the Catcher study, as given in Table 4.16. The collapsed table is a Format 1 table. It contains less information than the Format 2 table.

For an RBD it was recommended that a researcher perform a Fisher's test on each block before conducting the MH test (see p. 130). There were two reasons for this advice:

TABLE 4.16 The Collapsed Table for the Catcher Study

	Observed Counts			Row Proportions		
Location of Throw	S	F	Total	S	F	Total
In front of plate	17	8	25	0.68	0.32	1.00
Behind plate	10	15	25	0.40	0.60	1.00
Total	27	23	50			

1. Something interesting might be learned from the individual block analyses.
2. By comparing the direction and magnitude of the evidence in each block, the researcher might decide that the MH test is inappropriate.

Neither of these reasons is valid for the randomized pairs design. First, it can be shown that, because the block sizes are so small, the only possible values of **P** are 0.5 and 1.0. (In fact, for the third alternative, the only possible value is 1.0.) Thus, nothing of interest can be learned from the analysis of one block. Second, again because of the small block size, the pattern of evidence from block to block cannot be distinguished easily from chance variation. Thus, the only inference available is obtained from the MH test.

The evaluation of the MH test statistic is greatly simplified when a randomized block design is in fact a randomized pairs design. Recall from Equation 4.4 that the observed value of the MH test statistic is

$$z = \frac{u - E(U)}{\sqrt{\text{Var}(U)}}.$$

The details are not given, but it can be shown with some simple algebra that using the notation of the 2 × 2 table in Format 2,

$$u = a + b, \quad E(U) = a + 0.5(b + c), \quad \text{and} \quad \text{Var}(U) = 0.25(b + c).$$

Substituting these values into the above formula for z yields, after algebra,

$$z = \frac{b - c}{\sqrt{b + c}}. \tag{4.7}$$

For example, for the Catcher study, $b = 11$ and $c = 4$, giving

$$z = \frac{11 - 4}{\sqrt{11 + 4}} = 1.81.$$

Using the standard normal curve, the approximate P-value for the first alternative (Craig's choice) is the area to the right of 1.81, which equals 0.0351. Craig's data provide evidence that his son threw better from in front of home plate. If the null hypothesis were true, the probability of Craig's evidence or stronger evidence is only 0.0351. Because this probability is small, Craig's data are said to provide strong evidence in support of the alternative. In fact, the usual interpretation is to reject the null hypothesis in favor of the alternative because **P** is smaller than 0.05.

The MH test for paired data is usually called **McNemar's test.** Exact P-values for McNemar's test are given in Table A.4. To use this table, compute the values of

$$R = b + c \quad \text{and} \quad O = b. \tag{4.8}$$

Here are two examples of the use of Table A.4:

1. Suppose that $b = 8$ and $c = 3$; then $R = 11$ and $O = 8$. From Table A.4, **P** equals 0.113 for the first alternative, 0.967 for the second alternative, and 0.227 for the third alternative.

2. Suppose that $b = 2$ and $c = 6$; then $R = 8$ and $O = 2$. From Table A.4, **P** equals 0.965 for the first alternative, 0.145 for the second alternative, and 0.289 for the third alternative.

Note that Table A.4 can be used only if $b + c$ is equal to 14 or less. For larger values the standard normal curve approximation is usually adequate.

EXERCISES 4.3

1. John Nash, a student in my beginning statistics class, conducted an RPD to investigate whether his 10-month-old daughter walks better on the bed or on the floor. Here is his description of a trial:

    > [I place] Chelsea in a standing position, retreat to a distance of 2 to 3 feet while holding a shoe belonging to her in my teeth, and observe whether or not steps are taken toward the shoe in an attempt to grab it. A success occurs when at least one step is taken; a failure occurs if she falls before taking any steps.

    John's data are as follows:

Pair	1		2		3		4		5	
Trial:	1	2	3	4	5	6	7	8	9	10
Surface:	B	F	B	F	F	B	F	B	B	F
Response:	S	S	S	S	S	S	F	S	S	F

    (a) Why do you think John chose an RPD instead of a CRD?

    (b) Construct a 2 × 2 contingency table, in Format 2, to summarize these data. (Define the bed to be the first treatment.)

    (c) What is the proportion of successes with the first treatment? With the second treatment?

    (d) Use Table A.4 to obtain the exact P-value for the first alternative.

2. Mary Nichols-Pierce wrote in her report on her class project

    > For an aspiring novice pianist, it is important to be able to find middle C on the keyboard in the course of playing. Middle C is often played with either hand. It is important to locate middle C without continuously looking at the keyboard.

    Mary conducted an RPD with 20 trials; in each trial she attempted to locate middle C without looking at the keyboard. The treatments were her left and right hands. Mary's data are as follows:

Pair	1		2		3		4		5	
Trial:	1	2	3	4	5	6	7	8	9	10
Hand:	L	R	L	R	R	L	L	R	R	L
Response:	F	F	S	S	F	S	S	F	S	F

Pair	6		7		8		9		10	
Trial:	11	12	13	14	15	16	17	18	19	20
Hand:	R	L	R	L	L	R	R	L	L	R
Response:	F	S	F	F	F	S	F	F	S	S

(a) Why do you think Mary chose an RPD instead of a CRD?

(b) Construct a 2 × 2 contingency table, in Format 2, to summarize these data. (Define right hand to be the first treatment.)

(c) What is the proportion of successes with the first treatment? With the second treatment?

(d) Use Table A.4 to obtain the exact P-value for the third alternative.

3. Sarah Mason's daughter enjoys jumping rope. Sometimes she jumps a rope held in her hands and sometimes a rope attached to a stick. Sarah tested her daughter's jumping ability with an RPD with 50 trials. The first treatment was jumping the rope held in her hands, and the second treatment was jumping the rope attached to a stick. A success was defined as completing a predetermined number of jumps without a miss. Below is the 2 × 2 table summary of Sarah's data:

		Stick	
Hands	S	F	Total
S	10	14	24
F	1	0	1
Total	11	14	25

(a) Why do you think Sarah chose an RPD instead of a CRD?

(b) What is the proportion of successes with the first treatment? With the second treatment?

(c) Use the standard normal curve to obtain the approximate P-value for the first alternative.

4. Tirelo Modie enjoyed playing *djari* as a child in Botswana. As an adult, she decided to compare her ability with her right hand (the first treatment) and with her left hand. Each trial consisted of tossing the djari (similar to a shuttlecock) into the air. A trial resulted in a success if Tirelo was able to execute a particular maneuver on the descending djari with a stick. Tirelo performed an RPD with 50 pairs of trials. Here is a summary of her performance:

- In 15 pairs of trials she obtained a success with each hand.
- In 20 pairs of trials she obtained a failure with each hand.
- In 10 pairs of trials she obtained a success with her right hand but a failure with her left hand.
- In 5 pairs of trials she obtained a success with her left hand but a failure with her right hand.

(a) Construct the Format 2 table of observed frequencies. What is the proportion of successes on the first treatment? On the second treatment?

(b) Use the standard normal curve to obtain the approximate P-value for the first alternative.

5. Each trial of Margaret Algar-Gelembiuk's study was an attempt to tune a high E on her guitar. Margaret's first treatment was to use an A tuning fork, and her second treatment was to use an E tuning fork. Before performing the study, she suspected that the A tuning fork would be better than the E tuning fork. A battery-powered tuner was used to classify each trial a success or failure. (A value within 5 Hz of the target was classified a success.) Margaret performed an RPD on 25 pairs of trials. Her results were as follows:

- In 11 pairs of trials she obtained a success with each fork.
- In 10 pairs of trials she obtained a success with the A fork but a failure with the E fork.

- In 4 pairs of trials she obtained a success with the E fork but a failure with the A fork.

(a) Construct the Format 2 table of observed frequencies. What is the proportion of successes on the first treatment? On the second treatment?

(b) What is the appropriate alternative? Find the exact P-value.

6. Maisie enjoys catching a Frisbee in her mouth. Linda Evashevski, Maisie's human companion, performed an RPD with 50 trials to investigate whether Maisie is better at catching a 28-inch Frisbee (the first treatment) or a 36-inch Frisbee. Each trial was a toss of a Frisbee, and Maisie was credited with a success if she caught the Frisbee before it hit the ground. Here are Maisie's results:

- In six pairs of trials she obtained a success with each Frisbee.
- In four pairs of trials she obtained a failure with each Frisbee.
- In six pairs of trials she obtained a success with the smaller Frisbee but a failure with the larger Frisbee.

(a) Construct the Format 2 table of observed frequencies. What is the proportion of successes on the first treatment? On the second treatment?

(b) Before performing the study, Linda suspected that Maisie would find it easier to catch the larger Frisbee. Find the approximate P-value for the alternative that corresponds to Linda's suspicion.

7. Joe Harwell, a defenseman on a university hockey team, wanted to compare his accuracy at two ways of shooting a hockey puck. The first treatment was shooting a wrist shot from 5 feet inside the blue line, and the second treatment was shooting a slap shot from the same distance. If the shot went into the net, the trial was labeled a success. Joe performed an RPD on 25 pairs of trials. Here are his results:

- In eight pairs of shots he obtained a success with each shot.
- In four pairs of shots he obtained a failure with each shot.
- In eight pairs of shots he obtained a success with the wrist shot but a failure with the slap shot.

(a) Construct the Format 2 table of observed frequencies. What is the proportion of successes on the first treatment? On the second treatment?

(b) The conventional wisdom in hockey is that a wrist shot is more accurate. Is the conventional wisdom supported by this data? Compute the P-value for the appropriate alternative.

8. Chris Loomis conducted an RPD to investigate whether he hits a softball better in practice batting left-handed or right-handed. His trials were 25 pairs of swings. If the ball traveled out of the infield in the air (in fair territory), the swing was labeled a success. Anything else, including a miss, was labeled a failure. Chris obtained a total of 15 successes batting right-handed (the first treatment) and a total of 12 successes batting left-handed. In four pairs of swings, he had two successes.

(a) Construct the Format 2 table of observed frequencies. What is the proportion of successes on the first treatment? On the second treatment?

(b) Chris believed he was ambidextrous and hence chose the third alternative. Compute the approximate P-value.

(c) Pretend that Chris had performed a CRD and obtained the same results, 15 successes with the first treatment and 12 successes with the second treatment. Compute the approximate P-value for the third alternative. Compare this P-value with your answer to part (b) and comment.

9. Over the course of several days, Leigh Murphy performed an RPD with 50 pairs of trials on the meta-subject Attica, her canine companion. In the first treatment, Leigh stood by a window inside her home and *yelled*, "Squirrel, Attica!" In the second treatment Leigh stood by the same window and calmly remarked, "Hey Attica, squirrel." Attica's response was classified into one of two categories—a success if she got excited and a failure if she did not move. There were a total of 37 successes on the first treatment and a total of only 16 successes on the second treatment. In only five pairs of trials did Attica give two successes.

(a) Construct the Format 2 table of observed frequencies. What is the proportion of successes on the first treatment? On the second treatment?

(b) Leigh chose the first alternative; compute the approximate P-value.

10. After several years of observation, Jackie Schmidt believed that her canine companion Basia was very talented at catching popcorn thrown directly at her, but had some difficulty if Jackie's throw was off line. In order to investigate this further, Jackie conducted an RPD on 25 pairs of trials. In each trial Jackie tossed a popped kernel of corn to Basia. In the first treatment she tossed the kernel approximately 2 feet to Basia's right, and in the second treatment she threw the kernel approximately 2 feet to Basia's left. A trial was labeled a success if Basia caught the kernel before it hit the ground. Basia obtained a total of 16 successes with the first treatment; in addition, she caught both kernels in seven pairs of trials and missed both in four pairs.

   (a) Construct the Format 2 table of observed frequencies. What is the proportion of successes on the first treatment? On the second treatment?
   (b) Jackie chose the third alternative; compute the P-value.

11. Conduct an RPD for a sequence of 100 trials that investigates a question of interest to you.

   (a) Describe the study, including the subjects, response, treatments, and method of randomization. Why is this study of interest to you?
   (b) Present the data in a 2 × 2 table. Compute the sample proportion of successes with each treatment.
   (c) Conduct a hypothesis test for an alternative of your choice.

REFERENCES

1. C. Pedersen et al., "The Efficacy of Inosine Pranobex in Preventing the Acquired Immunodeficiency Syndrome in Patients with Human Immunodeficiency Virus Infection," *New England Journal of Medicine*, Vol. 322, June 21, 1990, pp. 1757–1763.

2. S. Kweder et al., "Inosine Pranobex—Is a Single Positive Trial Enough?" *New England Journal of Medicine*, Vol. 322, June 21, 1990, pp. 1807–1809.

3. P. A. Volberding et al., "Zidovudine in Asymptomatic Human Immunodeficiency Virus Infection," *New England Journal of Medicine*, Vol. 322, April 5, 1990, pp. 941–949.

4. M. A. Fischl et al., "The Efficacy of Azidothymidine (AZT) in the Treatment of Patients with AIDS and AIDS-Related Complex," *New England Journal of Medicine*, Vol. 317, July 23, 1987, pp. 185–191.

One-Population Models

- **5.1 SURVEYS**
- **5.2 PROBABILITY**
- **5.3 BERNOULLI TRIALS**
- **5.4 SOME PRACTICAL CONSIDERATIONS**

Chapter 5

Chapter 1 introduced the four components of an experimental design:
1. The selection of the subjects to be included in the study
2. The specification of the treatments to be compared
3. The specification of the response to be obtained from each subject
4. The specification of the method by which subjects are assigned to treatments

The first three chapters focused on a comparison of two treatments with a dichotomous response for studies in which the subjects are assigned to treaments by randomization. Chapter 1 considered data presentation and summary, and Chapters 2 and 3 considered the formal inference procedure of hypothesis testing. It was emphasized that the informal or formal conclusions of such studies can be applied only to the collection of subjects actually studied.

Researchers often generalize findings beyond the subjects in the study. Sometimes such generalizations are appropriate, and many times they are not. This chapter begins the study of these generalizations.

Chapter 1 introduced the idea of two types of studies: those with subjects that are distinct individuals and those with subjects that are trials. These two types of studies lead to two types of generalizations. For either type of generalization, the fundamental concept is the **population.** The population can be the collection of all distinct individuals of interest to the researcher. Alternatively, the population can be a mathematical model of the **process** that generates the outcomes of the trials.

The first type of population is called a **finite population** because it consists of a finite collection of individuals. For example, in 1986 the Wisconsin Department of Transportation was interested in the population of all persons having a

Wisconsin driver's license. The second type of population is called an **infinite population.** A mathematical model of the process that generates the outcomes of successive tosses of a coin is a simple example of an infinite population.

A finite population can be studied with either a census or a survey. A **census** examines all members of the population, whereas a **survey** examines just part of the population. The members of the population studied in a survey are called the **sample.** It is usually impractical to conduct a census of a finite population. For example, for the 1986 Wisconsin Driver Survey, the Department of Transportation did not have the resources to obtain the desired data from every licensed driver in the state.

An infinite population is a mathematical model for the process that generates the data, so the notion of examining the whole population is meaningless. The sample consists of the trials whose outcomes are actually observed.

The population is a concept that allows the researcher to generalize the knowledge gained from the study. In the case of a survey, the generalization consists of making inferences about the characteristics of the population of individuals based on the data from the individuals in the sample. In the case of a process, the generalization consists of using the observed trials to learn about unknown characteristics of the mathematical model or to predict the outcomes of future trials from the process.

Section 5.1 gives examples of surveys and defines the important notion of selecting a random sample from a finite population. Section 5.2 presents methods for computing probabilities for a random sample from a finite population, including the multiplication rule and binomial distribution. Section 5.3 introduces a simple and useful mathematical model for a process, namely, Bernoulli trials. It is shown that methods of computing probabilities for a random sample from a finite population can be applied, without modification, to problems involving Bernoulli trials. Finally, Section 5.4 raises some practical issues related to conducting a survey.

5.1 SURVEYS

A standard terminology is important for discussing finite populations. For each member of the population it is possible, in principle, to determine the values of a number of characteristics, or **variables.** For example, in the 1986 Wisconsin Driver Survey the 17 variables are the responses to 17 items on a questionnaire. Associated with a population, it is convenient to visualize a box of cards, called the **population box.** Each member of the population has a card in the box that contains the variables' values for that member. For example, the box for the 1986 Wisconsin Driver Survey contains a card for each person who was a licensed driver in Wisconsin in 1986, and each card contains the responses its driver would have given, if asked, to the 17 questionnaire items.

The values of the variables can be numbers or categories. The remainder of Part I of this book, Chapters 5 through 9, considers dichotomous variables. Part II, Chapters 10 and 11, considers multicategory variables, and Part III, Chapters 12 through 16, considers numerical variables. As a further simplification, Chapters 5, 6, 7, 10, 12, 14, 15, and 16 study variables one at a time. Chapters 8, 11, and 13 consider ways to describe and study the relation between two variables. Finally, a brief introduction to methods for simultaneously examining more than two variables per card is given in the optional Chapter 9. *In this chapter, attention is restricted to a single dichotomous variable on each card.*

As introduced in Chapter 1, it is convenient to label the dichotomous categories *success* and *failure*. Further, it is convenient to represent a success on a card by the number 1 and a failure by the number 0. The total number of cards in the box marked 1 is denoted by s (for success), and the total number of cards marked 0 is denoted by f (for failure). The total number of cards in the box is denoted by N. Clearly, since every card in the box is marked either 1 or 0, $s + f = N$.

Remember that the symbols s, f, and N have dual meanings. As defined above, they represent numbers of cards in the population box, but they also represent numbers of individuals in the population. Typically, it is easier to talk about cards in a box when one is developing the theory or performing computations. When one is interpreting the findings, however, it is better to use the terminology of the actual study.

At the planning stage of a survey, the values of s and f are unknown. These values are still unknown after the data are collected, unless the researcher conducts a census.

The box is characterized by its proportion of cards marked 1, which is denoted by $p = s/N$. Let q denote the proportion of failures in the population; $q = f/N$. Note that $p + q = 1$. By convention, attention will focus on p; but any statement about p yields an equivalent statement about q since they sum to 1. Note that this usage of p and q is consistent with the guidelines given in Chapter 1.

The basic question of statistical inference is

What do the results in the sample reveal about the value of p?

The answer to this question depends on how the sample was selected and, to a lesser degree, the size of the sample. The following example presents one of the most notorious surveys in American history. It demonstrates that having a large number of individuals in a sample is no guarantee of success!

EXAMPLE 5.1: *The Literary Digest Poll*

In 1936 the *Literary Digest* was interested in the population of all people who would vote in the 1936 U.S. presidential election. The sample was 2.4 million persons who responded to a mail questionnaire distributed by the magazine. A majority of the persons in the sample favored Alf Landon over Franklin Roosevelt for President. Consequently, the *Literary Digest* predicted that Landon would win, which turned out to be incorrect. The magnitude of their error was not small; in fact, the *Literary Digest* poll's prediction of Landon's share of the vote was 19 percentage points too high!

What went wrong? Why did the *Digest's* sample grossly overrepresent people who were planning to vote for Landon? The *Digest* drew its sample from lists of automobile and telephone owners—people who were relatively affluent by 1936 standards. Affluent persons were more likely to vote for Landon, the Republican candidate. ▲

As this example suggests, the goal of a researcher is to obtain a sample that is representative of the population. There is no way to ensure that the sample is representative, but if the sample is selected *at random* from the population, it is possible to quantify how representative the sample is likely to be. These notions are made precise later.

There are two related, but importantly different, methods of selecting a sample at random from a population. The first method mimics the process of randomization introduced in Chapter 1. The researcher selects n cards at random from the population box of N cards. The population members corresponding to the cards selected are the sample. This process of selecting n cards at random can be implemented in two ways:

- Reach into the box and select n cards at random.
- Reach into the box, select one card at random, and set it aside. Reach into the box, select one of the remaining cards at random, and set it aside. Continue in this manner until n cards have been selected.

The second version allows the researcher to consider the notions of the first card selected, the second card selected, and so on. The second version was not introduced in Chapter 2 because it adds nothing of value to the analysis and makes the brute force method even messier. The second version is introduced now because it motivates the following alternative way of selecting cards at random, namely,

Reach into the box and select one card at random, note its value, *place it back into the box*, and thoroughly mix the cards. Repeat this process until n cards have been selected.

This new method is called **random sampling with replacement,** and either version of the original method is called **random sampling without replacement.** The resultant sample is called a random sample with replacement or a random sample without replacement. The modifier *simple* is sometimes placed before *random sample* to distinguish it from, for example, a stratified random sample or a systematic random sample. This text does not consider other forms of random samples, and hence the extra modifier is not needed.

A natural question is

Which method of random sampling, with or without replacement, is better?

This question is investigated below. Not surprisingly, each method has some advantages.

Note that sampling without replacement ensures that n different cards are selected, whereas sampling with replacement makes it possible to select one or more cards more than once. For example, a sample of 100 cards drawn with replacement may yield only 95 different cards. Thus, from this perspective, sampling without replacement is superior to sampling the same number of cards with replacement because sampling with replacement might yield less information. But how much superior? It turns out that a good way to compare the sampling methods is by examining how they influence the probabilities of events. This idea is explored in the next two examples.

EXAMPLE 5.2: A Big Hypothetical Box

A hypothetical population consists of one million persons, of whom 60 percent are female and 40 percent are male. Suppose that 10 persons are selected at random from this population. Let A be the event that exactly 6 women and 4 men are selected. What is $P(A)$?

It can be shown that if sampling is performed at random without replacement, then $P(A) = 0.250824$. If, however, sampling is performed at random with replacement, then $P(A) = 0.250823$. The difference between these answers is 0.000001, only one in one million! ▲

EXAMPLE 5.3: A Tiny Hypothetical Box

A hypothetical population consists of 20 persons, of whom 60 percent are female and 40 percent are male. Suppose that 10 persons are selected at random from this population. Let A be the event that exactly 6 women and 4 men are selected. (This is the same definition of A given in the previous example.) What is $P(A)$?

It can be shown that if sampling is performed at random without replacement, the answer to this question is 0.350083. If, however, sampling is performed at random with replacement, the answer is 0.250823. The difference between these answers is 0.099260; approximately 1 in 10. ▲

The two previous examples illustrate a number of important points that are true in general:

- If the sample size is a small proportion of the population size, the probability of an event is approximately the same for either method of random sampling—with or without replacement. For example, in the big box the sample size divided by the population size is $10/1,000,000 = 0.00001$, and the probability of the event A is virtually unaffected by the sampling method used.

- If the sample size is a large proportion of the population size, the probability of any event can be very different for the two methods of random sampling. For example, in the tiny box the sample size divided by the population size is 10/20 = 0.5, and the probability of the event A is substantially different for the two sampling methods.

- For sampling with replacement, the number of cards in the box is irrelevant. All that matters is the proportion of cards marked 1. For instance, the event A of these examples has the same probability of occurring in the big and tiny boxes because in each box 60 percent of the cards are marked 1.

The first two points above refer to the sample size being a small or large proportion of the population size. As a practical matter, one needs a way to distinguish between these cases. The following is a commonly accepted general guideline:

> If the sample size n is less than 5 percent of the population size N, the probability of any event with random sampling with replacement is approximately equal to the probability of the event with random sampling without replacement.

The reason this guideline is useful is that computing probabilities is much easier when sampling is performed with replacement than when sampling is performed without replacement. A careful examination of this last claim is not possible in this book, but it is instructive to outline why it is true.

Researchers usually select a fairly large number of cards from the population box. For example, public opinion polls ordinarily have sample sizes of 1,000 or more persons. For the task of computing probabilities, it is conceptually easier to break a sample into its component parts, namely, the first card selected, the second card selected, and so on. When one is sampling with replacement, the box does not change from one selection to the next, but when one is sampling without replacement, the outcome of any selection changes the box for the next selection. It turns out that the mathematical arguments are much easier if the box never changes. Thus, often a researcher will select a random sample without replacement, but analyze the data as if the sampling were with replacement. According to the guideline, this approach is valid provided the sample size is less than 5 percent of the population size.

The following example shows that sometimes sampling with replacement can be preferred to sampling without replacement for other than mathematical reasons.

EXAMPLE 5.4: The Simulation Experiment

Recall the Colloquium study of the first three chapters. It was stated in Chapter 3 that there are 40,116,600 possible assignments of subjects to treatments. View each assignment as a member of a population. Thus, the population box has 40,116,600 cards in it, one for each assignment. Recall further that the value of the test statistic for the Colloquium study is $x = \frac{3}{7}$ and that the P-value for the first alternative is equal to $P(X \geq \frac{3}{7})$. Returning attention to the population box, let an assignment's card be marked 1 if the assignment would yield a value of X

greater than or equal to $\frac{3}{7}$, and let it be marked 0 otherwise. Then the proportion of successes in the box is

$$p = \frac{\text{The number of assignments that give } X \geq \frac{3}{7}}{\text{The total number of assignments}}.$$

This ratio is recognized as equaling $P(X \geq \frac{3}{7})$. Thus, the problem of computing the P-value for the Colloquium study, or any study of the first three chapters, can be phrased as a problem of computing p. As introduced in Section 3.2, a simulation experiment is a method for obtaining an approximate P-value when the exact answer is inaccessible. Recall that the runs of a simulation experiment are obtained as follows:

> The chance mechanism of assigning subjects to treatments is observed repeatedly under identical conditions; that is, after obtaining a particular assignment, the researcher returns the selected cards to the box, remixes the cards, and starts again. This process is continued until the desired number of runs has been obtained.

Upon reflection, it is clear that the runs of a simulation experiment, each of which yields one assignment, correspond to the selection of cards at random *with replacement* from the population box of 40 million plus possible assignments. Of course, the above algorithm could be modified to yield random sampling without replacement. It is not modified, however, because the necessary computer program would be much more complicated: time and memory would be consumed in keeping a record of which assignments have been selected in order to be certain there is no repeated examination of any assignment. ▲

In this book attention is restricted to the following two cases:
- The sample is selected at random with replacement.
- The sample is selected at random without replacement, and the sample size is less than 5 percent of the population size.

With this restriction, it is appropriate to assume, when performing computations, that sampling is performed at random with replacement.

5.2 PROBABILITY

5.2.1 Dichotomous Boxes

This section introduces the use of probability in surveys. In particular, the following problem is studied. A population box contains s cards marked 1 and f cards marked 0. The proportion of cards marked 1 is denoted by p and the proportion of cards marked 0 is denoted by q. A population box with this structure is called a **dichotomous box with proportion of successes equal to p and proportion of failures equal to q**. In statistical applications p and q are usually unknown; in this section, however, they are assumed to be known. The assumption that p and q are known is useful for building intuition. Chapter 6 illustrates how the results of this chapter can be used if p and q are unknown.

The researcher plans to select a random sample of n cards with replacement and wants to be able to compute the probabilities of various events associated with the outcome. It is convenient to begin by defining some random variables. Let X_1 denote the number on the first card selected, let X_2 denote the number on the second card selected, and so on, and let X_n denote the number on the nth (last) card selected. The purpose of the subscript is to identify when the card was selected. This section begins by computing probabilities for these random variables.

First, consider one random variable at a time. In particular, consider the number on the first card selected, X_1. Since each card in the box is marked either 1 or 0, those are the two possible values of X_1. Since a card is selected at random, each card in the box is equally likely to be selected. Applying the definition of probability for the equally likely case introduced in Section 2.3.1,

$$P(X_1 = 0) = \frac{\text{The number of cards that give } X_1 = 0}{\text{The number of cards in the box}} = \frac{f}{N} = q.$$

$$P(X_1 = 1) = \frac{\text{The number of cards that give } X_1 = 1}{\text{The number of cards in the box}} = \frac{s}{N} = p.$$

In words, the probability that the first card selected is marked 0 is equal to the proportion of cards in the population box marked 0, and the probability that it is marked 1 is equal to the proportion of cards in the population box marked 1. For example, suppose that the population is a dichotomous box with $p = 0.6$ and $q = 0.4$; then

$$P(X_1 = 0) = q = 0.4 \quad \text{and} \quad P(X_1 = 1) = p = 0.6.$$

Returning to an arbitrary dichotomous box, consider the number on the second card selected, X_2. Again, apply the definition of probability from Chapter 2, and remember that for sampling with replacement, the contents of the box at the second selection coincides with its original contents. The result is the following:

$$P(X_2 = 0) = \frac{\text{The number of cards that give } X_2 = 0}{\text{The number of cards in the box}} = \frac{f}{N} = q.$$

$$P(X_2 = 1) = \frac{\text{The number of cards that give } X_2 = 1}{\text{The number of cards in the box}} = \frac{s}{N} = p.$$

Note that the random variables X_1 and X_2 have identical sampling distributions; that is, the probability that either one equals 0 is q and the probability that either one equals 1 is p. The expression *have identical sampling distributions* is usually abbreviated as *are identically distributed*.

The above argument for the second card selected can be extended to the remaining cards selected. The result is as follows:

> The random variables $X_1, X_2, \ldots, X_n$ are identically distributed. More precisely, the probability that any particular one of these random variables equals 0 is q, and the probability that it equals 1 is p.

For example, suppose a researcher selects a random sample with replacement of 100 cards from a dichotomous box with $p = 0.6$ and $q = 0.4$; then, for example, $P(X_{37} = 1) = 0.6$ and $P(X_{82} = 0) = 0.4$.

5.2.2 The Multiplication Rule

The next task is to compute the probabilities that X_1 and X_2 take on given values simultaneously. In particular, the researcher wants to compute

$$P(X_1 = 0 \text{ and } X_2 = 0), P(X_1 = 0 \text{ and } X_2 = 1), P(X_1 = 1 \text{ and } X_2 = 0),$$

and $P(X_1 = 1 \text{ and } X_2 = 1)$.

In order to save space, the word *and* inside a probability statement is usually replaced by a comma; thus, the above expressions become

$$P(X_1 = 0, X_2 = 0), P(X_1 = 0, X_2 = 1), P(X_1 = 1, X_2 = 0),$$

and $P(X_1 = 1, X_2 = 1)$.

It turns out that considering two random variables is much more complicated than considering one random variable, and the ideas are best introduced with a brute force example.

Consider again a dichotomous box with $p = 0.6$ and $q = 0.4$, but now focus on a particular box, namely one with five cards. For convenience, name the three cards marked 1 as A, B, and C, and name the two cards marked 0 as D and E. Table 5.1 presents the 25 possible outcomes of the chance mechanism

TABLE 5.1 *All Possible Outcomes of Selecting Two Cards at Random with Replacement from a Population Consisting of Three Successes and Two Failures*

First Card	Second Card				
	A(1)	B(1)	C(1)	D(0)	E(0)
A(1)	AA(1,1)	AB(1,1)	AC(1,1)	AD(1,0)	AE(1,0)
B(1)	BA(1,1)	BB(1,1)	BC(1,1)	BD(1,0)	BE(1,0)
C(1)	CA(1,1)	CB(1,1)	CC(1,1)	CD(1,0)	CE(1,0)
D(0)	DA(0,1)	DB(0,1)	DC(0,1)	DD(0,0)	DE(0,0)
E(0)	EA(0,1)	EB(0,1)	EC(0,1)	ED(0,0)	EE(0,0)

of selecting two cards at random with replacement from the population box. To read this table please ignore, for the moment, the entries in parentheses and note that, for example, AB denotes the outcome that the first card selected is A and the second card selected is B. Note that the rows of the table correspond to the first card selected—for example, in all entries in the third row, C, card C is selected first—and the columns correspond to the second card selected. The numbers in parentheses are the values of the variables for the outcome; for example, outcome CD gives 1 for C and 0 for D and is written as (1,0) in the table. Since the two cards are selected at random with replacement from the population, it is natural to assume that the 25 outcomes in Table 5.1 are equally likely. The following results can be obtained by inspection of the table:

$$P(X_1 = 0, X_2 = 0) = \frac{\text{The number of outcomes that give } X_1 = 0 \text{ and } X_2 = 0}{\text{The number of possible outcomes}}$$

$$= \frac{4}{25} = 0.16.$$

$$P(X_1 = 0, X_2 = 1) = \frac{\text{The number of outcomes that give } X_1 = 0 \text{ and } X_2 = 1}{\text{The number of possible outcomes}}$$

$$= \frac{6}{25} = 0.24.$$

$$P(X_1 = 1, X_2 = 0) = \frac{\text{The number of outcomes that give } X_1 = 1 \text{ and } X_2 = 0}{\text{The number of possible outcomes}}$$

$$= \frac{6}{25} = 0.24.$$

$$P(X_1 = 1, X_2 = 1) = \frac{\text{The number of outcomes that give } X_1 = 1 \text{ and } X_2 = 1}{\text{The number of possible outcomes}}$$

$$= \frac{9}{25} = 0.36.$$

You should verify the following pairs of equations:

$P(X_1 = 0, X_2 = 0) = 0.16,$ $P(X_1 = 0)P(X_2 = 0) = 0.4(0.4) = 0.16.$
$P(X_1 = 0, X_2 = 1) = 0.24,$ $P(X_1 = 0)P(X_2 = 1) = 0.4(0.6) = 0.24.$
$P(X_1 = 1, X_2 = 0) = 0.24,$ $P(X_1 = 1)P(X_2 = 0) = 0.6(0.4) = 0.24.$
$P(X_1 = 1, X_2 = 1) = 0.36,$ $P(X_1 = 1)P(X_2 = 1) = 0.6(0.6) = 0.36.$

In words, the probability that X_1 and X_2 jointly take on any specified values is equal to the product of the individual probabilities that X_1 and X_2 take on those values. For example, the third pair of equations shows that the probability that

$X_1 = 1$ and $X_2 = 0$, which is 0.24, can be obtained as the product of the probability that $X_1 = 1$, which is 0.6, and the probability that $X_2 = 0$, which is 0.4. The four pairs of equations above are a special case of a general result called the **multiplication rule.** It is called the multiplication rule because the word *and* inside the probability statement is replaced by the operation of multiplication.

Note that the above argument for X_1 and X_2 can be applied to any two *different* random variables. For example, suppose $n \geq 100$ (just to make sure there *is* an X_{100}); then the above reasoning shows, for example, that

$$P(X_{32} = 1, X_{100} = 0) = P(X_{32} = 1)P(X_{100} = 0) = 0.6(0.4) = 0.24.$$

The requirement that the random variables be different is critical. For example, the event

$$(X_1 = 0 \text{ and } X_1 = 1)$$

is impossible and hence has probability 0 of occurring. An incorrect application of the multiplication rule would imply that the probability of this event equals

$$P(X_1 = 0)P(X_1 = 1) = pq = 0.6(0.4) = 0.24.$$

In addition, it is not obvious, but the multiplication rule can be extended to any number of *different* random variables. For example, assuming that $n \geq 4$,

$$P(X_1 = 1, X_2 = 0, X_3 = 1, X_4 = 0)$$
$$= P(X_1 = 1)P(X_2 = 0)P(X_3 = 1)P(X_4 = 0)$$
$$= 0.6(0.4)(0.6)(0.4) = 0.0576.$$

The utility of the multiplication rule comes from the fact that it enables one to compute the probability of an event that depends on two or more cards, which is difficult to obtain directly, as a product of probabilities of events involving just one card, each of which is easy to obtain.

It can be shown that the multiplication rule is true for any dichotomous box, not just the box considered in this brute force example. In fact, the multiplication rule is true for any population box, not just dichotomous boxes.

Below are several illustrations of the use of the multiplication rule.

EXAMPLE 5.5

Suppose that $n = 4$ cards are selected at random with replacement from a dichotomous box with $p = 0.5$ and $q = 0.5$. The analyst wants to know the probability that a card marked 1 is selected on every draw.

This problem involves selecting more than one card, so it is natural to consider using the multiplication rule. In order to use the multiplication rule, one must express the event of interest in terms of the individual selections:

$$P(\text{A card marked 1 is selected on every draw}) = P(X_1 = 1, X_2 = 1, X_3 = 1, X_4 = 1).$$

By the multiplication rule,

$$P(X_1 = 1, X_2 = 1, X_3 = 1, X_4 = 1)$$
$$= P(X_1 = 1)P(X_2 = 1)P(X_3 = 1)P(X_4 = 1)$$
$$= p \times p \times p \times p = p^4 = (0.5)^4 = 0.0625. \ \blacktriangle$$

It can be tedious to write numerous Xs with subscripts. A shorter route to the above answer is obtained by suppressing the Xs and writing the event of interest as $(1, 1, 1, 1)$, yielding

$$P(1, 1, 1, 1) = p \times p \times p \times p = p^4 = (0.5)^4 = 0.0625.$$

EXAMPLE 5.6

Suppose that $n = 3$ cards are selected at random with replacement from a dichotomous box with $p = 0.75$ and $q = 0.25$. Find the probability that a card marked 0 is selected on every draw.

The event of interest is $(0, 0, 0)$. By the multiplication rule,

$$P(0, 0, 0) = q \times q \times q = q^3 = (0.25)^3 = 0.0156. \ \blacktriangle$$

The clever use of the multiplication rule in conjunction with the addition rule, introduced in Section 2.4.2, can solve more complicated problems.

EXAMPLE 5.7

Suppose that $n = 5$ cards are selected at random with replacement from a dichotomous box with $p = 0.6$ and $q = 0.4$. Find the probability that a card marked 1 is selected at least four times in succession.

The event of interest is much more complicated than in the previous two examples; it is

$$(1, 1, 1, 1, 1 \text{ or } 1, 1, 1, 1, 0 \text{ or } 0, 1, 1, 1, 1).$$

The three outcomes connected by the word *or* are distinct; thus, by the addition rule:

$$P(1, 1, 1, 1, 1 \text{ or } 1, 1, 1, 1, 0 \text{ or } 0, 1, 1, 1, 1) = P(1, 1, 1, 1, 1) + P(1, 1, 1, 1, 0) + P(0, 1, 1, 1, 1).$$

Each of these three probabilities can be evaluated with the multiplication rule:

$$P(1, 1, 1, 1, 1) = p \times p \times p \times p \times p = p^5 = (0.6)^5 = 0.0778.$$
$$P(1, 1, 1, 1, 0) = p \times p \times p \times p \times q = p^4 q = (0.6)^4(0.4) = 0.0518.$$
$$P(0, 1, 1, 1, 1) = q \times p \times p \times p \times p = p^4 q = 0.0518.$$

Combining these results shows that the event of interest has probability

$$0.0778 + 0.0518 + 0.0518 = 0.1814. \ \blacktriangle$$

The random variables $X_1, X_2, \ldots, X_n$ studied in this section are an example of **independent random variables**. Selecting a random sample with replacement from a box is the most common way to obtain independent random variables, but there are other ways. For the most part, an investigation of the other ways is beyond the scope of this text, but this topic is addressed briefly in Chapters 8 and 11. From a practical viewpoint, the importance of independence rests in the following fact:

> If a collection of random variables is independent, the multiplication rule can be used for computing probabilities for any subcollection of them.

Combining the notion of independence with the earlier idea of identical distribution, one concludes that the random variables $X_1, X_2, \ldots, X_n$ are independent and identically distributed.

The multiplication rule is stated in terms of sampling with replacement from the same box. It also holds if each card is selected from a different box, as illustrated in the following example.

EXAMPLE 5.8

Suppose that one card is selected at random from a dichotomous box with $p = 0.7$ and $q = 0.3$ and a second card is selected at random from a dichotomous box with $p = 0.4$ and $q = 0.6$. Find the probability that both cards selected are marked 1.

P(The first card is marked 1, and the second card is marked 1)

$\quad = P$(The first card is marked 1) $\times P$(The second card is marked 1)

$\quad = 0.7(0.4) = 0.28$. ▲

5.2.3 The Binomial Distribution

Statisticians are often interested in combining the random variables $X_1, X_2, \ldots, X_n$ to create a new random variable. The most important way to combine them is by adding. Define a new random variable X by the equation

$$X = X_1 + X_2 + \cdots + X_n.$$

In words, X is the total of the numbers on the n cards selected. Since each card is marked 0 or 1, X can also be interpreted as the number of cards marked 1 in the sample. Since a 1 corresponds to a success, X can also be interpreted as the total number of successes in the sample. The text will move freely between these three interpretations of X. This section presents the sampling distribution of X. A complete derivation of the result requires some sophisticated mathematics; here, only the simple case of $n = 2$ is derived and the general formula is given.

For a sample of size $n = 2$ there are three possibilities for the value of X: 0, 1, or 2. In order to compute, for example, the probability that X equals 0, the event of interest must be written in terms of X_1 and X_2. More precisely,

$$P(X = 0) = P(X_1 + X_2 = 0) = P(X_1 = 0, X_2 = 0)$$
$$= P(X_1 = 0)P(X_2 = 0) = q \times q = q^2.$$

Similarly,

$$P(X = 2) = P(X_1 + X_2 = 2) = P(X_1 = 1, X_2 = 1)$$
$$= P(X_1 = 1)P(X_2 = 1) = p \times p = p^2.$$

The event $(X = 1)$ requires more care:

$$P(X = 1) = P(X_1 + X_2 = 1) = P(X_1 = 1, X_2 = 0 \text{ or } X_1 = 0, X_2 = 1).$$

By the addition rule, this last expression equals

$$P(X_1 = 1, X_2 = 0) + P(X_1 = 0, X_2 = 1).$$

Finally, applying the multiplication rule twice gives

$$P(X = 1) = p \times q + q \times p = 2pq.$$

These results are summarized in Table 5.2. This sampling distribution is called the **binomial (sampling or probability) distribution** for $n = 2$.

EXAMPLE 5.9

Suppose $n = 2$, $p = 0.7$, and $q = 0.3$. Then the binomial sampling distribution consists of the following equations:

$$P(X = 0) = q^2 = (0.3)^2 = 0.09.$$
$$P(X = 1) = 2pq = 2(0.7)(0.3) = 0.42.$$
$$P(X = 2) = p^2 = (0.7)^2 = 0.49. \blacktriangle$$

TABLE 5.2 *The Binomial Sampling Distribution for n = 2*

x	$P(X = x)$
0	q^2
1	$2pq$
2	p^2
Total	1

As stated above, for $n > 2$ the sampling distribution of X is given but not derived. First, some additional notation is required. For any positive integer r, define $r!$, read "r-factorial," by

$$r! = r \times (r-1) \times (r-2) \times \cdots \times 1. \tag{5.1}$$

In words, $r!$ is obtained as follows:

Start with r and multiply by successively smaller integers until 1 is reached.

A definition of $0!$ is useful. It is defined to equal 1:

$$0! = 1.$$

EXAMPLE 5.10

Below are some simple applications of the definition of $r!$:

$1! = 1, \quad 2! = 2 \times 1 = 2, \quad 3! = 3 \times 2 \times 1 = 6, \quad 4! = 4 \times 3 \times 2 \times 1 = 24,$
$5! = 5 \times 4 \times 3 \times 2 \times 1 = 120, \quad 6! = 6 \times 5 \times 4 \times 3 \times 2 \times 1 = 720.$ ▲

The sampling distribution of X is given by the formula

$$P(X = x) = \frac{n!}{x!(n-x)!} p^x q^{n-x} \quad \text{for } x = 0, 1, \ldots, n. \tag{5.2}$$

If $n = 2$, Equation 5.2 becomes

$$P(X = x) = \frac{2!}{x!(2-x)!} p^x q^{2-x} \quad \text{for } x = 0, 1, 2.$$

Successively replacing x by each of its values, 0, 1, and 2, gives three equations (remember that p^0, q^0, and $0!$ each equal 1):

$$P(X = 0) = \frac{2!}{0!2!} p^0 q^2 = \frac{2}{1(2)} q^2 = q^2.$$

$$P(X = 1) = \frac{2!}{1!1!} p^1 q^1 = \frac{2}{1(1)} pq = 2pq.$$

$$P(X = 2) = \frac{2!}{2!0!} p^2 q^0 = \frac{2}{2(1)} p^2 = p^2.$$

These equations agree with the entries in Table 5.2.

The collection of equations (one for each choice of x) contained in Equation 5.2 is referred to as the binomial (sampling or probability) distribution. To denote that the random variable X has the binomial (sampling or probability) distribution, one writes $X \sim \text{Bin}(n, p)$. The numbers n and p are called the **parameters** of the binomial distribution. Their numerical values reveal *which* binomial distribution is appropriate. Below are three examples of how the binomial distribution can be used.

EXAMPLE 5.11

Eight cards are selected at random with replacement from a dichotomous box with $p = 0.6$ and $q = 0.4$. Suppose the researcher wants to find the probability that the sum of the eight numbers selected equals 5. The goal is to compute $P(X = 5)$, so

$$n = 8, \quad x = 5, \quad \text{and} \quad n - x = 8 - 5 = 3.$$

Making these substitutions in Equation 5.2 gives

$$P(X = 5) = \frac{n!}{x!(n-x)!} p^x q^{n-x} = \frac{8!}{5!3!}(0.6)^5(0.4)^3.$$

It can be verified with a hand calculator that this number is 0.2787. ▲

EXAMPLE 5.12

Six cards are selected at random with replacement from a dichotomous box with $p = 0.7$ and $q = 0.3$. Suppose the researcher wants to find the probability that the sum of the six numbers selected equals 4. The goal is to compute $P(X = 4)$, so

$$n = 6, \quad x = 4, \quad \text{and} \quad n - x = 6 - 4 = 2.$$

Making these substitutions in Equation 5.2 gives

$$P(X = 4) = \frac{n!}{x!(n-x)!} p^x q^{n-x} = \frac{6!}{4!2!}(0.7)^4(0.3)^2.$$

It can be verified with a hand calculator that this number is 0.3241. ▲

EXAMPLE 5.13

Refer to Examples 5.2 and 5.3 (p. 149), the big and tiny hypothetical boxes, respectively. Ten cards are selected at random with replacement from a dichotomous box with $p = 0.6$ and $q = 0.4$. Suppose that the researcher wants to find the probability that the sum of the 10 numbers selected equals 6.

The goal is to compute $P(X = 6)$, so

$$n = 10, \quad x = 6, \quad \text{and} \quad n - x = 10 - 6 = 4.$$

Making these substitutions in Equation 5.2 gives

$$P(X = 6) = \frac{n!}{x!(n-x)!} p^x q^{n-x} = \frac{10!}{6!4!}(0.6)^6(0.4)^4.$$

It can be verified with a hand calculator that this number is 0.250823, as claimed earlier. ▲

TABLE 5.3 The Bin(25, 0.5) Distribution

x	$P(X = x)$	$P(X \leq x)$	$P(X \geq x)$
0	0.0000	0.0000	1.0000
1	0.0000	0.0000	1.0000
2	0.0000	0.0000	1.0000
3	0.0001	0.0001	1.0000
4	0.0004	0.0005	0.9999
5	0.0016	0.0020	0.9995
6	0.0053	0.0073	0.9980
7	0.0143	0.0216	0.9927
8	0.0322	0.0539	0.9784
9	0.0609	0.1148	0.9461
10	0.0974	0.2122	0.8852
11	0.1328	0.3450	0.7878
12	0.1550	0.5000	0.6550
13	0.1550	0.6550	0.5000
14	0.1328	0.7878	0.3450
15	0.0974	0.8852	0.2122
16	0.0609	0.9461	0.1148
17	0.0322	0.9784	0.0539
18	0.0143	0.9927	0.0216
19	0.0053	0.9980	0.0073
20	0.0016	0.9995	0.0020
21	0.0004	0.9999	0.0005
22	0.0001	1.0000	0.0001
23	0.0000	1.0000	0.0000
24	0.0000	1.0000	0.0000
25	0.0000	1.0000	0.0000
Total	1.0000		

If n is larger than 10, computing binomial probabilities by hand is tedious. Numerous published tables of binomial probabilities exist; see [1], for example. Many computer programs compute binomial probabilities, although they cannot handle values of n much larger than 100 because of computational difficulties. For example, the Bin(25, 0.5) distribution obtained with Minitab is presented in Table 5.3.

EXAMPLE 5.14

Suppose that the random variable X has a sampling distribution given by the binomial distribution with $n = 25$ and $p = 0.50$. Check that Table 5.3 yields the following equations:

$P(X = 10) = 0.0974$, $P(X = 15) = 0.0974$,   and $P(X \leq 12) = 0.5000$. ▲

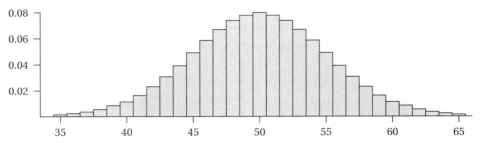

FIGURE 5.1 The Bin(100, 0.5) distribution.

The binomial sampling distribution can be presented visually, as well. Figures 5.1, 5.2, and 5.3 are probability histograms of three binomial distributions. An inspection of these figures indicates that the two distributions with $p = 0.5$ are symmetric and the distribution with $p = 0.2$ is somewhat asymmetric. This observation illustrates the following feature of all binomial distributions:

 The binomial distribution is symmetric if, and only if, $p = 0.5$. For $p = 0.5$ the point of symmetry is at $np = n(0.5)$.

Thus, the point of symmetry for the Bin(25, 0.5) distribution is at $np = 25(0.5) = 12.5$, and the point of symmetry of the Bin(100, 0.5) distribution is at $np = 100(0.5) = 50$. Both of these answers agree with a visual examination of the probability histograms. The Bin(100, 0.2) distribution is nearly symmetric, illustrating the following fact:

 For $p \neq 0.5$ the binomial distribution is nearly symmetric if both np and nq are at least 15.

For the Bin(100, 0.2) distribution, $np = 100(0.2) = 20$ and $nq = 100(0.8) = 80$, which are both larger than 15.

As another example, Figure 5.4 is the probability histogram of the Bin(50, 0.1) distribution. Note that $np = 50(0.1) = 5$. Because this value is smaller than 15, it is not surprising that the probability histogram is noticeably asymmetric.

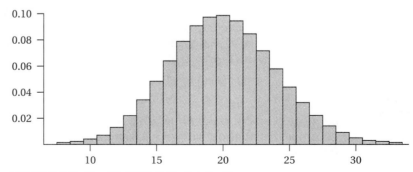

FIGURE 5.2 The Bin(100, 0.2) distribution.

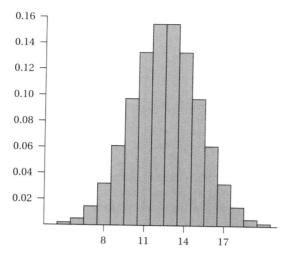

FIGURE 5.3 The Bin(25, 0.5) distribution.

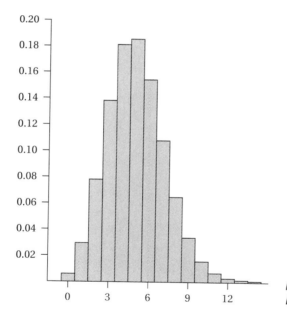

FIGURE 5.4 The Bin(50, 0.1) distribution.

In Chapter 3, for a number of different studies the probability histogram of the test statistic for Fisher's test was shown to have a single peak and to be symmetric or nearly symmetric. It was demonstrated in Section 3.4 that for such studies the standard normal curve can be used to obtain an approximate P-value, that is, to obtain an approximation to a probability that is either difficult or impossible to compute. The shapes of the probability histograms in

TABLE 5.4 Expected Values and Standard Deviations of Selected Binomial Distributions

Distribution	$E(X) = np$	$SD(X) = \sqrt{npq}$
Bin (100, 0.5)	50.0	5.0
Bin (100, 0.2)	20.0	4.0
Bin (25, 0.5)	12.5	2.5

Figures 5.1 through 5.3—single-peaked and symmetric or nearly symmetric—suggest (correctly, it turns out) that the standard normal curve can be used to obtain approximate binomial probabilites.

It can be shown that if $X \sim \text{Bin}(n, p)$, its expected value (mean) and standard deviation are given by the following two equations:

$$\mu = E(X) = np \tag{5.3}$$

and

$$\sigma = SD(X) = \sqrt{npq}. \tag{5.4}$$

The numbers obtained by applying Equations 5.3 and 5.4 to the Bin(100, 0.5), Bin(100, 0.2), and Bin(25, 0.5) distributions are listed in Table 5.4.

If $X \sim \text{Bin}(n, p)$, then

$$Z = \frac{X - np}{\sqrt{npq}}$$

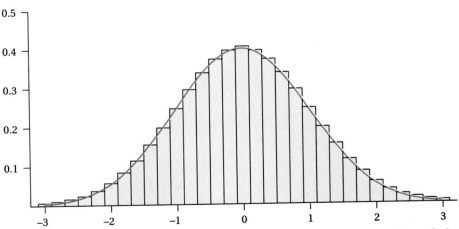

FIGURE 5.5 The standard normal curve and the standardized version of the Bin(100, 0.5) distribution.

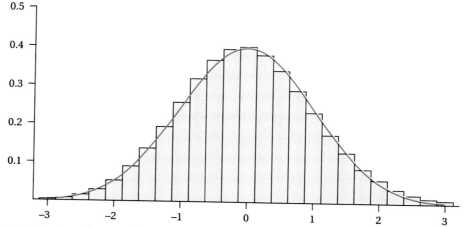

FIGURE 5.6 *The standard normal curve and the standardized version of the Bin(100, 0.2) distribution.*

is its standardized version. The probability histograms of the standardized versions of the Bin(100, 0.5), Bin(100, 0.2), and Bin(25, 0.5) distributions and the standard normal curve are presented in Figures 5.5, 5.6, and 5.7. Clearly, the standard normal curve is a very close approximation to each of these probability histograms. The general guideline is that this approximation is good if np and nq are both at least 15.

The mechanics of using the standard normal curve to obtain an approximation to a binomial probability are presented in Chapter 6.

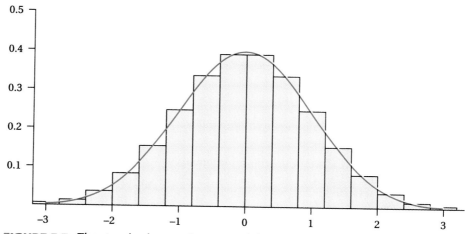

FIGURE 5.7 *The standard normal curve and the standardized version of the Bin(25, 0.5) distribution.*

EXERCISES 5.2

1. Two cards are selected at random with replacement from a dichotomous box with $p = 0.7$ and $q = 0.3$. Let X_1 equal the number on the first card selected and X_2 equal the number on the second card selected. Compute each of the following probabilities:

   (a) $P(X_1 = 1)$
   (b) $P(X_2 = 0)$
   (c) $P(X_1 = 1, X_2 = 0)$
   (d) $P(X_1 \geq X_2)$

2. Two cards are selected at random with replacement from a dichotomous box with $p = 0.2$ and $q = 0.8$. Let X_1 equal the number on the first card selected and X_2 equal the number on the second card selected. Compute each of the following probabilities:

   (a) $P(X_1 = 0)$
   (b) $P(X_2 = 1)$
   (c) $P(X_1 = 0, X_2 = 1)$
   (d) $P(X_1 = X_2)$

3. Four cards are selected at random with replacement from a dichotomous box with $p = 0.3$ and $q = 0.7$. Compute each of the following probabilities:

   (a) $P(1, 1, 0, 0)$ (Note: This notation was introduced in the text; it means that the first and second cards selected are labeled 1 and the third and fourth cards selected are labeled 0.)
   (b) $P(1, 0, 1, 0)$
   (c) The probability of obtaining an alternating sequence of zeros and ones

4. Six cards are selected at random with replacement from a dichotomous box with $p = 0.5$ and $q = 0.5$. Compute each of the following probabilities:

   (a) $P(1, 1, 1, 0, 0, 0)$
   (b) $P(1, 0, 1, 0, 1, 0)$
   (c) The probability of obtaining an alternating sequence of zeros and ones.

5. Suppose the random variables X and Y are independent. Given that
   $$P(X = 2) = 0.6 \text{ and } P(Y = 4) = 0.5,$$
   find $P(X = 2, Y = 4)$.

6. Suppose the random variables X and Y are independent. Given that
   $$P(X = 1) = 0.3 \text{ and } P(Y = 3) = 0.5,$$
   find $P(X = 1, Y = 3)$.

7. Two cards are selected at random with replacement from a dichotomous box with $p = 0.65$ and $q = 0.35$. Let X equal the sum of the two numbers selected. Compute each of the following probabilities:

   (a) $P(X = 0)$
   (b) $P(X = 1)$
   (c) $P(X = 2)$
   (d) Draw the probability histograms for X.

8. Two cards are selected at random with replacement from a dichotomous box with $p = 0.2$ and $q = 0.8$. Let X equal the sum of the two numbers selected. Compute each of the following probabilities:

   (a) $P(X = 0)$
   (b) $P(X = 1)$
   (c) $P(X = 2)$
   (d) Draw the probability histograms for X.

9. Six cards are selected at random with replacement from a dichotomous box with $p = 0.65$ and $q = 0.35$. Let X equal the sum of the six numbers selected. Compute each of the following probabilities:

   (a) $P(X = 3)$
   (b) $P(X = 4)$

10. Five cards are selected at random with replacement from a dichotomous box with $p = 0.2$ and $q = 0.8$. Let X equal the sum of the five numbers selected. Compute each of the following probabilities:

    (a) $P(X = 1)$
    (b) $P(X = 2)$

11. Evaluate the following:

    (a) $P(X = 3)$, for $X \sim \text{Bin}(6, 0.4)$
    (b) $P(X = 4)$, for $X \sim \text{Bin}(5, 0.7)$
    (c) $P(X = 2)$, for $X \sim \text{Bin}(7, 0.2)$

12. Evaluate the following:

    (a) $P(X = 2)$, for $X \sim \text{Bin}(6, 0.4)$
    (b) $P(X = 3)$, for $X \sim \text{Bin}(5, 0.7)$
    (c) $P(X = 3)$, for $X \sim \text{Bin}(7, 0.2)$

13. Suppose that $X \sim \text{Bin}(25, 0.5)$. Use Table 5.3 (p. 161) to obtain the following:

    (a) $P(X = 14)$
    (b) $P(X = 15)$
    (c) $P(X \leq 14)$
    (d) $P(X \leq 15)$
    (e) $P(X \geq 14)$
    (f) $P(X \geq 15)$

14. Suppose that $X \sim \text{Bin}(25, 0.5)$. Use Table 5.3 (p. 161) to obtain the following:

    (a) $P(X = 16)$
    (b) $P(X = 17)$
    (c) $P(X \leq 16)$
    (d) $P(X \leq 17)$
    (e) $P(X \geq 16)$
    (f) $P(X \geq 17)$

15. Find the standardized version of X for

    (a) $X \sim \text{Bin}(400, 0.5)$.
    (b) $X \sim \text{Bin}(200, 0.3)$.
    (c) $X \sim \text{Bin}(4000, 0.8)$.

16. Find the standardized version of X for

    (a) $X \sim \text{Bin}(400, 0.2)$.
    (b) $X \sim \text{Bin}(500, 0.4)$.
    (c) $X \sim \text{Bin}(1000, 0.6)$.

5.3 BERNOULLI TRIALS

This section begins the study of an infinite population—a mathematical model of a process that generates the outcomes of trials. This material is noticeably more abstract than any previous topic, so it is beneficial to start with an extremely simple example.

Many sporting events begin with the toss of a coin to decide which team chooses between possession of the ball and direction of attack. In fact, one could argue that tossing a coin is the canonical fair way to decide which of two individuals or groups is given an advantage or the right to make a choice. This section begins with an examination of the experiment of tossing a coin.

Suppose that a researcher is planning to toss a coin n times, where n is any positive integer. Let X_1 denote the outcome of the first toss, let X_2 denote the outcome of the second toss, and so on, and let X_n denote the outcome of the nth (last) toss. For convenience, let heads be called a success and labeled 1, and let tails be called a failure and labeled 0. As with a survey, it is fruitful to begin with an examination of the outcome of the first toss, X_1.

Let us now add the assumption that the coin is fair. To say that a coin is fair means that the two outcomes of the coin toss are equally likely to occur. This is the equally likely case, precisely the condition under which probability was introduced in Section 2.3.2. Thus,

$$P(X_1 = 0) = 0.5 \quad \text{and} \quad P(X_1 = 1) = 0.5.$$

This same reasoning can be applied to the remaining tosses; thus, the random variables $X_1, X_2, \ldots, X_n$ are identically distributed. For any of these random variables, the probability of obtaining the value 0 is $\frac{1}{2}$ and the probability of obtaining the value 1 is $\frac{1}{2}$. Note that one can consider a single toss of a fair coin to be equivalent to a single selection of a card at random from a dichotomous box with $p = 0.5$ and $q = 0.5$.

Next, consider the joint outcome of two or more tosses. For a survey, the analogous problem, that of computing probabilities for two or more cards, has a simple solution: assuming that the cards are selected at random with replacement, the researcher can use the multiplication rule. For a fair coin, if the researcher is *willing to assume* that the outcome of any toss has no effect on the outcome of any other toss, the outcomes of the coin tosses can be visualized as repeated sampling with replacement from a dichotomous box with $p = 0.5$ and $q = 0.5$. Hence, the multiplication rule can be used for repeated tosses of a fair coin. In the earlier terminology, the random variables $X_1, X_2, \ldots, X_n$ are independent. A picturesque way to describe independence for the tossing of a fair coin is to say that the coin has no memory.

Suppose the researcher is interested in the total number of heads in n tosses of a fair coin. Let X denote the total number of heads; clearly,

$$X = X_1 + X_2 + \cdots + X_n.$$

By the analogy with surveys, the sampling distribution of X is given by the Bin$(n, 0.5)$ distribution.

EXAMPLE 5.15

A researcher decides to toss a fair coin five times. Here are the probabilities of some selected events:

$$P(1, 0, 1, 0, 1) = 0.5(0.5)(0.5)(0.5)(0.5) = 0.0312.$$

$$P(1, 1, 1, 1, 1) = 0.5(0.5)(0.5)(0.5)(0.5) = 0.0312.$$

$$P(X = 5) = \frac{n!}{x!(n-x)!} p^x q^{n-x} = \frac{5!}{5!0!}(0.5)^5(0.5)^0 = 0.0312.$$

$$P(X = 3) = \frac{n!}{x!(n-x)!} p^x q^{n-x} = \frac{5!}{3!2!}(0.5)^3(0.5)^2 = 0.3125. \ \blacktriangle$$

5.3 Bernoulli Trials

The following are the important features of the fair coin-tossing example:

1. Each trial has two possible outcomes, a success (heads, or 1) or a failure (tails, or 0).
2. The probability of a success remains constant from trial to trial and is equal to $\frac{1}{2}$.
3. The outcomes of the trials are independent.

In words, the first feature means that the outcomes are dichotomous, the second means that they are identically distributed with successes and failures equally likely, and the third means that the multiplication rule can be used to compute probabilities of the outcomes of different trials.

For a second example, adopt everything about the coin example above *except* assume that the coin is weighted so that a head is three times as likely as a tail. For a fair coin a single toss can be visualized as a single selection from a box with one card marked 1 (for heads) and one card marked 0 (for tails). For the weighted coin described here, it is reasonable to visualize a single toss as a single selection from a box with *three* cards marked 1 (for heads) and one card marked 0 (for tails), since a head is three times as likely as a tail. Thus, the probability of a success on each trial is given by 0.75 instead of 0.5. This experiment has the three features above with the exception that the value 0.5 in feature 2 is replaced by 0.75.

EXAMPLE 5.16

A researcher decides to toss the weighted coin five times. Here are the probabilities of some selected events:

$$P(1, 0, 1, 0, 1) = 0.75(0.25)(0.75)(0.25)(0.75) = 0.0264.$$

$$P(1, 1, 1, 1, 1) = 0.75(0.75)(0.75)(0.75)(0.75) = 0.2373.$$

$$P(X = 5) = \frac{n!}{x!(n-x)!} p^x q^{n-x} = \frac{5!}{5!0!}(0.75)^5(0.25)^0 = 0.2373.$$

$$P(X = 3) = \frac{n!}{x!(n-x)!} p^x q^{n-x} = \frac{5!}{3!2!}(0.75)^3(0.25)^2 = 0.2637. \blacktriangle$$

The previous two examples can be generalized to give the assumptions of Bernoulli trials. Suppose a researcher plans to observe a sequence of n trials. Let X_1 denote the outcome of the first trial, let X_2 denote the outcome of the second trial, and so on, and let X_n denote the outcome of the nth (last) trial. The trials are called Bernoulli trials if the following three assumptions can be made:

- Assumption 1. Each trial results in one of two possible outcomes, which for convenience are labeled success and failure.

- Assumption 2. The probability of obtaining a success remains constant from trial to trial. This constant probability of success is denoted by the number p. The probability of a failure is denoted by q; clearly, $p + q = 1$.
- Assumption 3. The trials are independent.

As discussed earlier, the first assumption means that the trials must have a dichotomous outcome; the second means that the outcomes are identically distributed; and the third means that the multiplication rule can be used to compute probabilities of outcomes of different trials.

Let X denote the total number of successes in the n trials:

$$X = X_1 + X_2 + \cdots + X_n.$$

Probabilities are computed for X by using the binomial sampling distribution with parameters n and p.

EXAMPLE 5.17

A researcher decides to observe five Bernoulli trials with $p = 0.6$ and $q = 0.4$. Here are probabilities of selected events:

$$P(1, 0, 1, 0, 1) = 0.6(0.4)(0.6)(0.4)(0.6) = 0.0346.$$

$$P(1, 1, 1, 1, 1) = 0.6(0.6)(0.6)(0.6)(0.6) = 0.0778.$$

$$P(X = 5) = \frac{n!}{x!(n-x)!} p^x q^{n-x} = \frac{5!}{5!0!}(0.6)^5(0.4)^0 = 0.0778.$$

$$P(X = 3) = \frac{n!}{x!(n-x)!} p^x q^{n-x} = \frac{5!}{3!2!}(0.6)^3(0.4)^2 = 0.3456. \;\blacktriangle$$

Bernoulli trials are very useful for a mathematician. They allow the use of the multiplication rule and the binomial distribution to compute probabilities of events. It is very important, however, that a researcher not be corrupted by the allure of easy answers; the assumptions of Bernoulli trials should not be made without careful thought. If possible, relevant data should be examined too. The following examples illustrate these ideas.

EXAMPLE 5.18

Tom must take a 10-question true-false exam in physics. He has not studied for the exam and decides to select his answers randomly. (Fortunately, his physics teacher allows Tom to toss a fair coin during the exam.) Tom must score seven or more correct to pass the exam. What is the probability that he passes the exam?

To answer this question, some structure and assumptions are needed. Each question can be viewed as a dichotomous trial with outcomes correct (success) or incorrect (failure). Clearly, the questions satisfy the assumptions of Bernoulli trials with $p = 0.5$. Thus, Tom's total score, denoted by X, has a binomial distribution with $n = 10$ and $p = 0.5$. The probability that he passes is

$$P(X \geq 7) = P(X = 7) + P(X = 8) + P(X = 9) + P(X = 10).$$

If you are interested, you can use Equation 5.2 (p. 159) to verify that this sum equals 0.1719. Tom's future in physics looks bleak! ▲

The next example is a true story about a professor I know.

EXAMPLE 5.19

A few years ago, Milwaukee Brewers baseball player Paul Molitor had a 39-game hitting streak. (This means he had at least one hit in each of 39 consecutive games.) Before the game in which he would attempt to extend his streak to 40, a Madison radio announcer asked a professor, "What is the probability that Molitor will not get a hit in tonight's game?"

The professor reasoned as follows. It is common for a batter to have four at bats in a game, so assume there are $n = 4$ trials. Each at bat results in a hit (success) or no hit (failure). The probability of a hit p is not known, but as an approximation use 0.370, Molitor's batting average at the time. Next, assume that the trials (at bats) are independent. Finally, let X denote the total number of hits Molitor would obtain in the game. With these assumptions and notation, the answer to the original question is $P(X = 0)$, which can be computed easily:

$$P(X = 0) = \frac{n!}{x!(n-x)!} p^x q^{n-x} = \frac{4!}{0!4!}(0.370)^0 (0.630)^4 = 0.1575.$$

I disagreed with this answer; I had doubts about the assumption of independence, but my main concern was with the assumption that p remains constant from trial to trial. It is well known that some pitchers, such as Nolan Ryan, consistently yield low batting averages during their careers, whereas other pitchers consistently yield high batting averages during their (usually short) careers. In addition, many pitchers show tremendous variation in their performance from game to game. In short, I did not believe that the assumptions of Bernoulli trials were reasonable. If asked by the radio announcer, I would have said that I could not compute an exact probability or even an approximation that I trusted. Instead, I planned to listen to the game to find out what would actually happen! (I did listen, and Molitor failed to get a hit in four at bats.) It is better to realize and admit that one does not have enough information to answer a question than to make whatever assumptions are necessary to obtain an answer easily! ▲

The remaining two examples of this section develop tools for examining data to obtain insight into the validity of the assumptions of Bernoulli trials.

EXAMPLE 5.20

Each day the *Wisconsin State Journal*, a morning newspaper, forecasts the high temperature for the day for the city of Madison, Wisconsin. Data were obtained for each day during the period from March 21 to September 21, 1988, except for holidays when the paper was not published. Each day's forecast is arbitrarily labeled a success if it is within 2 degrees, in either direction, of the actual high and a failure otherwise. For example, if the forecasted high is 60 degrees, then the forecast is a success if the actual high is between 58 degrees and 62 degrees, inclusive, and it is a failure otherwise. The outcome of each day's forecast—a success or failure—can be viewed as a sequence of dichotomous trials. The processes that yield the outcomes are very complicated, involving the natural systems that cause the high temperature and the scientific systems that yield the forecasts. This example explores the question

Are the second and third assumptions of Bernoulli trials reasonable for this process?

The second assumption is that the probability of a successful forecast remained constant during the time of the study. As a longtime resident of Madison, I doubt the validity of this assumption. In particular, spring weather in Madison seems much more erratic than summer weather, so it might be more difficult to forecast accurately in the spring than in the summer. To investigate this possibility, the 178 trials were divided into a spring half and a summer half; the results are given in Table 5.5. This table shows that the proportion of successes was somewhat larger in the summer, but the difference is not very dramatic. I was surprised to see that the spring forecasts were nearly as good as the summer ones.

Table 5.6 divides the trials into sixths. This pattern is much more interesting! The proportion of successes is very low the first month, jumps substantially the second month, stays relatively constant through the end of the fourth month, and then drops again and remains fairly constant for the last two months.

TABLE 5.5 Comparison of the First and Second Halves of the Temperature Forecast Data

	Frequencies			Row Proportions		
Season	S	F	Total	S	F	Total
Spring	46	43	89	0.517	0.483	1.000
Summer	50	39	89	0.562	0.438	1.000
Total	96	82	178	0.539	0.461	1.000

TABLE 5.6 *Comparison of the Six Months of the Temperature Forecast Data*

	Frequencies			Row Proportions		
Month	S	F	Total	S	F	Total
First	10	19	29	0.345	0.655	1.000
Second	18	12	30	0.600	0.400	1.000
Third	18	12	30	0.600	0.400	1.000
Fourth	20	10	30	0.667	0.333	1.000
Fifth	15	15	30	0.500	0.500	1.000
Sixth	15	14	29	0.517	0.483	1.000
Total	96	82	178	0.539	0.461	1.000

Before getting too excited about this pattern, note that the number of trials in each month is small; the results of Chapter 11 will show that this pattern is not convincing.

With the data divided into so many pieces, a picture is helpful. Figure 5.8 is a plot of the proportion of successes versus the month; it is similar to the interaction graph introduced in Chapter 4, but it only has one curve, since there is only one source of data. The picture shows the pattern discussed above.

The third assumption of Bernoulli trials is that the outcomes of trials are independent. It seems reasonable to assume that the outcome of a forecast made in April is independent of the outcome of one made in July. It is not clear, however, that the outcome of one day's forecast is independent of the outcome of the next day's forecast. For example, consider the following scenario:

> Suppose Madison's weather is dominated by two kinds of systems. In the first weather system the temperature is stable and the high temperature is thus easy to forecast. In the second weather system the temperature fluctuates wildly and the high temperature is thus difficult to forecast. If these weather systems persist for several days, successive outcomes would not be independent—there would tend to be streaks of successes while the first system dominates and streaks of failures while the second system dominates.

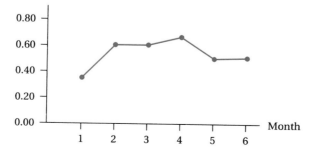

FIGURE 5.8 *Plot of proportion of successful temperature forecasts versus month.*

TABLE 5.7 Dependence of the Outcome of the Current Forecast on the Outcome of the Previous Forecast

Previous Forecast	Current Forecast					
	Frequencies			Row Proportions		
	S	F	Total	S	F	Total
S	54	42	96	0.562	0.438	1.000
F	41	40	81	0.506	0.494	1.000

Table 5.7 presents the relation between consecutive forecasts for the actual data. Of the successful forecasts, 56.2 percent are followed by a success, compared with 50.6 percent of the forecast failures. This difference is quite small. In summary, the data provide little evidence to suggest that the second and third assumptions of Bernoulli trials are unreasonable. ▲

EXAMPLE 5.21: The Tetris Study

Tetris is a video game that can be played on the Nintendo Entertainment System, a number of personal computers, or at an arcade. If you have never played Tetris, the following description of the game may be difficult to follow. Tetris rewards spatial reasoning and, not surprisingly, lightning reflexes. Essentially, shapes of one of seven possible configurations of blocks falls from the top of the screen. The player translates or rotates the falling shapes in an attempt to complete, with no gaps, a horizontal row of blocks. Each completed row of blocks disappears from the screen and blocks above the completed row, if there are any, drop down into the vacated space to allow room for more falling shapes. A player's score equals the number of rows completed before the screen overflows. After every 10 completed lines, the speed of the falling shapes increases, making the game more difficult. The game of Tetris is investigated further in Chapters 7, 10, 11, 15, and 16.

Each drop of a shape by the computer can be viewed as a trial. Each trial has seven possible outcomes and is best analyzed using the methods of Chapters 10 and 11. For now, the shapes are divided into two types: a straight row (my favorite, labeled a success) and any other shape (a failure). This example focuses on a very simple feature of the game, namely, whether the drops of shapes by the computer can be considered Bernoulli trials.

I observed 1,872 trials during eight plays of Tetris on my son's Nintendo system. The outcomes, by play, are given in Table 5.8 and graphed in Figure 5.9. There is some variation in the proportion of successes from play to play, but nothing remarkable. Thus, there is little evidence against the second assumption of Bernoulli trials.

By contrast, Table 5.9 shows that there is very strong evidence against the third assumption. Less than 1 percent of the successes were followed by a success,

TABLE 5.8 Comparison of Eight Plays of Tetris

	Frequencies			Row Proportions		
Play	S	F	Total	S	F	Total
First	22	171	193	0.114	0.886	1.000
Second	33	185	218	0.151	0.849	1.000
Third	36	215	251	0.143	0.857	1.000
Fourth	25	206	231	0.108	0.892	1.000
Fifth	30	198	228	0.132	0.868	1.000
Sixth	42	220	262	0.160	0.840	1.000
Seventh	33	215	248	0.133	0.867	1.000
Eighth	33	208	241	0.137	0.863	1.000
Total	254	1,618	1,872	0.136	0.864	1.000

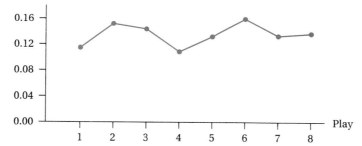

FIGURE 5.9 Plot of proportion of success for eight plays of Tetris.

TABLE 5.9 Dependence of the Outcome of the Current Shape on the Outcome of the Previous Shape for Tetris

	Current Shape					
	Frequencies			Row Proportions		
Previous Shape	S	F	Total	S	F	Total
S	2	251	253	0.008	0.992	1.000
F	249	1,362	1,611	0.155	0.845	1.000

EXERCISES 5.3

1. Assume that a student guesses on each question on a 25-item true-false exam. A score of 15 or more correct is needed to pass. Find the probability that the student passes. (Hint: Refer to Table 5.3, p. 161.)

2. Assume that a student guesses on each question on a 25-item true-false exam. A score of 17 or more correct is needed to pass. Find the probability that the student passes. (Hint: Refer to Table 5.3, p. 161.)

3. Assume that a student guesses on each question on a five-question multiple-choice exam. If each question has four different choices, find the probability that the student scores one or more correct. (Hint: Use the complement rule, introduced in Section 2.4.2.)

4. A multiple-choice exam consists of three questions. Question 1 has three possible answers, question 2 has four possible answers, and question 3 has five possible answers. If a student guesses on each question, find the probability that the student scores one or more correct. (Hint: Use the complement rule, introduced in Section 2.4.2.)

5. It can be tedious to obtain the table of counts that summarizes the relation between successive trials. (Table 5.7 is an example of such a table.) This exercise presents a way to reduce the tedium substantially. Consider the following artificial sequence of 10 dichotomous trials:

Trial	1	2	3	4	5	6	7	8	9	10
Outcome	S	S	F	S	F	S	S	F	F	F

It is easy to verify that this sequence has five successes and five failures. Because the sequence is short, the following table is easy to complete:

	Current		
Previous	S	F	Total
S			
F			
Total			

In particular, in trial pairs 1-2 and 6-7 a success is followed by a success. Similarly, in trial pairs 2-3, 4-5, and 7-8 a success is followed by a failure. Next, in trial pairs 3-4 and 5-6 a failure is followed by a success. Finally, in trial pairs 8-9 and 9-10 a failure is followed by a failure. These facts yield the following table:

	Current		
Previous	S	F	Total
S	2	3	5
F	2	2	4
Total	4	5	9

Note that the row total of the F row is only 4 even though five failures appear in the sequence. This occurs because the outcome of the last trial is a failure and that failure does not precede another trial in the sequence. Similarly, the S column total is 4 instead of 5 because the first trial is a success and that success does not follow another trial in the sequence. In summary, the row and column totals equal the observed frequencies of success and failure *except* that the last outcome is not counted in the row totals and the first outcome is not counted in the column totals. Thus, the row and column totals

```
S F F S S S F S F S S F S S S F S S S S
F S S F S S S F S S F F S S S S S S S F
F S F S S F S F F S S F S F S F S S S F
S F S S F S S S S S S S S S S S S S S F
S F S S S S F S F F S S S S S S F S S S
```

are very easy to obtain. Once they are obtained, the analyst needs to obtain only one count in the body of the table.

Repeat the above for the following sequence of dichotomous trials:

Trial	1	2	3	4	5	6	7	8	9	10
Outcome	F	S	S	S	F	F	S	F	F	S

6. In his report on his statistics project, Eric Statz wrote

> My dog Boone loves popcorn, so I thought if I threw pieces of popcorn for her to catch, the assumptions of Bernoulli trials would be reasonable. While sitting at the dining room table, I tossed one piece of popcorn every 5 seconds at approximately the same height and distance for her to catch.

The list of Eric's data is at the top of the page, in order from left to right.

(a) It can be verified easily that Boone achieved 33 successes in the first 50 trials and 38 successes in the last 50 trials, for a total of 71 successes and 29 failures in the study. Use these facts to investigate the validity of the second assumption of Bernoulli trials. Comment.

(b) Refer to the previous exercise. Because Boone's first and last trials were successes, the table at the bottom of this column can be partially completed as indicated.

	Current		
Previous	**S**	**F**	**Total**
S			70
F			29
Total	70	29	99

Next, verify that five times an *F* was followed by an *F*. Use this information to complete the above table. Compute its table of row proportions. Comment.

7. Linda Owen recorded the gender of the next 100 babies born at Meriter Hospital in Madison, Wisconsin, beginning on November 1, 1991. Explain why one might say it is obvious that the assumptions of Bernoulli trials are valid.

8. Giselle Lederman went to a (free) campus telephone and started calling phone numbers in a haphazard manner. Her trials consisted of calls that were answered. The trial was labeled a success if a woman answered the phone and a failure if a man answered. Do you think these are Bernoulli trials? Explain.

9. In her report on her statistics project, Kristen Joiner wrote

> I chose to test Muffin's preference for balls. She has two balls we often throw for her to fetch. One is small and blue, and the other is slightly larger, heavier, and red. My father rolled both balls for her with one hand at the same time; I recorded which ball she chose to chase first. Arbitrarily, the blue ball was labeled a success.

The list of Kristen's data is at the top of the following page, in order from left to right.

(a) The study consisted of 96 trials. Compare the outcomes of the first 48 trials to the outcomes of the last 48 trials.

(b) Investigate the relation between successive trials. (See Exercise 5 for a shortcut method of counting.)

10. Denise Howsare had never thrown darts until she performed her statistics project. Her study consisted of 100 throws, with a score of 13 or more labeled a success and a score of less than 13 (including missing the board) labeled a failure. Denise had 24 successes in her first 50 attempts and a total of 48 successes

```
S S S F F S S S S F F S S F S F S F F S F S
S S S F S S S F F F S S S F F F S F S F F F
S F S F S F S F F F F F F S F S F F F F S S
S S F S S F S F S S S S F F S F S F F F F S
S S S S S S F S S S S S S F F S
```

in the study. Her first trial was a failure and her last trial was a success. Finally, 28 of the successes were followed by another success. Use this information to investigate the assumptions of Bernoulli trials. (See Exercise 5.)

11. Margaret Algar-Gelembiuk hit 100 practice tennis "first serves" into the right court. During the first half of the study, she obtained 23 successes, and in the entire study she obtained 45 successes. Her first and last trials were failures and 30 of her failures were followed by another failure. Use this information to investigate the assumptions of Bernoulli trials. (See Exercise 5.)

12. Matt Roggensack attempted 100 golf putts from a distance of 10 feet. Matt had 26 successes in his first 50 putts and 32 successes in his last 50 putts. His first and last putts were successes, and 20 of his failures were followed by another failure. Use this information to investigate the assumptions of Bernoulli trials. (See Exercise 5.)

13. Rachael Rennard and Darby Schnarr recruited an assistant golf pro to assist them with their class project. The pro attempted 100 putts from a distance of approximately 5 feet. The pro had 32 successes in his first 50 putts and 34 successes in his last 50 putts. His first putt was a success, his last putt was a failure, and 25 of his failures were followed by a success. Use this information to investigate the assumptions of Bernoulli trials. (See Exercise 5.)

14. Mechele Pitt performed 100 dichotomous trials with her dog, Soochee, serving as her metasubject. At each trial, Soochee decided which of two toys to retrieve. Retrieving the toy with a squeaker was labeled a success. Soochee yielded 33 successes in the first 50 trials and 29 successes in the last 50 trials. The first trial was a success, the last trial was a failure, and nine of the failures were followed by another failure. Use this information to investigate the assumptions of Bernoulli trials. (See Exercise 5.)

15. Think of an experiment consisting of dichotomous trials for which

    - You can perform 100 trials in a fairly short period of time.
    - The assumptions of Bernoulli trials *might* be reasonable.
    - The value of p is likely to be in the range from 0.25 to 0.75 (that is, do not pick trials in which a success is either extremely easy or extremely difficult to obtain).

    Describe the experiment and conduct 100 trials. Use your results to investigate the validity of the assumptions of Bernoulli trials. Record the results of your trials in order and keep a copy for future reference.

5.4 SOME PRACTICAL CONSIDERATIONS

5.4.1 Problems with Sampling Human Populations

The Tetris study of the previous section illustrated the importance of examining relevant data in an effort to investigate whether the assumptions of Bernoulli trials are reasonable for a process. For a finite population the researcher faces a different challenge; namely, it can be extremely difficult to select a random sample of subjects from a population.

The first problem in a survey is the compilation of a list of all members of the population. As the 1990 and earlier federal censuses demonstrated, this is not an easy task for a population as large and diverse as that of the United States.

As another example, a medical researcher may be interested in the population of all persons who are HIV positive at a particular time. A total enumeration of this population is probably impossible; even after a herculean effort to obtain a nearly complete listing, the results would soon be outdated.

A second problem is that even if the researcher has a list of the entire population, the process of selecting n members at random for inclusion in the survey is formidable. This problem is becoming less serious in this age of inexpensive high-speed electronic computing, but it persists nonetheless.

If the first two problems have been solved, a third problem is actually locating the members of the sample. A fourth problem is that, once located, some members of the sample may refuse to participate in the study. These four problems are discussed in more detail for the 1986 Wisconsin Driver Survey.

Wisconsin had a computerized list of all licensed drivers, so the first and second problems were not substantial. The third and fourth problems, however, were very real. For example, a random sample of licensed drivers almost certainly would have been spread throughout the state. Sending interviewers to locate the sample members would have been time consuming and prohibitively expensive. Alternatively, mail or phone interviews likely would have had high rates of nonresponse from subjects.

Instead of a random sample of drivers, the Wisconsin researchers selected a number of driver's license examining stations at random from the collection of all stations and forced everyone who applied for a new license or a renewal during a one-month period to complete a questionnaire. (This eliminated nonresponse!) The Wisconsin Driver Survey was conducted during each of four years. After looking at the resulting data a number of different ways, I am convinced that no great distortion of reality occurs by assuming that the subjects in the surveys were selected at random from the populations.

The previous paragraph is not presented with the intent of convincing you that the Wisconsin Driver Survey was a good or poor study. Rather, the point is that most studies do not obtain a literal random sample of subjects. As a consumer of the results of research, one must develop the ability to assess the validity of conclusions. When presented with the results of a survey, the critical consumer will attempt to learn as much as possible about how the data were collected. All practicing statisticians can relate horror stories of researchers making the assumption of a random sample for mathematical convenience when it is clearly not appropriate. The *Literary Digest* poll discussed earlier in this chapter (p. 147) is an example of this practice. Here are some other examples; some of these will appear later in this book, and others will not.

- Assume that teenagers interviewed at shopping malls form a random sample of all teenagers.
- Assume that cadavers at a particular medical facility are a random sample of dead persons.
- Assume that college women at a particular university who see a gynecologist at the university health service are a random sample from the national population of college women.

5.4.2 Birthday Coincidences and Random Samples

Some time when you are in a group of 30 or more persons, ask everyone to state his or her birthday. You will likely find that at least two persons have the same birthday—same month and day, although possibly born in different years.

This birthday problem has been popular in beginning probability classes and texts for many years. The problem is usually stated as follows. Suppose that n persons are selected at random without replacement from a large population (the population size, N, is much bigger than n). Let A denote the event that at least two persons have the same birthday (the complement of A is the event that all n persons have different birthdays). With a slight extension of the ideas of this chapter, it can be shown that

$$P(A) = 1 - \frac{(366)(365)\cdots(367-n)}{366^n}. \qquad (5.5)$$

The derivation of this equation assumes that each of the 366 days of the year is equally likely to be the birthday of a person selected at random from the population. This assumption is clearly unnatural, since for most populations February 29 is much less likely than any of the other days. Some people prefer to ignore February 29, replacing each number in the numerator and denominator of Equation 5.5 by the number 1 smaller; it can be shown that this change has only a minor effect on the probability of the event A. In addition, it can be shown that if the assumption that all days are equally likely is replaced by different daily probabilities suggested by public health data on the number of babies born each day, the answers change very little from those given by Equation 5.5. The value of $P(A)$ can be computed easily with a hand calculator; selected results are given in Table 5.10. With 10 persons, for example, the probability of at least one match of birthdays is 0.1166, with 20 persons the probability is 0.4106, and for 23 persons it exceeds $\frac{1}{2}$. For 90 persons at least one match is a virtual certainty.

A number of other birthday coincidences are presented in [2]. They can be summarized as follows. Assume that n persons are selected at random without replacement from a much larger population, and assume further that every day of the year is equally likely to be the birthday of a person selected at random. Then the probability of any of the following events occurring is approximately $\frac{1}{2}$.

1. For $n = 7$, at least two persons have birthdays within a week of each other.

TABLE 5.10 *The Probability That at Least 2 Out of n Persons Have the Same Birthday.*

n	Prob.	n	Prob.	n	Prob.	n	Prob.	n	Prob.
2	0.0027	13	0.1939	24	0.5374	35	0.8135	46	0.9478
3	0.0082	14	0.2226	25	0.5677	36	0.8313	47	0.9544
4	0.0163	15	0.2523	26	0.5972	37	0.8479	48	0.9602
5	0.0271	16	0.2829	27	0.6258	38	0.8633	49	0.9654
6	0.0404	17	0.3143	28	0.6534	39	0.8775	50	0.9701
7	0.0561	18	0.3461	29	0.6799	40	0.8905	55	0.9861
8	0.0741	19	0.3783	30	0.7053	41	0.9025	60	0.9940
9	0.0944	20	0.4106	31	0.7295	42	0.9134	70	0.9991
10	0.1166	21	0.4428	32	0.7524	43	0.9234	80	0.9999
11	0.1408	22	0.4748	33	0.7740	44	0.9324	90	1.0000
12	0.1666	23	0.5063	34	0.7944	45	0.9405		

2. For $n = 14$, at least two persons have birthdays within a day of each other.
3. For $n = 18$, at least three persons have birthdays on the same day of the month, but not necessarily the same month. (For example, this event would occur if there were birthdays on March 1, July 1, and August 1 among the collection of 18 birthdays.)
4. For $n = 88$, at least three persons have the same birthday.
5. For $n = 187$, at least four persons have the same birthday.
6. For $n = 1,000$, at least nine persons have the same birthday.

One of the exercises at the end of this section asks you to select some people and investigate one of the items listed above. For example, you might determine the birthdays of $n = 7$ persons to see whether any two fall within a week of each other. Here are two ways in which you might collect data:

- Obtain a list of all residents of your city or a nearby city. Select seven names at random from the list, locate those persons, and determine their birthdays.
- Stand outside the library and question seven people selected haphazardly.

The first method is beyond reproach; it is a random sample, so you can be confident that the probability is approximately 50 percent that two persons will have a birthday within a week of each other. The first method is, however, a tremendous amount of work for one statistics exercise! A purist might argue that no probabilities can be computed for the second method because the subjects were not selected at random. I am comfortable, however, with saying there is approximately a 50 percent chance that the event of interest will occur because I believe that a person's reasons for being outside the library and being selected by you are unrelated to his or her date of birth. Of course, I may be wrong.

By contrast, suppose instead that you were asked to investigate attitudes in your city on some aspect of United States foreign policy. Then selecting persons haphazardly from outside the library could be very misleading. Attitudes on foreign policy may be related to various characteristics of the subject—age, education level, income level, poiitical philosophy and employment status, to name a few, that are themselves related to how likely it is that the person will be outside the library when the data are collected and also be selected by you.

Returning to the birthday problems, there are a number of bad ways to collect data—ways that will invalidate the answers given above. Here are two examples:

1. Select seven babies from the nursery in the maternity ward of a hospital.
2. Select seven persons standing in line to renew their driver's licenses.

For either of these methods, the chance of finding the desired match is substantially greater than 50 percent. The reason is obvious for the babies. For the other case, in many states a driver's license expires on the driver's birthday, and many people delay renewing the license as long as possible.

EXERCISES 5.4

1. Select one of the six events listed on p. 180. Sample an appropriate number of people (whose birthdays you do not already know!) such that the probability of a match is approximately 50 percent. Determine their birthdays and report whether or not a match occurs. Describe how you obtained your subjects.

2. Give a different bad way (not one of the two mentioned in the text) to sample for the birthday problem, that is, a method of sampling that is importantly different from random sampling.

REFERENCES

1. W. Beyer, ed., *Handbook of Tables for Probability and Statistics*, 2d ed. (Cleveland: Chemical Rubber Company, 1974), pp. 182–205.

2. P. Diaconis and F. Mosteller, "Methods for Studying Coincidences," *Journal of the American Statistical Association*, December, 1989, pp. 853–861.

Inference for a Population

- **6.1 INTRODUCTION**
- **6.2 CONFIDENCE INTERVALS**
- **6.3 MORE ON CONFIDENCE INTERVALS**
- **6.4 HYPOTHESIS TESTING**
- **6.5 PREDICTIONS***
- **6.6 HOW MANY ANALYSES?***

Chapter 6

6.1 INTRODUCTION

A researcher is planning to observe the value of a random variable X that is defined to be one of the following:

- The total number of successes in a random sample of size n from a population of distinct individuals
- The total number of successes in n Bernoulli trials

Recall from Chapter 5 (p. 159) that the sampling distribution of X is given by the binomial distribution.

In Chapter 5 the value of p was assumed to be known. With this information, the researcher is able to compute probabilities of the various possible values of X before collecting any data. This chapter takes a different viewpoint. It now is assumed that the value of p is unknown; thus, the researcher cannot compute the probabilities that X takes on given values. Instead, the researcher observes the value of X, denoted by x, and uses the number x to make an inference about the unknown value of p. The viewpoint of this chapter is much more realistic for most interesting scientific problems.

The value of p is rarely known for a finite population. If it were known, there would be little reason to conduct a survey. Occasionally, a researcher assumes knowledge of p for Bernoulli trials. For example, repeated tossing of a fair coin is modeled as Bernoulli trials with $p = 0.50$. In most interesting cases of Bernoulli trials, however, either p is unknown or the researcher wants to investigate whether the assumed value is correct. For example, a researcher might want to investigate whether a particular coin is actually fair. This chapter presents the three most common statistical methods for drawing conclusions about the unknown value of p: point estimation, confidence interval estimation, and hypothesis testing. In addition, the problem of predicting the future outcomes of a sequence of Bernoulli trials is considered.

The data are summarized by the sample size n and the observed value x of the random variable X. In words, x is the number of successes observed in the sample; note that $n - x$ is the number of failures observed in the sample.

The **point estimate** of the unknown p is denoted by $\hat{p}$ and is computed with the following formula:

$$\hat{p} = \frac{x}{n}. \tag{6.1}$$

Note that $\hat{p}$ equals the number of successes in the sample divided by the sample size. Thus, the proportion of successes in the sample, $\hat{p}$, is the point estimate of p, the proportion of successes in a finite population, or the probability of success on any trial for Bernoulli trials. It is convenient to define also the point estimate of $q = 1 - p$ to be

$$\hat{q} = 1 - \hat{p} = \frac{n - x}{n}.$$

Several examples illustrate these ideas.

EXAMPLE 6.1

The 1986 Wisconsin Driver Survey was introduced in **Chapter 5** (p. 145). Question 1 of the survey read

> How serious a problem do you think drunk driving is in Wisconsin at this time?

The driver had a choice of four possible responses: "extremely serious," "somewhat serious," "not very serious," and "not serious at all." For the purpose of this example, the last three choices are collapsed into one category, "not extremely serious." Label "extremely serious" a success and "not extremely serious" a failure, and assume that the data were obtained by selecting drivers at random from the population of Wisconsin drivers. (This assumption is discussed in Section 5.4.1.) The survey yielded $n = 1{,}677$ and $x = 877$, giving $\hat{p} = 877/1{,}677 = 0.523$. ▲

EXAMPLE 6.2

Suppose the assumptions of Bernoulli trials are reasonable for the temperature forecast study introduced in Example 5.20 (p. 172). As reported earlier, $n = 178$ and $x = 96$, giving $\hat{p} = 96/178 = 0.539$. ▲

EXAMPLE 6.3

The Colloquium study was introduced in Chapter 1 (see Example 1.1) and has been studied extensively in the preceding chapters. For Fisher's test and the first alternative, the P-value is equal to the probability of the event A:

$$A = \{\text{The test statistic is} \geq \tfrac{3}{7}\}.$$

As reported in Section 3.2, a simulation experiment of $n = 10,000$ runs yielded $x = 151$ occurrences of the event A. Thus, the point estimate of the P-value based on the simulation experiment is $\hat{p} = 151/10,000 = 0.0151$. ▲

The use of the word *point* in *point estimate* refers to the common mathematical practice of identifying each number with a point on the number line. Hence, a point estimate is a single-number estimate. An advantage of point estimates is the ease of understanding statements such as the following:

- It is estimated that 52.3 percent of all Wisconsin drivers in 1986 felt that drunk driving was an extremely serious problem.
- The estimated probability of a successful high temperature forecast in the *Wisconsin State Journal* is 0.539.

A disadvantage of point estimates is that they are usually incorrect. For example, the actual P-value of the Colloquium study is 0.0165, so the point estimate, 0.0151, although close to the actual value, is incorrect. In the next section confidence intervals are introduced as a way to deal with this disadvantage.

6.2 CONFIDENCE INTERVALS

This section shows how to compute a confidence interval and presents a sketch of the mathematical justification for confidence intervals. The next section discusses the interpretation of and some features of confidence intervals.

In order to obtain a 95 percent confidence interval for p, first compute

$$h = 1.96\sqrt{\frac{\hat{p}\hat{q}}{n}}.$$

Next, use the value of h to obtain

$$l = \hat{p} - h \quad \text{and} \quad u = \hat{p} + h.$$

The symbol h is for half-width, l is for lower bound, and u is for upper bound.

The 95 percent confidence interval for p is given by the interval of numbers $[l, u]$. In words, the researcher states, "I am 95 percent confident that $l \leq p \leq u$," or "I am 95 percent confident that p is between l and u."

EXAMPLE 6.4

For question 1 on the 1986 Wisconsin Driver Survey, $n = 1{,}677$ and $\hat{p} = 0.523$, giving

$$\hat{q} = 1 - \hat{p} = 1 - 0.523 = 0.477 \quad \text{and} \quad h = 1.96\sqrt{\frac{(0.523)(0.477)}{1677}} = 0.024.$$

Thus,

$$l = 0.523 - 0.024 = 0.499 \quad \text{and} \quad u = 0.523 + 0.024 = 0.547.$$

A 95 percent confidence interval for p is $[0.499, 0.547]$. In other words, the researcher is 95 percent confident that $0.499 \leq p \leq 0.547$, or, more simply, that p is between 49.9 percent and 54.7 percent. ▲

EXAMPLE 6.5

For the high-temperature forecasting study, $n = 178$ and $\hat{p} = 0.539$, giving

$$\hat{q} = 1 - \hat{p} = 1 - 0.539 = 0.461 \quad \text{and} \quad h = 1.96\sqrt{\frac{(0.539)(0.461)}{178}} = 0.073.$$

Thus,

$$l = 0.539 - 0.073 = 0.466 \quad \text{and} \quad u = 0.539 + 0.073 = 0.612.$$

A 95 percent confidence interval for p is $[0.466, 0.612]$. In other words, the researcher is 95 percent confident that $0.466 \leq p \leq 0.612$, or that p is between 46.6 percent and 61.2 percent. ▲

EXAMPLE 6.6

Consider again the point estimate of the P-value based on the simulation experiment for the first alternative for the Colloquium study. Recall that $n = 10{,}000$ and $\hat{p} = 0.0151$, giving

$$\hat{q} = 1 - \hat{p} = 1 - 0.0151 = 0.9849$$

and

$$h = 1.96\sqrt{\frac{(0.0151)(0.9849)}{10{,}000}} = 0.0024.$$

Thus,

$$l = 0.0151 - 0.0024 = 0.0127 \quad \text{and} \quad u = 0.0151 + 0.0024 = 0.0175.$$

A 95 percent confidence interval for p is $[0.0127, 0.0175]$. In other words, the researcher is 95 percent confident that the P-value is between 0.0127 and 0.0175. ▲

Recall that the exact P-value for the Colloquium study is 0.0165. This value falls between the boundaries, 0.0127 and 0.0175, of the above confidence interval;

whenever this happens statisticians say that the confidence interval is **correct.** The concept of correctness is useful, but it has its limitations. The analyst will never know if a particular interval is correct unless the entire population is studied. For example, the above confidence interval is known to be correct because it includes the exact P-value, which was obtained by looking at all 40,116,600 possible assignments of the 28 subjects to treatments. By contrast, for the 1986 Wisconsin Driver Survey and temperature forecasting examples, whether either confidence interval is correct is not known. It will be shown, however, in the following sections that, in a certain sense, each of these intervals has a 95 percent chance of being correct.

6.2.1 Derivation of the 95 Percent Confidence Interval

This section contains a large amount of algebraic reasoning and manipulation. It is not important to learn each detail, but it is important to understand the general outline of the approach. The basic idea is that the observable value of $\hat{p}$ is probably close to the unknown value of p, especially if the sample size is reasonably large. As the wording "probably close" suggests, the derivation of the confidence interval formula uses probability theory.

Remember that the computation of probabilities is associated with a chance mechanism. The chance mechanism for the current problem is one of the following:

- For a survey, selecting a random sample of size n from the population
- For Bernoulli trials, observing the outcomes of n trials

In addition, probabilities are computed before the result of the chance mechanism is observed.

The point estimate $\hat{p}$ is equal to the observed value, x, of the random variable X divided by the sample size, n. It is convenient to define a new random variable $\hat{P}$ by $\hat{P} = X/n$. Note the following facts about $\hat{P}$:

- It has a simple relationship to the random variable X; thus, probabilities computed for X easily yield probabilities for $\hat{P}$.
- The observed value of $\hat{P}$ is obtained by replacing X by its observed value; the result is x/n, which is recognized as $\hat{p}$. Thus, the point estimate $\hat{p}$ of the unknown p is simply the observed value of the random variable $\hat{P}$.

As defined earlier, $\hat{p}$ is called the point estimate of p; the random variable $\hat{P}$ is called the **point estimator** of p. Define the random variable $\hat{Q}$ equal to $1 - \hat{P}$; it is the point estimator of q, and its observed value is $\hat{q}$.

As discussed in Section 5.2.3, probabilities can be computed for X by using the binomial distribution. It turns out that for the type of probabilities required for a confidence interval, the binomial distribution is extremely messy to use. Fortunately, however, the computations are relatively easy if the standard normal curve approximation to the binomial distribution is used.

Recall that X has expected value and standard deviation given by
$$E(X) = np \quad \text{and} \quad SD(X) = \sqrt{npq}.$$
Thus, the standardized version of X is
$$Z = \frac{X - np}{\sqrt{npq}}.$$
The value of Z is not changed if its numerator and denominator are divided by n, yielding
$$Z = \frac{(X - np)/n}{(\sqrt{npq})/n} = \frac{X/n - np/n}{\sqrt{npq/n^2}} = \frac{\hat{P} - p}{\sqrt{pq/n}}.$$
The following derivations can be obtained from properties of the expected value and the standard deviation, but the details are beyond the scope of this text:
$$E(\hat{P}) = E\left(\frac{X}{n}\right) = \frac{E(X)}{n} = \frac{np}{n} = p.$$
$$SD(\hat{P}) = SD\left(\frac{X}{n}\right) = \frac{SD(X)}{n} = \frac{\sqrt{npq}}{n} = \sqrt{\frac{pq}{n}}.$$
Thus, the standardized version of $\hat{P}$ is
$$Z = \frac{\hat{P} - p}{\sqrt{pq/n}}. \tag{6.2}$$
Note that the standardized versions of X and $\hat{P}$ are the same.

Before proceeding, additional terminology is useful. The standard deviation of a point estimator is called its **standard error.** Thus, the standard error of $\hat{P}$ is $\sqrt{pq/n}$.

As argued in Section 5.2.3, probabilities for the standardized version of X, which is denoted by Z, can be approximated by the standard normal curve. In addition, from Table A.3, the area under the standard normal curve between -1.96 and $+1.96$ equals 0.95. Combining these results yields
$$P(-1.96 \leq Z \leq 1.96) \approx 0.95.$$
Next, replace the symbol Z by an equivalent form (Equation 6.2):
$$P\left(-1.96 \leq \frac{\hat{P} - p}{\sqrt{pq/n}} \leq 1.96\right) \approx 0.95.$$
Rearranging the terms in the last expression, implies that 0.95 is the approximate probability that
$$\hat{P} - 1.96\sqrt{\frac{pq}{n}} \leq p \leq \hat{P} + 1.96\sqrt{\frac{pq}{n}}. \tag{6.3}$$

The extremes of this string of inequalities look similar to l and u, the boundaries of the confidence interval, but two difficulties remain:

- The general guideline for using the standard normal curve approximation is that both np and nq equal or exceed 15. Since p and q are unknown, this condition cannot be checked. (Often this is not a serious problem in practice; if n is very large the researcher, using knowledge of the phenomenon being studied, might be certain that the condition is met.)
- Look at the extremes of Display 6.3. The numbers 1.96 and n are known; once data are obtained, the observed value of $\hat{P}$ can be computed; but the values of p and q are and will remain unknown.

In order to overcome both these difficulties, the Russian mathematician Slutsky replaced the unknown values of p and q in the extremes of Display 6.3 by their point estimators. He obtained

$$\hat{P} - 1.96\sqrt{\frac{\hat{P}\hat{Q}}{n}} \le p \le \hat{P} + 1.96\sqrt{\frac{\hat{P}\hat{Q}}{n}}. \tag{6.4}$$

Slutsky then proved that

- Probabilities based on the standard normal curve approximation remain valid after this exchange.
- The original condition of the general guideline in terms of the unknown p and q can be replaced by the condition that both $n\hat{p}$ and $n\hat{q}$ equal or exceed 15.

Thus, according to Slutsky's work, before data are collected the probability that the event described by Display 6.4 occurs is approximately 95 percent. After data are collected, the random variables $\hat{P}$ and $\hat{Q}$ are replaced by their observed values; the result is that Display 6.4 is transformed to

$$\hat{p} - 1.96\sqrt{\frac{\hat{p}\hat{q}}{n}} \le p \le \hat{p} + 1.96\sqrt{\frac{\hat{p}\hat{q}}{n}}. \tag{6.5}$$

This last expression is recognized as $l \le p \le u$; thus, the formula for the confidence interval has been derived.

For ease of exposition, the modifier *approximately* is usually dropped and one states that the probability or confidence is 95 percent.

6.2.2 Other Confidence Levels

Examine again the above derivation of the 95 percent confidence interval for p paying special attention to the role of the **confidence level** 95 percent. The value 95 percent was used in the first step to yield the values -1.96 and $+1.96$. Thereafter, no special properties of the numbers 95 percent, $+1.96$, or -1.96 were needed. Suppose the researcher wants 90 percent confidence instead of 95 percent; what would change? From Table A.3 the area under the standard normal curve between -1.645 and 1.645 equals 0.90. The earlier derivation of the 95 percent confidence interval can be repeated with every occurrence of 95 percent replaced by 90 percent, every occurrence of 1.96 replaced by 1.645, and every occurrence of -1.96 replaced by -1.645. These substitutions yield the following formula for the 90 percent confidence interval for p:

$$\hat{p} - 1.645\sqrt{\frac{\hat{p}\hat{q}}{n}} \leq p \leq \hat{p} + 1.645\sqrt{\frac{\hat{p}\hat{q}}{n}}.$$

In a similar manner, the 80 percent, 98 percent, and 99 percent confidence intervals for p can be obtained from the following facts:

- The area under the standard normal curve between -1.282 and 1.282 equals 0.80.
- The area under the standard normal curve between -2.326 and 2.326 equals 0.98.
- The area under the standard normal curve between -2.576 and 2.576 equals 0.99.

These five results can be summarized with one algorithm:

To obtain a confidence interval for p,

1. Select a confidence level.
2. Determine the value of z that corresponds to the confidence level from Table A.3 or Table 6.1.
3. The confidence interval is

$$\hat{p} - z\sqrt{\frac{\hat{p}\hat{q}}{n}} \leq p \leq \hat{p} + z\sqrt{\frac{\hat{p}\hat{q}}{n}}. \tag{6.6}$$

For brevity, the formula for the confidence interval is usually written as

$$\hat{p} \pm z\sqrt{\frac{\hat{p}\hat{q}}{n}}. \tag{6.7}$$

TABLE 6.1 *Values of z for a Confidence Interval Based on the Standard Normal Curve Approximation*

Confidence Level	z
80%	1.282
90%	1.645
95%	1.960
98%	2.326
99%	2.576

EXERCISES 6.2

1. On March 16–17, 1989, the *Newsweek* Poll interviewed what was called a national sample of 756 adults by telephone [1].

   (a) *Newsweek* magazine reported that 552 respondents agreed with the statement:

   > We should use fewer pesticides and chemicals to ensure safer food even if it means higher prices.

   Assuming that the 756 respondents were selected at random from all adults living in the United States, compute the point estimate of and the 95 percent confidence interval for the proportion of people in the population who would have agreed with the statement.

   (b) Of those surveyed, 340 reported that they "often or occasionally buy organic foods." With the assumptions above, compute the point estimate of and 95 percent confidence interval for the proportion of people in the population who often or occasionally buy organic foods.

   (c) Compare the responses to these two questions. In particular, discuss the issue of what people say compared with what they do.

2. On April 12–13, 1989, the *Newsweek* Poll interviewed what was called a national sample of 750 adults by telephone [2]. *Newsweek* magazine reported that 27 percent "think abortions should be legal under any circumstances." However, only 15 percent think abortions should be legal if the parents are unhappy with the gender of the fetus. Further, 18 percent "think abortions should be illegal under all circumstances," yet only 8 percent feel that abortions should be illegal if the woman's life is endangered. Discuss these inconsistencies in responses. What are some of the advantages of general questions? Of specific questions?

3. On March 9–10, 1989, the *Newsweek* Poll interviewed what was called a national sample of 756 adults by telephone [3]. *Newsweek* magazine reported that 469 respondents "approve of the way President George Bush is handling his job." Assuming that the 756 respondents were selected at random from all adults living in the United States, compute the point estimate of and the 90 percent confidence interval for the proportion of people in the population who would have responded that they approved, if they had been asked.

4. Exercise 14 on p. 27 presented the results of a study of the effectiveness of interferon in the treatment of hepatitis. Sixteen of 44 subjects who received interferon were cured. Assuming that these 44 subjects were selected at random from an appropriate population, compute the point estimate of and 80 percent confidence interval for the proportion in the population that would be cured with interferon treatment.

5. A study of chronic Crohn's disease was presented in Chapter 1 beginning on p. 20. Recall that 22 of 37 subjects given cyclosporine exhibited improvement by the end of the three months of treatment. Assuming that these 37 people were a random sample from the population of Crohn's disease sufferers who do not respond to the standard treatment, compute the point estimate of and 90 percent confidence interval for the proportion of the population that would obtain a success if given cyclosporine.

6. Exercise 16 on p. 27 introduced a study of different treatments for breast cancer.

   (a) Pretend that the women who received lumpectomy and irradiation were a random sample from a population, and analyze the data with a 95 percent confidence interval. Briefly interpret your interval.

   (b) Pretend that the women who received total mastectomy were a random sample from a population, and analyze the data with a 95 percent confidence interval. Briefly interpret your interval.

7. Exercise 17 on p. 27 introduced a study of different doses of lithium for the treatment of bipolar disorder. Pretend that the subjects who received the low dose were a random sample from a population, and analyze the data with a 95 percent confidence interval. Briefly interpret your interval.

8. Exercise 16 on p. 34 introduced a study of different shooting positions. On the assumption that Robert's shots from the kneeling position were Bernoulli trials, construct a 90 percent confidence interval for the probability of a success. Briefly interpret your interval.

9. The Three-Point Basket study was introduced in Chapter 1 (Example 1.6). The following table summarizes the results of the study:

Location	S	F	Total
In front	21	29	50
Left corner	20	30	50

   (a) Assuming that the shots from the front were 50 Bernoulli trials with an unknown probability of success, compute the point estimate of and 98 percent confidence interval for the probability of making a basket from the front.

   (b) Assuming that the shots from the left corner were 50 Bernoulli trials with an unknown probability of success, compute the point estimate of and 98 percent confidence interval for the probability of making a basket from the left corner.

10. Exercise 20 on p. 35 introduced a study of putting with or without a golf glove.

    (a) Assuming that Michael's putts while wearing a glove were Bernoulli trials, construct a 95 percent confidence interval for the probability of a success. Briefly interpret your interval.

    (b) Assuming that Michael's putts while not wearing a glove were Bernoulli trials, construct a 95 percent confidence interval for the probability of a success. Briefly interpret your interval.

11. Section 4.1 introduced a study of HIV-positive persons who did not have AIDS. If the data are combined across the different blocks, the results are 17 failures in 417 trials on the placebo (a failure is developing AIDS during the 24 weeks of the experiment). Assuming that these 417 subjects were selected at random from a population, compute the point estimate of and 95 percent confidence interval for the proportion in the population that would be successes at the end of 24 weeks if given the placebo.

12. A 1989 sample of 130 college women who visited a gynecologist at a particular (unnamed) university in the northeastern United States indicated that 113 were sexually experienced [4]. Assuming that this sample was selected at random from the population of all women at that university, compute the point estimate of and 95 percent confidence interval for the proportion in the population who were sexually experienced. Do you think it is reasonable to assume a random sample? Explain.

13. The following is part of the abstract of a French study on children born to HIV antibody–positive mothers [5]:

    Assessment of the risks of transmission of infection with human immunodeficiency virus type 1 (HIV-1) from mother to newborn is difficult, partly because of the persistence for up to a year of maternal antibodies transmitted passively to the infant. To determine the frequency of perinatal transmission of HIV infection, we studied from birth 308 infants born to seropositive women. Of 117 infants evaluated 18 months after birth, 32 were seropositive for HIV or had died of AIDS ($n = 6$); of the 32, only 2 remained asymptomatic. Another 76 infants were seronegative and free of symptoms, whereas nine were seronegative but had symptoms suggestive of HIV-1 infection.

    (a) Explain why the infants were not evaluated at age 6 months.

    (b) Why do you think only 117 of the 308 infants were evaluated at 18 months of age?

    (c) Assuming that the 117 infants are a random sample from an appropriate population, construct a 90 percent confidence interval for the proportion in the population who are seronegative and free of symptoms at age 18 months.

14. Refer to Exercise 12. Forty-six of the sexually experienced women reported that their partner "always or almost always" uses a condom during sexual intercourse, 66 reported that the partner "seldom or never" uses a condom, and 1 was uncertain. Ignoring the one uncertain woman and treating the sample as a random sample from the population of sexually experienced college women at that school, compute the point estimate of and 95 percent confidence interval for the proportion in the population whose partner always or almost always uses a condom. Do

you think it is reasonable to assume a random sample? Compare the assumption of a random sample in this and in Exercise 12.

15. Exercise 15 on p. 178 asked you to perform an experiment with 100 trials and a dichotomous response. Assuming that the trials were Bernoulli trials with an unknown probability of success, use your data to compute the point estimate of and 90 percent confidence interval for p.

16. French law permits voluntary interruption of pregnancy until amenorrhea has lasted 84 days. The law also specifies that a woman must allow one week for reflection between the time of her initial decision and the time of pregnancy interruption [6]. This exercise examines the results of a study on the effectiveness of RU 486. The abstract of the study reads, in part,

> In 2,115 women seeking voluntary termination of pregnancy after 49 days of amenorrhea or less, we studied the effect of a single 600 mg dose of mifepristone (RU 486), followed 36 to 48 hours later by the administration of one of two prostaglandin analogues, either gemeprost (1 mg) or sulprostone (0.25, 0.375, or 0.50 mg). Efficacy (success) was indicated by the complete expulsion of the conceptus without the need of an additional procedure. All other results were considered failures, and the pregnancy was terminated by a surgical method.

The report further stated, "The physician was free to determine the analogue and the dose to be used."

(a) The overall (ignoring type of analogue) efficacy rate in the study was 96.0 percent of 2,040 women (75 women did not return for follow-up). Assuming that the 2,040 women were a random sample from a population, analyze these data with a 95 percent confidence interval.

(b) The women on the four analogues differed considerably in time to expulsion, duration of bleeding, and presence and severity of pain. Given the design of the study, do you think it is valid to draw conclusions from a comparison of the four analogues? Give reasons for your answer.

6.3 MORE ON CONFIDENCE INTERVALS

In the previous section, the 95 percent confidence interval for the proportion of 1986 licensed drivers in Wisconsin who viewed drunk driving as an extremely serious problem was found to be [0.499, 0.547]. This section begins with an interpretation of what this means and some comments on its utility.

6.3.1 Interpretation of Confidence Intervals

Display 6.4 indicates that

$$P\left(\hat{P} - 1.96\sqrt{\frac{\hat{P}\hat{Q}}{n}} \leq p \leq \hat{P} + 1.96\sqrt{\frac{\hat{P}\hat{Q}}{n}}\right)$$

equals 0.95. Literally, this probability statement means that 95 percent of the possible samples of size 1,677 from the population of licensed drivers would yield a value of $\hat{p}$ (and $\hat{q} = 1 - \hat{p}$) such that

$$\hat{p} - 1.96\sqrt{\frac{\hat{p}\hat{q}}{n}} \leq p \leq \hat{p} + 1.96\sqrt{\frac{\hat{p}\hat{q}}{n}}.$$

In words, 95 percent of the possible samples would lead to an interval that actually contains the unknown p, or, alternatively, 95 percent of the possible samples would lead to an interval that is correct. The remaining 5 percent of the possible samples would yield an interval that does not contain the unknown p—an interval that is incorrect.

An individual researcher, of course, does not examine all possible samples. Instead, the researcher selects one sample and computes the confidence interval. For example, for the data given earlier for the Wisconsin Driver Survey the researcher computed the interval [0.499, 0.547]. Because the researcher does not know the value of p, he or she does not know whether this particular interval comes from one of the 95 percent of possible samples that yield a correct interval or from one of the 5 percent of possible samples that yield an incorrect interval. In short, the researcher does not know whether the interval is correct or incorrect.

You should not say that the *probability* that the interval [0.499, 0.547] is correct equals 0.95. A probability reflects uncertainty about the future operation of a chance mechanism. For a computed confidence interval, the chance mechanism has already operated and the interval is either correct or incorrect. As a result, statisticians use the phrase "95 percent confidence" to reflect the fact that the *method* used to obtain the interval has a 95 percent probability of yielding a correct interval. Another way to think of this is to invoke the long-run relative frequency interpretation of probability that was introduced in Section 2.4. If a researcher computes a large number of 95 percent confidence intervals, he or she knows that about 95 percent of the computed intervals are correct and about 5 percent are incorrect. There is no way, however (given that the researcher is ignorant of the values of the unknown parameters being estimated), to tell which intervals are correct and which are incorrect.

The ideas of the previous paragraph can be made more concrete by considering a simple artificial example generated by Minitab. Suppose the value of p for a population is 0.4. Ten researchers are interested in estimating p, and none of them knows that the true value is 0.4. Each researcher selects a random sample of size $n = 100$ and computes an 80 percent confidence interval. For example, the first researcher obtains $x = 47$ successes and finds the 80 percent confidence interval:

$$\hat{p} \pm 1.282\sqrt{\frac{\hat{p}\hat{q}}{n}} = 0.470 \pm 1.282\sqrt{\frac{(0.47)(0.53)}{100}} = 0.470 \pm 0.064 = [0.406, 0.534].$$

Similarly, the second researcher obtains $x = 31$ successes and finds that the 80 percent confidence interval is [0.251, 0.369]. (You might want to verify this interval.) The results for the 10 researchers are given in Table 6.2 and presented visually in Figure 6.1. Note the following features of these results:

- Given that $p = 0.4$, one can see that 8 of 10 (80 percent) of the 80 percent confidence intervals are correct and 2 of 10 (20 percent) are incorrect.
- From any researcher's perspective, p is unknown and there is no way of knowing whether or not a particular interval is correct. In particular, the

6.3 More on Confidence Intervals

TABLE 6.2 Ten 80 Percent Confidence Intervals for p Based on Random Samples of Size n = 100 for Which the Actual Value of p Is 0.4

Researcher	x	$\hat{p}$	l	u	Correct?
1	47	0.470	0.406	0.534	No
2	31	0.310	0.251	0.369	No
3	36	0.360	0.298	0.422	Yes
4	41	0.410	0.347	0.473	Yes
5	35	0.350	0.289	0.411	Yes
6	46	0.460	0.396	0.524	Yes
7	40	0.400	0.337	0.463	Yes
8	41	0.410	0.347	0.473	Yes
9	42	0.420	0.357	0.483	Yes
10	36	0.360	0.298	0.422	Yes

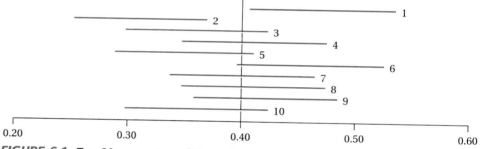

FIGURE 6.1 Ten 80 percent confidence intervals for p based on random samples of size n = 100. The actual value of p is 0.4. The number to the right of each interval identifies the researcher.

first and second researchers do not know that their intervals are incorrect, and the other eight researchers do not know that their intervals are correct. (Of course, if the researchers *compare their answers*, this remark can be modified. For example, because the first and second researchers' intervals do not overlap, it is clear that at least one of them must have an incorrect interval.)

The formula for the confidence interval is obtained by using the standard normal curve to approximate the sampling distribution of the standardized version of the point estimator $\hat{P}$. This seems reasonable because the standardized version of $\hat{P}$ is identical to the standardized version of X, and it was shown near the end of Section 5.2 that the sampling distribution of the latter can be approximated by the standard normal curve. The general guideline states that the approximation is good provided $n\hat{p}$ and $n\hat{q}$ both equal or exceed 15. But how good is the approximation? Actually, with a small amount of computation it is easy to evaluate the approximation for any selected special case. For example, suppose that, unknown to the researcher, $p = 0.4$, as in the above example. Suppose, in addition, that the researcher selects a random sample of size $n = 100$. The details are tedious, but it can be shown that the event that the 80 percent confidence interval includes the value 0.4 (it is correct) is equal to the event ($34 \leq X \leq 46$). With the help of Minitab, it can be shown that

$$P(34 \leq X \leq 46) = 0.816.$$

Thus, the exact confidence level is 81.6 percent if $p = 0.4$ and $n = 100$. In a similar way it can be shown that the exact confidence level is 80.8 percent if $n = 100$ and $p = 0.3$, and 79.8 percent if $n = 100$ and $p = 0.5$. Other values of n and p can be examined; the answers show that if the general guideline is met, the approximate confidence level is very close to the exact confidence level; that is, the approximation is quite good.

Now, let us turn to the utility of the information obtained from a confidence interval. The ideas are illustrated for the responses to question 1 on the 1986 Wisconsin Driver Survey.

Often it is useful to compare the information obtained from a confidence interval with two possible extreme procedures. At one extreme, the researchers could have decided to question every member of the population of drivers, not just the sample of $n = 1,677$ persons. A census would have revealed the exact value of p. If resources are unlimited, obtaining the value of p from a census is preferred to obtaining a confidence interval for p. In reality, however, resources are limited and a census is rarely a viable option. At the other extreme a researcher could decide to collect no data and simply report his or her opinion of the value of p. Compared with these extremes, a confidence interval gives an efficient and objective answer. Note that for Bernoulli trials the first extreme procedure, examining the whole population, is impossible since the population is a mathematical model of the process that generates the outcomes of the trials.

6.3.2 The Half-Width

A common reaction to the choice of the confidence level for a confidence interval is

> I want my confidence interval to be correct, so I will pick my confidence level to be as large as possible.

To investigate this notion it will be useful to construct five confidence intervals for p for the responses to question 1 on the 1986 Wisconsin Driver Survey. The results are given in Table 6.3 and presented visually in Figure 6.2. These displays demonstrate that as the confidence level increases, the value of the half-width, h, increases and the confidence interval becomes wider. This fact can be seen mathematically without reference to any particular example by noting that, from Table 6.1, as the confidence level increases, z increases. Thus, since h is proportional to z,

TABLE 6.3 *Five Confidence Intervals for the Proportion of Successes on Question 1 of the 1986 Wisconsin Driver Survey*

Confidence Level	h	Confidence Interval
80%	0.016	[0.507, 0.539]
90%	0.020	[0.503, 0.543]
95%	0.024	[0.499, 0.547]
98%	0.028	[0.495, 0.551]
99%	0.031	[0.492, 0.554]

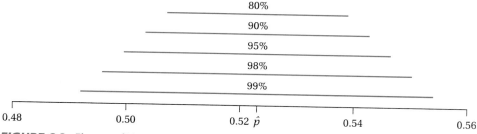

FIGURE 6.2 *Five confidence intervals for the proportion of successes on question 1 of the 1986 Wisconsin Driver Survey.*

$$h = z \times \sqrt{\frac{\hat{p}\hat{q}}{n}},$$

h increases as the confidence level increases. In summary, a higher confidence level achieves its higher chance of being correct at the expense of producing a wider interval. A wider confidence interval is a *less precise* statement about the value of p; in short, it is less useful. It is difficult to say exactly at what point a confidence interval becomes so wide that it is useless; in my opinion even the 99 percent confidence interval for question 1, which states that p is between 0.492 and 0.554, is sufficiently precise to be useful. By contrast, a confidence interval of, say, [0.25, 0.75] would be too wide to be useful. A value of p as small as 0.25 has very different practical ramifications than a value as large as 0.75. The smaller value shows little, and the larger value shows tremendous, public concern about drunk driving.

The formula for the half-width can be rewritten slightly to yield

$$h = z \times \frac{\sqrt{\hat{p}\hat{q}}}{\sqrt{n}}. \qquad (6.8)$$

The value of h depends on three numbers,

$$z, \hat{p}, \text{ and } n.$$

(Note that $\hat{q}$ also appears in Equation 6.8, but it equals $1 - \hat{p}$.) As discussed above, z depends on the analyst's choice of confidence level, with higher confidence levels leading to larger values of z and, hence, to larger values of h. The following example illustrates the effects of the other two terms, $\hat{p}$ and the sample size, n, on the half-width of a confidence interval.

EXAMPLE 6.7

Figure 6.3 plots the half-width, h, for a 95 percent confidence interval versus $\hat{p}$ for two different sample sizes, $n = 1{,}000$ and $n = 4{,}000$. At every value of $\hat{p}$, the curve for $n = 4{,}000$ is exactly half as tall as the curve for $n = 1{,}000$. This is a special case of the following two general facts:

1. For any fixed value of $\hat{p}$, the larger the value of n, the smaller the value of h.
2. For any fixed value of $\hat{p}$, a fourfold increase in the sample size results in only a twofold decrease in the half-width.

Both of these facts can be seen algebraically. The first is obvious because n appears in the denominator of h. The second fact is true because it is the square root of n that appears in the denominator and

$$\sqrt{4n} = \sqrt{4}\sqrt{n} = 2\sqrt{n}.$$

Thus, changing the sample size from n to $4n$ changes the denominator of h from $\sqrt{n}$ to $2\sqrt{n}$.

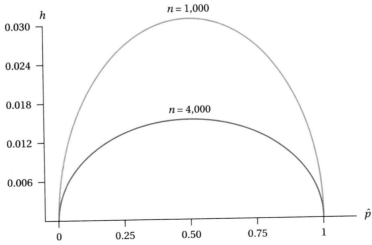

FIGURE 6.3 Plot of the half-width, h, versus $\hat{p}$ for a 95 percent confidence interval for $n = 1{,}000$ and $n = 4{,}000$.

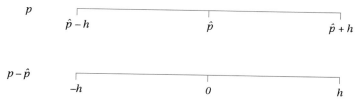

FIGURE 6.4 Visual demonstration that a confidence interval is correct if, and only if, the absolute value of $p - \hat{p}$ is less than or equal to h.

The effect of $\hat{p}$ is more complicated than the effect of n. As shown in Figure 6.3, for either value of n the largest value of h occurs when $\hat{p} = 0.5$. As the value of $\hat{p}$ moves away from one-half, in either direction, the half-width decreases. At first the decrease is slow; for $\hat{p}$ equal to 0.30 or 0.70, the value of h is only 8.4 percent smaller than it is for $\hat{p} = 0.50$. But as $\hat{p}$ approaches 0 or 1, the decrease is rapid; for $\hat{p}$ equal to 0.02 or 0.98, h is 71.9 percent smaller than it is for $\hat{p} = 0.50$. ▲

This section concludes with three remarks about confidence intervals. The first presents terminology that is used for all confidence intervals, not just those of this chapter. The second remark is applicable for any confidence interval that is derived by using the standard normal curve approximation. The third remark relates the findings of this chapter to terminology that is common in the mass media.

1. Sometimes the half-width, h, is referred to as an **error bound.** If the confidence interval is correct, then p lies in the interval $[\hat{p} - h, \hat{p} + h]$ and the difference between the known point estimate, $\hat{p}$, and the unknown true value of p is between $-h$ and $+h$ (see Figure 6.4). Thus, if the confidence interval is correct, h is a bound on the absolute value of the error in estimating p by $\hat{p}$.

2. The most popular choice for confidence level is 95 percent. This popularity is due, in part, to the following. It is common for researchers to report the point estimate, $\hat{p}$, and its **estimated standard error,** $(\hat{p}\hat{q}/n)^{1/2}$. A confidence interval can be obtained quickly, without recourse to a calculator or complicated arithmetic, by simply replacing z in the formula for the confidence interval by an integer value. For example, if a report states that $\hat{p} = 0.63$ and its estimated standard error is 0.07, then

    - The choice $z = 1$ gives $0.63 \pm 0.07 = [0.56, 0.70]$ for a confidence interval.
    - The choice $z = 2$ gives $0.63 \pm 0.14 = [0.49, 0.77]$ for a confidence interval.

- The choice $z = 3$ gives $0.63 \pm 0.21 = [0.42, 0.84]$ for a confidence interval.

It can be shown that using $z = 1$ gives only 68.3 percent confidence, which is too low, $z = 2$ gives 95.4 percent, and $z = 3$ gives 99.7 percent. It is generally agreed that the extra increase in confidence achieved by using $z = 3$ instead of $z = 2$ is not worth increasing h by 50 percent. (See, however, Section 6.6 for an exceptional case in which $z = 3$ may be preferred.) Thus, $z = 2$ is the popular choice, and many people use $z = 2$ and call it a 95 percent confidence interval, without worrying about the other 0.4 percent.

3. In the media, survey results are reported somewhat differently. One might read or hear, "Fifty-five percent of Americans support the president on this issue, with a margin of error of 3 percent." The numbers quoted above are $\hat{p}$ (0.55) and h (0.03) for a 95 percent confidence interval. If the results of several dichotomous responses are being reported, typically all of the $\hat{p}$'s and a single value for the sampling error are reported. This, of course, is incorrect; if there are different sample proportions or if they are based on different sample sizes, the sampling error of each sample proportion should be reported. Usually, the media simply report the largest of the individual sampling errors.

EXERCISES 6.3

1. A researcher selects 100 subjects at random from a population, observes 50 successes, and computes five confidence intervals. The five confidence levels are

    80%, 90%, 95%, 98%, and 99%,

    and the five intervals are

    $0.402 \leq p \leq 0.598$, $0.371 \leq p \leq 0.629$, $0.418 \leq p \leq 0.582$, $0.436 \leq p \leq 0.564$, and $0.384 \leq p \leq 0.616$.

    Match each interval with the correct level. (Hint: No computations are needed.)

2. Three researchers, Robin, Sally, and Teresa, independently select random samples from the same population. The sample sizes are 1,000 for Robin, 4,000 for Sally, and 250 for Teresa. Each researcher constructs a 95 percent confidence interval for her data. The half-widths of the three intervals are 0.015, 0.031 and 0.062. Match each half-width with its researcher. (Hint: No computations are needed.)

3. Daniel and Victor each select random samples of size 1,000 from different populations and construct 95 percent confidence intervals. The half-width of Daniel's interval is 0.030, and the half-width of Victor's interval is 0.025. Given that one of the values $\hat{p}$ is 0.20 and the other is 0.40, match the researcher with his value of $\hat{p}$. (Hint: No computations are needed.)

4. Monica and Sharon each select random samples of size 4,000 from different populations and construct 95 percent confidence intervals. The half-width of Monica's interval is 0.011, and the half-width of Sharon's interval is 0.015. Given that one of the values $\hat{p}$ is 0.15 and the other is 0.35, match the researcher with her value of $\hat{p}$. (Hint: No computations are needed.)

6.4 HYPOTHESIS TESTING

Recall from Chapter 2 (p. 42) that the four steps to a hypothesis test are

1. The hypotheses
2. The test statistic and its sampling distribution
3. The rule of evidence
4. The P-value

Following the presentation of the previous sections, the data are summarized by the sample size n and the observed number of successes x. Point and confidence interval estimates provide useful insight into the value of p. It is also instructive to conduct a hypothesis test if the following condition is satisfied:

> Of all the possible values for p there is one value, denoted by p_0, which, for some reason, is of special interest to the researcher.

Throughout this section the ideas of hypothesis testing are presented for a general problem involving one dichotomous variable, and they are illustrated with an extrasensory perception (ESP) experiment, described in the following example. The exercises will provide other examples of applications. Section 6.4.1 discusses the general case in which p_0 can take on general values, and Section 6.4.2 discusses the specific case in which $p_0 = 0.5$.

6.4.1 General Case

EXAMPLE 6.8: A Bernoulli Trials Model for ESP Testing

A researcher wants to investigate whether a metasubject (person) has ESP. A commonly used procedure consists of a sequence of trials. At each trial the researcher selects one card at random from a deck of five cards and the metasubject tries to determine, without sensory input, which of the five cards was selected. A correct determination is labeled a success and an incorrect one, a failure. It seems reasonable to tentatively assume the model of Bernoulli trials for this procedure, but in order to avoid prejudging the metasubject, the researcher assumes that p is unknown. Certain possible values of p have natural interpretations:

- If $p = 1$, the metasubject has perfect ESP.
- If $p = 0$, the metasubject has perfect reverse ESP.
- If $p = 0.20$, the metasubject performs as if guessing.
- If $p > 0.20$, the metasubject possesses (some) ESP.
- If $p < 0.20$, the metasubject possesses (some) reverse ESP. ▲

Step 1: The Hypotheses First, the researcher must determine the special value of interest, p_0. For example, suppose one wants to investigate whether a metasubject has ESP, using the procedure described above. The value of p is unknown, but the value 0.20 is particularly interesting, because if the metasubject is guessing, $p = 0.20$. Thus, for this problem, $p_0 = 0.20$ is the special value of interest. There are several points worth noting:

1. The choice of the special value of interest is the decision of the researcher; sometimes, as in the ESP example, the choice is natural, but other times it is not.
2. The special value of interest, p_0, is always known to the researcher. In contrast, the value of the parameter, p, is unknown.
3. The special value is chosen during the planning stage of the experiment, *before data are collected*. Its choice may be based upon theory or experience, but not on an examination of the data!

Once the special value of interest, p_0, has been determined, one specifies the hypotheses. The null hypothesis is that p equals the special value of interest, that is, $p = p_0$. There are three choices for the alternative hypothesis, namely $p > p_0$, $p < p_0$, or $p \neq p_0$. Following the terminology introduced in Section 2.2, we refer to these three alternatives as the first, second, and third alternatives, respectively. Combining the three choices for H_1 with the one choice for H_0 gives the following three choices for the pair of hypotheses:

$$H_0: p = p_0 \qquad H_0: p = p_0 \qquad H_0: p = p_0$$
$$H_1: p > p_0 \qquad H_1: p < p_0 \qquad H_1: p \neq p_0$$

For the ESP study, $p_0 = 0.20$, giving the following three choices for hypotheses:

$$H_0: p = 0.20 \qquad H_0: p = 0.20 \qquad H_0: p = 0.20$$
$$H_1: p > 0.20 \qquad H_1: p < 0.20 \qquad H_1: p \neq 0.20$$

Some of the considerations involved in the choice of an appropriate alternative hypothesis were discussed in Section 2.2. For ease of exposition, say that the choice of alternative reflects the purpose of the study. For example, in the ESP study suppose the purpose is to investigate whether the metasubject has ESP; then the choice of alternative hypothesis is $p > 0.20$. If the purpose is to investigate whether the metasubject has reverse ESP, the alternative is $p < 0.20$. Finally, if the purpose is to investigate whether the metasubject's performance is different (either better or worse) from guessing, $p \neq 0.20$ is the appropriate alternative hypothesis.

Step 2: The Test Statistic and Its Sampling Distribution Recall the following definition, given in Section 2.3:

> The test statistic is the number that summarizes the information in the data that is relevant to the problem of deciding between the hypotheses.

For a finite population, p is the proportion of successes in the population. Since the hypotheses depend on p, which is unknown, it is natural to summarize the data by computing the proportion of successes in the sample, $\hat{p}$. Similarly, for Bernoulli trials, p is the probability of success on any trial, and it is natural to summarize the data by computing the proportion of trials that yield successes, again $\hat{p}$. Thus, the natural choice for the test statistic is the random variable $\hat{P}$ that takes on the value $\hat{p}$. Working with $X = n\hat{P}$ is sometimes easier than with $\hat{P}$; the integer values of X are more convenient than the fractional values of $\hat{P}$, and all computer programs I know present probabilities for X. The results of this chapter are sometimes presented in terms of $\hat{P}$ so that their derivations are clear, but for purposes of computation, they are always presented in terms of X.

The researcher needs to know the sampling distribution of the test statistic assuming that the null hypothesis is true. From Section 5.2, X has the Bin(n, p) sampling distribution. If the null hypothesis is true, then $p = p_0$ and the sampling distribution of X is the Bin(n, p_0) sampling distribution. For example, suppose a researcher performs the ESP procedure described earlier with $n = 100$ trials. Then the null sampling distribution of X is Bin$(100, 0.2)$.

Step 3: The Rule of Evidence There are three rules of evidence, one for each possible alternative. Figure 6.5 presents the rules of evidence visually.

The first alternative states that the unknown p is larger than the specified p_0. The natural rule of evidence is

> The larger the value of $\hat{P}$, the stronger is the evidence in support of the alternative hypothesis that p is larger than p_0.

Since there is an increasing relationship between X and $\hat{P}$, this rule can be written as

> The larger the value of X, the stronger is the evidence in support of the alternative hypothesis that p is larger than p_0.

For example, consider the ESP procedure with $n = 100$ trials. Two possible outcomes are $X = 24$ and $X = 25$ successes. The rule of evidence that states that $X = 25$ provides stronger evidence that the subject has ESP than $X = 24$. Neither of these outcomes provides particularly convincing evidence in support of the alternative—see the computation of the P-value below—but $X = 25$ provides stronger evidence in support of the alternative than $X = 24$.

By similar reasoning, the rule of evidence for the second alternative is

> The smaller the value of $\hat{P}$, the stronger is the evidence in support of the alternative hypothesis that p is smaller than p_0.

Again, this rule can be written in terms of X

> The smaller the value of X, the stronger is the evidence in support of the alternative hypothesis that p is smaller than p_0.

For example, consider the ESP procedure with $n = 100$ trials. Two of the possible outcomes are $X = 11$ and $X = 10$ successes. The rule of evidence states that

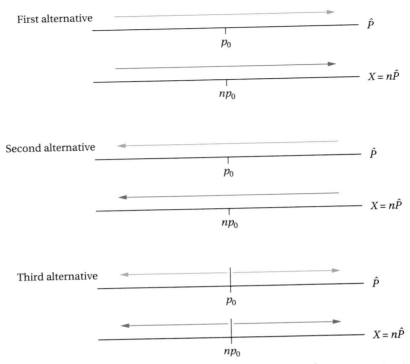

FIGURE 6.5 The rules of evidence in terms of either $\hat{P}$ or X. The horizontal axes indicate the values of the test statistic, and the arrows point in the direction of stronger evidence in support of the alternative.

$X = 10$ provides stronger evidence that the subject has reverse ESP than $X = 11$. Either of these outcomes provides statistically significant evidence in support of the alternative—see the computation of the P-value below.

There is some disagreement among statisticians about the rule of evidence for the third alternative, that $p \neq p_0$. The rule presented in this text gives answers similar to those of the other rules.

Recall that the expected value of $\hat{P}$ is p. If the null hypothesis is true, the expected value of $\hat{P}$ is p_0. The rule of evidence for the third alternative is

> The further the value of $\hat{P}$ from p_0, in either direction, the stronger is the evidence in support of the alternative hypothesis that p is different from p_0.

Since $X = n\hat{P}$, this rule can be written as

> The further the value of X from np_0, in either direction, the stronger is the evidence in support of the alternative hypothesis that p is different from p_0.

This last rule can be a bit confusing. Consider the ESP procedure with $n = 100$ trials; $np_0 = 100(0.2) = 20$, and the rule of evidence is

The further X is from 20, in either direction, the stronger is the evidence in support of the alternative hypothesis that p is different from 0.2.

For example, the values $X = 15$ and $X = 25$ are the same distance from 20; thus, they provide the same evidence in support of the third alternative. Values of X larger than 25 or smaller than 15 provide stronger evidence in support of the third alternative than either 15 or 25.

Step 4: The P-value Recall from Section 2.5 that the P-value is defined to be the (null) probability of the following event:

The test statistic achieves a value that gives either the same evidence as x in support of the alternative hypothesis or stronger evidence than x in support of the alternative hypothesis.

The P-value is computed under the assumption that the null hypothesis is true; for the current class of problems, this implies that probabilities for X are computed by using the binomial distribution with parameters n and p_0. Recall that p_0 is always known to the researcher; thus, under the assumption that the null hypothesis is true, probabilities can be computed for the test statistic X. The computation of the P-value will be demonstrated for ESP exams on several fictional persons.

EXAMPLE 6.9

Alice is given a 100-trial ESP exam. The purpose of the study is to investigate whether she performs better than she would by guessing. Thus, the hypotheses are

$$H_0: p = 0.20 \quad \text{and} \quad H_1: p > 0.20.$$

Suppose that Alice scores 28 correct on her exam. The rule of evidence states that any value of X larger than 28 would be stronger evidence than Alice's score; thus, the P-value is $P(X \geq 28)$, for $X \sim \text{Bin}(100, 0.2)$. Table 6.4 contains selected probabilities for the Bin(100,0.2) distribution. Reading from this table,

$$\mathbf{P} = P(X \geq 28) = 0.0342.$$

Alice's score, 28, is very strong evidence in support of the alternative because the probability is only 0.0342 of obtaining the same or stronger evidence. Alice's data are statistically significant for the first alternative. ▲

EXAMPLE 6.10

Brad is given a 100-trial ESP exam. The purpose of the study is to investigate whether he performs worse than he would by guessing. Thus, the hypotheses are

$$H_0: p = 0.20 \quad \text{and} \quad H_1: p < 0.20.$$

TABLE 6.4 Selected Probabilities for $X \sim \text{Bin}(100, 0.2)$.

x	$P(X \leq x)$	$P(X \geq x)$	x	$P(X \leq x)$	$P(X \geq x)$
5	0.0000	1.0000	22	0.7389	0.3460
6	0.0001	1.0000	23	0.8109	0.2611
7	0.0003	0.9999	24	0.8686	0.1891
8	0.0009	0.9997	25	0.9125	0.1314
9	0.0023	0.9991	26	0.9442	0.0875
10	0.0057	0.9977	27	0.9658	0.0558
11	0.0126	0.9943	28	0.9800	0.0342
12	0.0253	0.9874	29	0.9888	0.0200
13	0.0469	0.9747	30	0.9939	0.0112
14	0.0804	0.9531	31	0.9969	0.0061
15	0.1285	0.9196	32	0.9984	0.0031
16	0.1923	0.8715	33	0.9993	0.0016
17	0.2712	0.8077	34	0.9997	0.0007
18	0.3621	0.7288	35	0.9998	0.0003
19	0.4602	0.6379	36	0.9999	0.0002
20	0.5595	0.5398	37	1.0000	0.0001
21	0.6540	0.4405	38	1.0000	0.0000

Suppose that Brad scores 16 correct on his exam. The rule of evidence states that any value of X smaller than 16 would be stronger evidence than Brad's score; thus, the P-value is $P(X \leq 16)$, for $X \sim \text{Bin}(100, 0.2)$. Reading from Table 6.4,

$$\mathbf{P} = P(X \leq 16) = 0.1923.$$

Brad's data are not statistically significant for the second alternative. ▲

EXAMPLE 6.11

Carlos is given a 100-trial ESP exam. The purpose of the study is to discover whether he performs differently than he would by guessing. Thus, the hypotheses are

$$H_0: p = 0.20 \quad \text{and} \quad H_1: p \neq 0.20.$$

Suppose that Carlos scores 29 correct on his exam. In order to obtain the P-value, first compute

$$np_0 = 100(0.2) = 20,$$

the expected value of the test statistic if the null hypothesis is true. The observed value of the test statistic, 29, is 9 larger than its null expected value (20); thus, by the rule of evidence it provides the same evidence in support of the alternative hypothesis as the value 9 units below 20, which is $20 - 9 = 11$. Any value of the test statistic greater than or equal to 29 or less than or equal to 11 provides evidence in support of the alternative equal to or stronger than the evidence actually obtained. Therefore, from Table 6.4,

$$\mathbf{P} = P(X \geq 29) + P(X \leq 11) = 0.0200 + 0.0126 = 0.0326.$$

Carlos's data are statistically significant for the third alternative. ▲

Obtaining the Exact P-value The methods of the foregoing three examples can be generalized to any hypothesis test of the value of p. Let X denote the test statistic and x its observed value. The P-value can be obtained from the following two steps:

1. A table of $\text{Bin}(n, p_0)$ probabilities is needed. This table can be obtained with a computer program (for example, Minitab), from a published source (for example, [7]), or, if n is small, with a calculator. It is helpful if the table presents probabilities in the forms

$$P(X \geq x) \quad \text{and} \quad P(X \leq x).$$

2. Select the appropriate formula from the following:

For the first alternative: $\quad \mathbf{P} = P(X \geq x)$
For the second alternative: $\quad \mathbf{P} = P(X \leq x)$
For the third alternative: $\quad \mathbf{P} = P(X \geq x) + P(X \leq 2np_0 - x) \quad$ if $x > np_0$
$\qquad\qquad\qquad\qquad\qquad\quad \mathbf{P} = 1 \qquad\qquad\qquad\qquad\qquad\qquad$ if $x = np_0$
$\qquad\qquad\qquad\qquad\qquad\quad \mathbf{P} = P(X \leq x) + P(X \geq 2np_0 - x) \quad$ if $x < np_0$

Obtaining an Approximate P-value One can also use the standard normal curve to obtain an approximate P-value. Define the symbol q_0 as $q_0 = 1 - p_0$. The general guideline for using the standard normal curve approximation is that both np_0 and nq_0 are greater than or equal to 15. The method will be illustrated for the data given above for Alice, Brad, and Carlos, and then general formulas will be given.

Recall that for the test statistic X,

$$E(X) = np \quad \text{and} \quad SD(X) = \sqrt{npq}.$$

If the null hypothesis is true, p and q can be replaced in these equations by p_0 and q_0, respectively, yielding

$$E(X) = np_0 \quad \text{and} \quad SD(X) = \sqrt{np_0 q_0}.$$

Thus, if the null hypothesis is true, the standardized version of X is

$$Z = \frac{X - np_0}{\sqrt{np_0 q_0}}. \tag{6.9}$$

For example, consider the ESP procedure with $n = 100$ trials. If the null hypothesis is true, then the test statistic has expected value

$$np_0 = 100(0.2) = 20$$

and standard deviation

$$\sqrt{np_0 q_0} = \sqrt{100(0.2)(0.8)} = \sqrt{16} = 4.$$

EXAMPLE 6.12

Recall that Alice scored 28 correct on a 100-trial ESP exam. For the first alternative,

$$\mathbf{P} = P(X \geq 28).$$

This probability can be approximated by transforming X to its standardized version and then using the standard normal curve. First,

$$\mathbf{P} = P(X \geq 28) = P\left(\frac{X - 20}{4} \geq \frac{28 - 20}{4}\right) = P(Z \geq 2).$$

By the same argument given in Section 3.4, this last probability can be approximated by the area under the standard normal curve to the right of two. From Table A.2, this area equals 0.0228. For comparison, recall that the exact value of $\mathbf{P}$ is 0.0342. ▲

EXAMPLE 6.13

Recall that Brad scored 16 correct on his 100-trial ESP test. For the second alternative,

$$\mathbf{P} = P(X \leq 16) = P\left(\frac{X - 20}{4} \leq \frac{16 - 20}{4}\right) = P(Z \leq -1).$$

This last probability is approximated by the area under the standard normal curve to the left of -1. By symmetry, the area under the standard normal curve to the left of -1 equals the area under the curve to the right of $+1$; Table A.2 indicates that this latter area equals 0.1587. Thus, the approximate P-value equals 0.1587. For comparison, recall that the exact value of $\mathbf{P}$ is 0.1923. ▲

EXAMPLE 6.14

Recall that Carlos scored 29 correct on his 100-trial ESP test. For the third alternative,

$$\mathbf{P} = P(X \geq 29) + P(X \leq 11) = P\left(\frac{X - 20}{4} \geq \frac{29 - 20}{4}\right) + P\left(\frac{X - 20}{4} \leq \frac{11 - 20}{4}\right)$$

$$= P(Z \geq 2.25) + P(Z \leq -2.25).$$

These two terms are approximated by the areas under the standard normal curve to the right of 2.25 and to the left of -2.25, respectively. By symmetry, these areas are identical; thus, the approximation equals twice the area under the standard normal curve to the right of 2.25. Reading from Table A.2, the approximation is $2(0.0122) = 0.0244$. For comparison, recall that the exact value of $\mathbf{P}$ is 0.0326. ▲

One can obtain the standard normal curve approximation to the P-value for any problem by following the appropriate technique from one of the three previous examples. The method consists of the following two steps. First, let x denote the observed value of the test statistic X, and let

$$z = \frac{x - np_0}{\sqrt{np_0 q_0}}. \tag{6.10}$$

Note that z is the observed value of Z given in Equation 6.9. Second, the value of z gives the approximate P-value:

- For the first alternative, $p > p_0$, the approximate P-value is the area under the standard normal curve to the right of z.
- For the second alternative, $p < p_0$, the approximate P-value is the area under the standard normal curve to the right of $-z$.
- For the third alternative, $p \neq p_0$, the approximate P-value is twice the area under the standard normal curve to the right of $|z|$.

6.4.2 The Special Case of $p_0 = 0.5$

The case in which the special value of interest is equal to 0.5 is of special importance. Several hypothesis-testing problems introduced in this book are mathematically equivalent to this case.

If $p_0 = 0.5$, the three pairs of hypotheses are

$$\begin{array}{lll} H_0: p = 0.50 & H_0: p = 0.50 & H_0: p = 0.50 \\ H_1: p > 0.50 & H_1: p < 0.50 & H_1: p \neq 0.50 \end{array}$$

If $n \leq 14$, the P-value can be obtained immediately from Table A.4 for any alternative. Simply enter the table at $(R = n)$ and $(O = x)$ and read the P-value from the appropriate column.

EXAMPLE 6.15

A researcher wants to test the null hypothesis that $p = 0.5$ versus the alternative $p > 0.5$. A random sample of $n = 10$ yields $x = 8$ successes. Entering Table A.4 at $R = 10$ and $O = 8$ gives **P**= 0.055; the data are almost statistically significant. ▲

If $n > 14$, a table of binomial probabilities is needed for the exact P-value or the standard normal curve approximation can be used. If the researcher uses the standard normal curve to approximate the P-value, the formula for z can be rewritten as follows

$$z = \frac{x - np_0}{\sqrt{np_0 q_0}} = \frac{x - n(0.5)}{\sqrt{n(0.5)(0.5)}} = \frac{x - n(0.5)}{0.5\sqrt{n}} \times \frac{2}{2} = \frac{2x - n}{\sqrt{n}}.$$

Thus, *if $p_0 = 0.5$*, the observed value of the standardized version of the test statistic can be written as

$$z = \frac{2x - n}{\sqrt{n}}. \qquad (6.11)$$

EXAMPLE 6.16

A researcher wants to test the null hypothesis that $p = 0.5$ versus the alternative $p < 0.5$. A random sample of $n = 100$ yields $x = 39$ successes. Substituting into Equation 6.11 gives

$$z = \frac{2x - n}{\sqrt{n}} = \frac{2(39) - 100}{\sqrt{100}} = \frac{-22}{10} = -2.20.$$

According to the rule presented just below Equation 6.10, the approximate P-value is the area under the standard normal curve to the right of $-z = 2.20$. Reading from Table A.2, this area equals 0.0139. The data are statistically significant in support of the second alternative. ▲

The following example refers to material presented in Section 4.3, in the optional Chapter 4. If you skipped Chapter 4, you should proceed to the exercises at the end of this section.

EXAMPLE 6.17

It was stated above that Table A.4 gives the exact P-value for testing the null hypothesis that $p = 0.5$. The same table was used in Section 4.3 to obtain the exact P-value for McNemar's test. This example shows that this is no coincidence.

McNemar's test is used to compare two treatments with data obtained from a randomized pairs design. This presentation may prove a bit confusing because the use of the words *success* and *failure* and of the symbol n is different for McNemar's test than for the current section. First, the terminology of Chapter 4 is used. A study consists of n pairs of subjects with one member of each pair assigned, by randomization, to each treatment. The null hypothesis states that the two treatments are equally effective; the three possible alternatives state that the first treatment is superior to, inferior to, or different from the second treatment. There are four possible outcomes for each pair of subjects:

- A: Both subjects give successes.
- B: Only the subject on the first treatment gives a success.
- C: Only the subject on the second treatment gives a success.
- D: Neither subject gives a success.

Type A and D pairs give no insight into the relative merits of the treatments and are ignored by McNemar's test. Type B pairs give evidence that the first treatment is superior to the second, and type C pairs give evidence that the first treatment is inferior to the second. Recall that b and c denote the numbers of type B and type C pairs in the study, respectively. The exact P-value is obtained by entering Table A.4 at $R = b + c$ and $O = b$.

In the terminology of this chapter, let each type B or C pair be called a trial. Thus, the study consists of $n = b + c$ trials. These trials are dichotomous; let type B pairs be labeled successes and type C pairs be labeled failures. The total number of successes in the study is $x = b$. In addition, it seems reasonable to assume that the trials are independent. Suppose the Chapter 4, null hypothesis is true. Then each trial is equally likely to give a success or a failure. (If the treatments are identical, the treatment that wins is determined entirely by the chance process of randomization.) Thus, these trials are Bernoulli trials with $p = 0.5$.

Before turning to the alternatives, let us summarize what has been established. The null hypothesis for McNemar's test can be viewed as the null hypothesis $p = 0.5$ of this chapter with the identifications that n of this chapter equals $b + c$ of Chapter 4, and x of this chapter equals b of Chapter 4. Thus, the Chapter 4 directive to enter Table A.4 at $R = b + c$ and $O = b$ is in perfect agreement with the current chapter's directive to enter the table at $R = n$ and $O = x$. All that remains is to show that the alternatives from the two chapters match.

If the first alternative of Chapter 4 is true, the first treatment is superior to the second. Given that a pair of subjects yields exactly one (Chapter 4) success—that is, given that either pair type B or C occurs—it is more likely that type B occurs. In other words, a (Chapter 6) success is more likely than a (Chapter 6) failure, meaning $p > q$ or, equivalently, that $p > 0.5$. Thus, the first alternative for McNemar's test corresponds to the first alternative for this section's test. Similarly, the second and third alternatives for McNemar's test can be shown to correspond to the second and third alternatives of the test of this section, respectively.

Finally, for the standard normal curve approximation to the P-value, this section directs the researcher to compute

$$z = \frac{2x - n}{\sqrt{n}}.$$

Making the substitutions $x = b$ and $n = b + c$ gives

$$z = \frac{2b - (b + c)}{\sqrt{b + c}} = \frac{b - c}{\sqrt{b + c}},$$

which is in agreement with the result given in Chapter 4 for McNemar's test. ▲

EXERCISES 6.4

1. Daniel, Oscar, and Victor are given a 1,000-trial ESP test. Daniel scores 220 correct, Oscar scores 225 correct, and Victor scores 170 correct.
   (a) Who has the strongest evidence in support of the first alternative?
   (b) Who has the strongest evidence in support of the second alternative?
   (c) Who has the strongest evidence in support of the third alternative?

2. Henrietta, Monica, and Sharon are given a 500-trial ESP test. Henrietta scores 110 correct, Monica scores 105 correct, and Sharon scores 70 correct.
   (a) Who has the strongest evidence in support of the first alternative?
   (b) Who has the strongest evidence in support of the second alternative?
   (c) Who has the strongest evidence in support of the third alternative?

3. Pedro is given a 400-trial ESP test and scores 90 correct. Find the approximate P-value if
   (a) The alternative is that he has positive ESP.
   (b) The alternative is that he performs differently than he would by guessing.

4. Quinn is given a 1,000-trial ESP test and scores 160 correct. Find the approximate P-value if
   (a) The alternative is that he has positive ESP.
   (b) The alternative is that he performs differently than he would by guessing.

5. Test $H_0: p = 0.50$ versus $H_1: p > 0.50$ with $n = 10$. Find the P-value if the observed value of X is
   (a) $x = 7$.
   (b) $x = 9$.
   (c) $x = 5$.

6. Test $H_0: p = 0.50$ versus $H_1: p < 0.50$ with $n = 8$. Find the P-value if the observed value of X is
   (a) $x = 2$.
   (b) $x = 0$.
   (c) $x = 5$.

7. Test $H_0: p = 0.50$ versus $H_1: p \neq 0.50$ with $n = 8$. Find the P-value if the observed value of X is
   (a) $x = 2$.
   (b) $x = 0$.
   (c) $x = 5$.

8. Test $H_0: p = 0.50$ versus $H_1: p \neq 0.50$ with $n = 10$. Find the P-value if the observed value of X is
   (a) $x = 7$.
   (b) $x = 9$.
   (c) $x = 5$.

9. Refer to Exercise 3 on p. 193. Test the null hypothesis that the percentage of the U.S. population that approved of President Bush's handling of his job was equal to 0.50 versus the alternative that it was larger. (Hint: Recall that 469 of 756 sampled approved.)

10. Linda Owen recorded the gender of the first 100 babies born at Meriter Hospital in Madison, Wisconsin, beginning on November 1, 1991, and counted 52 females and 48 males. Assuming that these births represent Bernoulli trials, test the null hypothesis that the probability of a female baby is 0.50 versus the third alternative.

11. Exercise 9 on p. 177 introduced a study of Kristen Joiner's dog. In a sequence of 96 trials, Muffin chased her blue ball 57 times and her red ball 39 times. Assuming that the trials are Bernoulli trials, test the null hypothesis that the probability that Muffin chases the blue ball is 0.50 versus the third alternative.

12. Giselle Lederman went to a (free) campus telephone and started calling phone numbers in a haphazard manner. Her trials consisted of calls that were answered. The trial was labeled a success if a woman answered the phone and a failure if a man answered. She obtained 64 successes in a sample of size 100. Assuming that the trials are Bernoulli trials, test the null hypothesis that the probability that a female answers is 0.50 versus the alternative that the probability is larger than 0.50.

13. Data obtained using cytogenetic methods over the past 20 years suggest that the origin of the extra chromosome 21 in trisomy 21 (Down's syndrome) is maternal in 80 percent of the cases and paternal in 20 percent of the cases [8]. Unfortunately, in many cases, cytogenetic methods are not conclusive. Recent advances in using DNA polymorphisms have caused researchers to reexamine the 80 percent–20 percent split. In particular, data on 193 children with trisomy 21 yield 184 cases of maternal origin and 9 cases of paternal origin. Use these new data to test the null hypothesis that the origin is maternal in 80 percent of the cases versus the third alternative.

6.5 PREDICTIONS*

The material in this section is not used in the remainder of the book. The uninterested reader may skip ahead to Section 6.6. (Section 6.6 is also optional, so you may want to skip ahead to Chapter 7.)

The methods of this section are true for any random variable with a binomial sampling distribution; in practice, however, they usually are applied only to sequences of Bernoulli trials. Predictions deal with future experimentation. The researcher is planning to observe the value of a random variable, Y, that has a binomial distribution with parameters m and p. (Note that the sample size for Y is m, not n.) The goal is to predict the value of Y before it is observed. There are two types of predictions: point predictions and interval predictions. There are several analogies with point and interval estimation. A point prediction is a single number, whereas an interval prediction is a range of values. In most interesting problems, point predictions have a very low probability of being correct. Prediction intervals give up the precision of a point prediction to achieve a certain probability of being correct. There is, however, an important difference between estimation and prediction. With estimation the researcher never knows whether a point or interval estimate is correct (unless the entire population is studied), but for prediction the eventual observation of Y will reveal whether or not the prediction is correct.

In practice, p is usually unknown. It is easier, though, to present the results by first considering the case in which p is known.

Throughout this section assume that mp and mq are both at least 15, implying that the standardized version of the sampling distribution of Y can be well approximated by the standard normal curve.

6.5.1 The Case in Which p Is Known

The random variable Y has a binomial distribution; thus, $E(Y) = mp$.

The point prediction of the value of Y is simply its expected value, mp, rounded off to the nearest integer, if necessary.

EXAMPLE 6.18

A fair coin will be tossed 100 times. Let Y denote the number of heads obtained. The sampling distribution of Y is the binomial distribution with $m = 100$ and $p = 0.5$. The expected value of Y is $mp = 100(0.5) = 50$. Because the expected value is an integer, there is no need to determine the nearest integer. Thus, the point prediction of Y is 50.

The coin was tossed 100 times, and $Y = 46$ heads were obtained; thus, the point prediction is incorrect. It is not surprising that the point prediction is incorrect: the probability that $Y = 50$ can be shown to equal 0.0796. ▲

EXAMPLE 6.19

Teri will be given the ESP exam with $n = 100$. Predict the number of correct responses assuming that Teri is guessing.

Let Y denote the number of correct responses. The sampling distribution of Y is the binomial distribution with $m = 100$ and $p = 0.2$. The expected value of Y is $mp = 100(0.2) = 20$. Because the expected value is an integer, the point prediction of Y is 20.

With a computer playing the role of Teri, I performed the experiment and obtained $Y = 23$ correct responses; thus, the point prediction is incorrect. It is not surprising that the point prediction is incorrect, because the probability that $Y = 20$ can be shown to equal 0.0993. ▲

The point prediction of Y is simply its expected value. The interval prediction takes into account its variation as measured by the standard deviation. The result will be stated and illustrated, but not derived.

Let Y have a binomial distribution with known parameters m and p. A prediction interval for Y is obtained by computing

$$mp \pm z\sqrt{mpq}, \qquad (6.12)$$

where z is determined by the researcher's choice of the probability of a correct prediction; values of z are given in Table A.3.

Typically, the values computed with Formula 6.12 are not integers; each value should be rounded off to the nearest integer. The result is the prediction interval.

Formula 6.12 will be applied to the two examples given earlier in this section.

EXAMPLE 6.20

A fair coin will be tossed 100 times. Obtain a 95 percent prediction interval for the number of heads that will be obtained.

The choice of 95 percent gives $z = 1.96$, and the assumption of fairness implies that $p = 0.5$ and $q = 0.5$. Substituting these numbers and $m = 100$ into Formula 6.12 gives

$$mp \pm z\sqrt{mpq} = 100(0.5) \pm 1.96\sqrt{100(0.5)(0.5)} = 50 \pm 9.8,$$

which gives the values 40.2 and 59.8. After rounding, the 95 percent prediction interval is [40, 60].

It was reported earlier that the eventual value of Y was 46. Thus, the prediction interval is correct. ▲

EXAMPLE 6.21

Teri is given the ESP exam with 100 trials. Obtain a 90 percent prediction interval for the number of correct responses assuming that Teri is guessing.

The choice of 90 percent gives $z = 1.645$, and the assumption of guessing implies $p = 0.2$ and $q = 0.8$. Substituting these numbers and $m = 100$ into Formula 6.12 gives

$$mp \pm z\sqrt{mpq} = 100(0.2) \pm 1.645\sqrt{100(0.2)(0.8)} = 20 \pm 6.6,$$

which gives the values 13.4 and 26.6. After rounding, the 90 percent prediction interval is [13,27]. It was reported earlier that the eventual value of Y was 23. Thus, the prediction interval is correct. ▲

6.5.2 The Case in Which p Is Unknown

In this case the point prediction of the previous section, mp, cannot be used because p is unknown. If the researcher has no previous data, this problem cannot be solved. Suppose, therefore, that the researcher has previously observed n trials from the process under study and counted the number of successes in those trials. A set of previous data in a prediction problem is sometimes called a **training set.** Let x denote the number of successes in the training set; then x is the observed value of a random variable, X, that has a binomial distribution with n trials and probability of success on each trial given by p. Although the value of X does not reveal the exact value of p, it does provide the researcher with some objective information about the value of p. Let $\hat{p}$ represent the proportion of successes in the training set; that is, $\hat{p} = x/n$. The point prediction of Y is obtained by taking its expected value, $E(Y) = mp$, and replacing the unknown p by its estimate from the training set, $\hat{p}$.

In summary, the point prediction of Y is

$$m\hat{p} = m \cdot \frac{x}{n}. \qquad (6.13)$$

If this number is not an integer, round it off to the nearest integer.

EXAMPLE 6.22

Example 5.20 examined the daily high-temperature forecasts for Madison, Wisconsin, that appeared in the *Wisconsin State Journal* during the spring and summer of 1988. Recall that a forecast is labeled a success if it is within 2° of the actual high. The newspaper presented 178 predictions, which resulted in 96 successes and 82 failures. The assumptions of Bernoulli trials were tentatively adopted, and several inspections of the data suggested that those assumptions were reasonable. These data will serve as the training set, giving

$$n = 178, \quad x = 96, \quad \text{and} \quad \hat{p} = 96/178 = 0.539.$$

Suppose that on September 21, 1988, a researcher defined Y to equal the number of successes in the next 100 forecasts. The point prediction of Y is

$$m\hat{p} = 100(0.539) = 53.9;$$

rounding this answer gives 54 as the point prediction. Four months later, it was found that only 44 of those 100 predictions were successful. ▲

Below, a formula is given for the prediction interval for Y when p is unknown. The formula can be derived by using some algebra and simple properties of the variance, but the proof is beyond the scope of this text.

Let Y have a binomial distribution with known parameter m and unknown parameter p. Assume that there is a training set (previous data on the process) that yielded x successes in n trials, with x and $n - x$ each greater than or equal to 15. Define

$$\hat{p} = \frac{x}{n}.$$

A prediction interval for the value of Y is obtained by computing

$$m\hat{p} \pm z\sqrt{m\hat{p}\hat{q}}\sqrt{1 + \frac{m}{n}}, \tag{6.14}$$

where z is determined by the researcher's choice of the probability of a correct prediction; as has been mentioned before, values of z are given in Table A.3. Typically, the values computed with Formula 6.14 are not integers. As in the case in which p is known, these values should be rounded off to the nearest integer to give the prediction interval.

Use of this formula is illustrated below.

EXAMPLE 6.23

Refer to the previous example. Find a 95 percent prediction interval for the number of successes in the next 100 forecasts.

The value of p is unknown, but there is a training set for which $x = 96$, $n = 178$, and $\hat{p} = 0.539$. The choice of 95 percent implies that $z = 1.96$. Substituting these values and $m = 100$ into Formula 6.14 gives

$$m\hat{p} \pm z\sqrt{m\hat{p}\hat{q}}\sqrt{1 + \frac{m}{n}} = 100(0.539) \pm 1.96\sqrt{100(0.539)(0.461)}\sqrt{1 + \frac{100}{178}}$$

$$= 53.9 \pm 1.96(4.98)(1.25) = 53.9 \pm 12.2,$$

which gives the values 41.7 and 66.1. After rounding, the 95 percent prediction interval for Y is [42, 66]. Recall that the observed value of Y was 44; the prediction interval is correct. ▲

EXERCISES 6.5

1. A person says

    > The law of averages states that if I toss a fair coin 10,000 times, the number of heads obtained will be extremely close to 5,000.

    Construct a 95 percent prediction interval for the number of heads in 10,000 tosses of a fair coin. What does your answer say about the above interpretation of the law of averages?

2. Exercise 6 on p. 177 described a study in which Eric Statz tossed popcorn to his dog Boone. In the first 50 trials, Boone obtained 33 successes. On the assumption of Bernoulli trials, use these data to answer the following questions about the last 50 trials.

    (a) What is the point prediction of the number of successes Boone would achieve in the last 50 trials?

    (b) Construct a 90 percent prediction interval for the number of successes Boone would achieve in the last 50 trials.

    (c) Given that Boone actually obtained 38 successes in the last 50 trials, comment on your previous two answers.

3. Give the point prediction and 90 percent prediction intervals for the number of female babies among the next 200 babies born at Meriter Hospital in Madison, Wisconsin.

    (a) Assume that each birth is equally likely to be a boy or a girl.

    (b) Assume that the probability of a girl is unknown, but that there is a training set of 52 female and 48 male births.

4. Exercise 12 on p. 214 described a study in which the gender of the person answering a phone was determined for 100 phone calls. Sixty-four calls were answered by women and 36 by men. Suppose that the researcher made an additional 75 phone calls.

    (a) What is the point prediction of the number of the 75 calls that would be answered by a female?

    (b) Construct a 80 percent prediction interval for the number of the 75 calls that would be answered by a female.

5. Exercise 9 on p. 177 described a study in which Kristen Joiner gave her dog two choices of a ball to chase. In the first 48 trials, Muffin chased the blue ball 27 times and the red ball 21 times. Assuming that the trials were Bernoulli trials, use these data to answer the following questions about the last 48 trials:

    (a) What is the point prediction of the number of times Muffin would chase the blue ball in the last 48 trials?

    (b) Construct a 90 percent prediction interval for the number of times Muffin would chase the blue ball in the last 48 trials.

    (c) Given that Muffin actually chased the blue ball 30 times in the last 48 trials, comment on your previous two answers.

6. Exercise 10 on p. 177 described a study in which Denise Howsare threw 100 darts. In the first 50 trials, Denise obtained 24 successes. On the assumption of Bernoulli trials, use these data to answer the following questions about the last 50 tosses:

    (a) What is the point prediction of the number of successes Denise would obtain in the last 50 trials?

    (b) Construct a 90 percent prediction interval for the number of successes Denise would obtain in the last 50 trials.

    (c) Given that Denise actually obtained 24 successes in the last 50 trials, comment on your previous two answers.

7. Exercise 15 on p. 178 asked you to perform an experiment with 100 trials and a dichotomous response.
   (a) Use your results to compute a point prediction of and a 90 percent prediction interval for the number of successes you will achieve if you perform 100 more trials.
   (b) Perform 100 more trials, count the number of successes, and compare the number with your prediction. Comment.

6.6 HOW MANY ANALYSES?*

This section is optional, so the uninterested reader may proceed to Chapter 7.

Most statisticians believe that the interpretation given to a confidence interval or a P-value should be modified if more than one set of data or more than one variable per subject are analyzed. There are two classes of problems, independent analyses and dependent analyses; examples of each follow.

EXAMPLE 6.24: Independent Analyses

In earlier examples, a 100-trial ESP exam was given to fictional subjects Alice and Brad. It is reasonable to assume that knowledge of Alice's performance gives no information about Brad's performance. Because of this assumption, it is said that any analysis of Alice's data is **independent** of any analysis of Brad's data. ▲

Mathematically, independent analyses can arise from random samples from different population boxes.

EXAMPLE 6.25: Dependent Analyses

These ideas apply to any survey; the ideas are illustrated for the 1986 Wisconsin Driver Survey. Recall that question 1 measured drivers' attitudes on the seriousness of drunk driving. Question 14 asked the respondent to report how often he or she drank alcoholic beverages.

Based on the sample of 1,677 drivers one can analyze the responses to question 1 and the responses to question 14. In contrast to the previous example, here it is *not* reasonable to assume that knowing the sample results on one of these questions gives no information about the sample results on the other. For example, perhaps nondrinkers are more likely than heavy drinkers to perceive the drunk-driving problem as extremely serious. Thus, if by chance nondrinkers are overrepresented in the random sample, it is likely that persons who believe that the problem of drunk driving is extremely serious are overrepresented, as well. Of course, the above conjecture on the relationship between drinking and attitude may be incorrect. The point is that it is not reasonable to assume there is no relationship; the presence or absence of a relationship can and should be decided by looking at data as illustrated in Sections 8.3 and 8.4. (Incidentally, the data overwhelmingly support the conjecture.) Statisticians say that analysis of question 1 is not independent of (or is **dependent** upon) analysis of question 14. ▲

Mathematically, dependent analyses can arise from one random sample of cards from one box with each card containing the values of many variables. The different analyses correspond to examining different variables.

6.6.1 Hypothesis Testing, Independent Analyses

This is a fairly complicated topic. The ideas are introduced for a variation on a familiar problem.

Suppose that each of 200 students is given a 100-trial ESP exam. For each student the null hypothesis, that $p = 0.20$, is tested against the first alternative, that $p > 0.20$. Thus, there will be a total of 200 hypothesis tests. These 200 analyses are independent analyses as defined earlier in this section. Here are results for the experiment of testing 200 students, with the P-values obtained from Table 6.4 (p. 208):

- One student scored 32 correct, giving **P** = 0.0031.
- One student scored 30 correct, giving **P** = 0.0112.
- One student scored 29 correct, giving **P** = 0.0200.
- Three students scored 28 correct, giving **P** = 0.0342.
- The remaining 194 students scored 27 or fewer correct, giving a variety of P-values, all of which are larger than 0.05.

Six of the 200 tests have given results that are statistically significant. But, before concluding that these six students have ESP, consider the following argument.

Assume for the moment that all 200 students were guessing. Each student's exam score can be viewed as a single Bernoulli trial with success defined as a score of 28 or more and a failure defined as a score of 27 or fewer. From Table 6.4, the probability of success is 0.0342. Let Y denote the number of students who score 28 or more on their ESP exam. Note that Y equals the number of students who achieve statistically significant results. Clearly, Y has a binomial distribution with $n = 200$ trials and $p = 0.0342$. Thus, $E(Y) = 200(0.0342) = 6.84$. In words, assuming that every one of the 200 students is guessing, the expected value of Y is 6.84. Thus, it is not at all remarkable that six people would obtain statistically significant results. A researcher might reasonably conclude that the experimental results do not discredit the assumption that everyone was guessing.

Whereas it would be unwise to conclude, as suggested earlier, that six students have ESP, it is unfair to conclude that no one has ESP. *Perhaps* one student, for example the student who scored 32 correct, has ESP; it is not fair to deny that accomplishment simply because the student had the misfortune of being tested with 199 persons who were guessing! Ideally, the experimenter would give another ESP exam to each of the six persons who obtained the statistically significant results. The second exams would show one of the following:

- One or more persons continue to perform significantly better than they would by guessing.
- All six subjects' performances deteriorate to something indistinguishable from guessing (or worse).

If further examination is not possible, the results of the original study should be interpreted with a great deal of caution.

If a researcher stops the above analysis with the conclusion that six students have ESP without attempting, even informally, to adjust for the fact that these six students were screened from a much larger group of subjects, then that researcher is said to be committing the **focusing-on-a-winner-fallacy.** The fallacy occurs whenever the researcher performs a number of tests and interprets the significant tests as if they were the only ones performed. The fallacy has many guises, as will be demonstrated by examples throughout this book.

The above ESP exam results, which, incidentally, were generated by Minitab with each student guessing, can also demonstrate another example of the fallacy. Suppose that instead of giving the exam to 200 different students, the same student took the exam 200 times. An unscrupulous researcher might report only the most favorable result and state

> On a 100-trial ESP exam, the student scored 32 correct to yield a P-value of 0.0031.

This researcher is focusing on the best outcome (the winner) and ignoring the others.

Sometimes the statistician must be a detective to discover the presence of the fallacy. For example, suppose a person states

> I took the 100-trial ESP exam and scored 32 correct; my P-value is 0.0031, and I am convinced that I have ESP!

Before drawing any conclusions, it would be wise to determine whether the exam was given to anyone else or whether this person took the exam more than once.

There is a formula that is particularly useful for adjusting for the focusing-on-a-winner fallacy. It can be applied to any problem involving independent analyses, not just hypothesis testing, so it will be stated in some generality.

Suppose that cards will be selected from a box to determine whether or not an event A occurs. Let the probability that the event A occurs be denoted by $P(A)$. Next, suppose that there are $(m - 1)$ other boxes, each identical to the first, and that from each box the same experiment will be performed to see whether or not A occurs. (Thus, the basic experiment is performed m times.)

To fix ideas, consider the ESP experiment described above. On the assumption that a student is guessing, the ESP exam can be modeled as the selection of 100 cards at random with replacement from a dichotomous box with $p = 0.2$ and $q = 0.8$. Let A be the event that the student scores 28 or more correct. As shown earlier, $P(A) = 0.0342$. If 200 students are taking the ESP exam and they are all guessing, the entire examination of 200 students can be modeled as the selection of 100 cards at random with replacement from each of $m = 200$ identical boxes.

Returning to the general framework, let B be the event that the event A occurs at least once in the m different experiments. It can be shown that

$$P(B) = 1 - [1 - P(A)]^m. \qquad (6.15)$$

The use of this equation will be demonstrated for the ESP exam on 200 subjects. Let A be the event that a particular subject scores 28 or more correct; $P(A) = 0.0342$. For $m = 200$ subjects, the probability that at least one subject scores 28 or more correct is

$$P(B) = 1 - (1 - 0.0342)^{200} = 1 - (0.9658)^{200}.$$

With a scientific calculator this can be evaluated as

$$1 - (0.9658)^{200} = 1 - 0.0009 = 0.9991.$$

Thus, even if every subject is guessing, there is a 99.9 percent chance that at least one subject will obtain statistically significant results. Next, let A be the event that a subject scores 32 or more correct; for a particular subject, $P(A) = 0.0031$. For $m = 200$ subjects, the probability that at least one subject scores 32 or more correct is

$$P(B) = 1 - (1 - 0.0031)^{200} = 1 - (0.9969)^{200} = 1 - 0.5374 = 0.4626.$$

Thus, even if every student is guessing, there is a 46 percent chance that at least one student will score 32 or more correct.

6.6.2 Hypothesis Testing, Dependent Analyses

Suppose that a random sample of subjects is selected from a population and each subject completes an m-item questionnaire, where m can be any positive integer. Suppose further that a hypothesis test is performed for each item, yielding m values of **P**. Let Y denote the number of times the test results are statistically significant, that is, the number of times that the P-value is less than or equal to 0.05. For example, perhaps $m = 200$ tests are performed and $Y = 9$ of them are statistically significant.

As with independent analyses, the usual goal of the analyst is to consider how Y would behave if, in fact, all null hypotheses were correct. For the independent analyses case, quite informative results, including Equation 6.15, were obtained because the multiplication rule could be used. For dependent analyses the multiplication rule cannot be used; there is, however, one useful result:

$$E(Y) \approx (0.05)m. \tag{6.16}$$

This display contains an approximation sign, $\approx$, instead of an equal sign because the probability of a statistically significant result is only approximately 0.05. In fact, if the researcher uses an exact test, the probability of obtaining a statistically significant result is usually less than 0.05. (For example, for a 100-trial ESP test the data are statistically significant if, and only if, the test statistic equals or exceeds 28, an event shown to have probability 0.0342.)

Consider the example given above with $m = 200$ and $Y = 9$. On the assumption that all 200 null hypotheses are true,

$$E(Y) \approx (0.05)m = 0.05(200) = 10.$$

Thus, obtaining nine significant results does not discredit the assumption that all 200 null hypotheses are correct.

6.6.3 Confidence Intervals

Both the independent and dependent analyses situations will be considered in this section. Suppose the researcher plans to compute m confidence intervals, each with the same confidence level. Let α represent the probability that any particular confidence interval is incorrect. For example, a 95 percent confidence interval gives $\alpha = 0.05$, a 99 percent confidence interval gives $\alpha = 0.01$, and a 99.74 percent confidence interval (corresponding to $z = 3$ in the formula for a confidence interval) gives $\alpha = 0.0026$.

The following two results are true for either analysis type.

Suppose m confidence intervals are to be constructed, each with the same probability, α, of being incorrect. Let Y denote the number of confidence intervals that will be incorrect. Then

$$E(Y) = \alpha m. \tag{6.17}$$

The second result is called Bonferroni's result.

Suppose that m confidence intervals are to be constructed, each with the same probability, α, of being incorrect. The probability that all m confidence intervals will be correct is at least

$$1 - \alpha m.$$

If the confidence intervals are computed from independent analyses, Bonferroni's result can be refined:

Suppose that m confidence intervals are to be constructed, each with the same probability, α, of being incorrect. If the analyses are independent, the probability that all m confidence intervals will be correct is equal to

$$(1 - \alpha)^m. \tag{6.18}$$

Following are some examples of the use of these three results.

EXAMPLE 6.26

Suppose that $m = 10$ confidence intervals are to be constructed, each with confidence level equal to 99 percent ($\alpha = 0.01$). Equation 6.17 shows that the expected number of incorrect intervals is

$$0.01 \times 10 = 0.1.$$

Bonferroni's result states that the probability that all 10 confidence intervals will be correct is at least

$$1 - 10 \times 0.01 = 1 - 0.10 = 0.90.$$

For this example, the latter result gives a more interesting answer. If the analyses are independent, the probability that all 10 confidence intervals will be correct is equal to

$$(1 - 0.01)^{10} = (0.99)^{10} = 0.9044.$$

Note the relationship between these last two answers: Bonferroni's result states that the probability is at least 0.90, and the latter states that it equals 0.9044. ▲

EXAMPLE 6.27

Suppose that $m = 100$ confidence intervals will be constructed, each with confidence level equal to 99 percent ($\alpha = 0.01$). Equation 6.17 states that the expected number of incorrect intervals is

$$0.01 \times 100 = 1.$$

Bonferroni's result states that the probability that all 100 confidence intervals will be correct is at least

$$1 - 100 \times 0.01 = 1 - 1 = 0.$$

This is a totally useless answer because probabilities are always at least 0. For this example the former result obviously gives a more interesting answer. If the analyses are independent, the probability that all 100 confidence intervals will be correct is equal to

$$(1 - 0.01)^{100} = (0.99)^{100} = 0.3660,$$

by Formula 6.18. This is a big improvement over the answer obtained with Bonferroni's result. ▲

In some studies the researcher wants all confidence intervals to be correct. In such cases, the answers obtained from Bonferroni's result and Formula 6.18 for the previous example are disappointing. The only way to improve these answers is to increase the confidence level of each interval. Of course, this has the drawback of making each interval wider. A 99 percent confidence interval corresponds to $z = 2.576$; if z is increased to 3, the confidence level increases to 99.74 percent and α decreases to 0.0026. Substituting this smaller value of α into Bonferroni's result for the problem of the previous example, $m = 100$ confidence intervals, gives

$$1 - 100 \times 0.0026 = 1 - 0.26 = 0.74.$$

Thus, the probability is at least 74 percent that all 100 confidence intervals will be correct. Similarly, if the analyses are independent, Formula 6.18 yields

$$(1 - 0.0026)^{100} = (0.9974)^{100} = 0.7708,$$

as the probability that all 100 intervals are correct. Both of these answers are big improvements over the results for $\alpha = 0.01$. The price of this improvement is that every confidence interval has become 16.4 percent wider because $z = 2.576$ has been replaced by $z = 3$.

EXERCISES 6.6

1. Suppose that $m = 5$ persons are given a 100-trial ESP test. If all five persons are guessing, find the probability that at least one person scores 29 or more correct.

2. A researcher computes 1,000 confidence intervals, each with confidence level equal to 95 percent. What is the expected number of incorrect confidence intervals?

3. A researcher computes ten 95 percent confidence intervals.
   (a) Use Bonferroni's result to obtain a lower bound on the probability that all 10 intervals are correct.
   (b) Suppose the intervals are the result of independent analyses. Use Equation 6.18 to obtain the probability that all 10 intervals are correct.
   (c) Discuss the difference between these answers.

4. A friend announces, "I obtained 10 consecutive heads when I tossed a coin; what is the probability of that happening?" Comment.

5. A newspaper article reported that a married couple, each born on October 31, had a baby delivered on October 31 by a doctor born on October 31. The article stated that the probability of this event is

$$\left(\frac{1}{365}\right)^4 = \frac{1}{17{,}748{,}900{,}625},$$

approximately one in 18 billion.
   (a) Explain the reasoning behind the paper's answer.
   (b) Criticize the paper's answer.

REFERENCES

1. "Anxiety in the Market," *Newsweek*, March 27, 1989, p. 22.

2. "Americans and Abortion," *Newsweek*, April 24, 1989, p. 39.

3. "The Honeymoon's Still On: A *Newsweek* Poll," *Newsweek*, March 20, 1989, p. 27.
4. B. A. DeBuono et al., "Sexual Behavior of College Women in 1975, 1986, and 1989," *New England Journal of Medicine*, Vol. 322, March 22, 1990, pp. 821–825.
5. S. Blanche et al., "A Prospective Study of Infants Born to Women Seropositive for Human Immunodeficiency Virus Type I," *New England Journal of Medicine*, Vol. 320, June 22, 1989, pp. 1643–1648.
6. L. Silvestre et al., "Voluntary Interruption of Pregnancy with Mifepristone (RU 486) and a Prostaglandin Analogue: A Large Scale French Experience," *New England Journal of Medicine*, Vol. 322, March 8, 1990, pp. 645–648.
7. W. Beyer, ed., *Handbook of Tables for Probability and Statistics*, 2d ed. (Cleveland: Chemical Rubber Company, 1974).
8. S. Antonarakis et al., "Parental Origin of the Extra Chromosome in Trisomy 21 As Indicated by Analysis of DNA Polymorphisms," *New England Journal of Medicine*, Vol. 324, March 28, 1991, pp. 872–876.

Comparing Two Populations

7.1 **THE FOUR TYPES OF STUDIES**
7.2 **INFERENCE**
7.3 **INTERPRETATION OF STUDIES**

Chapter 7

Chapters 5 and 6 studied a single population with a dichotomous response variable. The population could be a collection of distinct individuals, called a finite population, or a mathematical model for a process that generates the outcomes of trials, called an infinite population. The only infinite populations considered were those that arise from the assumptions of Bernoulli trials. For Bernoulli trials or a finite population, the population can be taken to be a dichotomous box with proportion of successes equal to p and proportion of failures equal to q, where $p + q = 1$. The population is determined completely by the number p.

This chapter extends the main techniques of Chapter 6—point estimation, confidence interval estimation, and hypothesis testing—to a comparison of two populations. The methods of this chapter can be applied to a variety of scientific inquiries. Although the mathematical theory and computations are the same for all studies, the interpretation of the results depends heavily on the type of study. It is shown in this chapter that four types of studies are obtained by combining the two types of populations—a finite population or Bernoulli trials—with the important dichotomy of controlled versus observational studies.

7.1 THE FOUR TYPES OF STUDIES

This section presents a number of examples that introduce the four types of studies, namely, controlled or observational studies on either finite populations or Bernoulli trials. First, the Dating study is an example of an observational study on finite populations.

EXAMPLE 7.1: Dating Study

The first population comprises undergraduate men at the University of Wisconsin-Madison, and the second population comprises undergraduate men at Texas A&M University. Each man's response is his answer to the following question:

> If a woman is interested in dating you, do you generally prefer for her to ask you out, to hint that she wants to go out with you, or to wait for you to act.

The response "ask" is labeled a success, and the other responses are labeled failures. The purpose of the study is to compare the proportion of successes at Wisconsin with the proportion of successes at Texas A&M [1]. ▲

The Chronic Crohn's Disease study is a controlled study with finite populations.

EXAMPLE 7.2: Chronic Crohn's Disease Study

Recall the study of chronic Crohn's disease, introduced in Chapter 1 (Examples 1.4 and 1.5). Consider the collection of all persons who have chronic Crohn's disease and are resistant to or intolerant of the standard therapy. Both the first and the second populations consist of these persons. For the first population the researcher imagines that each person in the population is given cyclosporine, and for the second population the researcher imagines that each person in the population is given the placebo. Following the presentation of Chapter 1, the response could be improvement at the end of therapy, a success, or no improvement at the end of therapy, a failure. The purpose of the study is to compare the proportions of successes in the two populations. ▲

The exercises and future exposition give additional examples of two finite populations. The two examples above illustrate the following general facts:

- In an observational study the set of individuals who comprise the first population is different from the set of individuals who comprise the second population. (See also the discussion of two-points-in-time studies beginning on p. 243.) The researcher has no control over the population to which any individual belongs. The researcher simply *observes* the population status of each individual in the study. For example, in the Dating study the researchers had no control over whether a particular man in their study was an undergraduate at Wisconsin or at Texas A&M.
- A controlled study consists of a single **superpopulation** of individuals with a particular characteristic. In the above example the superpopulation is all persons who suffer from chronic Crohn's disease and are resistant to or intolerant of the standard therapy. The superpopulation generates two populations that correspond to the two treatments of interest. In the

above example, these populations correspond to cyclosporine and the placebo. Note that the two populations consist of exactly the same individuals, namely, the individuals in the superpopulation. The researcher does not control membership in the superpopulation, but once a sample of individuals is selected for study, the researcher does control which treatment is received by each individual in the study. As discussed in Chapter 1, a good way to exercise this control is through the process of randomization. These ideas are discussed further in Section 7.1.1.

- It is sometimes useful to imagine how an observational study could be changed into a controlled study. For the Dating study one might proceed as follows. Start with the superpopulation of all male high school seniors who plan to attend college. A researcher selects a random sample of, say, 200 of these men and randomly assigns 100 of them to attend Wisconsin and the other 100 to attend Texas A&M. Of course, this controlled study would be highly unethical.

The Three-Point Basket study is a controlled study on Bernoulli trials.

EXAMPLE 7.3: Three-Point Basket Study

As discussed in the Three-Point Basket study of Chapter 1 (Example 1.6), Clyde Gaines was interested in comparing his ability to shoot three-point jump shots from two locations: in front of the basket and the left corner. Assume that Clyde's shots from in front as well as his shots from the left corner satisfy the assumptions of Bernoulli trials. Note, however, that one does not assume that each location has the same probability of success: the purpose of the study is to compare these two probabilities. This is a controlled study because the researcher had the power to assign each trial (shot) either to the first population—in front of the basket—or to the second population—the left corner. Clyde exercised this power by using randomization. ▲

Finally, the High-Temperature Forecast study is an observational study on Bernoulli trials.

EXAMPLE 7.4: High-Temperature Forecast Study

Section 5.3 introduced methods for examining data from trials to investigate informally whether or not the assumptions of Bernoulli trials are reasonable. In this chapter the conclusions of these examinations are made formal. For example, data were obtained on the high-temperature forecasts in Madison, Wisconsin, for a six-month period, spring and summer of 1988 (Example 5.20). The second assumption of Bernoulli trials is that the probability of success remained constant during the six months. The validity of this assumption can be explored by breaking the original data into two sequences, one for spring and one for summer. Assume that the spring 1988 trials as well as the summer 1988 trials satisfy the assumptions of Bernoulli trials. Note again that one does not assume that each season has the same probability of success, for the purpose of the study is to compare these two probabilities. This is an observational study because the researcher could not assign a particular trial to either population. For example, the trial on May 3, 1988, was observed to be a spring trial—the researcher could not assign it to be a summer day. ▲

7.1.1 Sampling

This section presents the methods of sampling for the four types of studies introduced above.

In an observational study on finite populations, the researcher first decides on the number of individuals to be sampled from each population. Let n_1 denote the number to be sampled from the first population and let n_2 denote the number to be sampled from the second population. This generalizes the notation used in Chapter 6. Previously, n denoted the sample size; now it is necessary to introduce subscripts to distinguish between the two sample sizes. Once the sample sizes are chosen, the researcher selects a random sample of size n_1 from the first population and a random sample of size n_2 from the second population.

Suppose a researcher wants to investigate the relationship between a dichotomous response and the gender of the subject. For a specific example, suppose a researcher wants to compare the attitudes of male and female drivers on the seriousness of drinking and driving. Following the procedure of the previous paragraph, the researcher would select a random sample of size n_1 from the population of female drivers and a random sample of size n_2 from the population of male drivers. Frequently in practice, however, the researcher does not have a list of the populations of female and male drivers, but only a list of the superpopulation of all drivers. In such a situation, the researcher could select a random sample of n persons from the superpopulation of all drivers and then classify each sampled person as female or male. Let n_1 and n_2 denote the numbers of females and males, respectively, that are obtained in the random sample from the superpopulation of all drivers. Note that if a researcher selects random samples from the individual (that is, female and male) populations, the numbers of females and males in the sample are set by the researcher, but if the researcher selects a random sample from the superpopulation, the values of n_1 and n_2 are determined by the chance mechanism of sampling and are therefore not known in advance of sampling. It can be shown mathematically that the distinction between having n_1 and n_2 selected by the researcher and having them determined by sampling from the superpopulation is unimportant. In simpler terms, if a researcher selects, for example, 200 individuals from the superpopulation of all drivers and obtains $n_1 = 105$ females and $n_2 = 95$ males, it is valid to pretend that the data were obtained by selecting a random sample of size 105 from the population of female drivers and a random sample of size 95 from the population of male drivers.

Recall that a controlled study on finite populations begins with a superpopulation of subjects. The responses in the first and second populations correspond to the responses the members of the superpopulation would give if they were exposed to the first and second treatments, respectively. The researcher selects the sample sizes, n_1 and n_2. Let n denote the sum of these two sample sizes: $n = n_1 + n_2$. Next, the researcher selects a random sample of size n from the superpopulation; n_1 of the individuals in the sample are assigned to the first treatment by randomization, and the remaining n_2 are assigned to the second treatment.

For a controlled study on Bernoulli trials, the researcher first selects the values of the sample sizes, n_1 and n_2. Again, let n denote the sum of these two sample sizes: $n = n_1 + n_2$. The researcher plans to observe n trials and assigns, by randomization, n_1 of the trials to the first population and the remaining n_2 trials to the second population.

Finally, for an observational study on Bernoulli trials, the researcher first selects the values of the sample sizes, n_1 and n_2. The researcher observes n_1 trials from the first population and n_2 trials from the second population. As for an observational study on finite populations, the trials could arise from classifying the observations from a superpopulation of trials. This idea is illustrated later, in Example 7.14.

If one of the above methods of sampling is used, the resultant samples are called **independent random samples.**

7.1.2 Data Presentation and Summary

In the Dating study the researchers selected a random sample of $n_1 = 107$ from the population of undergraduate men at Wisconsin and a random sample of $n_2 = 100$ from the population of undergraduate men at Texas A&M. Sixty men at Wisconsin and 31 men at Texas A&M said they prefer to be asked, and the remaining men prefer that women hint or wait. Table 7.1 presents the observed counts and row proportions for these data.

Recall that the data for the Chronic Crohn's Disease study were presented in Chapter 1 (Example 1.4). For convenience, the data are reproduced in Table 7.2. For the purpose of this chapter, add the assumption that the 71 persons in the

TABLE 7.1 *Responses to the Dating Study*

| | Observed Frequencies | | | Row Proportions | | |
| | Preference | | | Preference | | |
Population	Ask	Other	Total	Ask	Other	Total
Wisconsin	60	47	107	0.561	0.439	1.000
Texas A&M	31	69	100	0.310	0.690	1.000
Total	91	116	207			

TABLE 7.2 Results of the Chronic Crohn's Disease Study

| | Observed Frequencies | | | Row Proportions | | |
| | Clinical Result | | | Clinical Result | | |
Population	Improved	Not Improved	Total	Improved	Not Improved	Total
Cyclosporine	22	15	37	0.595	0.405	1.000
Placebo	11	23	34	0.324	0.676	1.000
Total	33	38	71			

study were selected at random from the superpopulation of potential subjects. This assumption is admittedly a fantasy, and it is discussed in Section 7.3.3.

The data for the Three-Point Basket study were presented in Chapter 1 (Example 1.6) and are reproduced in Table 7.3. As stated earlier, in this chapter it is assumed that the shots from the two locations are Bernoulli trials. Finally, the data for the High-Temperature Forecast study, presented in Chapter 5 (Example 5.20), are reproduced in Table 7.4. As stated earlier, it is assumed that the forecasts for each season are Bernoulli trials.

Each of the tables discussed above looks like a 2×2 contingency table in Format 1 except that the rows designate populations rather than treatments. To avoid a cumbersome proliferation of types of tables, the tables of this chapter also are called Format 1. The general notation for a Format 1 table and its table of row proportions is given in Table 7.5. Note that this agrees exactly with the notation introduced in Section 1.2.

A dichotomous population is characterized by its value of p. For a finite population, p equals the proportion of population members that are successes; for Bernoulli trials, p equals the probability of success on each trial. In the studies of this chapter there are two populations, so it is necessary to introduce more complicated notation to distinguish between the two p's. Let p_1 denote the proportion of successes (or probability of success) for the first population

TABLE 7.3 Results of the Three-Point Basket Study

| | Observed Frequencies | | | Row Proportions | | |
| | Result | | | Result | | |
Population	Basket	Miss	Total	Basket	Miss	Total
In front	21	29	50	0.420	0.580	1.000
Left corner	20	30	50	0.400	0.600	1.000
Total	41	59	100			

TABLE 7.4 Results of the High-Temperature Forecast Study

	Observed Frequencies			Row Proportions		
	Result			Result		
Population	S	F	Total	S	F	Total
Spring	46	43	89	0.517	0.483	1.000
Summer	50	39	89	0.562	0.438	1.000
Total	96	82	178			

TABLE 7.5 Notation for a 2× 2 Table in Format 1 and Its Table of Row Proportions

	Observed Frequencies			Row Proportions		
	Response			Response		
Population	S	F	Total	S	F	Total
1	a	b	n_1	$\hat{p}_1$	$\hat{q}_1$	1
2	c	d	n_2	$\hat{p}_2$	$\hat{q}_2$	1
Total	m_1	m_2	n			

and let p_2 denote the proportion of successes (or probability of success) for the second population. Define $q_1 = 1 - p_1$ and $q_2 = 1 - p_2$; note that q_1 and q_2 are the proportions of failures (probabilities of failure) for the first and second populations, respectively. The statistical inference procedures introduced in the next section focus on the values of p_1 and p_2.

7.2 INFERENCE

Inference is appropriate whenever p_1 and p_2 are unknown to the researcher.

The point estimates of p_1, q_1, p_2, and q_2 are given by their sample versions $\hat{p}_1$, $\hat{q}_1$, $\hat{p}_2$, and $\hat{q}_2$, respectively. The point estimates can be read directly from the table of row proportions.

One compares populations by comparing the values of p_1 and p_2. A common way to compare these two numbers is by computing their difference, $p_1 - p_2$. The point estimate of $p_1 - p_2$ is obtained by replacing each unknown quantity by its point estimate; the result is $\hat{p}_1 - \hat{p}_2$.

EXAMPLE 7.5

This example computes the point estimate of $p_1 - p_2$ for each of the four studies of this chapter.

- For the Dating study, from Table 7.1,

$$\hat{p}_1 - \hat{p}_2 = 0.561 - 0.310 = 0.251.$$

In words, it is estimated that Wisconsin undergraduate men are 25.1 percentage points more likely than Texas A&M undergraduate men to give the ask response.

- For the Chronic Crohn's Disease study, from Table 7.2,

$$\hat{p}_1 - \hat{p}_2 = 0.595 - 0.324 = 0.271.$$

In words, if given to the entire population, cyclosporine compared to the placebo would increase the proportion of improved cases by an estimated 27.1 percentage points.

- For the Three-Point Basket study, from Table 7.3,

$$\hat{p}_1 - \hat{p}_2 = 0.420 - 0.400 = 0.020.$$

In words, it is estimated that Clyde's probability of making a three-point basket was 2 percentage points higher from in front of the basket than from the left corner.

- For the High-Temperature Forecast study, from Table 7.4,

$$\hat{p}_1 - \hat{p}_2 = 0.517 - 0.562 = -0.045.$$

In words, it is estimated that the probability of a successful forecast was 4.5 percentage points lower in spring than in summer, 1988. ▲

After computing the point estimate of $p_1 - p_2$, the next goal usually is to compute a confidence interval for this quantity. The formula will be given and its use demonstrated. The interested reader is referred to the (optional) next section for an outline of the mathematical reasoning that yields the formula.

With the assumptions and notation of this chapter, a confidence interval for $p_1 - p_2$ is given by

$$(\hat{p}_1 - \hat{p}_2) \pm z \sqrt{\frac{\hat{p}_1 \hat{q}_1}{n_1} + \frac{\hat{p}_2 \hat{q}_2}{n_2}} \qquad (7.1)$$

where z is determined by the confidence level and can be found in Table A.3. The general guideline is that this formula can be used if each of $a, b, c,$ and d in the 2×2 contingency table is equal to 10 or more, or, equivalently, if each of $n_1 \hat{p}_1, n_1 \hat{q}_1, n_2 \hat{p}_2,$ and $n_2 \hat{q}_2$ is equal to 10 or more.

It can be shown that simple algebra transforms Formula 7.1 into the following formula, which is written in terms of observed frequencies instead of point estimates:

$$\left(\frac{a}{n_1} - \frac{c}{n_2}\right) \pm z \times \sqrt{\frac{ab}{n_1^3} + \frac{cd}{n_2^3}}. \tag{7.2}$$

The use of Formulas 7.1 and 7.2 is illustrated below for the examples of this chapter.

EXAMPLE 7.6

For the Dating study, from Table 7.1,

$\hat{p}_1 = 0.561$, $\hat{q}_1 = 0.439$, $n_1 = 107$, $\hat{p}_2 = 0.310$, $\hat{q}_2 = 0.690$, and $n_2 = 100$.

Thus, the 95 percent confidence interval for $p_1 - p_2$ is

$$(0.561 - 0.310) \pm 1.96 \sqrt{\frac{(0.561)(0.439)}{107} + \frac{(0.310)(0.690)}{100}}$$

$$= 0.251 \pm 1.96(0.0666) = 0.251 \pm 0.131 = [0.120, 0.382].$$

The endpoints of this interval are both positive; thus, all the values in the interval are positive. In words, at the 95 percent confidence level one can conclude that the proportion of Wisconsin men who prefer to be asked, p_1, is larger than the proportion of Texas A & M men who prefer to be asked, p_2. Again at the 95 percent confidence level, the difference between these proportions is at least 0.120 and at most 0.382. ▲

EXAMPLE 7.7

For the Chronic Crohn's Disease study, Formula 7.2 can be used to obtain the 95 percent confidence interval for $p_1 - p_2$. From Table 7.2,

$a = 22$, $b = 15$, $n_1 = 37$, $c = 11$, $d = 23$, and $n_2 = 34$.

Thus, the 95 percent confidence interval for $p_1 - p_2$ is

$$\left(\frac{22}{37} - \frac{11}{34}\right) \pm 1.96 \sqrt{\frac{(22)(15)}{(37)^3} + \frac{(11)(23)}{(34)^3}} = 0.271 \pm 1.96(0.1138)$$

$$= 0.271 \pm 0.223 = [0.048, 0.494].$$

All the numbers in this confidence interval are positive. In words, at the 95 percent confidence level one can conclude that the proportion of persons who would improve if given cyclosporine exceeds the proportion who would improve if given the placebo. Again at the 95 percent confidence level, the difference between these proportions is at least 0.048 and at most 0.494. This interval is disappointingly wide; from a practical viewpoint, a cyclosporine advantage of only 0.048 is very different from an advantage of 0.494. ▲

EXAMPLE 7.8

For the Three-Point Basket study, from Table 7.3,
$$a = 21, \ b = 29, \ n_1 = 50, \ c = 20, \ d = 30, \text{ and } n_2 = 50.$$
Thus, the 95 percent confidence interval for $p_1 - p_2$ is

$$\left(\frac{21}{50} - \frac{20}{50}\right) \pm 1.96 \sqrt{\frac{(21)(29)}{(50)^3} + \frac{(20)(30)}{(50)^3}} = 0.020 \pm 1.96(0.0983)$$

$$= 0.020 \pm 0.193 = [-0.173, 0.213].$$

This interval includes 0; thus, at the 95 percent confidence level the two success probabilities might be equal. This interval is very wide; at one extreme (the upper end point), Clyde's probability of success from the front might be as much as 21.3 percentage points higher, but at the other extreme, his probability of success from the left corner might be as much as 17.3 percentage points higher. ▲

EXAMPLE 7.9

For the High-Temperature Forecast study, from Table 7.4,
$$\hat{p}_1 = 0.517, \ \hat{q}_1 = 0.483, \ n_1 = 89, \ \hat{p}_2 = 0.562, \ \hat{q}_2 = 0.438, \text{ and } n_2 = 89.$$
Thus, the 95 percent confidence interval for $p_1 - p_2$ is

$$(0.517 - 0.562) \pm 1.96 \sqrt{\frac{(0.517)(0.483)}{89} + \frac{(0.562)(0.438)}{89}}$$

$$= -0.045 \pm 1.96(0.0746) = -0.045 \pm 0.146 = [-0.191, 0.101].$$

As in the previous example, the interval includes 0; thus, at the 95 percent confidence level the two success probabilities might be equal. In addition, at one extreme, spring forecasts might have probability of success as much as 10.1 percentage points higher, but at the other extreme, summer forecasts might have probability of success as much as 19.1 percentage points higher. ▲

7.2.1 Derivation of the Confidence Interval Formula*

As mentioned earlier, this section is optional and is intended for the reader who enjoys algebra.

You may want to review the derivation of the confidence interval for a single p that was presented in Section 6.2.1. Applying that approach to the current problem, the point estimates $\hat{p}_1, \hat{q}_1, \hat{p}_2,$ and $\hat{q}_2$ are viewed as the observed values of the point estimators (random variables) $\hat{P}_1, \hat{Q}_1, \hat{P}_2,$ and $\hat{Q}_2$, respectively. It was stated that the sampling distribution of $\hat{P}$ has the following properties:

$$E(\hat{P}) = p \quad \text{and} \quad \text{Var}(\hat{P}) = \frac{pq}{n}.$$

(Actually, the result was stated in terms of the standard deviation, but the variance is more convenient for the argument below.) By simply including the subscripts of this chapter, these equations yield

$$E(\hat{P}_1) = p_1, \quad \text{Var}(\hat{P}_1) = \frac{p_1 q_1}{n_1}, \quad E(\hat{P}_2) = p_2, \quad \text{and} \quad \text{Var}(\hat{P}_2) = \frac{p_2 q_2}{n_2}.$$

A general mathematical result states that the expected value of the difference of two random variables is equal to the difference of the individual expected values. Applied to the current problem, this result implies that

$$E(\hat{P}_1 - \hat{P}_2) = E(\hat{P}_1) - E(\hat{P}_2) = p_1 - p_2.$$

Note that the value of $\hat{P}_1$ is computed from the data from the first population and the value of $\hat{P}_2$ is computed from the data from the second population. It can be shown that because the samples are independent, these random variables are independent, too. It can be shown that if two random variables are independent, the variance of their difference is equal to the *sum* (not the difference!) of the individual variances. Applied to the current problem, this implies that

$$\text{Var}(\hat{P}_1 - \hat{P}_2) = \text{Var}(\hat{P}_1) + \text{Var}(\hat{P}_2) = \frac{p_1 q_1}{n_1} + \frac{p_2 q_2}{n_2}.$$

These formulas for the expected value and variance enable one to calculate the standardized version of $\hat{P}_1 - \hat{P}_2$. It can be shown that the probability histogram of this standardized version can be well approximated by the standard normal curve. Applying the same algebraic arguments that were used in Chapter 6, one obtains, for example, for 95 percent confidence,

$$P\left(\hat{P}_1 - \hat{P}_2 - 1.96\sqrt{\frac{p_1 q_1}{n_1} + \frac{p_2 q_2}{n_2}} \leq p_1 - p_2 \leq \hat{P}_1 - \hat{P}_2 + 1.96\sqrt{\frac{p_1 q_1}{n_1} + \frac{p_2 q_2}{n_2}}\right)$$
$$\approx 0.95.$$

As in Chapter 6, the work of Slutsky can be used to show that the unknown proportions in the extreme terms of this string of inequalities can be replaced by their point estimators. After data are obtained, the point estimators are replaced by their observed values, the point estimates; the result is Formula 7.1 for the special case of 95 percent confidence and $z = 1.96$. Other confidence levels are handled by replacing 1.96 with the appropriate value of z.

7.2.2 Hypothesis Testing

For the problem of this chapter, the null hypothesis states that $p_1 = p_2$. For finite populations, this means that the proportion of individuals that are successes is the same in each population; for Bernoulli trials, this means that the probability of success on each trial is the same in each population. There are three possible choices for the alternative hypothesis:

$$H_1: p_1 > p_2, \quad H_1: p_1 < p_2, \quad \text{or} \quad H_1: p_1 \neq p_2,$$

referred to as the first, second, or third alternative, respectively. If a success is the more desirable response, the first alternative states that the first population is superior to the second population, the second alternative states that the first population is inferior to the second population, and the third alternative states that the first population is either superior or inferior to the second population.

The similarity between this null and these three alternative hypotheses and the ones in Chapter 2 is striking. Thus, it may not be too surprising to learn that the remaining three steps of the current hypothesis test are identical to steps 2, 3, and 4 of Fisher's test presented in Chapters 2 and 3. (Or it may be surprising; after all, in the earlier analysis probability arose from randomization, whereas in the current case it arises from the random sampling of populations.) Thus, all the earlier results apply to the current problem. For example, if $n_1 = n_2$ and the common value is less than or equal to 14, Table A.1 yields the exact P-value. Alternatively, one can can obtain an approximate P-value using either a simulation experiment or the standard normal curve. For the latter method, recall that the approximate P-value is obtained by completing these two steps:

1. Compute z:

$$z = \frac{\sqrt{n-1}(ad-bc)}{\sqrt{n_1 n_2 m_1 m_2}}.$$

2. Approximate the P-value:
   - For the first alternative, $p_1 > p_2$, the approximate P-value equals the area under the standard normal curve to the right of z.
   - For the second alternative, $p_1 < p_2$, the approximate P-value equals the area under the standard normal curve to the right of $-z$.
   - For the third alternative, $p_1 \neq p_2$, the approximate P-value equals twice the area under the standard normal curve to the right of $|z|$.

Recall the following rule given in Chapter 3:

Let r_1 equal the smaller of n_1 and n_2, and let r_2 equal the smaller of m_1 and m_2. The general guideline for using the standard normal curve approximation for Fisher's test is that

$$\frac{r_1 r_2}{n} \geq 5.$$

This general guideline also is appropriate for the test of this chapter.

Before illustrating this method for the four studies of this chapter, this paragraph discusses an advanced topic for the interested reader. The formula for z given above is sometimes modified by replacing the term $\sqrt{n-1}$ in the numerator by $\sqrt{n}$. If n is reasonably large, this change affects the value of z very little, and hence the approximate P-value does not change dramatically. Either formula for z gives an approximate P-value, and it can be shown that neither one consistently gives a better approximation. Some people prefer $\sqrt{n-1}$ because it corresponds to the exact standardization as discussed in Chapter 3. Others prefer $\sqrt{n}$ because with this choice the formula for z is algebraically identical to

$$(\hat{p}_1 - \hat{p}_2)/\sqrt{\frac{\hat{p}\hat{q}}{n_1} + \frac{\hat{p}\hat{q}}{n_2}},$$

where $\hat{p}$ equals the sample proportion of successes in the combined samples,

$$\hat{p} = \frac{a+c}{n_1 + n_2} = \frac{m_1}{n},$$

and $\hat{q} = 1 - \hat{p}$. This latter formula for z has the disadvantage of being messier to compute, presenting a serious potential for inaccuracy due to round-off error. Thus, the original formula for z (with $\sqrt{n-1}$) will be used in this text.

EXAMPLE 7.10

For the Dating study the researchers chose the third alternative because they were interested in finding any difference between the two populations. From Table 7.1,

$$a = 60, \ b = 47, \ n_1 = 107, \ c = 31, \ d = 69, \ n_2 = 100, \ m_1 = 91,$$
$$m_2 = 116, \text{ and } n = 207.$$

Substituting these numbers into the formula for z gives

$$z = \frac{\sqrt{206}(60 \times 69 - 47 \times 31)}{\sqrt{107(100)(91)(116)}} = 3.62.$$

The area under the standard normal curve to the right of 3.62 is smaller than 0.0002; thus, the approximate P-value for the third alternative is smaller than 2(0.0002), or 0.0004. The data are highly statistically significant; it seems clear that the populations do not have the same proportions of successes. ▲

EXAMPLE 7.11

A hypothesis test was conducted in Chapter 3 for the Chronic Crohn's Disease study. The value of z was found to be 2.26, and the approximate P-value for the first alternative was shown to equal 0.0119. The data are statistically significant; cyclosporine therapy appears to be superior to the placebo. ▲

EXAMPLE 7.12

A hypothesis test was conducted in Chapter 3 for the Three-Point Basket study. The value of z was found to be 0.20 and the approximate P-value for the third alternative was shown to equal 0.8414. The data are not statistically significant; one should continue to believe that Clyde is equally accurate at shooting from in front of the basket and from the left corner. ▲

EXAMPLE 7.13

For the High-Temperature Forecast study I chose the third alternative because I am interested in finding any difference between spring and summer. A case could also be made for the second alternative because of my opinion that spring weather is more erratic. From Table 7.4,

$$a = 46, \ b = 43, \ n_1 = 89, \ c = 50, \ d = 39, \ n_2 = 89, \ m_1 = 96,$$
$$m_2 = 82, \text{ and } n = 178.$$

Substituting these numbers into the formula for z gives

$$z = \frac{\sqrt{177}(46 \times 39 - 43 \times 50)}{\sqrt{89(89)(96)(82)}} = -0.60.$$

For the third alternative the absolute value of z is needed. The area under the standard normal curve to the right of $|z| = 0.60$ is 0.2743. Thus, the approximate P-value for the third alternative is 2(0.2743), or 0.5486. It is quite clear that there is little evidence in support of the alternative of a difference in success rates between seasons. ▲

The analysis of the previous example suggests that it may be reasonable to believe that the second condition for Bernoulli trials holds for the high-temperature forecasts for the entire sequence of six months. The analysis certainly does not *prove* that the second assumption is valid: only one possible departure from the assumption has been investigated. The next example sheds some light on the third assumption, that of independence.

EXAMPLE 7.14

Table 7.6, presented in Chapter 5, is reproduced here for convenience. These data can be used to conduct a formal hypothesis test of the third assumption of Bernoulli trials for the six months of forecast data. The data consist of 178 trials, and each trial except the first is preceded by a trial. Each trial after the first can be viewed as coming from one of two populations. If its previous trial is a success,

TABLE 7.6 Dependence of the Outcome of the Current Forecast on the Outcome of the Previous Forecast

	Current Forecast					
	Frequencies			Row Proportions		
Previous Forecast	S	F	Total	S	F	Total
Success	54	42	96	0.562	0.438	1.000
Failure	41	40	81	0.506	0.494	1.000
Total	95	82	177			

it is selected from the first population, but if its previous trial is a failure, it is selected from the second population. If the trials are indeed independent, the result of the previous trial has no influence on the current trial and these two populations are identical—that is, $p_1 = p_2$. On the other hand, if $p_1 \neq p_2$, the trials are not independent and are not Bernoulli trials. Thus, the independence of trials is explored by testing the null hypothesis that $p_1 = p_2$ versus the third alternative, that $p_1 \neq p_2$.

The observed value of the standardized version of the test statistic is

$$z = \frac{\sqrt{n-1}(ad-bc)}{\sqrt{n_1 n_2 m_1 m_2}} = \frac{\sqrt{176}(54 \times 40 - 42 \times 41)}{\sqrt{96(81)(95)(82)}} = 0.75.$$

The area under the standard normal curve to the right of 0.75 equals 0.2266. Thus, for the third alternative, the approximate P-value equals 2(0.2266), or 0.4532. Based on this examination of the data, there is little reason to doubt the assumption of independence. Note that this is an observational study because the experimenter cannot control which population a particular trial is assigned to—the assignment depends on the previous day's outcome, which the experimenter only observes. In fact, the data come from the superpopulation of daily forecasts, and each trial is classified to a population according to the previous day's result.

Further insight can be gained by constructing a 95 percent confidence interval for $p_1 - p_2$,

$$\left(\frac{a}{n_1} - \frac{c}{n_2}\right) \pm z \times \sqrt{\frac{ab}{n_1^3} + \frac{cd}{n_2^3}} = \left(\frac{54}{96} - \frac{41}{81}\right) \pm 1.96\sqrt{\frac{54(42)}{(96)^3} + \frac{41(40)}{(81)^3}}$$

$$= 0.056 \pm 1.96(0.0752) = 0.056 \pm 0.147 = [-0.091, 0.203].$$

At one extreme, a success may be 20.3 percentage points more likely after a success than after a failure. At the other extreme, a success may be 9.1 percentage points less likely after a success than after a failure. ▲

The following example, taken from the Wisconsin Driver Survey, introduces an important class of observational studies on finite populations. They are referred to as **two-points-in-time studies.**

EXAMPLE 7.15

The populations are

- Population 1: All Wisconsin licensed drivers in 1983
- Population 2: All Wisconsin licensed drivers in 1982

Independent random samples were selected from these populations. Each person in the study was asked question 5 on the survey:

Have you heard the term "implied consent" used in connection with drunken driving?

("Implied consent" means that everyone who obtains a Wisconsin driver's license implicitly agrees to a breath test of blood alcohol content on demand. Failure to submit to a test results in the immediate loss of the driver's license. Of course, the results of the test may be challenged in court later. Question 5 was placed on the survey to obtain a rough measure of awareness of this term among the population. Another question on the survey dealt with the driver's understanding of implied consent.) The possible responses were yes and no. Table 7.7 presents the data from the study. Note that $p_1 - p_2$ equals the increase from 1982 to 1983 in the proportion of all drivers who would answer yes. The point estimate of this difference is $\hat{p}_1 - \hat{p}_2$; reading from the table, this difference is $0.457 - 0.397 = 0.060$. Thus, the estimated increase in positive responses is 6 percentage points. A 95 percent confidence interval for the difference is

$$(\hat{p}_1 - \hat{p}_2) \pm z\sqrt{\frac{\hat{p}_1 \hat{q}_1}{n_1} + \frac{\hat{p}_2 \hat{q}_2}{n_2}} = 0.060 \pm 1.96\sqrt{\frac{0.457(0.543)}{1{,}072} + \frac{0.397(0.603)}{1{,}589}}$$

$$= 0.060 \pm 1.96(0.0195) = 0.060 \pm 0.038 = [0.022, 0.098].$$

At the 95 percent confidence level, the increase in positive response from 1982 to 1983 was between 2.2 and 9.8 percentage points.

The state's researchers chose the first alternative because they were looking for an increase in positive responses as a result of an advertising campaign. The observed value of the standardized version of the test statistic is

$$z = \frac{\sqrt{n-1}(ad-bc)}{\sqrt{n_1 n_2 m_1 m_2}} = \frac{\sqrt{2{,}660}(490 \times 958 - 582 \times 632)}{\sqrt{1{,}072(1{,}589)(1{,}121)(1{,}540)}} = 3.07.$$

The area under the standard normal curve to the right of 3.07 is 0.0011. Thus, the approximate P-value for the first alternative equals 0.0011. The data are highly statistically significant; the data strongly support the alternative that the proportion of positive responses increased from 1982 to 1983 in the population of Wisconsin licensed drivers. ▲

TABLE 7.7 *Responses to Question 5 by Year for the Wisconsin Driver Survey*

	Frequencies			Row Proportions		
	Response			Response		
Year	Yes	No	Total	Yes	No	Total
1983	490	582	1,072	0.457	0.543	1.000
1982	631	958	1,589	0.397	0.603	1.000
Total	1,121	1,540	2,661			

Before the previous example, it was stated that a two-points-in-time study is an observational study. The example did not actually address that claim. The primary reason these studies are viewed as observational is revealed in the next section, where the results of the various studies are interpreted.

7.3 INTERPRETATION OF STUDIES

This chapter has considered six studies in detail: studies on dating, Crohn's disease, implied consent, and shooting baskets, and two studies on high-temperature forecasts. In this section the results of these studies are given detailed interpretations. Some general remarks are made.

For a controlled study on finite populations, the two populations are the same individuals and the populations differ only in the treatment received. Thus, if the populations are different, the difference can be attributed to a difference between treatments. (Remember that conclusions based on data always have some chance of being incorrect!) For the study of chronic Crohn's disease, for example, a researcher can conclude that cyclosporine is better than the placebo. The situation for an observational study, however, is quite different.

The study on dating shows quite conclusively that the populations are different. The researcher must take care, however, not to read any more into the results of the study. There is no information in the study to suggest *why* the populations differ. In particular, there is no basis for concluding that the experience of being at Wisconsin has a different effect on men's preferences than the experience of being at Texas A&M. More precisely, there are likely many identifiable differences between these populations of men that have nothing to do with their respective universities. A list of possible differences between these populations that could account, at least in part, for the differences in preferences might include the following:

- Most obviously, most Texas A&M men grew up in Texas and most Wisconsin men grew up in Wisconsin. Perhaps the cultural, ethnic, or religious differences between people in these states influence the men's preferences.

- Again rather obviously, Texas A&M men tend to date Texas women and Wisconsin men tend to date Wisconsin women. Perhaps the difference in male preferences results from differences in female preferences or behaviors in the two states.

- Perhaps a greater proportion of Texas A&M men major in technical fields, and perhaps men in technical fields have different preferences than men in nontechnical fields.

The differences between universities is said to be **confounded** with other differences between the groups of men, including but not limited to those differences listed above. Thus, a researcher cannot validly give a causal interpretation to the relationship in the data. In particular, one cannot conclude that the difference in preferences is a result of differences between the colleges attended.

A similar problem exists for a two-points-in-time study. For the study of implied consent, for example, it was quite clear that the proportion of positive responses increased from 1982 to 1983. It is impossible, however, to determine what caused the increase. The increase could have been the result of any or all of the following:

- The state ran an extensive advertising campaign to educate the public about the drinking and driving law.
- There was increased state and local enforcement of the drunk driving law.
- Mothers Against Drunk Driving and Students Against Drunk Driving promoted educational efforts to increase awareness about drinking and driving.

The Wisconsin Department of Transportation had been directed by law to educate the public on a variety of issues related to drunk driving. Thus, officials at the Department of Transportation preferred the first explanation given above, but they were honest enough to concede that, based on their survey of drivers, it was impossible to assign credit to any particular source.

Similar remarks are true when the populations are Bernoulli trials. For example, if Clyde Gaines had shot significantly better from one location, a researcher could conclude that the change of location caused the difference in performance. By contrast, if the spring high-temperature forecasts had been statistically significantly inferior to the summer ones, there could be a number of possible explanations unrelated to the season, including the following:

- Perhaps the hardware or software for collecting meteorological data improved in summer, making forecasts more accurate.
- Perhaps those involved in making forecasts improved from experience or training.
- Perhaps those making forecasts in spring were replaced in summer by more skilled forecasters.

Thus, in summary, a significant difference between populations yields a causal interpretation for a controlled study but not for an observational study. It is possible, however, that there is a causal relationship in an observational study. The point is that an observational study cannot rule out other possible explanations. For example, it was clear for many years that the population of heavy smokers had a higher rate of lung cancer than the population of nonsmokers. Some smoking proponents noted that such studies were observational and that there could therefore be one or more important confounding factors. As far as one knew, it was possible, for example, that a gene predisposes some people both to smoking and to developing lung cancer. Thus, stopping smoking would not change their genetic makeup—they would simply be more irritable when they developed lung cancer! Eventually, controlled studies on animals and a better understanding of the biology of cancer led most scientists to conclude that there is a causal relationship between smoking and lung cancer.

7.3.1 Practical Importance

The remarks of this section apply to most situations in which statistical inference is used, not just to problems that fit the framework of this chapter. The ideas are illustrated first with the Three-Point Basket study, and then some general comments are made.

The null hypothesis states that $p_1 = p_2$, which also can be written $p_1 - p_2 = 0$. The third alternative states that $p_1 - p_2 \neq 0$. It is instructive to consider some possible values for this difference. It is possible that $p_1 - p_2$ equals 0.50. In words, Clyde has 50-percentage-point-higher probability of making a three-point shot from in front than from the left corner. Fifty percentage points are a huge difference in basketball! It is possible, however, that $p_1 - p_2$ equals 0.001. In words, if Clyde attempts 1,000 shots from in front and 1,000 shots from the left corner, he can expect to make one more basket from in front of the basket than from the left corner! A difference as small as 0.001 has no practical importance in basketball.

These two numerical examples suggest the following strategy. Before conducting a study, the researcher should decide which values of $p_1 - p_2$ would be of **practical importance**. For example, for the sake of argument let us assume that before performing his study Clyde Gaines decided that if

$$p_1 - p_2 \geq 0.06 \quad \text{or} \quad p_1 - p_2 \leq -0.06,$$

then the difference in his skill at the two locations has practical importance.

The difficulty, of course, is that even after collecting data the researcher does not know the exact value of $p_1 - p_2$. The above ideas, however, can still be useful. Considerable insight can be gained by examining the point estimate and the 95 percent (or some other level if you prefer) confidence interval. For the Three-Point Basket study, the point estimate of $p_1 - p_2$ is 0.02. Note that *if this estimate is correct*—that is, if the actual value of $p_1 - p_2$ equals 0.02—then the difference in skill between locations is not of practical importance. The 95 percent confidence interval, however, tells a very different, less clear, story. It ranges from a low of −17.3 percentage points to a high of 21.3 percentage points. It includes all the values, from −0.06 to 0.06, that are not of practical importance, but it also includes a wide range of values that are of practical importance. Thus, in this sense, the study is not very conclusive. The problem is that the confidence interval is too wide. As noted in Section 6.3, a confidence interval can be made narrower by collecting more data. The effect of increasing the sample size is illustrated for a fictitious study, called the Ironman Clyde study.

TABLE 7.8 *Practical Importance of the Six Studies in This Chapter*

Study	Definition of Practical Importance	95% Confidence Interval	Conclusion of Study
Dating	$p_1 - p_2 \geq 0.10$ or $p_1 - p_2 \leq -0.10$	[0.120, 0.382]	Practical importance
Chronic Crohn's Disease	$p_1 - p_2 \geq 0.10$	[0.048, 0.494]	Cannot tell
Three-Point Basket	$p_1 - p_2 \geq 0.06$ or $p_1 - p_2 \leq -0.06$	[−0.173, 0.213]	Cannot tell
High-Temperature Forecast			
Spring vs. summer	$p_1 - p_2 \geq 0.10$ or $p_1 - p_2 \leq -0.10$	[−0.191, 0.101]	Cannot tell
Consecutive trials	$p_1 - p_2 \geq 0.10$ or $p_1 - p_2 \leq -0.10$	[−0.091, 0.203]	Cannot tell
Implied Consent	$p_1 - p_2 \geq 0.05$	[0.022, 0.098]	Cannot tell

Ironman Clyde shot 5,000 jump shots(!) from each location and had $\hat{p}_1 = 0.42$ and $\hat{p}_2 = 0.40$; note that these are the same values as in the actual study. It can be shown (see Exercise 1) that the 95 percent confidence interval for $p_1 - p_2$ is [0.001, 0.039]. All of the values in this interval are positive, so Ironman Clyde is a better shooter from in front of the basket than from the left corner. (The same conclusion can be drawn from a hypothesis testing analysis—see Exercise 1.) Note, however, that none of the values in the confidence interval are of practical importance to Clyde. Thus, at the 95 percent confidence level, the researcher can conclude that although Ironman Clyde may shoot somewhat better from in front than from the left corner, the difference is not of any practical importance.

Table 7.8 presents an assessment of practical importance for the six studies considered in this chapter. Note that the definitions of practical importance, given in the second column, are mine; it would be better to have experts in the respective research areas make the definitions, but my definitions will suffice for illustration. You can verify that the entries in the third column agree with the answers given earlier in this chapter and that the conclusions in the fourth column are consistent with the information in the other columns. This table tells an unpopular story: often the conclusion of a study is that the researcher cannot tell if there is practical importance. In these instances the researcher needs more data to answer this question.

7.3.2 Statistical Power*

The material in this section is optional. It will not be needed later in this text.

In the previous subsection it was noted that after the completion of a study a researcher might discover that not enough data were collected. The concept of statistical power, or power for short, provides one way to anticipate such a problem before data are collected. In this short section I attempt only to provide a brief introduction to power. The ideas will be illustrated with the Three-Point Basket study.

Power is the probability that the null hypothesis will be rejected. Recall that the null hypothesis is rejected if, and only if, the P-value is less than or equal to 0.05. It turns out that the computation of power becomes very messy if one considers the exact P-value; thus, for simplicity the P-value will be computed by using the standard normal curve approximation. For ease of exposition, no reference will be made to the approximate nature of the probabilities computed using the standard normal curve.

Recall that for the Three-Point Basket study the alternative hypothesis is the third alternative, $p_1 \neq p_2$. For the third alternative, the P-value is equal to twice the area under the standard normal curve to the right of $|z|$, where

$$z = \frac{\sqrt{n-1}(ad-bc)}{\sqrt{n_1 n_2 m_1 m_2}}$$

is the test statistic. Table A.2 indicates that the area under the standard normal curve to the right of 1.96 is 0.025. Thus, twice the area under the standard normal curve to the right of 1.96 is 0.05. It follows that the approximate P-value is less than or equal to 0.05 if, and only if, $|z| \geq 1.96$. Thus, the event that the null hypothesis is rejected can be viewed as the event that the absolute value of the test statistic is greater than or equal to 1.96. The probability of this event will now be computed under a variety of situations.

Thus far in this text, probabilities have been computed under the assumption that the null hypothesis is true. If the null hypothesis is true, the sampling distribution of the test statistic can be approximated by the standard normal curve. Thus, if the null hypothesis is true, the probability of rejecting the null hypothesis equals twice the area under the standard normal curve to the right of 1.96. This area is 0.05. This is a desirable property of Fisher's test: if the null hypothesis is true, there is only a 5 percent chance that the test will err and reject the null hypothesis. (Note: Nothing specific to Fisher's test is needed in the above argument. In fact, for any hypothesis-testing procedure, if the researcher uses the principle of rejecting the null hypothesis if the P-value is less than or equal to 0.05, then—at least approximately—the probability of rejecting a true null hypothesis equals 0.05.)

Suppose, however, that the null hypothesis is not true. In this situation it would be desirable to reject the null hypothesis in favor of the alternative. A natural question is

> What is the probability that the test will reject the null hypothesis if, in fact, the null hypothesis is false?

This is a difficult question to answer for two reasons. First, there are many ways the null hypothesis could be false (more on this below). Second, given the exact way in which the null hypothesis is false, the computation of this probability can be difficult.

For example, in the Three-Point Basket study one possible way for the null hypothesis to be false is that $p_1 = 0.45$ and $p_2 = 0.40$. In words, it is possible that each of Clyde's shots from the front has probability 0.45 of being a success and each of his shots from the left corner has probability 0.40 of being a success. If, in fact, $p_1 = 0.45$ and $p_2 = 0.40$, what is the probability that Fisher's test will make a correct decision by rejecting the null hypothesis? In other words, what is the probability that the absolute value of the test statistic will equal or exceed 1.96? This probability is difficult to compute, so I decided to approximate it with a simulation experiment. Each run of the simulation experiment is rather complicated and will be described below.

First, I had Minitab generate 50 Bernoulli trials with $p_1 = 0.45$; the result was 24 successes and 26 failures. Second, I had Minitab generate 50 Bernoulli trials with $p_2 = 0.40$; the result was 18 successes and 32 failures. These results yield the following 2×2 contingency table of generated data:

Location	Basket	Miss	Total
In front	24	26	50
Left corner	18	32	50
Total	42	58	100

Third, I had Minitab compute the value of the test statistic for these data. The result is

$$z = \frac{\sqrt{99}(24 \times 32 - 26 \times 18)}{\sqrt{50(50)(42)(58)}} = 1.21.$$

Thus, the first run of the simulation experiment yields the decision not to reject the null hypothesis because the absolute value of the test statistic is smaller than 1.96.

The above procedure was repeated until a total of 1,000 runs were obtained. Of the 1,000 runs, only 101 generated data sets led to the (correct) decision of rejecting the null hypothesis. The estimated power of the test (the estimated probability of rejecting the null hypothesis) if $p_1 = 0.45$ and $p_2 = 0.40$ is $101/1{,}000 = 0.101$.

The point of the above analysis is the following. If, in fact, $p_1 = 0.45$ and $p_2 = 0.40$, the probability that the Three-Point Basket study would *correctly conclude* that the null hypothesis is false is estimated to be only 0.101. This is a disappointingly small number!

To gain further insight, the above 1,000-run simulation experiment was repeated for a total of four choices of p_1 and three choices of n. In each case,

TABLE 7.9 *Estimated Power for the Three-Point Basket Study Based on a 1,000-Run Simulation Experiment*

		n	
p_1^*	100	200	400
0.45	0.101	0.109	0.186
0.50	0.198	0.315	0.538
0.55	0.356	0.566	0.873
0.60	0.559	0.842	0.986

* For each case in this table, $p_2 = 0.40$.

a balanced CRD was used. The results of the simulation experiment are presented in Table 7.9. Reading from this table, if $p_1 = 0.45$ and $p_2 = 0.40$, and if the study had $n = 400$ shots, the estimated power is only 0.186. Even for such a large study, the chance that Fisher's test will correctly detect the difference in Clyde's ability is meager. The actual study, which had $n = 100$, had an estimated 55.9 percent chance of detecting that $p_1 = 0.60$ and $p_2 = 0.40$. I would be very surprised, however, if a talented basketball player exhibited such a huge difference, 20 percentage points, in his ability from the two locations (you, of course, may reasonably disagree with me).

7.3.3 The Assumption of Independent Random Samples

It was stated earlier in this chapter that the men in the Dating study were selected at random from their respective populations. That is not literally true. In fact, the men were all students in an introductory psychology class at their university, and each psychology class requires students to enroll as subjects in one or more studies. (This is, as you may know from experience, a common practice at universities.) In their paper, the researchers did not claim to have independent random samples, but they used the methods of this chapter, which require that assumption. I am not being critical of the researchers; their approach is not uncommon, and it is implicit in the scientific community that such studies should be interpreted more carefully because the subjects are not true random samples.

Section 5.4.1 discussed the selection of subjects for the 1986 Wisconsin Driver Survey. Similar procedures were used in 1982 and 1983; thus, the study of the familiarity with the term *implied consent* was not based on literally independent random samples.

The 71 persons in the Chronic Crohn's Disease study were selected in large part because they were conveniently available to the researchers. They were not literally a random sample from a superpopulation. In fact, since physicians generally are interested in a superpopulation that includes future patients, a literally random sample in a medical study is usually impossible—a sample cannot include individuals who are not in the superpopulation at the time of the study!

Despite the fact that literally independent random samples are rare, researchers believe that much knowledge and insight can be gained by using the results of this and future chapters. Be on guard, however, for the next *Literary Digest* poll, whatever its guise might be!

Occasionally a person discovers that the data have not been obtained from independent random samples and, in fact, that the departure from the assumption is serious enough to warrant not using the methods of this chapter. What can be salvaged? The contingency table of observed frequencies and its row proportions still can be used to describe the data. For an observational study, nothing else should be done. For a controlled study with randomization, however, the results of the first three chapters of this book can be used. For example, the P-values for the Chronic Crohn's Disease and Three-Point Basket studies are the same whether or not one assumes independent random samples. (The scopes of the conclusions differ, of course.)

EXERCISES FOR CHAPTER 7

1. Assume that independent random samples with $n_1 = 5,000$ and $n_2 = 5,000$ yield $\hat{p}_1 = 0.420$ and $\hat{p}_2 = 0.400$. (These are the hypothetical Ironman Clyde data discussed in the narrative.)

   (a) Construct the 95 percent confidence interval for $p_1 - p_2$.

   (b) Find the approximate P-value for testing $p_1 = p_2$ versus the third alternative.

2. Exercises 12 and 14 on p. 194 detailed a 1989 survey of college women. The sample of 112 sexually experienced college women indicated that 46 used condoms "always or almost always" during sexual intercourse. In a similar study at the same school three years earlier, 30 of 129 sexually experienced college women gave the same answer [2].

   (a) Which of the four types of studies is this?

   (b) On the assumption that these were independent random samples, analyze this data with a 95 percent confidence interval.

   (c) On the assumption that these were independent random samples, conduct a hypothesis test with the alternative that the use of condoms increased during the three year period.

   (d) Write a brief summary of the results of the study. Include, if appropriate, a discussion of confounding, practical importance, and the assumption of independent random samples.

3. *Newsweek* reported that the Gallup Organization interviewed a national sample of 756 adults by telephone on March 9–10, 1989 [3]. Sixty-two percent of those surveyed responded that they approved when asked

   Do you approve or disapprove of the way George Bush is handling his job?

   In a similar poll one month earlier, 55 percent responded that they approved. Assume that the earlier poll had the same sample size (*Newsweek* did not report the earlier sample size).

   (a) Which of the four types of studies is this?

(b) Assuming that these were independent random samples, analyze this data with a 95 percent confidence interval.

(c) Assuming that these were independent random samples, conduct a hypothesis test with the alternative that the opinion changed.

(d) Write a brief summary of the results of the study. Include, if appropriate, a discussion of confounding, practical importance, and the assumption of independent random samples.

4. Exercise 10 on p. 177 introduced a study of Denise Howsare throwing 100 darts. In the process of investigating the third condition for Bernoulli trials, the following table was obtained:

	Current Trial		
Previous Trial	S	F	Total
S	28	19	47
F	20	32	52
Total	48	51	99

Refer to Example 7.14 (p. 242).

(a) Test the null hypothesis that $p_1 = p_2$ versus the third alternative.

(b) Construct a 95 percent for $p_1 - p_2$.

(c) Briefly discuss what you have learned from parts (a) and (b).

5. Exercise 5 on p. 33 described the following CRD. Pam Crawford's son threw a ball at a target 50 times, 25 times with each hand. He obtained 12 successes with his right hand and 10 with his left.

(a) Which of the four types of studies is this?

(b) Assuming that these were independent random samples, analyze these data with a 95 percent confidence interval.

6. Exercise 16 on p. 27 introduced a study that compared two treatments for breast cancer. Pretend that the subjects constituted a random sample from the superpopulation of all women with breast cancer. Construct a 95 percent confidence interval that compares the treatments. Interpret your answer.

7. Example 5.21 (p. 174) reported the results of a study of the video game Tetris. In the process of investigating the third condition for Bernoulli trials, the following table was obtained:

	Current Shape		
Previous Shape	S	F	Total
Success	2	251	253
Failure	249	1,362	1,611

Refer to Example 7.14 (p. 242).

(a) Test the null hypothesis that $p_1 = p_2$ versus the third alternative.

(b) Briefly discuss what you have learned.

8. Exercise 20 on p. 35 introduced a study of putts attempted with and without a golf glove. Assuming that each method gives Bernoulli trials, construct a 95 percent confidence interval to compare the methods. Interpret your answer.

9. Exercise 11 on p. 178 introduced a study of Margaret Algar-Gelembiuk making 100 practice first serves in tennis. In the process of investigating the third condition for Bernoulli trials, the following table was obtained:

	Current Trial		
Previous Trial	S	F	Total
S	21	24	45
F	24	30	54
Total	45	54	99

Refer to Example 7.14 (p. 242).

(a) Test the null hypothesis that $p_1 = p_2$ versus the third alternative.

(b) Construct a 95 percent confidence interval for $p_1 - p_2$.

(c) Briefly discuss what you have learned from parts (a) and (b).

10. Question 7 on the Wisconsin Driver Survey measured the respondent's knowledge of the legal limit of blood alcohol concentration in Wisconsin (0.10 percent). In 1982, 50.1 percent of 1,589 respondents gave the correct answer compared with 56.2 percent of 1,072 in 1983. Construct a 95 percent confidence interval for the increase in knowledge from 1982 to 1983.

11. Question 12 on the 1983 and 1984 Wisconsin Driver surveys asked, "Have you heard anything about Wisconsin's recent drunken driving law?" In 1983, 64.8 percent of the 1,072 persons questioned responded yes, and in 1984, 59.0 percent of the 2,632 persons questioned responded yes. Construct a 95 percent confidence interval for the decrease in positive responses in the population from 1983 to 1984. Briefly comment on your findings. Include, if appropriate, a discussion of confounding, practical importance, and the assumption of independent random samples.

12. I collected data on the first 14 weeks of the 1990 major league baseball season. In the American League the team that was leading at the end of the sixth inning went on to win 458 out of 527 games. In the National League the team that was leading at the end of the sixth inning went on to win 390 out of 450 games. Assuming that these data are independent random samples from two populations, construct a 95 percent confidence interval for $p_1 - p_2$ (let the American League be the first population). Write a brief summary of your findings. Include, if appropriate, a discussion of confounding, practical importance, and the assumption of independent random samples.

13. Refer to Exercise 11. Respondents who had heard of the law were given a five-question quiz. The first question read

> Is there a mandatory 90-day driver license suspension for an operating while intoxicated conviction?

In 1983, 65.0 percent of the 695 persons who answered gave the correct answer (yes). In 1984, 66.4 percent of the 1,552 persons who answered gave the correct answer. Construct a 95 percent confidence interval for the increase in knowledge in the population from 1983 to 1984. Briefly comment on your findings. Include, if appropriate, a discussion of confounding, practical importance, and the assumption of independent random samples.

14. Refer to Exercise 12. In the American League, the team ahead at the end of the seventh inning went on to win 484 out of 529 games. In the National League, the team that was leading at the end of the seventh inning went on to win 423 out of 467 games. Assuming that these data are independent random samples from two populations, construct a 95 percent confidence interval for $p_1 - p_2$ (let the American League be the first population). Write a brief summary of your findings. Include, if appropriate, a discussion of confounding, practical importance, and the assumption of independent random samples.

15. Refer to Exercise 11. Respondents who had heard of the law were given a five-question quiz. The second question read

> Does it lower the blood alcohol concentration for intoxication to 0.08 percent?

In 1983, 65.9 percent of the 695 persons who answered gave the correct answer (no). In 1984, 60.9 percent of the 1,552 persons who answered gave the correct answer. Construct a 95 percent confidence interval for the increase in knowledge in the population from 1983 to 1984. Briefly comment on your findings. Include, if appropriate, a discussion of confounding, practical importance, and the assumption of independent random samples.

16. Refer to Exercise 12. In the American League the team ahead at the end of the eighth inning went on to win 515 out of 543 games. In the National League the team that was leading at the end of the eighth inning went on to win 458 out of 477 games.

(a) Assuming that these data are independent random samples from two populations, construct a 95 percent confidence interval for $p_1 - p_2$ (let the American League be the first population).

(b) Write a brief summary of your findings. Include, if appropriate, a discussion of confounding, practical importance, and the assumption of independent random samples.

(c) (Hypothetical) At the end of the eighth inning in a National League game in which the score is 15-0, a television announcer states

> Don't go away: Statistics tells us that there is a 4 percent chance that the trailing team will come back and win.

What do you think?

(d) (Hypothetical) At the end of the eighth inning in a National League game in which the score is 10-9, a television announcer states

> Statistics tells us that there is only a 4 percent chance that the trailing team will win.

What do you think?

17. Exercise 21 on p. 35 asked you to perform a balanced CRD consisting of 100 trials. Assuming that both treatments give data that are Bernoulli trials, construct a 95 percent confidence interval to compare the treatments. Interpret your answer.

18. This exercise refers to the material presented in the optional section entitled Statistical Power (Sec. 7.3.2). Refer to Table 7.9.

(a) Suppose that before conducting the Three-Point Basket study, Clyde Gaines had said, "If, in fact, $p_1 = 0.50$ and $p_2 = 0.40$, I want my test to have at least a 30 percent chance of correctly rejecting the null hypothesis." What advice would you give Clyde?

(b) Suppose that before conducting the study, Gaines had said, "If, in fact, $p_1 = 0.50$ and $p_2 = 0.40$, I want my test to have at least an 80 percent chance of correctly rejecting the null hypothesis." What advice would you give Clyde?

REFERENCES

1. C. Muehlenhard and E. Miller, "Traditional and Nontraditional Men's Responses to Women's Dating Initiation," *Behavior Modification*, July 1988, pp. 385–403.

2. B. A. DeBuono et al., "Sexual Behavior of College Women in 1975, 1986, and 1989," *New England Journal of Medicine*, Vol. 322, March 22, 1990, pp. 821–825.

3. "The Honeymoon's Still On: A *Newsweek* Poll," March 20, 1989, p. 27.

Two Dichotomous Responses

8.1 INTRODUCTION

8.2 CONDITIONAL PROBABILITY AND SCREENING TESTS

8.3 ESTIMATING PROBABILITIES AND CONDITIONAL PROBABILITIES

8.4 HYPOTHESIS TESTING AND CONFIDENCE INTERVALS

8.5 COMPARING p_A WITH p_B

Chapter 8

8.1 INTRODUCTION

Chapter 5 introduced the concept of a population for one dichotomous variable per individual or trial. Such a population can be represented by a single number p. Chapter 6 presented methods of statistical inference for the common situation when p is unknown. This chapter extends some of the material of Chapters 5 and 6 to studies in which there are two dichotomous variables per subject.

A population model for two dichotomous variables can arise for a collection of individuals (a finite population) or as a mathematical model for a process that generates two dichotomous variables per trial. It is easier to introduce the ideas and notation for a finite population; an example of a mathematical model is presented later in the chapter.

Recall from Chapter 5 that a finite population can be visualized as a box of cards, with one card for each member of the population. In Chapter 5 each card had a single piece of information—either a 1 or a 0—and the box could be summarized by the proportion of cards marked 1, which is denoted by p. The problem of the current chapter is more complicated. Let the categories of the first variable be denoted by A and A^c, and let the categories of the second variable be denoted by B and B^c. Four types of cards are in the box, corresponding to the different possible combinations of A or A^c with B or B^c. We make the following definitions:

- N_{AB} denotes the number of cards in the box that are marked A and B.
- N_{AB^c} denotes the number of cards in the box that are marked A and B^c.

TABLE 8.1 Table of Population Counts

	B	B^c	Total
A	N_{AB}	N_{AB^c}	N_A
A^c	N_{A^cB}	$N_{A^cB^c}$	N_{A^c}
Total	N_B	N_{B^c}	N

- N_{A^cB} denotes the number of cards in the box that are marked A^c and B.
- $N_{A^cB^c}$ denotes the number of cards in the box that are marked A^c and B^c.

Let $N = N_{AB} + N_{AB^c} + N_{A^cB} + N_{A^cB^c}$ denote the total number of cards in the box. We further define the following:

- p_{AB} denotes the proportion of cards in the box that are marked A and B.
- p_{AB^c} denotes the proportion of cards in the box that are marked A and B^c.
- p_{A^cB} denotes the proportion of cards in the box that are marked A^c and B.
- $p_{A^cB^c}$ denotes the proportion of cards in the box that are marked A^c and B^c.

It is convenient to present the population counts in a table, as illustrated in Table 8.1. This table implicitly defines four more Ns; for example, $N_A = N_{AB} + N_{AB^c}$ denotes the number of cards in the box that are labeled A. Table 8.1 is called the **table of population counts** for the finite population. If every number in a table of population counts is divided by N, the result is called the **table of population proportions,** which is given in Table 8.2. This table implicitly defines four more p's; for example, $p_{A^c} = p_{A^cB} + p_{A^cB^c}$ is the proportion of cards in the box that are marked A^c. The population proportions also can be interpreted as probabilities for the chance mechanism of selecting one card at random from the box. Suppose, for example, that a researcher selects one card at random from a population box. Then the probability that the card selected has an A and a B on it is

$$\frac{\text{The number of cards that are labeled } A \text{ and } B}{\text{The total number of cards in the box}} = \frac{N_{AB}}{N} = p_{AB}.$$

Thus, the table of population proportions also is called the table of probabilities.

The following section introduces an important application of these ideas, screening tests for a disease or condition.

TABLE 8.2 Table of Population Proportions or Probabilities

	B	B^c	Total
A	p_{AB}	p_{AB^c}	p_A
A^c	p_{A^cB}	$p_{A^cB^c}$	p_{A^c}
Total	p_B	p_{B^c}	1

8.2 CONDITIONAL PROBABILITY AND SCREENING TESTS

Early diagnostic, or screening, tests are an important component of modern health care. Many colleges require students to have a tuberculosis skin test before admission. Routine Pap tests are recommended for early detection of cervical cancer. The Red Cross screens donated blood for the HIV antibody, and health centers test persons for the HIV antibody. A mammogram can provide early detection of breast cancer. A screening test gives either a positive or negative response: a positive response is taken to mean that the condition of interest is present, and a negative response is taken to mean that the condition of interest is absent. For example, a positive Pap test is taken to mean the woman has cervical cancer. I say "taken to mean" because screening tests can be mistaken, a fact that often is overlooked in discussions of the individual and societal implications of recommending or requiring screening tests. This section uses the general framework of the previous section to study such screening tests. In the process of studying screening tests, the important notion of **conditional probability** is introduced.

A population consists of N persons. Two characteristics of each person are of interest. First, there is the person's actual health status, in which the condition of interest is either present, denoted by A, or absent, denoted by A^c; second, there is the person's response to the screening test, either positive, denoted by B, or negative, denoted by B^c. It is helpful to describe in words each of the entries in Table 8.2:

- p_A is the proportion of the population who have the condition.
- p_{A^c} is the proportion of the population who do not have the condition.
- p_B is the proportion of the population who would give a positive response to the screening test (test positive).
- p_{B^c} is the proportion of the population who would give a negative response to the screening test (test negative).
- p_{AB} is the proportion of the population who have the condition and would test positive; it is the proportion of the population who would receive a **correct positive** test result.

- p_{AB^c} is the proportion of the population who have the condition but would test negative; it is the proportion of the population who would receive a **false negative** test result.
- p_{A^cB} is the proportion of the population who do not have the condition but would test positive; it is the proportion of the population who would receive a **false positive** test result.
- $p_{A^cB^c}$ is the proportion of the population who do not have the condition and would test negative; it is the proportion of the population who would receive a **correct negative** test result.

This section considers how to use these probabilities to judge the utility of a screening test. The topic begins with some simple observations.

False negatives and false positives are mistakes, so a good screening test will have small values of p_{AB^c} and p_{A^cB}. One must realize that in most situations the practical implications of the two possible errors are quite different. Consider the tuberculosis skin test. The result of a false negative is that an ill person is not treated, possibly leading to the person's becoming more seriously ill or infecting other persons. A false positive means that a person without tuberculosis is labeled diseased. In practice, a positive skin test is followed by a chest x-ray, which typically will unearth any false positives. Thus, for the person, the practical implications of a false positive are temporary unwarranted anxiety and the health risk of receiving an unneeded chest x-ray. Clearly, the false negative is a more serious error. The exercises ask you to evaluate the seriousness of the two possible errors for the other conditions presented earlier.

In order to develop these ideas further, let us consider some specific values for the various probabilities. Table 8.3 presents population counts and proportions for a hypothetical condition and screening test. (Hypothetical values are used because it is clearly impossible to determine exact probabilities for the examples discussed earlier.) Inspection of this table reveals a number of facts, including the following:

- $p_A = 0.100$; in words, 10.0 percent of the population have the condition.
- $p_B = 0.120$; in words, 12.0 percent of the population would give a positive response to the screening test.

TABLE 8.3 *Hypothetical Population Counts and Proportions for a Condition and Its First Screening Test*

| | Screening Test | | | | Screening Test | | |
Condition	Positive (B)	Negative (B^c)	Total	Condition	Positive (B)	Negative (B^c)	Total
Present (A)	12	88	100	Present (A)	0.012	0.088	0.100
Absent (A^c)	108	792	900	Absent (A^c)	0.108	0.792	0.900
Total	120	880	1,000	Total	0.120	0.880	1.000

- $p_{AB^c} = 0.088$; in words, 8.8 percent of the population would obtain a false negative test result.
- $p_{A^cB} = 0.108$; in words, 10.8 percent of the population would obtain a false positive test result.

This is not an effective screening test; conditional probability, which is defined next, shows just how bad it is.

Imagine the chance mechanism of selecting one person at random from this population. The probability that the person selected has the condition is $p_A = 0.100$. The notion of conditional probability enables an analyst to answer the following question:

> Suppose this randomly selected person is given the screening test and yields a positive response. Given this additional information, what is the probability that the person has the condition?

The answer to this question is written symbolically as $p_{A|B}$, which is read, "the probability that A occurs given that B occurs"; in the current context this is "the probability that the person has the condition given that his or her screening test is positive." One answers this question most easily by examining the population counts in Table 8.3.

Given that the person's screening test is positive, it follows that the person selected is among the 120 in the first (B) column of the table. Twelve of these 120 persons have the condition (A). Thus,

$$p_{A|B} = \frac{12}{120} = 0.100.$$

This is the same number as p_A; in words, knowledge that the screening test is positive *does not change* the probability that the person has the condition.

One also can consider the following question:

> Suppose this randomly selected person is given the screening test and yields a negative response. What is the probability that the person has the condition?

The answer to this question is $p_{A|B^c}$. Repeating the argument given above for $p_{A|B}$, one finds

$$p_{A|B^c} = \frac{88}{880} = 0.100.$$

To summarize, the probability that a randomly selected person has the condition is 0.1; given that the person yields a positive result on the screening test, this probability is still 0.1; and given that the person yields a negative result on the screening test, this probability is still 0.1. Clearly, this screening test is of no value.

Table 8.4 presents population counts and proportions for the same hypothetical condition but for a different screening test. Imagine the chance mechanism of selecting one person at random from this population. The probability that the person selected has the condition is $p_A = 0.100$. Following the method given earlier,

$$p_{A|B} = \frac{95}{104} = 0.913 \quad \text{and} \quad p_{A|B^c} = \frac{5}{896} = 0.006.$$

In words, without the results of the screening test, the probability that a randomly selected person has the condition is 10 percent; given that the screening test is positive, this probability is 91.3 percent; and given that it is negative, this probability is 0.6 percent. Thus, this second screening test, although not absolutely conclusive, is informative.

It is convenient to derive formulas for conditional probabilities in terms of the general notation presented in Tables 8.1 and 8.2. First,

$$p_{A|B} = \frac{N_{AB}}{N_B}; \tag{8.1}$$

in words, the conditional probability of A given B is the proportion of all individuals with characteristic B who also have characteristic A. Recall that

$$p_{AB} = \frac{N_{AB}}{N} \quad \text{and} \quad p_B = \frac{N_B}{N}.$$

Equation 8.1 can be written as

$$p_{A|B} = \frac{N_{AB}}{N_B} = \frac{N_{AB}/N}{N_B/N} = \frac{p_{AB}}{p_B}.$$

For future reference,

$$p_{A|B} = \frac{p_{AB}}{p_B}. \tag{8.2}$$

TABLE 8.4 Hypothetical Population Counts and Proportions for a Condition and Its Second Screening Test

	Screening Test				Screening Test		
Condition	Positive (B)	Negative (B^c)	Total	Condition	Positive (B)	Negative (B^c)	Total
Present (A)	95	5	100	Present (A)	0.095	0.005	0.100
Absent (A^c)	9	891	900	Absent (A^c)	0.009	0.891	0.900
Total	104	896	1,000	Total	0.104	0.896	1.000

In words, the conditional probability of A given B is the ratio of the proportion with both characteristics to the proportion with the latter characteristic. This verbal interpretation can be used to obtain easily any conditional probability of interest. For example, the conditional probability of A given B^c is the ratio of the proportion with both characteristics (AB^c) to the proportion with the latter characteristic (B^c). Thus,

$$p_{A|B^c} = \frac{p_{AB^c}}{p_{B^c}}.$$

It is also of interest to compute the conditional probability of B given A. Adapting Equation 8.2 yields

$$p_{B|A} = \frac{p_{AB}}{p_A}.$$

For the second screening test for the hypothetical condition,

$$p_{B|A} = \frac{p_{AB}}{p_A} = \frac{0.095}{0.100} = 0.95.$$

In words, given that a randomly selected person has the condition, the probability that the person (also) yields a positive response to the screening test is 95 percent.

A common error is to confuse p_{AB}, $p_{A|B}$ and $p_{B|A}$. As computed above, the values of these probabilities for the second screening test are 0.095, 0.913, and 0.950, respectively. The first of these, p_{AB}, is the proportion of the entire population who correctly test positive; $p_{A|B}$ is the proportion of those testing positive who have the condition; and $p_{B|A}$ is the proportion of those having the condition who test positive. Abusing the language and taking the perspective of the randomly selected person, $p_{A|B}$ is the probability that I have the condition given that my test is positive, and $p_{B|A}$ is the probability that my test is positive given that I have the condition. A member of the population is primarily interested in $p_{A|B}$, and a researcher or public health official is more interested in $p_{B|A}$.

The remarks of the previous paragraph can be extended to false positives (A^c occurring with B), false negatives (A occurring with B^c), and correct negatives (A^c occurring with B^c). For the second screening test and false positives,

$$p_{A^cB} = 0.0090, \quad p_{A^c|B} = \frac{p_{A^cB}}{p_B} = \frac{0.009}{0.104} = 0.0865,$$

$$\text{and} \quad p_{B|A^c} = \frac{p_{A^cB}}{p_{A^c}} = \frac{0.009}{0.900} = 0.0100.$$

These numbers are quite different (see the exercises for even more extreme examples). I have seen each of these numbers referred to as "the" false positive rate in different studies. Thus, it is incumbent upon consumers of research results to take care to learn which formula for "the" false positive rate is being used. Similarly, for the second screening test and false negatives,

$$p_{AB^c} = 0.0050, \quad p_{A|B^c} = \frac{p_{AB^c}}{p_{B^c}} = \frac{0.005}{0.896} = 0.0056,$$

$$\text{and} \quad p_{B^c|A} = \frac{p_{AB^c}}{p_A} = \frac{0.005}{0.100} = 0.0500.$$

Finally, for the second screening test and correct negatives,

$$p_{A^cB^c} = 0.8910, \quad p_{A^c|B^c} = \frac{p_{A^cB^c}}{p_{B^c}} = \frac{0.891}{0.896} = 0.9944,$$

$$\text{and} \quad p_{B^c|A^c} = \frac{p_{A^cB^c}}{p_{A^c}} = \frac{0.891}{0.900} = 0.9900.$$

For any screening test, $p_{B|A}$ is called the **sensitivity** of the test and $p_{B^c|A^c}$ is called the **specificity** of the test. As computed above, the sensitivity of the second screening test 0.95 and the specificity is 0.99.

Refer to the table of population proportions for the first screening test given in Table 8.3. Notice that

$$p_A p_B = 0.1(0.12) = 0.012 \quad \text{and} \quad p_{AB} = 0.012.$$

Thus, $p_{AB} = p_A p_B$; in words, the proportion with characteristics A and B equals the proportion with characteristic A multiplied by the proportion with characteristic B. In short, the multiplication rule holds for A and B; thus, A and B are referred to as **independent events.** It can be shown that if A and B are independent events, the other three combinations of A or A^c with B or B^c are independent as well. If one establishes that A and B are independent, it follows that A^c and B, for example, are independent; that is, $p_{A^cB} = p_{A^c}p_B$. If A and B are independent, the first variable (which takes on value A or A^c) is said to be independent of the second variable (which takes on value B or B^c). This agrees with the definition of independent random variables given in Chapter 5 (see p. 157).

If A and B are independent, then

$$p_{A|B} = \frac{p_{AB}}{p_B} = \frac{p_A p_B}{p_B} = p_A.$$

In words, the conditional probability of A given B equals the **unconditional probability** of A. (Unconditional probability is just probability; the modifier *unconditional* is added for emphasis to distinguish it from conditional probability.) This argument can be extended to yield the facts listed below. If interested, you may verify them.

- If the two variables are independent, the conditional probability of A given either B or B^c equals the unconditional probability of A.
- If the two variables are independent, the conditional probability of A^c given either B or B^c equals the unconditional probability of A^c.
- If the two variables are independent, the conditional probability of B given either A or A^c equals the unconditional probability of B.

- If the two variables are independent, the conditional probability of B^c given either A or A^c equals the unconditional probability of B^c.

Thus, if the variables are independent, there is no point in computing conditional probabilities.

Conversely, if A and B are not independent, that is, if they are **dependent**, then

$$p_{AB} \neq p_A p_B.$$

$$p_{A|B} = \frac{p_{AB}}{p_B} \neq \frac{p_A p_B}{p_B} = p_A.$$

In words, the conditional probability of A given B does not equal the unconditional probability of A. Again, this argument can be extended to conclude that *for dependent events, the conditional probabilities are always different from the unconditional probabilities.*

When given a table of population proportions, always check first for independence. Consider the following table of population proportions, for example:

	B	B^c	Total
A	0.12	0.28	0.40
A^c	0.18	0.42	0.60
Total	0.30	0.70	1.00

Here $p_A p_B = 0.4(0.3) = 0.12$, and $p_{AB} = 0.12$. Thus, A and B are independent, the variables are independent, and there is no reason to compute any conditional probabilities.

By contrast, if the table of population proportions is that given in Table 8.5, then $p_A p_B = 0.4(0.3) = 0.12$, but $p_{AB} = 0.18$. Thus, A and B are dependent, the characteristics are dependent, and the conditional probabilities may give interesting insights into the structure of the population.

TABLE 8.5 *An Example of Dependent Variables*

Proportion of Population				Conditional Probability of B or B^c Given A or A^c				Conditional Probability of A or A^c Given B or B^c			
	B	B^c	Total		B	B^c	Total		B	B^c	Unconditional
A	0.18	0.22	0.40	A	0.45	0.55	1.00	A	0.60	0.31	0.40
A^c	0.12	0.48	0.60	A^c	0.20	0.80	1.00	A^c	0.40	0.69	0.60
Total	0.30	0.70	1.00	Unconditional	0.30	0.70	1.00	Total	1.00	1.00	1.00

An easy way exists to obtain all of the conditional probabilities without repeated application of Equation 8.2. Simply divide each entry in the table of population proportions by its row total; the resultant table contains the conditional probabilities of B or B^c given A or A^c. For example, divide each entry in the first (A) row of Table 8.5 by 0.40, divide each entry in the second (A^c) row by 0.60, and divide each entry in the third (Total) row by 1. The result of these divisions is given in the second section of this table. Reading from this table, one finds, for example, that $p_{B|A} = 0.45$. The row labeled "Unconditional" presents the unconditional probabilities of B and B^c for comparison with the conditional values.

Similarly, if each entry in the table of population proportions is divided by its column total, the resultant table contains all conditional probabilities of the As given the Bs. The result of this operation for the current example is given in the third section of Table 8.5. Reading from this table, one finds, for example, that $p_{A|B} = 0.60$. The column labeled "Unconditional" presents the unconditional probabilities of A and A^c for comparison with the conditional values.

Remember that dividing by row totals gives the conditional probabilities of Bs given As, and dividing by column totals gives the conditional probabilities of As given Bs. If you have the resultant table but do not know how the division was performed, look at the row and column totals. If the row totals are all equal to 1, the table of population proportions has been divided by its row totals and the resultant table contains the conditional probabilities of Bs given As. If the column totals are all equal to 1, the table of population proportions has been divided by its column totals, and the resultant table contains the conditional probabilities of As given Bs.

If the events A and B are independent, then $p_{AB} = p_A p_B$. This is the multiplication rule, which was introduced in Section 5.2. There is a generalization of this rule that can be applied to any two events, including dependent events.

Take the application of the formula for conditional probability

$$p_{B|A} = \frac{p_{AB}}{p_A},$$

and multiply each side of this equation by p_A; the result is

$$p_A p_{B|A} = p_{AB}.$$

Interchange the left and right sides of this equation to yield

$$p_{AB} = p_A p_{B|A}. \tag{8.3}$$

In words, the probability that A and B occur is equal to the probability that A occurs multiplied by the conditional probability that B occurs given that A occurs.

8.2 Conditional Probability and Screening Tests

In a similar way, the equation

$$p_{A|B} = \frac{p_{AB}}{p_B}$$

yields

$$p_{AB} = p_B p_{A|B}. \tag{8.4}$$

Equations 8.3 and 8.4 state that the probability that two events occur is equal to the probability that either event occurs multiplied by the conditional probability the remaining event occurs, given that the other event occurs. This is called the **multiplication rule for conditional probabilities.**

In practice, use the rule that helps solve the problem at hand. For example, given that $p_B = 0.5$ and $p_{A|B} = 0.7$, Equation 8.4, but not Equation 8.3, can be used to obtain $p_{AB} = 0.5(0.7) = 0.35$.

EXAMPLE 8.1: Building a Table of Probabilities

Suppose you are given that $p_A = 0.24$, $p_{B|A} = 0.75$, and $p_{B|A^c} = 0.50$. This information allows you to complete the table of population proportions. First, note that by the complement rule, $p_{A^c} = 0.76$. Next, by Equation 8.3,

$$p_{AB} = p_A p_{B|A} = 0.24(0.75) = 0.18.$$

The multiplication rule for conditional probabilities also yields

$$p_{A^c B} = p_{A^c} p_{B|A^c} = 0.76(0.50) = 0.38.$$

These results can be placed in the table of population proportions, yielding the following:

	B	B^c	Total
A	0.18		0.24
A^c	0.38		0.76
Total			1.00

Addition and subtraction transform this incomplete table to the complete table of population proportions:

	B	B^c	Total
A	0.18	0.06	0.24
A^c	0.38	0.38	0.76
Total	0.56	0.44	1.00

▲

EXERCISES 8.2

1. What do the terms *correct positive, correct negative, false positive,* and *false negative* mean for each of the following examples? Compare and contrast the consequences of false positive and false negative screening test results.
   (a) A Pap test
   (b) A mammogram

2. What do the terms *correct positive, correct negative, false positive,* and *false negative* mean for each of the following examples? Compare and contrast the consequences of false positive and false negative screening test results.
   (a) An HIV antibody test for donated blood
   (b) An HIV antibody test for diagnosing a person

3. In a hypothetical population, 360 persons have both characteristics A and B, 140 persons have both characteristics A and B^c, 240 persons have both characteristics A^c and B, and 260 persons have both characteristics A^c and B^c.
   (a) Construct the table of population counts.
   (b) Construct the table of population proportions.
   (c) Construct the table of conditional probabilities of the Bs given the As.
   (d) Construct the table of conditional probabilities of the As given the Bs.
   (e) Are the variables statistically independent?

4. In a hypothetical population of 100 persons, 60 persons have characteristic A, 40 persons have characteristic B, and 30 persons have both these characteristics.
   (a) Construct the table of population counts.
   (b) Construct the table of population proportions.
   (c) Construct the table of conditional probabilities of the Bs given the As.
   (d) Construct the table of conditional probabilities of the As given the Bs.
   (e) Are the variables statistically independent?

5. In a hypothetical population of 1,000 persons, 500 persons have characteristic A, 400 persons have characteristic B, and 200 persons have both of these characteristics.
   (a) Construct the table of population counts.
   (b) Construct the table of population proportions.
   (c) Construct the table of conditional probabilities of the Bs given the As.
   (d) Construct the table of conditional probabilities of the As given the Bs.
   (e) Are the variables statistically independent?

6. In a hypothetical population of 500 persons, 200 persons have characteristic A, 150 persons have characteristic B, and 60 persons have both of these characteristics.
   (a) Construct the table of population counts.
   (b) Construct the table of population proportions.
   (c) Construct the table of conditional probabilities of the Bs given the As.
   (d) Construct the table of conditional probabilities of the As given the Bs.
   (e) Are the variables statistically independent?

7. In a hypothetical population, $p_A = 0.3$, $p_B = 0.4$, and the variables are independent. Construct the table of probabilities.

8. In a hypothetical population, $p_A = 0.7$ and the variables are independent. What is the value of $p_{A|B}$? What is the value of $p_{A^c|B^c}$?

9. In a hypothetical population, $p_A = 0.2$, $p_{B|A} = 0.8$, and $p_{B|A^c} = 0.6$.
   (a) Use the multiplication rule for conditional probabilities to construct the table of population proportions.
   (b) Construct the table of the conditional probabilities of the As given the Bs.

10. In a hypothetical population, $p_A = 0.3$, $p_{B|A} = 0.7$, and $p_{B|A^c} = 0.4$.
    (a) Use the multiplication rule for conditional probabilities to construct the table of population proportions.
    (b) Construct the table of the conditional probabilities of the As given the Bs.

11. For a particular screening test,

$p_A = 0.01$, $p_{B|A} = 0.95$, and $p_{B^c|A^c} = 0.99$.

(a) Use the multiplication rule for conditional probabilities to construct the table of population proportions.
(b) Construct the table of the conditional probabilities of the Bs given the As.
(c) Construct the table of the conditional probabilities of the As given the Bs.
(d) Refer to the three tables you have just constructed. Identify the three false positive rates and compare their numerical values. Comment.
(e) Refer to the three tables you have just constructed. Identify the three false negative rates and compare their numerical values. Comment.

12. For a particular screening test,

$p_A = 0.001$, $p_{B|A} = 0.95$, and $p_{B^c|A^c} = 0.99$.

(a) Use the multiplication rule for conditional probabilities to construct the table of population proportions.
(b) Construct the table of the conditional probabilities of the Bs given the As.
(c) Construct the table of the conditional probabilities of the As given the Bs.
(d) Refer to the three tables you have just constructed. Identify the three false positive rates and compare their numerical values. Comment.
(e) Refer to the three tables you have just constructed. Identify the three false negative rates and compare their numerical values. Comment.

13. This exercise is motivated by a 1987 report on the screening test for the HIV antibody [1]. A sample of blood is drawn from a person and subjected to an enzyme immunoassay, which can be either reactive or nonreactive. If the result is nonreactive, the screening test is negative. If, however, the result is reactive, another sample of blood is analyzed with the more expensive Western blot test. In that case, the result of the screening test is that of the Western blot test. In short, the screening test is positive if the enzyme immunoassay is reactive and the Western blot test is positive. The screening test is negative if the enzyme immunoassay is nonreactive or the Western blot test, if given, is negative. Of course, none of the population proportions are known, nor are they easy to estimate. In such cases, researchers settle for studying the consequences of specific educated guesses of the values of the probabilities. In the report mentioned, the authors used the values $p_A = 0.0001$, $p_{AB^c} = 0$, and $p_{B|A^c} = 0.00005$.

(a) Write a few sentences that explain, in simple language, the meaning of each of these three assumed values.
(b) Use the multiplication rule for conditional probabilities to construct the table of population proportions.
(c) Construct the table of the conditional probabilities of the Bs given the As.
(d) Construct the table of the conditional probabilities of the As given the Bs.
(e) Refer to the three tables you have just constructed. Identify the three false positive rates and compare their numerical values. Comment.
(f) Refer to the three tables you have just constructed. Identify the three false negative rates and compare their numerical values. Comment.

(The interested reader may want to refer also to [2] and [3].)

8.3 ESTIMATING PROBABILITIES AND CONDITIONAL PROBABILITIES

The previous section demonstrated how to compute conditional probabilities from a table of population proportions. In most statistical applications the entries in the table of population proportions are unknown. Thus, the researcher must

TABLE 8.6 Observed Frequencies of Responses on Gender and Attitude for the 1986 Wisconsin Driver Survey.

Gender	The Drunk Driving Problem Is		Total
	Extremely Serious (B)	Not Extremely Serious (B^c)	
Female (A)	419	332	751
Male (A^c)	357	396	753
Total	776	728	1,504

settle for estimates of the population proportions and conditional probabilities. The ideas will be introduced with an example.

In the 1986 Wisconsin Driver Survey, several variables were measured on each subject, including

- Gender
- Attitude toward drunken driving

Attitude was measured by the response to question 1, "How serious a problem do you think drunk driving is in Wisconsin?" Possible responses were "extremely serious" and "not extremely serious."

The results of the survey are given in Table 8.6. (The actual sample size was not 1,504, the total for this table, but 1,689. Unfortunately, 179 subjects did not respond to the question on gender and 6 others did not provide their attitude. All analyses that follow are based on the assumption that ignoring this nonresponse does not bias the results. Certainly, some subjects found the gender question to be intrusive, but there is reason to believe that the physical design of the questionnaire made it easy to overlook the request for gender.) Notice that this is *not* the table of population counts; it is the **table of observed frequencies**.

In order to have a reasonable estimate of the table of row proportions, assume that the 1,504 persons surveyed were selected at random from the population. If each entry in Table 8.6 is divided by the sample size, 1,504, the result is the **table of relative frequencies**, which is given in Table 8.7. Each entry in the table of relative frequencies is the point estimate of the corresponding entry in the table

TABLE 8.7 Table of Relative Frequencies of Responses on Gender and Attitude for the 1986 Wisconsin Driver Survey

Gender	The Drunk Driving Problem Is		Total
	Extremely Serious (B)	Not Extremely Serious (B^c)	
Female (A)	0.279	0.221	0.499
Male (A^c)	0.237	0.263	0.501
Total	0.516	0.484	1.000

of population proportions. For example, 0.279 is the point estimate of p_{AB}, the proportion of the population that is female and responds "extremely serious." Similarly, 0.263 is the point estimate of $p_{A^c B^c}$, the proportion of the population that is male and responds "not extremely serious," and 0.516 is the point estimate of p_B, the proportion of the population that responds "extremely serious."

If each entry in the table of relative frequencies (or, equivalently, in the table of observed frequencies) is divided by its row total, each entry in the resultant table is the point estimate of the corresponding entry in the table of conditional probabilities of the Bs given the As. Table 8.8 was obtained in this way. Suppose a person is selected at random from the population of all licensed drivers in Wisconsin in 1986. Reading from Table 8.8, one finds that the point estimate of the probability that the selected person would respond "extremely serious" (p_B) is 0.516. If it is given, however, that the selected person is female, the estimated probability that she would respond "extremely serious" ($p_{B|A}$) is 0.558. Finally, if it is given that the selected person is male, the estimated probability that he would give the response "extremely serious" ($p_{B|A^c}$) is 0.474.

Finally, if each entry in the table of relative frequencies (or, equivalently, in the table of observed frequencies) is divided by its column total, each entry in the resultant table is the point estimate of the corresponding entry in the table of conditional probabilities of the As given the Bs. Table 8.9 was obtained in this way. Suppose a person is selected at random from the population of all licensed drivers in Wisconsin in 1986. Reading from Table 8.9, one finds that the point estimate of the probability that the selected person is female (p_A) is 0.499. If it is given, however, that the selected person responded "extremely serious," the estimated probability that the person is female ($p_{A|B}$) is 0.540. Finally, if it is given that the selected person did not respond "extremely serious," the estimated probability that the person is female ($p_{A|B^c}$) is 0.456.

The above ideas can be applied to the table of observed frequencies for a random sample from any finite population to obtain point estimates of the population proportions and all conditional probabilities.

Table 8.6 is an example of a 2 × 2 contingency table in Format 3. The general notation for a Format 3 table is given in Table 8.10. The notation for the entries in the table agree with the notation of the earlier tables. The different formats are distinguished by the row and column labels.

TABLE 8.8 Table of Point Estimates of the Conditional Probabilities of Attitude Given Gender for the 1986 Wisconsin Driver Survey

Gender	The Drunk Driving Problem Is		Total
	Extremely Serious (B)	Not Extremely Serious (B^c)	
Female (A)	0.558	0.442	1.000
Male (A^c)	0.474	0.526	1.000
Unconditional	0.516	0.484	1.000

TABLE 8.9 *Table of Point Estimates of the Conditional Probabilities of Gender Given Attitude for the 1986 Wisconsin Driver Survey*

Gender	The Drunk Driving Problem Is		Unconditional
	Extremely Serious (B)	Not Extremely Serious (B^c)	
Female (A)	0.540	0.456	0.499
Male (A^c)	0.460	0.544	0.501
Total	1.000	1.000	1.000

The methods of this chapter can also be applied to a sequence of trials provided they satisfy the following conditions. You might want to compare these conditions with those for Bernoulli trials given in Section 5.3.

1. Each trial yields the values of two dichotomous variables. The categories of the first variable are denoted by A and A^c, and the categories of the second variable are denoted by B and B^c. Thus, each trial results in one of four possible outcomes:

$$AB, \; AB^c, \; A^cB, \; \text{or} \; A^cB^c.$$

2. The probabilities of the four possible outcomes,

$$p_{AB}, \; p_{AB^c}, \; p_{A^cB}, \; \text{and} \; p_{A^cB^c},$$

remain constant from trial to trial.

3. The trials are independent.

These are referred to as the assumptions of **bivariate Bernoulli trials.** (Bivariate Bernoulli trials are a special case of **multinomial trials,** defined in Chapter 10.) Note that the third condition, independence, does not mean that A and B are independent; it means that the results of different trials are independent. For example, the probability that the first trial gives AB and the second trial gives A^cB is equal to $p_{AB}p_{A^cB}$ by the multiplication rule for independent trials. The ideas will be illustrated with an example.

Larry Bird, formerly a professional basketball player, attempted two free throws on 338 occasions in games during the 1980–81 and 1981–82 basketball

TABLE 8.10 *2 × 2 Contingency Table in Format 3*

	B	B^c	Total
A	a	b	n_1
A^c	c	d	n_2
Total	m_1	m_2	n

8.3 Estimating Probabilities and Conditional Probabilities

TABLE 8.11 *Observed Frequencies for 338 Pairs of Free Throws Shot by Larry Bird*

First Shot	Second Shot		Total
	B (S)	B^c (F)	
A (S)	251	34	285
A^c (F)	48	5	53
Total	299	39	338

seasons. Each pair of free throws is a trial. The first dichotomous response is the outcome of the first free throw, and the second dichotomous response is the outcome of the second free throw. Define the following events:

- A: The first free throw is a success.
- A^c: The first free throw is a failure.
- B: The second free throw is a success.
- B^c: The second free throw is a failure.

Bird's 338 pairs of attempts are assumed to satisfy the conditions for bivariate Bernoulli trials. The results of Bird's shooting are presented in Table 8.11 [4]. It is useful to examine the information in this table carefully. Reading from Table 8.11, one learns, for example, that on 251 trials Bird made both free throws, on 5 trials he missed both, on 34 trials he made only the first, and on 48 trials he made only the second. Dividing each entry in the table of observed frequencies, Table 8.11, by the total number of trials, $n = 338$, yields the table of relative frequencies, Table 8.12. This latter table shows, for example, that Larry Bird made 84.3 percent of his first free throws and 88.5 percent of his second free throws, and that on 74.3 percent of the trials he made both free throws.

It is also possible, but not necessary, to compute two tables of point estimates of the conditional probabilities. This study's goal was to discover whether the probability of success on the second shot depends on the outcome of the first shot. In other words, interest lies in studying the conditional probabilities of the Bs

TABLE 8.12 *Relative Frequencies for 338 Pairs of Free Throws Shot by Larry Bird*

First Shot	Second Shot		Total
	B (S)	B^c (F)	
A (S)	0.743	0.100	0.843
A^c (F)	0.142	0.015	0.157
Total	0.885	0.115	1.000

TABLE 8.13 Point Estimates of the Conditional Probabilities of the Outcome of Larry Bird's Second Free Throw Given the Outcome of His First Free Throw

First Shot	Second Shot		Total
	B (S)	B^c (F)	
A (S)	0.881	0.119	1.000
A^c (F)	0.906	0.094	1.000
Unconditional	0.885	0.115	1.000

given the As. If each entry in Table 8.11 or 8.12 is divided by its row total, the result is the table of point estimates of the desired conditional probabilities. The results of these divisions are given in Table 8.13. As mentioned previously, Bird made 88.5 percent of his second shots: after a successful first shot, he made 88.1 percent of his second shots, and after a failed first shot, he made 90.6 percent of his second shots. Bird's proportion of successes on the second shot was 2.5 percentage points lower after a hit than after a miss.

8.4 HYPOTHESIS TESTING AND CONFIDENCE INTERVALS

Consider the 1986 Wisconsin Driver Survey study of gender and attitude, presented above. Table 8.8 shows that the estimates of the conditional probabilities are not equal to the estimates of the unconditional probabilities. For example, the estimates of $p_{B|A}$, $p_{B|A^c}$, and p_B are, respectively, 0.558, 0.474, and 0.516. In this section hypothesis testing and confidence interval procedures will be developed that enable the analyst to use the point estimates of the conditional probabilities to infer characteristics of the true (population) conditional probabilities.

8.4.1 Hypothesis Testing

The null hypothesis states that the variables are independent. Symbolically, the null hypothesis implies that $p_{B|A} = p_B$ and $p_{B|A^c} = p_B$. Thus:

 The null hypothesis can be written as

$$H_0: p_{B|A} = p_{B|A^c}.$$

There are three choices for the alternative, each reflecting a different form of dependence in the population:

$$H_1: p_{B|A} > p_{B|A^c}, \qquad H_1: p_{B|A} < p_{B|A^c}, \quad \text{or} \quad H_1: p_{B|A} \neq p_{B|A^c}.$$

These alternatives are referred to as the first, second, and third alternatives, respectively. Note the following two points:

1. The null and three alternative hypotheses are written in terms of B given A or A^c; they are mathematically equivalent, respectively, to the following hypotheses, in which the roles of A and B are interchanged:

$$H_0: p_{A|B} = p_{A|B^c}.$$

$$H_1: p_{A|B} > p_{A|B^c}, \quad H_1: p_{A|B} < p_{A|B^c}, \quad \text{or} \quad H_1: p_{A|B} \neq p_{A|B^c}.$$

2. Some statisticians advocate using the third alternative exclusively, because it includes both directions of dependence.

Steps 2, 3, and 4 of the test for these hypotheses coincide with steps 2, 3, and 4 of Fisher's test, which were introduced in Chapter 2. The reason for this correspondence is discussed in Chapter 11. From the standard normal curve one can obtain an approximate P-value:

First, compute z:

$$z = \frac{\sqrt{n-1}(ad - bc)}{\sqrt{n_1 n_2 m_1 m_2}}. \tag{8.5}$$

Second, approximate the P-value:

- For the first alternative, the approximate P-value is the area under the standard normal curve to the right of z.
- For the second alternative, the approximate P-value is the area under the standard normal curve to the right of $-z$.
- For the third alternative, the approximate P-value is twice the area under the standard normal curve to the right of $|z|$.

The appropriate area is found in Table A.2. The following rule was given in Chapter 3:

Let r_1 equal the smaller of n_1 and n_2, and let r_2 equal the smaller of m_1 and m_2. The general guideline for using the standard normal curve approximation for Fisher's test is that

$$\frac{r_1 r_2}{n} \geq 5.$$

This general guideline also is appropriate for the test of this section.

Consider the study of gender and attitude on the 1986 Wisconsin Driver Survey. Suppose the researcher is interested in finding any kind of pattern; then the third alternative is appropriate. From Table 8.6,

$a = 419$, $b = 332$, $c = 357$, $d = 396$, $n_1 = 751$,

$n_2 = 753$, $m_1 = 776$, $m_2 = 728$, and $n = 1{,}504$.

Substituting these values into Equation 8.5 gives

$$z = \frac{\sqrt{1{,}503}(419 \times 396 - 332 \times 357)}{\sqrt{751(753)(776)(728)}} = 3.25.$$

Thus, the approximate P-value is twice the area under the standard normal curve to the right of 3.25. Using Table A.2, this value is 2(0.0006), or 0.0012. These data are highly statistically significant; it seems clear that gender and attitude are dependent variables.

As another example, consider the study of Larry Bird's free-throw shooting. If the researcher is interested in finding any kind of dependence between the outcomes of the two shots, the third alternative is appropriate. From Table 8.11,

$a = 251$, $b = 34$, $c = 48$, $d = 5$, $n_1 = 285$,

$n_2 = 53$, $m_1 = 299$, $m_2 = 39$, and $n = 338$.

Substituting these values into Equation 8.5 gives

$$z = \frac{\sqrt{337}(251 \times 5 - 34 \times 48)}{\sqrt{285(53)(299)(39)}} = -0.52.$$

Thus, the approximate P-value is twice the area under the standard normal curve to the right of 0.52. Using Table A.2, this value is 2(0.3015), or 0.6030. These data are not statistically significant, and there is little reason to doubt the null hypothesis that the outcome of Bird's second shot was independent of the outcome of his first shot.

8.4.2 Confidence Intervals

This section presents and illustrates the formulas for obtaining a confidence interval for

$$p_{B|A} - p_{B|A^c} \quad \text{and} \quad p_{A|B} - p_{A|B^c}.$$

Consider again the study of gender and attitude. The analysis of this chapter has been based on the assumption that the 1,504 persons in the study were a random sample from the population of Wisconsin licensed drivers in 1986. Imagine the following alternative way of performing this study:

1. Divide the population of Wisconsin licensed drivers into two subpopulations. The first subpopulation consists of female drivers, and the second subpopulation consists of male drivers.
2. Let the attitude indicated by choice of "extremely serious" be labeled a successful response and let the attitude indicated by choice of "not extremely serious" be labeled a failed response.

3. Let p_1 denote the proportion of successes in the subpopulation of female drivers and let p_2 denote the proportion of successes in the subpopulation of male drivers.
4. Obtain data by selecting independent random samples from the two subpopulations.

With this alternate formulation, Formula 7.2 (p. 237) can be used to obtain a confidence interval for $p_1 - p_2$. Next, note that $p_{B|A}$ of this chapter is the same number as p_1: they both denote the proportion of females who would have chosen "extremely serious." Similarly, $p_{B|A^c}$ is the same number as p_2: they both denote the proportion of males who would have chosen "extremely serious." All that differs between the original problem of this chapter and the alternate formulation is the method of sampling. It can be shown that if a random sample is selected from the population of all licensed drivers, then it is mathematically valid to *pretend* that the sampling method consists of independent random samples from the two subpopulations. (Exercise 4 shows that the converse is not true—namely, if the data arise as independent random samples from two subpopulations, it can be very misleading to pretend that they arose from a random sample from the original population.) The utility of this result is that it implies that Formula 7.2 can yield a confidence interval for $p_{B|A} - p_{B|A^c}$. For ease of reference, the formula is reproduced here.

The confidence interval for $p_{B|A} - p_{B|A^c}$ is given by

$$\left(\frac{a}{n_1} - \frac{c}{n_2}\right) \pm z \times \sqrt{\frac{ab}{n_1^3} + \frac{cd}{n_2^3}} \qquad (8.6)$$

where the confidence level is used to obtain z from Table A.3. The general guideline for using this formula is that a, b, c, and d should each be greater than or equal to 10.

For the relationship between gender and attitude on the 1986 Wisconsin Driver Survey,

$$a = 419, \; b = 332, \; c = 357, \; d = 396, \; n_1 = 751, \text{ and } n_2 = 753.$$

Substituting these values into Equation 8.6 gives, for the 95 percent confidence level,

$$\left(\frac{419}{751} - \frac{357}{753}\right) \pm 1.96 \times \sqrt{\frac{419(332)}{(751)^3} + \frac{357(396)}{(753)^3}}$$

$$= (0.558 - 0.474) \pm 1.96(0.02568) = 0.084 \pm 0.050 = [0.034, 0.134].$$

In words, the 95 percent confidence interval for the difference between the population proportions of females and males who respond "extremely serious" is [0.034, 0.134]. At the 95 percent confidence level a researcher can assert that the proportion of females in the population who give the response "extremely serious" is between 3.4 and 13.4 percentage points higher than the proportion of males in the population who give the same response.

The data for Larry Bird give $d = 5$; thus, the general guideline for using Formula 8.6 is violated. During the same two-year period of study, another player, Nate Archibald, shot 321 pairs of free throws. His table of observed frequencies has values

$$a = 203, \ b = 42, \ c = 62, \ d = 14, \ n_1 = 245, \text{ and } n_2 = 76.$$

Substituting these values into Equation 8.6 gives, for the 95 percent confidence level,

$$\left(\frac{203}{245} - \frac{62}{76}\right) \pm 1.96 \times \sqrt{\frac{203(42)}{(245)^3} + \frac{62(14)}{(76)^3}} = (0.829 - 0.816) \pm 1.96(0.05057)$$
$$= 0.013 \pm 0.099 = [-0.086, 0.112].$$

The point estimate, 0.013, indicates that Archibald's proportion of successes was 1.3 percentage points higher after a hit than after a miss for the two-year period under study. The confidence interval indicates that, at the 95 percent level, Archibald could have been as much as 11.2 percentage points better after a hit than after a miss; or, at the other extreme, he could have been as much as 8.6 percentage points worse after a hit than after a miss. The interval is too wide to be of much practical value.

The roles of A and B can be reversed in the above argument to obtain the confidence interval for $p_{A|B} - p_{A|B^c}$.

The confidence interval for $p_{A|B} - p_{A|B^c}$ is given by

$$\left(\frac{a}{m_1} - \frac{b}{m_2}\right) \pm z \times \sqrt{\frac{ac}{m_1^3} + \frac{bd}{m_2^3}} \tag{8.7}$$

where the confidence level is used to obtain z from Table A.3. The general guideline for using this formula is that a, b, c, and d should each be greater than or equal to 10.

EXERCISES FOR SECTIONS 8.3 AND 8.4

1. In the 1984 Wisconsin Driver Survey, 264 of 1,140 women and 214 of 1,200 men surveyed said they did not drink alcohol.

    (a) Present this data in a 2×2 table. (Let A denote "female" and B denote "does not drink alcohol.")

    (b) Compute the table of relative frequencies.

    (c) Compute the table of point estimates of conditional probabilities of drinking status given gender. Compare briefly the point estimates of $p_{B|A}$, $p_{B|A^c}$ and p_B; use words, not symbols.

    (d) Compute the table of point estimates of conditional probabilities of gender given drinking status. Compare briefly the point estimates of $p_{A|B}$, $p_{A|B^c}$ and p_A; use words, not symbols.

(e) Are the characteristics statistically independent in the population? Use the third alternative.

(f) Construct a 95 percent confidence interval for $p_{B|A} - p_{B|A^c}$.

(g) Briefly discuss what you have learned from these computations.

2. Question 7 on the Wisconsin Driver Survey measured knowledge of the legal blood alcohol concentration limit. The results for the 1984 survey are as follows:

	Answer		
Gender	Correct	Incorrect	Total
Female	634	498	1,132
Male	787	410	1,197
Total	1,419	908	2,329

(a) Construct the table of sample relative frequencies.

(b) Construct the table of point estimates of the conditional probabilities of answer given gender. Briefly explain in words, not symbols, what is revealed.

(c) Test the null hypothesis of independence versus the third alternative.

(d) Construct a 95 percent confidence interval for the difference in the proportions of correct answers for females and males in the population.

(e) Briefly discuss what you have learned from these computations.

3. Question 11(d) on the Wisconsin Driver Survey asked the person to rate the effectiveness of mandatory jail terms as a method for "keeping people from driving after (having) too much to drink." The possible responses were "most important," "fairly important," "somewhat important," and "not very important." For the purpose of this analysis, consider two choices, "most important" and "not most important." The results for the 1984 survey are as follows:

	Mandatory Jail Terms Are		
Gender	Most Important	Not Most Important	Total
Female	476	560	1,036
Male	468	623	1,091
Total	944	1,183	2,127

(a) Construct the table of sample relative frequencies.

(b) Construct the table of point estimates of the conditional probabilities of opinion given gender. Briefly explain in words, not symbols, what is revealed.

(c) Test the null hypothesis of independence versus the third alternative.

(d) Construct a 95 percent confidence interval for the difference in the proportions of "most important" responses for females and males in the population.

4. Below is a 2 × 2 table for a hypothetical population of 1,000 persons:

	B	B^c	Total
A	10	10	20
A^c	40	940	980
Total	50	950	1,000

(a) Construct the table of probabilities.

(b) Construct the table of the conditional probabilities of the Bs given the As.

(c) Construct the table of the conditional probabilities of the As given the Bs.

An experimenter selects random samples of size 20 each without replacement from the subpopulations of persons with characteristics A and A^c. (Note that this provides a complete census of the subpopulation with characteristic A.) The results are as follows:

	B	B^c	Total
A	10	10	20
A^c	1	19	20
Total	11	29	40

An analyst decides to pretend that the data were obtained by selection of a random sample of size 40 from the original population of 1,000 persons.

(d) Compute this analyst's estimate of the table of probabilities.

(e) Compute this analyst's estimate of the table of conditional probabilities of the Bs given the As.

(f) Compute this analyst's estimate of the table of conditional probabilities of the As given the Bs.

(g) Compare the six tables you have constructed.

5. A one-year study was conducted at five hospitals in the Seattle area. Every person who sought care for a bicycle-related injury at one of the hospitals was included as a subject in the study. (Actually, 13.9 percent of the accident victims chose not to participate.) Two variables were measured for each subject:

- Whether or not a bicycle helmet was being worn at the time of the accident
- Whether or not the injury included a head injury

The results of the study are given in the following table [5]:

	Head Injury?		
Helmet Worn?	Yes	No	Total
Yes	17	103	120
No	218	330	548
Total	235	433	668

(a) Discuss whether or not these data can be assumed to be a random sample from some population. If they can, specify the population.

(b) Assuming that the data are a random sample, construct the table of point estimates of the conditional probabilities of type of injury given helmet use. Discuss the results.

(c) Still assuming that the data are a random sample, test the null hypothesis of independence versus the alternative that the probability of a head injury is lower for those who wear helmets.

(d) Again making the assumption of a random sample, construct a 95 percent confidence interval for $p_{B|A} - p_{B|A^c}$. Briefly interpret the answer.

(e) What are some of the strengths and weaknesses of this study?

(f) A researcher states, "I estimate that 120/668 = 0.180, or 18 percent, of all bicycle riders in Seattle wear helmets." Comment.

6. A newspaper article entitled, "Study Adds to Alcoholism Gene Theory" [6] stated, in part, the following:

> The gene pinpointed in the new study has two alternative forms.... The researchers looked at both alternative forms of the gene—the "A-1 allele" and the "A-2 allele"—in brain matter from the cadavers of 70 subjects, 35 alcoholics and 35 non-alcoholics. ... The A-1 allele was present in 69 percent of the alcoholics but only in 20 percent of the non-alcoholics, the researchers said.
>
> Such a high correlation was surprising, given that alcoholism comes in a number of types and is almost certain to have a number of causes, the researchers said.
>
> "A large majority of alcoholics in the present study had experienced repeated treatment failures in their alcoholic rehabilitation and the cause of death was primarily attributed to the chronic damaging effects of alcohol on their bodily systems," the researchers wrote.
>
> It is possible the A-1 allele is associated with a particular subtype of virulent alcoholism in which the person fails to respond to treatment, they wrote.

View this data as two responses on each subject, the first response being alcoholism status, with A denoting alcoholic, and the second response being the gene form, with B denoting the A-1 allele.

(a) Construct the table of sample frequencies.

(b) The article was not very informative about how the 70 cadavers were selected for study. Since this is clearly a path-breaking study, it is likely that not much effort was made to obtain a random sample. Ignoring that problem for now, which of the following do you think is the most reasonable assumption about the method of sampling? Explain your answer.

i. A random sample was selected from the population of all cadavers.

ii. Random samples were selected from the population of cadavers with the A-1 allele and the population with the A-2 allele.

iii. Random samples were selected from the population of cadavers for which the cause of death was "primarily attributed to the chronic damaging effects of alcohol on their bodily systems" and the population of nonalcoholics.

(c) Suppose that the third sampling method discussed in part (b) was, in fact, used. Test the null hypothesis of independence versus the third alternative. Explain why the third alternative is appropriate.

7. The data in Exercise 5 suggest that wearing a helmet reduces the chance of head injury. The most serious head injuries involve injury to the brain. The table at the top of the next column examines the 235 persons who had head injuries.

(a) Assuming that these data are a random sample, compute the table of point estimates of the conditional probabilities of type of injury given helmet use. Discuss the results.

Helmet Worn?	Brain Injury?		Total
	Yes	No	
Yes	4	13	17
No	95	123	218
Total	99	136	235

(b) Assuming that these data are a random sample, test the null hypothesis of independence versus the alternative that the probability of a brain injury is lower for those who wear helmets.

8. Refer to Exercise 6. Suppose random samples were selected from the population of cadavers for which the cause of death was "primarily attributed to the chronic damaging effects of alcohol on their bodily systems" and the population of nonalcoholics.

(a) Suppose a scientist proposes using the gene form as a screening test for the type of alcoholism described in the article. Ignoring the problem of obtaining brain matter from a living person, which of the three false positive rates defined in the text can be estimated from the data? Estimate all false positive rates you have identified.

(b) Repeat the previous item for the false negative rates.

(c) (Hypothetical) Suppose it were discovered that the 35 alcoholics in the study were related to each other. Would this fact change your interpretation of the results of the study?

8.5 COMPARING p_A WITH p_B

For ease of exposition refer to outcomes A and B as successes on the first and second variables, respectively. Similarly, refer to outcomes A^c and B^c as failures on the first and second variables, respectively.

Consider the study of Larry Bird presented in the previous section. The analysis indicated that there is not convincing evidence for rejecting the null hypothesis that the results of the two shots are statistically independent. Another natural question is

> Did Larry Bird have the same probability of success on the first and second shots?

In symbols, this question asks whether p_A equals p_B. From Table 8.12 the point estimates of p_A and p_B are, respectively, 0.843 and 0.885. These point estimates suggest that Bird's probability of success was larger for the second shot than it was for the first shot. Generally:

One can test
$$H_0: p_A = p_B$$
versus any of three possible choices for the alternative hypothesis,
$$H_1: p_A > p_B, \quad H_1: p_A < p_B, \quad \text{or} \quad H_1: p_A \neq p_B.$$

As usual, these are referred to as the first, second, and third alternatives, respectively. It is argued below that this test is mathematically equivalent to a test studied in Section 6.4.2.

If the null hypothesis is true,
$$p_A = p_B. \tag{8.8}$$
Reading from the table of probabilities, Table 8.2 (p. 259), one sees that
$$p_A = p_{AB} + p_{AB^c} \quad \text{and} \quad p_B = p_{AB} + p_{A^cB}.$$
Substituting these two equations into Equation 8.8 gives $p_{AB} + p_{AB^c} = p_{AB} + p_{A^cB}$. After canceling, $p_{AB^c} = p_{A^cB}$ if the null hypothesis is true. Thus, the null hypothesis depends only on the probabilities of the outcomes AB^c and A^cB—namely, the trials that result in exactly one success. In particular, the probabilities of the outcomes AB and A^cB^c are irrelevant. It is therefore not too surprising to discover that the hypothesis test ignores a and d in the table of observed frequencies. After all, these numbers count the number of occurrences of the irrelevant outcomes. Instead, the test focuses on b and c.

Let $m = b + c$. The actual study consists of n trials (pairs of free throws), but only m of these trials are relevant for the question at hand, testing whether p_A equals p_B. Let $Y_1, Y_2, \ldots, Y_m$, be the random variables corresponding to these m relevant observations. Note that these random variables satisfy the conditions for Bernoulli trials:

- Each relevant trial yields a dichotomous outcome, either AB^c or A^cB.
- Let p be the probability that a relevant trial yields outcome AB^c. The assumption that the original n observations are a random sample implies that p remains constant from relevant trial to relevant trial. (It can be shown that p equals $p_{AB^c}/(p_{AB^c} + p_{A^cB})$, but this identity is not needed.)
- The assumption of a random sample implies that the n observations are independent; thus, the m relevant trials are independent, too.

If the null hypothesis is true, $p_{AB^c} = p_{A^cB}$. In words, the two possible outcomes of a relevant trial are equally likely. Thus, if the null hypothesis is true, $p = 0.5$.

Suppose the first alternative, $p_A > p_B$, is true. By the above reasoning it follows that $p_{AB^c} > p_{A^cB}$. In words, the first possible outcome of a relevant trial is more likely than the second. Thus, if the first alternative hypothesis is true, $p > 0.5$. Similarly, the second alternative means $p < 0.5$, and the third alternative means $p \neq 0.5$. To summarize, the hypothesis-testing problem of this section is mathematically equivalent to the test of whether $p = 0.5$ given in Section 6.4.2. The salient features of that test are the following:

1. If $b + c \leq 14$, then the exact P-value can be obtained by entering Table A.4 at

$$R = b + c \quad \text{and} \quad O = b.$$

2. If $b + c > 14$, first compute

$$z = \frac{b - c}{\sqrt{b + c}}.$$

Then,

- The approximate P-value for the first alternative, $p_A > p_B$, equals the area under the standard normal curve to the right of z.
- The approximate P-value for the second alternative, $p_A < p_B$, equals the area under the standard normal curve to the right of $-z$.
- The approximate P-value for the third alternative, $p_A \neq p_B$, is equal to twice the area under the standard normal curve to the right of $|z|$.

For the study of Larry Bird, for example, Table 8.11 gives

$$b = 34 \quad \text{and} \quad c = 48.$$

Thus,

$$z = \frac{34 - 48}{\sqrt{34 + 48}} = -1.55.$$

The approximate P-value for the third alternative, that his probability of making a shot changed from the first shot to the second, is twice the area under the standard normal curve to the right of 1.55. Using Table A.2, one finds this area to be 2(0.0606), or 0.1212. The data are not statistically significant, and one should continue to believe that Bird had the same probability of making each shot.

If b and c are both greater than or equal to 10, a confidence interval for $p_A - p_B$ can be obtained. The point estimate of p_A is

$$\frac{n_1}{n} = \frac{a + b}{n},$$

and the point estimate of p_B is

$$\frac{m_1}{n} = \frac{a + c}{n}.$$

Thus, the point estimate of $p_A - p_B$ is

$$\frac{a+b}{n} - \frac{a+c}{n} = \frac{b-c}{n}.$$

Formula 8.9 is a special case of a formula that is presented in Chapter 16:

The confidence interval for $p_A - p_B$ is

$$\frac{b-c}{n} \pm z \times \sqrt{\frac{n(b+c) - (b-c)^2}{(n-1)n^2}} \quad (8.9)$$

where the confidence level is used to obtain z from Table A.3. The general guideline for using this formula is that both b and c should be greater than or equal to 10.

The use of Formula 8.9 will be illustrated with the above data for Larry Bird. Recall that

$$b = 34, \quad c = 48, \quad \text{and} \quad n = 338.$$

Thus, a 95 percent confidence interval for $p_A - p_B$ is

$$\frac{34-48}{338} \pm 1.96 \times \sqrt{\frac{338(34+48) - (34-48)^2}{337(338)^2}} = -0.041 \pm 1.96(0.02674)$$

$$= -0.041 \pm 0.052 = [-0.093, 0.011].$$

The confidence interval includes 0, so at the 95 percent level it is possible that the two probabilities are identical. On the other hand, the interval is fairly wide; Bird was an excellent free-throw shooter, and a half-width of 5.2 percentage points is substantial.

Consider again the study of gender and attitude in the 1986 Wisconsin Driver Survey, the results of which are presented in Table 8.6. In this study, p_A is the proportion of females in the population and p_B is the proportion of population members who would respond "extremely serious" to question 1 on the survey. Obviously, there is little interest in comparing p_A with p_B, so the methods of this section would not be used.

EXERCISES 8.5

The exercises of this subsection all involve different aspects of the responses to question 11 on the 1984 Wisconsin Driver Survey:

Which do you feel would be most effective in keeping people from driving after too much to drink? Please rank each in order of importance.

The five proposals were
- S: Strong enforcement
- R: Revoking the driver's license
- E: More education
- J: Mandatory jail terms
- F: Higher fines

For each proposal, the subject chose from these four responses: most important, fairly important, somewhat important, and not very important. For the analyses in this chapter, the last three possible responses will be collapsed into one category, yielding a dichotomous response with categories "most important" and "not most important."

The five proposals can be analyzed separately using the methods of Chapter 6. The following display gives for each proposal the number of respondents, n (not all subjects responded to all proposals), and the number, x, who selected the response "most important."

	Proposal				
	S	R	E	J	F
x	1,553	1,357	980	1,024	1,079
n	2,396	2,379	2,240	2,317	2,324

In addition, for each pair of proposals, a 2 × 2 table of sample frequencies can be created. For example, if the rows denote the response to proposal S and the columns the response to proposal R, the table will enable an analyst to study how attitude toward strong enforcement is related to attitude toward revoking the driver license. The table for proposals S and R is as follows:

	Revoking Driver License		
Strong Enforcement	Most Important	Not Most Important	Total
Most important	959	529	1,488
Not most important	357	451	808
Total	1,316	980	2,296

For the questions below, you do not need this entire table; the values $b = 529$, $c = 357$, and $n = 2,296$ are sufficient.

It would require too much space to list the 10 tables that arise from all pairwise comparisons of proposals. Instead, Table 8.14 presents selected information for the remaining nine tables. Each row in Table 8.14 gives information from a different 2 × 2 table. The first column, "Row," gives the proposal that determines the rows of the 2 × 2 table, and the second column, "Column," gives the column proposal. For all 2 × 2 tables the symbols A and B denote the response "most important" and A^c and B^c denote "not most important." The remaining three columns of Table 8.14 give the values of b, c, and n for the 2 × 2 table.

1. Use the appropriate formulas from Chapter 6 to find the point estimate of and 99 percent confidence interval for the proportion of drivers in the population who considered strong enforcement to be a most important proposal.

2. Use the appropriate formulas from Chapter 6 to find the point estimate of and 99 percent confidence interval for the proportion of drivers in the population who considered revoking the driver's license to be a most important proposal.

3. Use the appropriate formulas from Chapter 6 to find the point estimate of and 99 percent confidence interval for the proportion of drivers in the population who considered more education to be a most important proposal.

4. Use the appropriate formulas from Chapter 6 to find the point estimate of and 99 percent confidence interval for the proportion of drivers in the population who considered mandatory jail terms to be a most important proposal.

5. Use the appropriate formulas from Chapter 6 to find the point estimate of and 99 percent confidence interval for the proportion of drivers in the population who considered higher fines to be a most important proposal.

TABLE 8.14. Summary Statistics for Responses to Question 11 on the 1984 Wisconsin Driver Survey

Row	Column	b	c	n
S	E	757	281	2,171
S	J	753	295	2,252
S	F	625	219	2,265
R	E	713	400	2,175
R	J	589	299	2,261
R	F	511	283	2,267
E	J	508	526	2,141
E	F	428	528	2,146
J	F	324	381	2,237

6. Test the null hypothesis that the proportion of drivers who would have rated strong enforcement a most important proposal equals the proportion of all drivers who would have rated revoking the driver's license a most important proposal. Use the third alternative. Construct a 99 percent confidence interval for the proportion of all drivers who would have rated strong enforcement a most important proposal minus the proportion of all drivers who would have rated revoking the driver's license a most important proposal.

7. Test the null hypothesis that the proportion of drivers who would have rated strong enforcement a most important proposal equals the proportion of all drivers who would have rated more education a most important proposal. Use the third alternative. Construct a 99 percent confidence interval for the proportion of all drivers who would have rated strong enforcement a most important proposal minus the proportion of all drivers who would have rated more education a most important proposal.

8. Test the null hypothesis that the proportion of drivers who would have rated strong enforcement a most important proposal equals the proportion of all drivers who would have rated mandatory jail terms a most important proposal. Use the third alternative. Construct a 99 percent confidence interval for the proportion of all drivers who would have rated strong enforcement a most important proposal minus the proportion of all drivers who would have rated mandatory jail terms a most important proposal.

9. Test the null hypothesis that the proportion of drivers who would have rated strong enforcement a most important proposal equals the proportion of all drivers who would have rated higher fines a most important proposal. Use the third alternative. Construct a 99 percent confidence interval for the proportion of all drivers who would have rated strong enforcement a most important proposal minus the proportion of all drivers who would have rated higher fines a most important proposal.

10. Test the null hypothesis that the proportion of drivers who would have rated revoking the driver's license a most important proposal equals the proportion of all drivers who would have rated more education a most important proposal. Use the third alternative. Construct a 99 percent confidence interval for the proportion of all drivers who would have rated revoking the driver's license a most important proposal minus the proportion of all drivers who would have rated more education a most important proposal.

11. Test the null hypothesis that the proportion of drivers who would have rated revoking the driver's license a most important proposal equals the proportion of all drivers who would have rated mandatory jail terms a most important proposal. Use the third alternative. Construct a 99 percent confidence interval for the proportion of all drivers who would have rated revoking the driver's license a most important proposal minus the proportion of all drivers who would have rated mandatory jail terms a most important proposal.

12. Test the null hypothesis that the proportion of drivers who would have rated revoking the driver's license a most important proposal equals the proportion of all drivers who would have rated higher fines a most important proposal. Use the third alternative. Construct a 99 percent confidence interval for the proportion of all drivers who would have rated revoking the driver's license a most important proposal minus the proportion of all drivers who would have rated higher fines a most important proposal.

13. Test the null hypothesis that the proportion of drivers who would have rated more education a most important proposal equals the proportion of all drivers who would have rated mandatory jail terms a most important proposal. Use the third alternative. Construct a 99 percent confidence interval for the proportion of all drivers who would have rated more education a most important proposal minus the proportion of all drivers who would have rated mandatory jail terms a most important proposal.

14. Test the null hypothesis that the proportion of drivers who would have rated more education a most important proposal equals the proportion of all drivers who would have rated higher fines a most important proposal. Use the third alternative. Construct a 99 percent confidence interval for the proportion of all drivers who would have rated more education a most important proposal minus the proportion of all drivers who would have rated higher fines a most important proposal.

15. Test the null hypothesis that the proportion of drivers who would have rated mandatory jail terms a most important proposal equals the proportion of all drivers who would have rated higher fines a most important proposal. Use the third alternative. Construct a 99 percent confidence interval for the proportion of all drivers who would have rated mandatory jail terms a most important proposal minus the proportion of all drivers who would have rated higher fines a most important proposal.

REFERENCES

1. K. B. Meyer and S. G. Pauker, "Sounding Board: Screening for HIV: Can We Afford the False Positive Rate?" *New England Journal of Medicine*, Vol. 317, July 23, 1987, pp. 238–241.

2. L. B. Ellwein, "Decision Analysis of the HTLV-III Screening Test for Blood Donors," *AIDS Research*, February, 1986, pp. 5–17.

3. G. Friedland, "Early Treatment for HIV: The Time Has Come," *New England Journal of Medicine*, Vol. 322, April 5, 1990, pp. 1000–1002.

4. A. Tversky and T. Gilovich, "The Cold Facts about the 'Hot Hand' in Basketball," *Chance: New Directions for Statistics and Computing*, Winter, 1989, pp. 16–21.

5. R. S. Thompson et al., "A Case-Control Study of the Effectiveness of Bicycle Safety Helmets," *New England Journal of Medicine*, Vol. 320, May 25, 1989, pp. 1361–1366.

6. B. C. Coleman, "Study Adds to Alcoholism Gene Theory," *Wisconsin State Journal*, April 18, 1990, p. 1A.

Adjusting for a Factor

9.1 THE HAZARDS OF COLLAPSING TABLES
9.2 HYPOTHESIS TESTING

Chapter 9

9.1 THE HAZARDS OF COLLAPSING TABLES

9.1.1 Simpson's Paradox

Table 9.1 presents three sets of tables. Each set consists of tables of observed frequencies and row proportions for free-throw data. The first set is familiar; it presents the free-throw data for Larry Bird for the 1980–81 and 1981–1982 seasons and was analyzed in Chapter 8. The second set of tables is analogous data for another member of these Boston Celtics teams, Rick Robey [1]. In the current example, primary interest rests in the dependence of the second shot on the outcome of the first shot. It is more convenient to use the terminology of Chapter 7 rather than that of Chapter 8. As discussed in Chapter 8, either approach is valid mathematically.

Thus, the second shot is viewed as the response and the first shot defines the populations. More precisely, label a second-shot hit a success and a second-shot miss a failure. A response (outcome of the second shot) is observed to be from the first population if the first shot is a hit, and it is observed to be from the second population if the first shot is a miss. The populations are infinite, and the responses from each population are assumed to satisfy the conditions for Bernoulli trials. For example, after a hit (first population) Larry Bird's responses (second shot outcomes) are assumed to be Bernoulli trials with unknown probability of success p_{1B}. (The numeral 1 is, as usual, for the first population; the B is for *Bird*.) Similarly, after a miss Bird's responses are assumed to be Bernoulli trials with unknown probability of success p_{2B}. For Robey, the

TABLE 9.1 Observed Frequencies and Row Proportions for Pairs of Free Throws

	Observed Frequencies				Row Proportions		
	Hit Second Shot	Miss Second Shot	Total		Hit Second Shot	Miss Second Shot	Total
	Larry Bird						
Hit First Shot	251	34	285	Hit First Shot	0.881	0.119	1.000
Miss First Shot	48	5	53	Miss First Shot	0.906	0.094	1.000
Total	299	39	338				
	Rick Robey						
Hit First Shot	54	37	91	Hit First Shot	0.593	0.407	1.000
Miss First Shot	49	31	80	Miss First Shot	0.612	0.388	1.000
Total	103	68	171				
	Collapsed Table						
Hit First Shot	305	71	376	Hit First Shot	0.811	0.189	1.000
Miss First Shot	97	36	133	Miss First Shot	0.729	0.271	1.000
Total	402	107	509				

unknown probabilities of successes for the first and second populations are denoted by p_{1R} and p_{2R}, respectively. The following point estimates of the unknown probabilities are obtained from Table 9.1:

$$\hat{p}_{1B} = 0.881, \quad \hat{p}_{2B} = 0.906, \quad \hat{p}_{1R} = 0.593, \quad \text{and} \quad \hat{p}_{2R} = 0.612.$$

Note that for each player the proportion of second free throws made after a hit was lower than the proportion made after a miss. For Bird, the difference is $0.881 - 0.906 = -0.025$, or 2.5 percentage points lower after a hit; for Robey, the difference is $0.593 - 0.612 = -0.019$, or 1.9 percentage points lower after a hit.

The third set of tables begins with the **collapsed table**. The collapsed table is obtained by combining the data for the two players; that is, by *collapsing over* players. Using the standard notation, for the Bird table $a = 251$ and for the Robey table $a = 54$; thus, for the collapsed table, $a = 251 + 54 = 305$. The other entries of the collapsed table $(b, c, \ldots)$ are obtained in the same way—simply add the corresponding entries from the Bird and Robey tables.

The row proportions of the collapsed table reveal an interesting relationship. In the collapsed table, the proportion of second free throws made after a hit is much larger than the proportion made after a miss. More precisely, the difference, denoted $\hat{p}_{1C} - \hat{p}_{2C}$, is equal to $0.811 - 0.729 = 0.082$, or the proportion is 8.2 percentage points larger after a hit. In particular, note that although each player individually performed somewhat better after a miss, in the collapsed table the pattern is reversed! This reversal is called **Simpson's paradox.** The first task in this chapter is to investigate Simpson's paradox.

To understand Simpson's paradox, it is convenient to define the important concept of a weighted average. You are familiar with the notion of an arithmetic average.

Let $x_1, x_2, \ldots, x_k$ be k numbers. The arithmetic average of these numbers is simply

$$\frac{x_1 + x_2 + \cdots + x_k}{k}.$$

For example, if $k = 2$, $x_1 = 0$, and $x_2 = 10$, the arithmetic average of x_1 and x_2 is $(0 + 10)/2 = 5$. Numbers $w_1, w_2, \ldots, w_k$ are called **weights** if they satisfy the following two properties:

- Each number is nonnegative.
- $w_1 + w_2 + \cdots + w_k = 1$.

The **weighted average** of $x_1, x_2, \ldots, x_k$ with respect to the weights $w_1, w_2, \ldots, w_k$ is

$$x_1 w_1 + x_2 w_2 + \cdots + x_k w_k.$$

For the earlier example, with $k = 2$, suppose the weights are $w_1 = 0.2$ and $w_2 = 0.8$; then the weighted average of $x_1 = 0$ and $x_2 = 10$ with respect to $w_1 = 0.2$ and $w_2 = 0.8$ is

$$x_1 w_1 + x_2 w_2 = 0(0.2) + 10(0.8) = 0 + 8 = 8.$$

Figure 9.1 shows that the weighted average can be viewed as a center of gravity. In this figure, the two x's, 0 and 10, are plotted on the horizontal axis. Centered

FIGURE 9.1 Weighted average of 0 and 10 using weights 0.2 and 0.8, respectively. The weighted average, 8, equals the center of gravity of the rectangles.

at each value of x is a rectangle with height equal to the weight associated with the value. Each rectangle should have the same width, but the width can be any number. For aesthetic purposes, the widths are usually drawn small enough to prevent any overlap of the rectangles. The weighted average is the center of gravity of the rectangles.

Returning to the general case of k numbers and k weights, consider the special case in which all of the weights are equal. Since the weights add to 1, the common value of the k weights is $1/k$. In this case the weighted average of the numbers $x_1, x_2, \ldots, x_k$ is

$$x_1 w_1 + x_2 w_2 + \cdots + x_k w_k = \frac{x_1}{k} + \frac{x_2}{k} + \cdots + \frac{x_k}{k} = \frac{x_1 + x_2 + \cdots + x_k}{k}.$$

This last expression is recognized as the arithmetic average of the numbers. Thus, the arithmetic average is the special case of the weighted average that occurs if the weights are equal. The arithmetic average sometimes is called the **unweighted** or **equally weighted average.**

Consider the row proportion $\hat{p}_{1C} = 0.811$ in the collapsed table of the free-throw data. It is computed, of course, by dividing the number of successes in the first row of the collapsed table, 305, by the row total, 376; $305/376 = 0.811$. It is useful to write the numerator as $305 = 251 + 54$. This apportionment mirrors the players' contributions to the numerator: Bird had 251 successes on the first population, and Robey had 54. Next, write

$$0.811 = \frac{305}{376} = \frac{251}{376} + \frac{54}{376} = \frac{251}{285} \times \frac{285}{376} + \frac{54}{91} \times \frac{91}{376}$$
$$= 0.881(0.758) + 0.593(0.242). \tag{9.2}$$

Let $w_{1B} = 0.758$ and $w_{1R} = 0.242$. In words, of the 376 successful first shots attempted by either player, w_{1B} is the proportion attempted by Bird and w_{1R} is the proportion attempted by Robey. (Thus, the numeral 1 is for the first row—success on the first shot—of the table, and the letter identifies the source.) Note that these two w's are nonnegative and sum to 1, so they are weights. With this and the earlier notation, Equation 9.2 can be written as

$$\hat{p}_{1C} = \hat{p}_{1B} w_{1B} + \hat{p}_{1R} w_{1R}. \tag{9.3}$$

In words, the sample proportion of successes in the first row of the collapsed table equals the weighted average of the sample proportions of successes in the component tables. The weights are given by the proportion of observations contributed by each table.

This argument can be applied equally well to the second row of the collapsed table, yielding

$$\hat{p}_{2C} = \hat{p}_{2B} w_{2B} + \hat{p}_{2R} w_{2R}. \tag{9.4}$$

The weights are computed as before; w_{2B} is the proportion of the second sample in the collapsed table that comes from Bird; namely, $53/133 = 0.398$. Similarly,

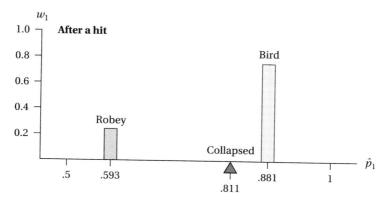

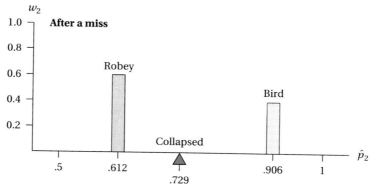

FIGURE 9.2 Visual explanation of Simpson's paradox for the Free Throw study.

w_{2R} is the proportion of the second sample in the collapsed table that comes from Robey; namely, $w_{2R} = 80/133 = 0.602$.

Figure 9.2 illustrates Simpson's paradox for these free-throw data. The top picture is for the samples (Bird and Robey) from the first populations (after a hit), and the bottom picture is for the samples (Bird and Robey) from the second populations (after a miss). On the horizontal axes are plotted the $\hat{p}$'s, and the heights of the rectangles equal the weights. First, note that both Bird's and Robey's rectangles in the bottom picture are shifted slightly to the right of their rectangles in the top picture. This shift reflects the fact that each player had a higher sample proportion of successes after a miss than after a hit. Next, focus on the weights. In the collapsed sample from the first populations, about three-quarters of the data come from Bird and one-quarter from Robey. In the collapsed sample from the second populations, however, only about 40 percent of the data come from Bird and the remaining 60 percent is from Robey. In the collapsed sample from the first populations, the sample proportion of successes, 0.811, is much closer to Bird's value, 0.881, than to Robey's value, 0.593, because of the larger weight associated with Bird's value. By contrast, in the collapsed sample from the second populations the sample proportion, 0.729, is pulled down by the large weight associated with Robey's value. In words, the collapsed sample from the second populations has a lower proportion of successes than the collapsed sample from the first populations because a larger share of its data comes from Robey, who was a much poorer free-throw shooter than Bird.

Here is another way to interpret the above analysis of the free-throw data. A person seeing only the collapsed table would draw the conclusion that second free throws that follow hits have a substantially higher proportion of successes than second free throws that follow misses. If, however, this person would **adjust** for differences between players (by analyzing each player separately) an opposite conclusion would be reached: for either player, second free throws that follow misses have a somewhat higher proportion of successes than second free throws that follow hits. Do not conclude that the adjusted analysis represents the truth. Perhaps a further adjustment would lead to a different conclusion. (For example, it might be important to adjust for whether the basketball game was played at home or on the road.)

This example illustrates the most distressing feature of an observational study. Once the analysis is completed, the researcher must remain aware that adjusting for a factor could lead to a drastically different conclusion. This feature occurs with all types of data, dichotomous, multicategory, and numerical.

The above ideas can be illustrated with an observational study on finite populations, too. In Chapter 8, for example, the relationship between attitude and gender was explored for the 1986 Wisconsin Driver Survey. As with our study of free throws, it is more convenient to use the terminology of Chapter 7. The first population consists of female drivers, and the second population consists of male drivers. The response is the answer to question 1 on the survey: the answer "extremely serious" is labeled a success and the answer "not extremely serious" is labeled a failure. The observed frequencies are presented in Table 8.6 (p. 270). The table of row proportions indicates that a higher proportion of females than males gave the successful response. The hypothesis testing and confidence interval analyses strongly suggest that there is a relationship in the populations, too. This 2 × 2 table, however, can be viewed as a collapsed table. The researcher must decide how to divide this collapsed table into component tables. In other words, the researcher must select a factor.

There are no absolute rules for choosing a factor for study. One characteristic of a good researcher is the ability to choose factors well. Here is how I proceed. In the current study, the analysis shows that women and men have different attitudes on the seriousness of drinking and driving. I ask myself, "What other characteristic of a person, besides gender, is likely to influence attitude on this issue?" My answer is the person's frequency of drinking alcoholic beverages. Thus, I divided the collapsed table into component tables based on self-reported drinking frequency. The component tables and collapsed tables are given in

TABLE 9.2 *Observed Frequencies and Row Proportions for the Relationship between Attitude and Gender by Self-Reported Drinking Frequency*

	Observed Frequencies				Row Proportions		
	Response				Response		
Gender	S	F	Total	Gender	S	F	Total
Light Drinkers							
Female	234	133	367	Female	0.638	0.362	1.000
Male	151	100	251	Male	0.602	0.398	1.000
Total	385	233	618				
Moderate Drinkers							
Female	133	142	275	Female	0.484	0.516	1.000
Male	115	146	261	Male	0.441	0.559	1.000
Total	248	288	536				
Heavy Drinkers							
Female	42	54	96	Female	0.438	0.562	1.000
Male	83	146	229	Male	0.362	0.638	1.000
Total	125	200	325				
Collapsed Table							
Female	409	329	738	Female	0.554	0.446	1.000
Male	349	392	741	Male	0.471	0.529	1.000
Total	758	721	1,479				

Table 9.2. The "light drinkers" are persons who reported that they drank alcohol rarely or never; the "moderate drinkers" reported that they drank alcohol a few times a month; and the "heavy drinkers" reported that they drank alcohol more than once a week. A comparison of Table 8.6 with the collapsed table in Table 9.2 shows that the latter is missing 25 persons, because 13 women and 12 men did not respond to the question on drinking frequency on the questionnaire.

Table 9.3 highlights an interesting pattern in these data. The difference between the female and male proportions of successes in the collapsed table, 0.083, exceeds the corresponding difference in each of the component tables. In words, collapsing magnifies the gender difference. This pattern is presented

TABLE 9.3 Comparison of Proportions of Successes for the Study of Gender and Attitude

Table	$\hat{p}_1$	$\hat{p}_2$	$\hat{p}_1 - \hat{p}_2$
Light drinkers	0.638	0.602	0.036
Moderate drinkers	0.484	0.441	0.043
Heavy drinkers	0.438	0.362	0.076
Collapsed	0.554	0.471	0.083

visually in Figure 9.3. The vertical distance between dots equals the difference between the female and male proportions of successes, and the largest distance occurs in the collapsed table. Figure 9.3 is an **interaction graph** and was introduced in the optional Chapter 4, in Section 4.1. The interaction graph also shows visually that, for either gender, as self-reported drinking frequency increases, the proportion of successes declines.

Figure 9.4 shows why the gender difference in the collapsed table, 0.083, is larger than any of the gender differences in the component tables. For each gender, heavy drinkers have the lowest proportion of successes, and the heights of the rectangles in the figure show that the males in the sample have a higher proportion of heavy drinkers than the females in the sample do.

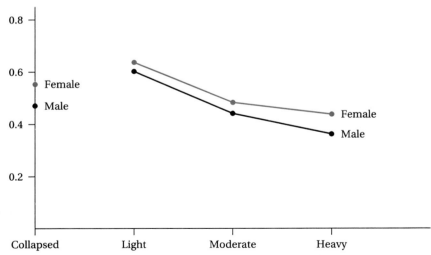

FIGURE 9.3 Interaction graph of the proportion of successful responses by gender by drinking frequency.

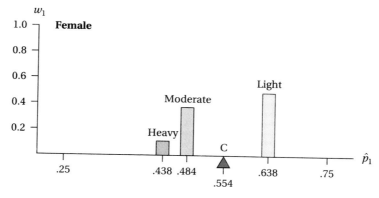

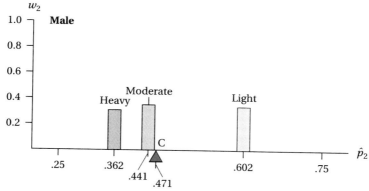

FIGURE 9.4 Visual presentation of the relationship between the proportion of successes in the component tables and the collapsed table (C) for study of gender and attitude.

9.1.2 Standardized Rates

Consider the question

> In the 1986 Wisconsin Driver Survey, how do the attitudes of women and men compare on the seriousness of drinking and driving?

One answer is obtained by looking at the collapsed table:

> The proportion of female successes in the sample is 8.3 percentage points larger than the proportion of male successes in the sample.

If the purpose of the research is only to compare the two genders, this is a perfectly good answer. Suppose, however, that the researcher wants to gain some insight into why women and men differ. In this case, the above answer can be improved by adding the following:

- For the light drinkers, the proportion of successes among the females in the sample is 3.6 percentage points higher than the proportion of successes among the males in the sample.
- For the moderate drinkers, the proportion of successes among the females in the sample is 4.3 percentage points higher than the proportion of successes among the males in the sample.
- For the heavy drinkers, the proportion of successes among the females in the sample is 7.6 percentage points higher than the proportion of successes among the males in the sample.

Thus, if one takes into account drinking frequency, the observed difference between women and men decreases from the collapsed table's value of 8.3 percentage points; the decrease is small for the heavy drinkers and substantial for the others.

The above three-part answer, one for each component table, is more difficult to remember and interpret than a single-number comparison. This difficulty is amplified when there are a large number of component tables. Computing a **standardized rate** is a way to avoid this difficulty.

Table 9.4 presents the relative frequency distributions of the self-reported drinking frequency of the female and male drivers. The frequencies are obtained from Table 9.2; the relative frequencies are, of course, the weights that appear in Figure 9.4.

TABLE 9.4 Relative Frequency Distributions of Self-Reported Drinking Frequency for Females and Males

Drinking Category	Frequency	Relative Frequency
Females		
Light	367	$w_{1L} = 0.497$
Moderate	275	$w_{1M} = 0.373$
Heavy	96	$w_{1H} = 0.130$
Total	738	1.000
Males		
Light	251	$w_{2L} = 0.339$
Moderate	261	$w_{2M} = 0.352$
Heavy	229	$w_{2H} = 0.309$
Total	741	1.000

For the free-throw data, Equation 9.3 showed that the proportion of successes in the collapsed table can be obtained by computing a weighted average of the proportions of successes in the component tables. By similar reasoning, the proportion of female successes in the collapsed table, $\hat{p}_{1C}$, can be computed as

$$\hat{p}_{1C} = \hat{p}_{1L} w_{1L} + \hat{p}_{1M} w_{1M} + \hat{p}_{1H} w_{1H}.$$

Substituting numbers from Tables 9.3 and 9.4 into this equation gives

$$\hat{p}_{1C} = 0.638(0.497) + 0.484(0.373) + 0.438(0.130) = 0.554.$$

In a similar way, the proportion of successes for males in the collapsed table is

$$\hat{p}_{2C} = \hat{p}_{2L} w_{2L} + \hat{p}_{2M} w_{2M} + \hat{p}_{2H} w_{2H}.$$

Substituting numbers from Tables 9.3 and 9.4 into this equation gives

$$\hat{p}_{2C} = 0.602(0.339) + 0.441(0.352) + 0.362(0.309) = 0.471.$$

The idea of a standardized rate is to estimate the proportion of successes there would have been in the female sample *if the women had had the same weights—relative frequencies of self-reported drinking categories—as the men.* This number is obtained by computing the weighted average of the female proportions of successes *using the male weights,* and is denoted by $\hat{p}_{1S}$ (S is for *standardized*). In particular,

$$\hat{p}_{1S} = \hat{p}_{1L} w_{2L} + \hat{p}_{1M} w_{2M} + \hat{p}_{1H} w_{2H}. \tag{9.5}$$

Substituting into this equation gives

$$\hat{p}_{1S} = 0.638(0.339) + 0.484(0.352) + 0.438(0.309) = 0.522.$$

The roles of males and females can be reversed in the above, yielding an estimate of the proportion of successes there would have been in the male sample *if the men had had the same weights—relative frequencies of self-reported drinking categories—as the women.* This number is obtained by computing the weighted average of the male proportions of successes using the female weights, and is denoted by $\hat{p}_{2S}$. The formula is

$$\hat{p}_{2S} = \hat{p}_{2L} w_{1L} + \hat{p}_{2M} w_{1M} + \hat{p}_{2H} w_{1H}. \tag{9.6}$$

Substituting into this equation gives

$$\hat{p}_{2S} = 0.602(0.497) + 0.441(0.373) + 0.362(0.130) = 0.511.$$

The above computations can be summarized as follows:

- For the actual data, the difference in proportion of successes between females and males is $0.554 - 0.471 = 0.083$.
- If the women in the survey drank as much as the men in the survey, the women's estimated proportion of successes would drop from the observed

0.554 to 0.522, a decline of 3.2 percentage points. Another way to state this is that if the women in the survey drank as much as the men in the survey, it is estimated that the difference between female and male success proportions would be $0.522 - 0.471 = 0.051$.

- If the men in the survey drank as little as the women in the survey, the men's estimated proportion of successes would increase from the observed 0.471 to 0.511, a change of 4.0 percentage points. Another way to state this is that if the men in the survey drank as little as the women in the survey, it is estimated that the difference between female and male success proportions would be $0.554 - 0.511 = 0.043$.

An extremely unfortunate characteristic of standardized rates is that the two differences above, 0.051 and 0.043, are not the same number; the interpretation of standardized rates would be easier if they were. In any event, somewhat less than half of the difference between females and males in the original data disappears if one adjusts for differences in drinking frequency.

9.2 HYPOTHESIS TESTING

The material of the previous section is descriptive. This section presents a population model for the data of this chapter. The ideas are introduced with the study of gender and attitude discussed earlier.

There are two populations of interest, female and male licensed drivers. The three levels of the factor chosen by the researcher, self-reported drinking frequency, divides these populations into three distinct **subpopulations**. For the current study, the three subpopulations of the female population are female light drinkers, female moderate drinkers, and female heavy drinkers. Similarly, the three subpopulations of the male population are male light drinkers, male moderate drinkers, and male heavy drinkers. The subpopulations of female and male heavy drinkers form a **matched pair of subpopulations,** as do the subpopulations of female and male moderate drinkers, and the subpopulations of female and male light drinkers. Note that the subpopulations of female heavy drinkers and male light drinkers, for example, do not form a matched pair of subpopulations.

The data are assumed to be independent random samples from the two populations. (The Mantel-Haenszel test of this section also is valid if the data are independent random samples from all of the subpopulations.)

There are three strategies for analyzing such data with a hypothesis test. The first strategy is to perform Fisher's test on the collapsed table, as described in Chapter 7. Specifically, let p_1 denote the proportion of successes in the first population and let p_2 denote the proportion of successes in the second population. The researcher wants to test the null hypothesis that $p_1 = p_2$. There are, of course, three choices for the alternative hypothesis. For the study of gender and attitude, I will choose the third alternative, that the populations of women and men have different proportions of successes. From Table 9.2,

$$a = 409,\ b = 329,\ c = 349,\ d = 392,\ n_1 = 738,\ n_2 = 741,$$
$$m_1 = 758,\ m_2 = 721,\ \text{and}\ n = 1{,}479.$$

To obtain the approximate P-value, compute

$$z = \frac{\sqrt{n-1}(ad-bc)}{\sqrt{n_1 n_2 m_1 m_2}} = \frac{\sqrt{1{,}478}(409 \times 392 - 329 \times 349)}{\sqrt{738(741)(758)(721)}} = 3.20.$$

For the third alternative, the approximate P-value equals twice the area under the standard normal curve to the right of 3.20. From Table A.2, this area equals 2(0.0007), or 0.0014. The observed difference in attitudes between genders is highly statistically significant.

The second strategy is to perform Fisher's test on the component tables, that is, for each matched pair of subpopulations. Anticipating the third strategy below, it is convenient to use an alternative formula for z. Instead of computing

$$z = \frac{\sqrt{n-1}(ad-bc)}{\sqrt{n_1 n_2 m_1 m_2}},$$

for the component table under study, use the formula

$$z = \frac{a - E(A)}{\sqrt{\text{Var}(A)}}. \tag{9.7}$$

In this equation a denotes, as usual, the entry in the first row and first column of the component table being analyzed. The other values in this equation are computed by

$$E(A) = \frac{n_1 m_1}{n} \quad \text{and} \quad \text{Var}(A) = \frac{n_1 n_2 m_1 m_2}{n^2(n-1)}.$$

(The equivalence of these two formulas for z was discussed in Section 4.2.)

The values of a, $E(A)$, $\text{Var}(A)$, and z and the approximate P-value for the third alternative are given in Table 9.5 for the study of gender and attitude. For now, ignore the (bottom) row labeled "Combined." The conclusions of these individual analyses are quite different from the conclusion based on the analysis of the collapsed table. In each component table, there is evidence that the female subpopulation has a higher proportion of successes than the male subpopulation. The evidence, however, is not close to being statistically significant in any matched pair of subpopulations, as the smallest approximate P-value is 0.2040.

The third strategy is complicated; to avoid an unnecessary notational explosion, the hypotheses are stated in words rather than symbols. This third strategy

TABLE 9.5 Computations for the Second and Third Analysis Strategies for the Study of Gender and Attitude

Subpopulation	a	E(A)	Var(A)	z	Approximate P*
Light drinkers	234	228.63	35.067	0.91	0.3628
Moderate drinkers	133	127.24	33.353	1.00	0.3174
Heavy drinkers	42	36.92	16.060	1.27	0.2040
Combined	409	392.79	84.480	1.76	0.0784

*The approximate P-values are computed for the third alternative.

for hypothesis testing is appropriate only if the researcher believes that the *direction* of the differences between matched pairs of subpopulations is consistent across the different matched pairs. More precisely, suppose the researcher wants to test the null hypothesis that within each matched pair of subpopulations, females and males have the same proportion of successes. This null hypothesis implies that the subpopulations of female and male light drinkers have the same proportion of successes, the subpopulations of female and male moderate drinkers have the same proportion of successes, and the subpopulations of female and male heavy drinkers have the same proportion of successes. This null hypothesis does not say anything about there being any other relationships between these proportions. For example, this null hypothesis does *not* say that the subpopulations of female light drinkers and female moderate drinkers have the same proportion of successes. In fact, recall that I chose self-reported drinking as my factor because I suspected that drinking frequency influenced attitude.

For the third strategy, there are three possible choices for the alternative hypothesis:

- The first alternative states that females have a higher proportion of successes than males in every matched pair of subpopulations.
- The second alternative states that females have a lower proportion of successes than males in every matched pair of subpopulations.
- The third alternative states that females have either a higher proportion of successes than males in every matched pair of subpopulations or a lower proportion of successes than males in every matched pair of subpopulations.

The test statistic for this third strategy is written in terms of the statistics obtained for the second strategy, Fisher's test for each component table. In particular, let

- $U = \sum A$
- $u = \sum a$, the observed value of U

- $E(U) = \sum E(A)$
- $\text{Var}(U) = \sum \text{Var}(A)$

where the sums are taken over the different component tables in the study. For the study of gender and attitude, for example, these sums are taken over the three tables corresponding to the three categories of self-reported drinking frequency. These sums are presented in the bottom row of Table 9.5. In this table,

$$u = 409, \quad E(U) = 392.79, \quad \text{and} \quad \text{Var}(U) = 84.480.$$

The test statistic is the standardized version of U and is given by the formula

$$Z = \frac{U - E(U)}{\sqrt{\text{Var}(U)}}.$$

The observed value of the test statistic is

$$z = \frac{u - E(U)}{\sqrt{\text{Var}(U)}}.$$

For the study of gender and attitude,

$$z = \frac{409 - 392.79}{\sqrt{84.480}} = 1.76.$$

This value appears in the bottom row of Table 9.5. It can be shown that if the null hypothesis is true, the sampling distribution of the test statistic can be approximated well by the standard normal curve. The approximate P-value is given by the familiar rule:

- For the first alternative, the approximate P-value equals the area under the standard normal curve to the right of z.
- For the second alternative, the approximate P-value equals the area under the standard normal curve to the right of $-z$.
- For the third alternative, the approximate P-value equals twice the area under the standard normal curve to the right of $|z|$.

In the current example, the approximate P-value for the third alternative is $2(0.0392) = 0.0784$. Thus, combining the information across matched pairs of subpopulations leads to a result that is not quite statistically significant.

The hypothesis test of the third strategy is called the Mantel-Haenszel test and is abbreviated as the MH test. It was introduced in Chapter 4 (Section 4.2) for a randomized block design.

It is natural to wonder which of the three analysis strategies is preferred for the study of gender and attitude. My answer is that it depends on the goals of the research. A researcher who only wants to compare females and males should use the first strategy and analyze the collapsed table. A researcher who wants to compare females and males after adjusting for differences in drinking habits

TABLE 9.6 Computations for the Second and Third Analysis Strategies for the Free Throw Study

Subpopulation	a	E(A)	Var(A)	z	Approximate P*
Larry Bird	251	252.12	4.575	−0.52	0.6030
Rick Robey	54	54.81	10.257	−0.25	0.8026
Combined	305	306.93	14.832	−0.50	0.6170

*The approximate P-values are for the third alternative.

should use either the second or the third strategy. If it seems reasonable to suspect that the direction of differences between matched pairs of subpopulations is consistent across matched pairs, I recommend the third strategy—generally one learns more from an appropriate combination of information from different sources than from examining each source individually.

This chapter concludes with a familiar study and a lengthy examination of a new study. Each example is contained in its own section.

9.2.1 Free Throws

In this section the free-throw data for Larry Bird and Rick Robey (presented in Table 9.1) are analyzed with a number of hypothesis tests. In these tests, the third alternative is used.

In the collapsed table,

$$a = 305, \; b = 71, \; c = 97, \; d = 36, \; n_1 = 376, \; n_2 = 133,$$
$$m_1 = 402, \; m_2 = 107, \text{ and } n = 509.$$

To obtain the approximate P-value, first compute

$$z = \frac{\sqrt{n-1}(ad-bc)}{\sqrt{n_1 n_2 m_1 m_2}} = \frac{\sqrt{508}(305 \times 36 - 71 \times 97)}{\sqrt{376(133)(402)(107)}} = 1.99.$$

For the third alternative, the approximate P-value is 2(0.0233) = 0.0466. The observed difference, that the players shot better after a hit, is statistically significant.

The computations for the second and third strategies are presented in Table 9.6. The second analysis strategy indicates that each player shot somewhat better after a miss, but that neither player came close to achieving statistical significance. The MH test reaches the same conclusion.

9.2.2 Elder Abuse

L. Vinton [2] studied the 428 cases of elder abuse reported to county departments of social services in Wisconsin in 1986. "Abuse" encompasses physical, medicinal, emotional, and financial mistreatment. After a case of abuse was reported to officials, a social worker interviewed the victim and offered an appropriate service. Vinton's primary interest was in determining to what extent the victim's response to the offer of service was related to the gender of the abuser. The response has two categories: the offer of services could be accepted, a success, or refused, a failure. The first population corresponds to victims of female abusers, and the second population corresponds to victims of male abusers.

There are many ways to view Vinton's data, including the following:

1. The data are the entire population of reported cases of elder abuse in Wisconsin in 1986.
2. The data are a random sample from the process that generated reported cases of elder abuse in Wisconsin in 1986. For example, each senior citizen in Wisconsin in 1986 was a candidate to be the victim in a reported case of abuse. The process by which certain seniors end up in this data set is, to say the least, complicated.
3. The data are a random sample from the entire population of all reported elder abuse cases in Wisconsin for a number of years including 1986. This is analogous to the assumption that subjects in a medical study form a random sample from the population of current and future possible patients.
4. The data are a random sample from all cases of elder abuse in Wisconsin in 1986 (both reported and unreported cases).

From the first of these perspectives, descriptive methods of studying the data are appropriate, but inference is not needed because the entire population has been sampled. The other three perspectives make both descriptive and inferential procedures appropriate. The first perspective is beyond reproach. In my opinion, the second and third perspectives are sufficiently reasonable to be useful. The fourth perspective, however, is unreasonable: I do not believe that reported cases are representative of unreported cases. You, of course, can reasonably disagree with my opinion. (Unfortunately, in practice researchers do not always explain the perspectives that justify their inferences.) In this section the data are examined by descriptive and inferential procedures; you can decide if the latter are reasonable.

In addition to the populations and response, for each case Vinton measured a number of factors. Three of them, all dichotomous, will be considered in this section:

- Disability status of abused (disabled or not)
- Care giver status of abuser (primary care giver or not)
- Relationship of abuser to abused (primary relative or not)

Vinton's data are presented in Table 9.7, which contains cells for each of the 32 combinations of the values taken on by the five dichotomous variables. The number in each cell counts the number of cases in that combination. For

TABLE 9.7 Data for the Elder Abuse Study

	Primary Care Giver						Not Primary Care Giver					
	Primary Relative			Not Primary Relative			Primary Relative			Not Primary Relatvie		
Abuser	A*	R*	T	A*	R*	T	A*	R*	T	A*	R*	T
	Disabled											
Female	36	14	50	29	7	36	18	9	27	13	7	20
Male	47	38	85	7	3	10	42	21	63	27	10	37
Total	83	52	135	36	10	46	60	30	90	40	17	57
Component Table		1			2			3			4	
	Not Disabled											
Female	0	1	1	0	0	0	10	6	16	10	5	15
Male	2	2	4	0	1	1	25	19	44	8	11	19
Total	2	3	5	0	1	1	35	25	60	18	16	34
Component Table		5			6			7			8	

*The response A is for *accepted social services*, and R is for *refused social services*.

example, the 36 in the upper left corner of the table means that there were 36 cases in which the victim was disabled, the offer of social services was accepted by the victim, and the abuser was a primary care giver, a primary relative, and a female.

It is difficult to assimilate and understand the information in this table. It is helpful to obtain a quick picture of the data set by examining the five variables individually. This is achieved by counting the number of cases for each value of each variable and then summarizing the counts by computing relative frequencies. The results are as follows:

- Sixty-four percent (274 of 428) of the victims accepted services.
- Sixty-one percent (263) of the abusers were male.
- Sixty-eight percent (290) of the abusers were primary relatives.
- Forty-four percent (187) of the abusers were primary care givers.
- Seventy-seven percent (328) of the victims were disabled.

Vinton conjectured that victims of male abusers would be less likely to accept social services than victims of female abusers, possibly out of fear of retribution by the abuser. This conjecture is examined first for the collapsed table, which

TABLE 9.8 *Collapsed Table for the Elder Abuse Study*

	Observed Frequencies				Row Percentages		
Abuser	Accept	Refuse	Total	Abuser	Accept	Refuse	Total
Female	116	49	165	Female	0.703	0.297	1.000
Male	158	105	263	Male	0.601	0.399	1.000
Total	274	154	428	Total	0.640	0.360	1.000

ignores the three factors. The collapsed table and its row proportions are given in Table 9.8. There is evidence in the data supporting Vinton's conjecture: a higher proportion of victims of female abusers accepted services than victims of male abusers. The point estimate of the population difference is $\hat{p}_1 - \hat{p}_2 = 0.703 - 0.601 = 0.102$, slightly more than 10 percentage points.

In addition, Vinton's conjecture suggests conducting a hypothesis test of the null hypothesis, $p_1 = p_2$, versus the first alternative, $p_1 > p_2$. The observed value of the test statistic is

$$z = \frac{\sqrt{n-1}(ad-bc)}{\sqrt{n_1 n_2 m_1 m_2}} = \frac{\sqrt{427}(116 \times 105 - 49 \times 158)}{\sqrt{165(263)(274)(154)}} = 2.14,$$

and the approximate P-value, which is the area under the standard normal curve to the right of 2.14, is 0.0162. Thus, the collapsed table supports Vinton's conjecture. The difference in the sample proportion of successes is large, and the test is statistically significant.

Next, the data are examined in each of the eight component tables formed by all possible combinations of the three factors. In Table 9.7, the first three columns of numbers constitute the 2 × 2 component table for disabled victims who were abused by a primary care giver who was a primary relative, the next three columns of numbers constitute the component table for disabled victims who were abused by a primary care giver who was not a primary relative, and so on. The eight component tables are numbered in Table 9.7 for convenience. The most obvious feature of these eight component tables is that very few cases are counted in tables 5 and 6—tables that represent victims who were not disabled and were abused by a primary care giver. Table 9.9 presents a number of useful summaries of the eight component tables:

- The first column identifies the component table.
- The second column lists the values of $\hat{p}_1$, the proportion of victims of female abusers who accepted services.
- The third column lists the values of $\hat{p}_2$, the proportion of victims of male abusers who accepted services.
- The fourth column lists the values of $\hat{p}_1 - \hat{p}_2$. Note that there is no entry for tables 5 and 6 in the second, third, and fourth columns because of the paucity of data.

TABLE 9.9 Summary Statistics for the Component Tables of the Elder Abuse Study

Table	$\hat{p}_1$	$\hat{p}_2$	$\hat{p}_1 - \hat{p}_2$	a	E(a)	Var(a)	z	Approximate P*
1	0.72	0.55	0.17	36	30.7	7.511	1.93	0.0268
2	0.81	0.70	0.11	29	28.2	1.361	0.69	0.2451
3	0.67	0.67	0.00	18	18.0	4.247	0.00	0.5000
4	0.65	0.73	−0.08	13	14.0	2.766	−0.60	0.7257
5	—	—	—	0	0.4	0.240	—	—
6	—	—	—	0	0.0	0.000	—	—
7	0.62	0.57	0.05	10	9.3	2.900	0.41	0.3409
8	0.67	0.42	0.25	10	7.9	2.152	1.43	0.0764
MH test:	—	—	—	116	108.5	21.177	1.63	0.0516

*The approximate P-values are for the first alternative.

- The fifth, sixth, and seventh columns provide the values of a, $E(A)$, and $Var(A)$, which are needed for hypothesis testing.
- The eighth column lists the observed value of the test statistic z for each table. There are no entries for tables 5 and 6 because the general guideline for using the standard normal curve to approximate the P-value is not even close to being met because of the lack of data. The value of z in the bottom row is for the MH test.
- The ninth column gives the standard normal curve approximation to the P-values for the first alternative.

You are encouraged to verify the numbers in Table 9.9. An interaction graph is given in Figure 9.5.

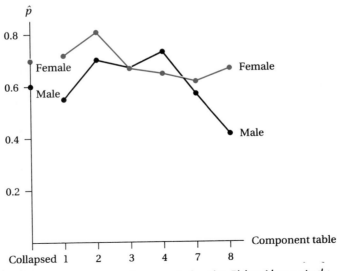

FIGURE 9.5 Interaction graph for the Elder Abuse study.

A number of features of the data are revealed in Table 9.7, Figure 9.5, and Table 9.9. You will be asked to discuss them in the exercises.

EXERCISES FOR CHAPTER 9

1. Question 7 on the Wisconsin Driver Survey measured knowledge of the legal blood alcohol concentration limit. Let a correct answer be a success. Several exercises will examine the relationship between knowledge and gender for the 1984 survey. The collapsed table is as follows:

	S	F	Total
Female	634	498	1,132
Male	787	410	1,197
Total	1,419	908	2,329

    Let subpopulation 1 be drivers who would have reported drinking alcohol several times a week; similarly, let subpopulations 2, 3, and 4 be drivers who would have reported drinking alcohol several times a month, less often, and not at all, respectively. (You may have noticed that the 1984 categories are different from those used in 1986. The change of categories made interyear comparisons difficult and will be discussed in Chapter 11.) The results are as follows:

	Subpopulation					
	1			2		
	S	F	T	S	F	T
Female	66	34	100	131	94	225
Male	160	69	229	254	120	374
Total	226	103	329	385	214	599

	Subpopulation					
	3			4		
	S	F	T	S	F	T
Female	297	246	543	140	124	264
Male	233	148	381	140	73	213
Total	530	394	924	280	197	477

    (a) Construct the table of row percentages for the collapsed table.
    (b) Construct the table of row proportions for each component table.
    (c) Construct the interaction graph of the proportion of successes, including the points for the collapsed table. Comment on what this picture reveals.

2. Question 11(d) on the Wisconsin Driver Survey asked for a rating of the effectiveness of mandatory jail terms as a method for "keeping people from driving after (having) too much to drink." The possible responses were "most important," "fairly important," "somewhat important," and "not very important." Define "most important" to be a success and any other response a failure. Several exercises will examine the relationship between response to this question and gender for the 1984 survey. The collapsed table is as follows:

	S	F	Total
Female	476	560	1,036
Male	468	623	1,091
Total	944	1,183	2,127

Let subpopulation 1 be drivers who would have reported drinking alcohol several times a week; similarly, let subpopulations 2, 3, and 4 be drivers who would have reported drinking alcohol several times a month, less often, and not at all, respectively. The results are as follows:

	Subpopulation					
	1			2		
	S	F	T	S	F	T
Female	28	62	90	85	125	210
Male	85	124	209	132	213	345
Total	113	186	299	217	338	555

	Subpopulation					
	3			4		
	S	F	T	S	F	T
Female	243	258	501	120	115	235
Male	143	204	347	108	82	190
Total	386	462	848	228	197	425

(a) Construct the table of row proportions for the collapsed table.

(b) Construct the table of row proportions for each component table.

(c) Construct the interaction graph of the proportion of successes, including the points for the collapsed table. Comment on what this picture reveals.

3. Refer to Exercise 1.

   (a) Use the standardized rate to estimate the proportion of women who would have known the legal limit for blood alcohol concentration if women drank as much as men.

   (b) Use the standardized rate to estimate the proportion of men who would have known the legal limit for blood alcohol concentration if men drank as little as women.

   (c) Comment on your findings in parts (a) and (b).

4. Refer to Exercise 2.

   (a) Use the standardized rate to estimate the proportion of women who would have given a successful response if women drank as much as men.

   (b) Use the standardized rate to estimate the proportion of men who would have given a successful response if men drank as little as women.

   (c) Comment on your findings in parts (a) and (b).

5. Refer to Exercise 1. Use the collapsed table to test the null hypothesis that knowledge does not depend on gender; use the third alternative.

6. Refer to Exercise 2. Use the collapsed table to test the null hypothesis that response does not depend on gender; use the third alternative.

7. Refer to Exercise 1. Complete the entries in the following display:

Subpopulation	a	$E(A)$	$Var(A)$	z
1	66	68.69	15.015	−0.69
2	131			−2.40
3		311.46		
4			28.760	
Combined				

Use the above display to test, in each matched pair of subpopulations, the null hypothesis that knowledge does not depend on gender versus the third alternative. Comment on your findings.

8. Refer to Exercise 2. Complete the entries in the following display:

Subgroup	a	$E(A)$	$Var(A)$	z
1	28	34.01	14.840	−1.56
2	85	82.11	31.140	0.52
3				
4				
Combined				

Use the above display to test, in each matched pair of subpopulations, the null hypothesis that response does not depend on gender versus the third alternative. Comment on your findings.

	Female Abusers		*Male Abusers*	
Component Table	Frequency	Relative Frequency	Frequency	Relative Frequency
1	50			
2	36			
3	27			
4	20			
5, 6	1			
7	16			
8	15			
Total	165			

9. Use the display in Exercise 7 to perform the MH test with the third alternative. Compare and contrast the three test strategies:

   - Analysis of the collapsed table (Exercise 5)
   - Separate analyses of each matched pair of subpopulations (Exercise 7)
   - Combining evidence across matched pairs of subpopulations (this exercise)

10. Use the display in Exercise 8 to perform the MH test with the third alternative. Compare and contrast the three test strategies:

    - Analysis of the collapsed table (Exercise 6)
    - Separate analyses of each matched pair of subpopulations (Exercise 8)
    - Combining evidence across matched pairs of subpopulations (this exercise)

Exercises 11–13 refer to the study of elder abuse discussed in the text. You will need to refer to the tables and figures presented in the text.

11. Complete the table at the top of the page. (Note that the component tables 5 and 6 have been combined.)

12. Write a few sentences describing the interaction graph, Figure 9.5. Note that component tables 1 and 8 are mirror images of each other; similarly, component tables 2 and 7, and so on, are mirror images of each other.

13. Refer to Exercise 11. Obtain the standardized rate estimate of the proportion of successes for victims of female abusers if they had had the same distribution of perpetrators as victims of male abusers. Obtain the standardized rate estimate of the proportion of successes for victims of male abusers if they had had the same distribution of perpetrators as victims of female abusers. Compare your answers to the values in the collapsed table. Comment.

REFERENCES

1. A. Tversky and T. Gilovich, "The Cold Facts about the Hot Hand in Basketball," *Chance: New Directions for Statistics and Computing*, Winter, 1989, pp. 16–21.

2. L. Vinton, "Correlates of Abused Elders' Anticipated Use of Services," Ph.D. diss., University of Wisconsin-Madison, 1988.

Part II

MULTICATEGORY RESPONSES

Chapter 10 **One Multicategory Response**

Chapter 11 **Tests of Homogeneity and Independence**

One Multicategory Response

10.1 DATA PRESENTATION AND SUMMARY

10.2 THE POPULATION MODEL AND ESTIMATION

10.3 THE CHI-SQUARED GOODNESS OF FIT TEST

Chapter 10

Part I of this book examined dichotomous responses: response variables that can take on one of exactly two possible values. The remainder of the book extends many of the results of Part I to more complicated responses. In addition, the more complicated responses require some new methods for description and analysis. There are two types of these more complicated responses, and they are illustrated in the following two examples.

EXAMPLE 10.1

Question 9 on the 1982 Wisconsin Driver Survey read

Which do you think is most intoxicating?

The subjects were given a choice of four answers, "can of beer," "glass of wine," "mixed drink," or "about the same." This choice of answers is a **multicategory response** with four categories. ▲

EXAMPLE 10.2

Sally goes to a golf driving range and hits a ball off a tee with a 3-iron. The response is the distance the ball travels in yards. This is a **numerical response,** so named because the response is a number. ▲

Chapter 10 extends the results of Chapters 5 and 6 for a single dichotomous response and one population, to a single multicategory response and one population. Chapter 11 extends the results of Chapters 1 through 3 and Chapter 7 for a single dichotomous response and two treatments or populations, to a single multicategory response or more than two treatments or populations. Chapter 11 also extends the results of Chapter 8 for two dichotomous responses and a single population to the case where either or both of the responses are multicategory. For brevity and simplicity, the results of Chapters 4 and 9 are not extended to a multicategory response. Chapters 12 through 16 extend the results of Part I to one or two numerical responses per subject.

It is important to realize that in many situations a response can be dichotomous, multicategory, or numerical, depending upon the preference of the researcher. For example, let the response be the age of an adult person. Age is usually reported by rounding down to the nearest year, a number, and is naturally viewed as a numerical response. Alternatively, a researcher may want to treat age as a multicategory response with categories such as

18–24, 25–34, 35–44, 45–54, 55–64, and 65 or above.

Finally, age could be dichotomous, with categories such as "old enough to be president of the United States" (35 or older) and "not old enough to be president of the United States."

10.1 DATA PRESENTATION AND SUMMARY

Table 10.1 presents the frequency and relative frequency distributions of the responses to question 9 on the 1982 Wisconsin Driver Survey. Notice that 1,574 persons responded to the question, with over half (55.4 percent) selecting the response "about the same." The second most popular response was that a mixed drink is most intoxicating. Relatively few persons believed that either beer or wine is most intoxicating. Figure 10.1 presents a bar chart of the relative frequency distribution.

TABLE 10.1 Responses to "Which Do You Think Is Most Intoxicating?" (Question 9) on the 1982 Wisconsin Driver Survey

Response	Frequency	Relative Frequency
About the same	872	0.554
Mixed drink	588	0.374
Can of beer	75	0.048
Glass of wine	39	0.025
Total	1,574	1.001

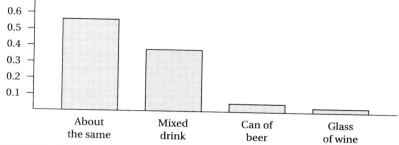

FIGURE 10.1 Bar chart of the relative frequency distribution of the responses of question 9 on the 1982 Wisconsin Driver Survey.

Question 3 on the 1982 Wisconsin Driver Survey read

If a person drives while drunk, what is his/her chance of being stopped by the police?

The researchers could have asked for the percentage chance of being stopped, a numerical response. Instead, they opted for a multicategory response with five possible values: "very low," "low," "about 50:50," "high," and "very high." The actual chance of being stopped was, not surprisingly, unknown, but Wisconsin State Patrol officials believed that the chance was less than 1 percent. The purpose of this question was to measure the perception of Wisconsin drivers. In the 1982 survey 1,530 people responded to this question. The data are presented in Table 10.2 and Figure 10.2. A quick examination of the data shows that the sampled drivers' perceptions of the chance of being stopped far exceed the police estimate. This discrepancy suggests that drivers may overestimate the effectiveness of police enforcement or underestimate the frequency of drunk driving.

TABLE 10.2 Responses to "If a Person Drives While Drunk, What Is His/Her Chance of Being Stopped by the Police?" (Question 3) on the 1982 Wisconsin Driver Survey

Response	Frequency	Relative Frequency
Very low	160	0.105
Low	407	0.266
About 50:50	699	0.457
High	172	0.112
Very High	92	0.060
Total	1,530	1.000

There is an important difference between the two examples presented in this section. The responses to question 9—"can of beer," "glass of wine," "mixed drink," and "about the same"—are not ordered, but the responses to question 3—"very low," "low," "about 50:50," "high," and "very high"—are ordered. More precisely, a multicategory response is **ordered** if the following condition is met:

For any three possible response categories, one of the categories naturally is viewed as being between the other two.

Select any three possible responses to question 3, such as "low," "high," and "very high." It is natural to view "high" as falling between the "low" and "very high" responses. For question 9, however, there is no natural way to order the responses "can of beer," "glass of wine," and "mixed drink."

Another way to view an ordered multicategory response is that there is an *underlying variable* that the researcher wants to measure, and the categories reflect how much of the underlying variable is possessed by the subject. For question 3, for example, the underlying variable is the perceived chance of being stopped, and the ordered response categories—very low, low, about 50:50, high, and very high—reflect increasing values of the perceived chance.

For an ordered multicategory response, the relative frequency distribution and its bar chart should follow the ordering of the categories, as illustrated in Table 10.2 and Figure 10.2. For an unordered multicategory response, however, the relative frequency distribution and its bar chart present first the category with the highest frequency (or, equivalently, relative frequency), followed by the category with the second highest frequency, and so on, as illustrated in Table 10.1 and Figure 10.1.

The methods presented in Chapters 10 and 11 can be used for ordered or unordered multicategory responses. Numerous books present methods appropriate for ordered multicategory responses but inappropriate for unordered responses. The book by Alan Agresti [1] is an excellent reference for these methods. Finally, ordered categorical data are sometimes analyzed using the methods for numerical data that will be introduced in Chapter 12 and studied in Part III of this book.

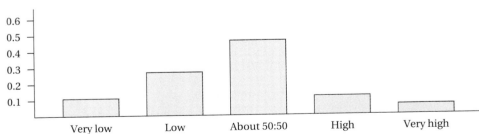

FIGURE 10.2 Bar chart of the relative frequency distribution of the responses of question 3 on the 1982 Wisconsin Driver Survey.

EXAMPLE 10.3: The Daily Game

The data in the above two examples arose from a study of different subjects. The following is an example of multicategory data that are obtained by observing a sequence of trials. This example also demonstrates how easy it is to be confused as to whether a response is ordered or unordered.

The Wisconsin state lottery includes a daily jackpot game. A 50-cent ticket allows a player to select six different integers between 1 and 36, inclusive. Each evening, lottery officials select six different integers between 1 and 36. Any player whose six numbers match the state's selection wins $250,000. The state's method for selecting the winning six numbers is advertised to be fair (that is, the possible outcomes are equally likely). There are two natural ways to study a lottery. First, on the assumption that the state's method of selecting numbers *actually is* fair, the probability that any particular ticket wins is 1 in 1,947,792, because it can be shown that there are 1,947,792 different ways to select six numbers from 36 numbers. Second, a researcher can critically examine the assumption of fairness by observing and analyzing the outcomes of many of the daily drawings. This second approach is the topic of this example.

Data were obtained for 235 daily drawings, from July 14, 1991, to March 4, 1992. The real goal of the study was to investigate whether the 1,947,792 possible outcomes for the six numbers selected are equally likely. The data reveal that 235 of the possible outcomes occurred once each and the remaining 1,947,557 outcomes did not occur. This result provides virtually no insight into the real goal. One approach would be to observe about 20 million outcomes—which would require about 55,000 years—to see whether some outcomes occur much more often than others. Since this approach is clearly impossible, I will instead examine the $235 \times 6 = 1,410$ numbers that were obtained on the 235 days. If the state's selection procedure is fair, then except for chance variation, each of the 36 numbers should have been selected the same number of times.

View each individual number drawn by the state to be a response. The possible values of the response are the integers $1, 2, \ldots, 36$. It is perhaps natural to assume implicitly that categories corresponding to numbers are ordered. For this example, however, the numbers are simply labels placed on 36 balls in a drum. The ball numbered 2, for example, is not between the balls numbered 1 and 3. Another way to look at it is that the lottery game would remain unchanged if the 36 balls were labeled with 36 different nonsense symbols instead of numbers, yet no one would argue that nonsense symbols are ordered.

Figures 10.3, 10.4, and 10.5 present three bar charts of the frequency distribution for the 1,410 balls selected. Figure 10.3 presents the frequency for ball 1 first, followed by the frequency for ball 2, which is followed by the frequency for ball 3, and so on. According to the guidelines given in this chapter, this ordering would be used by an analyst who views the categories (ball numbers) as ordered. Figure 10.4, on the other hand, presents the frequency for ball 15 first because it was selected most often, followed by ball 31 because it was selected second most

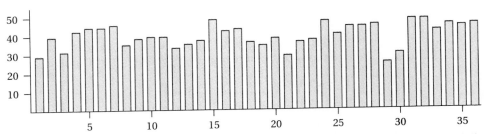

FIGURE 10.3 Bar chart of frequency distribution of the numbers selected in 235 daily lottery games.

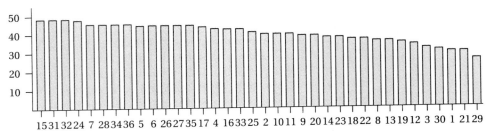

FIGURE 10.4 Bar chart of the frequency distribution of the numbers selected in 235 daily lottery games.

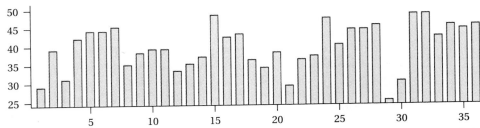

FIGURE 10.5 Potentially misleading bar chart of the frequency distribution of the numbers selected in 235 daily lottery games.

often, and so on. This figure would be used by an analyst, such as myself, who views the categories as unordered. Figure 10.3 suggests more variation in the frequencies than Figure 10.4 because of its arbitrary ordering of the categories, which gives an up-and-down pattern to the bars. (A test of the null hypothesis of fairness presented later in this chapter shows that the pattern in the data is not statistically significant.)

More serious than the discrepancy between Figures 10.3 and 10.4 is a gross distortion of the data that could result from examining Figure 10.5. For example, in the 235 drawings, ball 28 and ball 29 were selected 45 and 25 times, respectively. Thus, it is accurate to say that the frequency of ball 28 exceeds the frequency of ball 29 by 20, or to say that ball 28 was selected 80 percent more often than ball 29 (45/25 = 1.8, so 45 is 80 percent larger than 25). Figure 10.5 cuts off the first 24 occurrences of each ball from the chart. The result is that the rectangle above ball 28 is 21 times taller than the adjacent rectangle above ball 29, which could easily be misinterpreted as indicating that ball 28 occurred 21 times as often as ball 29! Exercise 6 below explores other features of this picture.

EXERCISES 10.1

1. In the 1986 Wisconsin Driver Survey, in response to the question

    If a person drives while drunk, what is his/her chance of being stopped by the police?

    46 subjects responded "very low," 338 responded "low," 752 responded "about 50:50," 378 responded "high," and 166 responded "very high." Present these data with a relative frequency table and a bar chart. Compare, in words, the 1982 responses (Table 10.2 and Figure 10.2) with the 1986 responses.

2. In the 1984 Wisconsin Driver Survey, drivers were asked the question

    Which do you think is most intoxicating?

    Eighty-nine subjects selected "can of beer," 74 selected "glass of wine," 940 selected "mixed drink," and 1,482 selected "about the same." Summarize these data with a relative frequency table and a bar chart. Compare, in words, the 1982 responses (Table 10.1 and Figure 10.1) with the 1984 responses.

3. My canine companion Casey enjoys eating dog biscuits. Her favorite brand comes in seven flavors, bacon, cheese, liver, meat, milk, poultry, and vegetable. I have noticed that she responds least enthusiastically when I give her a vegetable-flavored treat. This observation inspired me to study the distribution of flavors in a box of dog biscuits. I purchased a box of biscuits and sorted the 120 biscuits by flavor. Let each biscuit be a subject and let the response be its flavor. My results were as follows:

    bacon 19, cheese 17, liver 20, meat 11, milk 21, poultry 12, and vegetable 20.

    (a) How many categories does the multicategory variable have?
    (b) Are the categories ordered or unordered?
    (c) Construct the relative frequency distribution for these data.
    (d) Construct a bar chart for the relative frequency distribution for these data.

4. Trix cereal comes in five fruit flavors, each with its own distinctive shape. A friend of mine was interested in the distribution of the flavors. He emptied a large box of the cereal and sorted the pieces according to flavor. Let each piece of cereal be a subject and let the response be its flavor. His results were as follows:

    grape 420, lemon 530, lime 610, orange 585, and raspberry 470.

    (a) How many categories does the multicategory variable have?
    (b) Are the categories ordered or unordered?
    (c) Construct the relative frequency distribution for these data.
    (d) Construct a bar chart for the relative frequency distribution for these data.

5. The video game Tetris was introduced in Chapter 5 (p. 174). The progress of the game is driven by a sequence of trials. Each trial is the introduction at the top of the screen of a falling shape made up of four squares. There are seven possible shapes for each trial to assume; arbitrarily label these shapes A, B, C,

D, E, F, and G. A sequence of 1,872 trials yielded the following frequencies for each shape:

A 245, B 278, C 257, D 272, E 281, F 285, and G 254.

(a) Construct the relative frequency distribution for the data (take the categories to be unordered).
(b) Construct a bar chart for the relative frequency distribution for the data.

6. Refer to Figures 10.3 and 10.5.

(a) How do the vertical axes of these bar charts differ? Give at least one important consequence of this difference.
(b) I have examined several brochures promoting a number of different lotteries across North America. Many of the brochures included a bar chart of the frequencies of selected numbers over some period of time. *Every* such bar chart was distorted in the same way that Figure 10.5 is distorted. It is my experience that many, if not most, bar charts that appear in the print media are distorted in the same way. Find a bar chart in the print media that has this distortion, photocopy it, and submit it for this exercise.
(c) Write a plausible conjecture for why this distortion is so popular.

10.2 THE POPULATION MODEL AND ESTIMATION

Following the development of Chapter 5, a finite population represents a collection of distinct individuals and an infinite population represents a mathematical model for the process that generates the outcomes of trials. It is easier to start with a finite population.

Let k represent the number of possible values (number of categories) of the response variable. The case $k = 2$ was studied in Part I of this book; this chapter will restrict attention to the case $k > 2$. For the examples on the Wisconsin Driver Survey given in the previous section, $k = 5$ for the response to question 3 and $k = 4$ for the response to question 9. Number the categories $1, 2, \ldots, k$. This numbering is arbitrary, but if the categories are ordered it is standard to have the numbers assigned to the categories follow the same order as the categories. For example, make the following assignments for the above studies:

For question 3: $1 =$ "very low," $2 =$ "low," $3 =$ "about 50:50," $4 =$ "high," and $5 =$ "very high."

For question 9: $1 =$ "about the same," $2 =$ "mixed drink," $3 =$ "can of beer," and $4 =$ "glass of wine."

Imagine a population box with one card in it for each member of the population. A population member's card contains one number, the number of the category of the individual's response. Let

$N_1 = $ the number of cards in the box marked 1.
$N_2 = $ the number of cards in the box marked 2.
$\vdots$
$N_k = $ the number of cards in the box marked k.

Further, let N equal the total number of cards in the box: $N = N_1 + N_2 + \cdots + N_k$. For the response to question 3 on the Wisconsin Driver Survey, for example,

- N_1 denotes the number of licensed drivers in Wisconsin who would respond "very low" if questioned.
- N_2 denotes the number of licensed drivers in Wisconsin who would respond "low" if questioned.
- N_3 denotes the number of licensed drivers in Wisconsin who would respond "about 50:50" if questioned.
- N_4 denotes the number of licensed drivers in Wisconsin who would respond "high" if questioned.
- N_5 denotes the number of licensed drivers in Wisconsin who would respond "very high" if questioned.
- N denotes the number of licensed drivers in Wisconsin.

Let $p_1 = N_1/N$, $p_2 = N_2/N$, ..., $p_k = N_k/N$. Clearly, p_1 equals the proportion of cards in the box marked 1, p_2 equals the proportion of cards in the box marked 2, and so on. For the response to question 3 on the Wisconsin Driver Survey, for example,

- p_1 denotes the proportion of licensed drivers in Wisconsin who would respond "very low" if questioned.
- p_2 denotes the proportion of licensed drivers in Wisconsin who would respond "low" if questioned.
- p_3 denotes the proportion of licensed drivers in Wisconsin who would respond "about 50:50" if questioned.
- p_4 denotes the proportion of licensed drivers in Wisconsin who would respond "high" if questioned.
- p_5 denotes the proportion of licensed drivers in Wisconsin who would respond "very high" if questioned.

Alternatively, $p_1, p_2, \ldots, p_k$ can be interpreted as probabilities:

Consider the experiment of selecting one card at random from the box. Here p_1 is the probability that the card selected is marked 1, p_2 is the probability that the card selected is marked 2, and so on.

Knowledge of the values of $p_1, p_2, \ldots, p_k$ yields a total understanding of the population. Unfortunately, such knowledge is impossible without a census of the population.

Again, following the development in Chapter 5, the sample consists of the members of the population on which data are collected. Two important ways to obtain a sample are a random sample with replacement and a random sample without replacement. The definitions of these methods of sampling that were given in Chapter 5 apply, without modification, to the current situation. Again, as in Chapter 5, it can be shown that if the sample size is less than 5 percent of the population size, then there is no important difference between random sampling with and without replacement. For a random sample with replacement a researcher can use the multiplication rule, as in Chapter 5, to compute the probabilities of certain complicated events. Because the multiplication rule can be used, the random variables corresponding to the numbers on the cards selected are called independent random variables. (Note: You may want to review the presentation of these ideas in Chapter 5. As stated before, this chapter extends the results of Chapters 5 and 6 from a dichotomous to a multicategory response. This chapter focuses on statistical methods for analyzing data and provides few details of the mathematical development.)

10.2.1 Multinomial Trials

There are many possible mathematical models for a sequence of trials. Only the model of multinomial trials will be studied in this text, however, because it is the simplest interesting model.

Recall the definition of Bernoulli trials given in Chapter 5:

Suppose an experiment consists of n trials. The trials are Bernoulli trials if the following three conditions are satisfied:

1. Each trial has two possible outcomes—a success or a failure.
2. The probability of success remains constant from trial to trial. This constant probability is denoted by p.
3. The trials are independent; in particular, the multiplication rule can be used.

Multinomial trials are a natural extension of this definition.

Suppose an experiment consists of n trials. The trials are called multinomial trials if the following three conditions are satisfied:

1. Each trial has $k > 2$ possible outcomes, which are denoted $1, 2, \ldots, k$.
2. The probabilities of the different outcomes remain constant from trial to trial. These constant probabilities are denoted $p_1, p_2, \ldots, p_k$. These probabilities may be known or unknown, and, in either case, they sum to 1.
3. The trials are independent; in particular, the multiplication rule can be used.

Bivariate Bernoulli trials, introduced in Section 8.3, are a special case of multinomial trials with $k = 4$ categories, AB, AB^c, A^cB, and A^cB^c.

EXAMPLE 10.4

Consider the experiment of tossing a die repeatedly and recording the number obtained each time. Are these multinomial trials? First of all, the word "tossing" is implicitly assumed to mean that the die is thrown with sufficient force to cause a good amount of bouncing before it settles. This is achieved in gambling casinos by requiring the player to bounce the die off a (not too) distant wall. The first condition is satisfied, with $k = 6$ possible outcomes. The second condition is that the probability of each possible outcome remains constant. In words, this assumption is that the die's composition does not change during the course of experimentation, which seems reasonable. Finally, the third assumption is that the die does not have a memory, which also seems reasonable. Remember that it is not necessary to know the probability of each of the possible outcomes in order to have multinomial trials, but if one is willing to assume the die is balanced, or fair, the six outcomes are equally likely to occur. As discussed in Section 2.3, if the outcomes are equally likely to occur, the probability of any particular outcome is equal to 1 divided by the number of possible outcomes, that is, 1 divided by 6, or $\frac{1}{6}$. ▲

In a development similar to that for Bernoulli trials, it can be shown that observing a sequence of multinomial trials is mathematically equivalent to selecting cards at random, with replacement, from an appropriate box of cards. For instance, suppose the analyst assumes that the die in the previous example is fair. Then successive casts of the die can be visualized as sampling at random, with replacement, from a box containing cards numbered 1, 2, 3, 4, 5, and 6. In summary, for a multicategory response the same mathematical arguments apply to finite populations and multinomial trials. Thus, while developing mathematical results, one can ignore whether a population is finite or composed of multinomial trials.

EXAMPLE 10.5

Gregor Johann Mendel (1822–1884) performed many important experiments in biology. In one study he harvested 556 peas from 15 double hybrid plants. Of his peas, 315 were round and yellow, 108 were round and green, 101 were wrinkled and yellow, and 32 were wrinkled and green. Consider each pea to be a trial. The first condition for multinomial trials is satisfied, with $k = 4$ possible types of peas. You should consider whether or not the second and third assumptions of multinomial trials seem reasonable for Mendel's experiment. Biologists regularly make these assumptions for data like Mendel's. Figure 10.6 presents the bar chart for Mendel's data. These pea data are analyzed further later in this chapter. ▲

10.2.2 Estimation

Following the ideas of Chapter 6 for one dichotomous variable, this chapter presents three types of statistical inference for the k proportions $p_1, p_2, \ldots, p_k$:
- Point estimation
- Confidence interval estimation
- Hypothesis testing

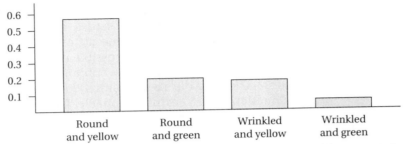

FIGURE 10.6 Bar chart of the relative frequency distribution of the types of peas grown by Mendel.

The data consist of a sample of n subjects assumed to have been selected at random with replacement from the population. The assumption that the sample was selected at random should be examined critically for each application of these methods. The responses of the n subjects studied are summarized by simple counting. Define the random variables

O_1 = the number of times the response 1 occurs in the sample.
O_2 = the number of times the response 2 occurs in the sample.
$\vdots$
O_k = the number of times the response k occurs in the sample.

(The O is for *observed*.) By tradition, observed values of these random variables also are denoted by uppercase letters. The sum of these observed frequencies equals the sample size:

$$O_1 + O_2 + \cdots + O_k = n.$$

The point estimators of the population proportions are the sample proportions (sample relative frequencies) $\hat{P}_1 = O_1/n$, $\hat{P}_2 = O_2/n, \ldots, \hat{P}_k = O_k/n$.

For the two questions from the Wisconsin Driver Survey studied in this section, the point estimates of the population proportions are given by the sample proportions that appear in the relative frequency columns of Tables 10.1 and 10.2 (p. 316). For example, the point estimate of the proportion of the 1982 Wisconsin licensed drivers who, if questioned, would have answered "mixed drink" to question 9 is $\hat{p}_2 = 0.374$. As another example, the point estimate of the proportion of the same population who would have answered "about 50:50" to question 3 is $\hat{p}_3 = 0.457$. For Mendel's pea data, the point estimate of the probability of obtaining a pea that is round and yellow is $315/556 = 0.567$.

The methods of Chapter 6 can be used to obtain a confidence interval or conduct a hypothesis test for any individual population proportion. For example, suppose a researcher wants a confidence interval for the proportion of 1982 Wisconsin licensed drivers who would have responded "very low" to question 9. To proceed, consider the response "very low" to be a success and any other response to be a failure. Thus, from Table 10.1 the data consist of $x = 160$ successes in a sample of size $n = 1{,}530$. The sample proportions of successes and failures are

$$\hat{p} = \frac{x}{n} = \frac{160}{1{,}530} = 0.105 \quad \text{and} \quad \hat{q} = 1 - \hat{p} = 1 - 0.105 = 0.895.$$

These numbers can be substituted into Formula 6.7,

$$\hat{p} \pm z\sqrt{\frac{\hat{p}\hat{q}}{n}}.$$

For a 99 percent confidence interval, for example, $z = 2.576$ and the confidence interval is

$$0.105 \pm 2.576\sqrt{\frac{(0.105)(0.895)}{1{,}530}} = 0.105 \pm 0.020 = [0.085, 0.125].$$

In words, at the 99 percent confidence level one can conclude that between 8.5 percent and 12.5 percent of the population of licensed drivers would have responded "very low" to question 9.

The above argument can be applied to each of the five population proportions for question 9. Below are the 99 percent confidence intervals for these proportions.

- For p_2, the proportion who would have responded "low," the confidence interval is

$$0.266 \pm 2.576\sqrt{\frac{(0.266)(0.734)}{1{,}530}} = 0.266 \pm 0.029 = [0.237, 0.295].$$

- For p_3, the proportion who would have responded "about 50:50," the confidence level is

$$0.457 \pm 2.576\sqrt{\frac{(0.457)(0.543)}{1{,}530}} = 0.457 \pm 0.033 = [0.424, 0.490].$$

- For p_4, the proportion who would have responded "high," the confidence interval is

$$0.112 \pm 2.576\sqrt{\frac{(0.112)(0.888)}{1{,}530}} = 0.112 \pm 0.021 = [0.091, 0.133].$$

- Finally, for p_5, the proportion who would have responded "very high," the confidence interval is

$$0.060 \pm 2.576\sqrt{\frac{(0.060)(0.940)}{1{,}530}} = 0.060 \pm 0.016 = [0.044, 0.076].$$

The interpretation of these intervals *individually* is familiar. For example, the analyst is 99 percent confident that the unknown p_1 is between 0.085 and 0.125. The interpretation of these intervals *simultaneously*, or *collectively*, however, is not so straightforward. Difficulties arise both in the interpretation of the values in the intervals and in the assessment of the confidence level.

Viewed together, there is an inconsistency in these five intervals. To see this, note that the confidence intervals for p_1, p_2, p_3, p_4, and p_5 include the values 0.125, 0.295, 0.490, 0.133, and 0.076, respectively (these values are the five upper endpoints). These five values sum to 1.119, whereas the unknown proportions must sum to 1. In other words, these five numbers taken individually are plausible values of the unknown proportions of licensed drivers, but collectively they cannot possibly be correct! (It is not easy to remove this inconsistency, and in this text I will not attempt to do so.)

Before data are collected, the probability that any individual interval will be correct is 0.99. Suppose, however, that the researcher wants to know the probability that all five intervals will be correct. Using the terminology of Section 6.6, these five confidence intervals are an example of a dependent analysis, and although the probability that all five intervals will be correct is unknown, by Bonferroni's result this unknown probability is 0.95 or greater.

In contrast with confidence intervals, the next section shows that it is relatively easy to perform a hypothesis test that considers all k proportions simultaneously. But first it is convenient to digress and study the family of chi-squared curves.

10.2.3 The Chi-squared Curves

The standard normal curve was introduced in Chapter 3 as a tool for computing approximate probabilities. The standard normal curve is not the only approximating curve used by statisticians. There are a number of curves, similar enough in their characteristics to be called a family of curves, that will be useful throughout this chapter and the next. This family of curves is called the **chi-**

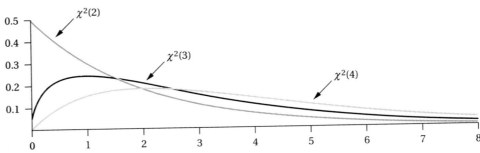

FIGURE 10.7 Three chi-squared curves.

squared curves. Three chi-squared curves are presented in Figure 10.7. This figure illustrates two general properties of all chi-squared curves:

- There is no area under any chi-squared curve to the left of 0. (This partly explains the name; the word "squared" in "chi-squared" serves as a reminder that negative numbers are not possible.)
- Each chi-squared curve is asymmetric.

A further property of these curves is

- The total area under any chi-squared curve is equal to 1.

Chi-squared curves are characterized by the parameter called the *number of degrees of freedom*, which is abbreviated df. The number of degrees of freedom can be any positive integer. Thus, there is a chi-squared curve with one degree of freedom, a chi-squared curve with two degrees of freedom, and so on. Figure 10.7 presents chi-squared curves with 2, 3, and 4 degrees of freedom. A chi-squared curve with 5 degrees of freedom, for example, is denoted by $\chi^2(5)$. The symbol χ is the lowercase Greek letter chi. In general, for any positive integer m, $\chi^2(m)$ denotes the chi-squared curve with m degrees of freedom.

Table A.5 in the Appendix provides selected areas for chi-squared curves with 30 or fewer degrees of freedom. Due to space limitations, there is only one row of numbers in Table A.5 for each chi-squared curve, compared with two pages in Table A.2 for the standard normal curve. The few numbers per curve given in Table A.5, however, are particularly useful for the hypothesis-testing applications needed in this text.

Use of Appendix Table A.5. Table A.5 gives areas under chi-squared curves to the right of selected numbers. The number of degrees of freedom for the chi-squared curve determines the row of the table to be used. For example, suppose one wants to find the area under the chi-squared curve with two degrees of freedom to the right of 4.61. To find this area, read across row 2 of Table A.5 until the entry 4.61 is located, as illustrated below, in a reproduction of a portion of Table A.5.

		Area to the Right		
df	...	0.50	0.10	0.05
1	...	.	.	.
2	...	1.39	4.61	5.99

The value 4.61 resides in the column headed by 0.10. The area to the right of 4.61 under the chi-squared curve with two degrees of freedom is 0.10. This result is presented graphically in Figure 10.8.

As another example, read across row 5 of Table A.5 to the entry 11.07. The number 11.07 is in the column headed 0.05. Thus, the area to the right of 11.07 under the chi-squared curve with five degrees of freedom is equal to 0.05. For a

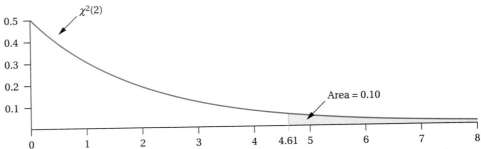

FIGURE 10.8 Visual demonstration that the area to the right of 4.61 under the chi-squared curve with two degrees of freedom equals 0.10.

final example, read across row 10 to the entry 23.21. The number 23.21 is in the column headed 0.01: the area to the right of 23.21 under the chi-squared curve with 10 degrees of freedom is equal to 0.01. The following example gives more illustrations of the use of Table A.5.

EXAMPLE 10.6

Use Table A.5 to verify the following answers:

1. Find the area to the right of 1.39 under the chi-squared curve with two degrees of freedom.
   Locate the value 1.39 in row 2 of Table A.5. The value 1.39 is in the column headed 0.50; thus, the area to the right of 1.39 is 0.50.

2. Find the area to the right of 3.50 under the chi-squared curve with two degrees of freedom.
   The value 3.50 does not appear in row 2 of Table A.5. As a result, the area to its right cannot be obtained, but the following bound on the area is very useful. The method introduced here is called the invisible column solution. Read across the column headings in Table A.5 from left to right and note that the numbers decrease. Read across any other row of the table from left to right and note that the numbers increase. Reading across row 2, note that if 3.50 were in the row it would lie between the entries 1.39 and 4.61. Thus, imagine an extra invisible column in Table A.5 with the entry 3.50 in row 2. The column heading of this invisible column will be the desired area (see Table 10.3). Unfortunately, the heading of the invisible column cannot be read, but the heading would lie between the headings of the two visible columns to either side of it, namely 0.50 and 0.10. Thus, the area is somewhere between 0.10 and 0.50 (it is standard to write the smaller number first). Figure 10.9 presents this argument visually.

3. Find the area to the right of 21.62 under the chi-squared curve with 10 degrees of freedom.

TABLE 10.3 Invisible Column Approach to Finding the Area to the Right of 3.50 under the Chi-squared Curve with Two Degrees of Freedom

df	...	0.50	?	0.10
1	...			
2	...	1.39	3.50	4.61

The value 21.62 does not appear in row 10 of Table A.5. If 21.62 were in the row, it would fall between the entries 20.48 and 23.21. The column headings for the values 20.48 and 23.21 are 0.025 and 0.01, respectively. Thus, the desired area is between 0.01 and 0.025.

4. Find the area to the right of 12.00 under the chi-squared curve with two degrees of freedom.

 The invisible column approach is needed again. Unlike the previous two problems, here 12.00 does not fall between two entries in row 2. Instead, it is larger than every entry in the row. Thus, the invisible column with entry 12.00 lies to the right of the last column in the table, and hence its column heading is smaller than all column headings (remember that column headings decrease to the right). It follows that the area to the right of 12.00 is less than 0.01.

5. Find the area to the right of 2.00 under the chi-squared curve with 12 degrees of freedom.

 The invisible column approach is needed again. This time the value, 2.00, is smaller than every entry in row 12 of Table A.5. Thus, the area to the right of 2.00 is larger than 0.99. ▲

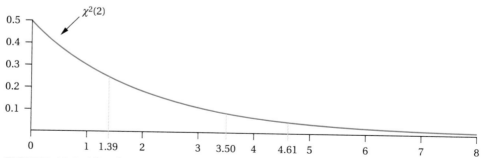

FIGURE 10.9 Visual presentation of the validity of the invisible column approach. According to Table A.5, the area to the right of 4.61 is 0.10; similarly, the area to the right of 1.39 is 0.50. Clearly, the area to the right of 3.50 is between these two areas. Thus, the area to the right of 3.50 is between 0.10 and 0.50.

The invisible column approach and, in fact, Table A.5 itself become more outdated each day. Many computer programs and even pocket calculators can compute areas under a chi-squared curve. For example, with the help of Minitab, I found that

- The area to the right of 3.50 under the chi-squared curve with two degrees of freedom is 0.1738.
- The area to the right of 21.62 under the chi-squared curve with ten degrees of freedom is 0.0171.
- The area to the right of 12.00 under the chi-squared curve with two degrees of freedom is 0.0025.
- The area to the right of 2.00 under the chi-squared curve with 12 degrees of freedom is 0.9994.

Compare each of these areas to the bounds obtained in Example 10.6 from Table A.5.

There is a useful connection between the chi-squared curve with one degree of freedom and the standard normal curve. Suppose Z is a random variable with sampling distribution given by the standard normal curve. Then the sampling distribution of Z^2 is given by the chi-squared curve with one degree of freedom. An example may help make this result more believable. If the sampling distribution of Z is given by the standard normal curve, then $P(Z \geq 1.96$ or $Z \leq -1.96)$ equals the area under the standard normal curve to the right of 1.96 plus the area under the standard normal curve to the left of -1.96. By symmetry, this sum is equal to twice the area under the standard normal curve to the right of 1.96. According to Table A.2, this area is 2(0.0025), or 0.05. Alternatively,

$$P(Z \geq 1.96 \text{ or } Z \leq -1.96) = P(Z^2 \geq (1.96)^2) = P(Z^2 \geq 3.84).$$

By the above result, this last probability is equal to the area to the right of 3.84 under the chi-squared curve with one degree of freedom. Reading from Table A.5, one finds this area to be 0.05, which agrees with the answer from the standard normal curve.

The practical value of this result is that Table A.2, the table of areas under the standard normal curve, can be used to compute areas under the chi-squared curve with one degree of freedom. This is helpful because Table A.2 contains much more information than row 1 of Table 5. The reader who enjoys algebra can use this result to prove the following:

Let χ^2 denote any positive number. Let χ equal the positive square root of χ^2. The area to the right of χ^2 under the chi-squared curve with one degree of freedom is equal to twice the area under the standard normal curve to the right of χ.

The use of this result is illustrated in the following example.

EXAMPLE 10.7

Verify the following answers:

1. Find the area to the right of 9.00 under the chi-squared curve with one degree of freedom.

   The square root of 9.00 is 3.00. Thus, the desired area equals twice the area under the standard normal curve to the right of 3.00. According to Table A.2, this area is 2(0.0013), or 0.0026. By contrast, according to Table A.5, the area is less than 0.01. The answers from the two tables are consistent, but the answer obtained by using the standard normal curve is more precise.

2. Find the area to the right of 4.25 under the chi-squared curve with one degree of freedom.

   The square root of 4.25 is 2.06. Thus, the desired area equals twice the area under the standard normal curve to the right of 2.06. According to Table A.2, this area is 2(0.0197), or 0.0394. By contrast, according to Table A.5, the area is bounded by 0.025 and 0.05. The answers from the two tables are consistent, but the answer obtained by using the standard normal curve is more precise. ▲

EXERCISES 10.2

1. Compute 95 percent confidence intervals for the four population proportions for Mendel's pea data given in Example 10.5 (p. 325).

2. Compute 95 percent confidence intervals for the four population proportions for the responses to question 9 on the 1982 Wisconsin Driver Survey data given in Table 10.1 (p. 316).

3. Use Table A.5 to find the area to the right of 4.57 under the chi-squared curve with 11 degrees of freedom.

4. Use Table A.5 to find the area to the right of 7.04 under the chi-squared curve with 13 degrees of freedom.

5. Use Table A.5 to find the area to the right of 7.55 under the chi-squared curve with eight degrees of freedom.

6. Use Table A.5 to find the area to the right of 29.00 under the chi-squared curve with 20 degrees of freedom.

7. Use Table A.5 to find the area to the right of 48.00 under the chi-squared curve with 29 degrees of freedom.

8. Use Table A.5 to find the area to the right of 17.00 under the chi-squared curve with five degrees of freedom.

9. Use Table A.5 to find the area to the right of 10.00 under the chi-squared curve with two degrees of freedom.

10. Use Table A.5 to find the area to the right of 7.00 under the chi-squared curve with 19 degrees of freedom.

11. Use Table A.2 to find the area to the right of 7.00 under the chi-squared curve with one degree of freedom. Compare your answer with that obtained with Table A.5.

12. Use Table A.2 to find the area to the right of 2.00 under the chi-squared curve with one degree of freedom. Compare your answer with that obtained with Table A.5.

10.3 THE CHI-SQUARED GOODNESS OF FIT TEST

The hypothesis test developed in this section is called the chi-squared goodness of fit test.

The first step in a hypothesis test is the specification of the hypotheses. Recall from Section 6.4 that for a dichotomous response, the researcher must select a special value of interest for p, which is denoted p_0. For the problem of this chapter, the researcher must select k special values of interest, one for each proportion, $p_1, p_2, \ldots, p_k$. The method of making this selection is demonstrated below with some examples. Note that the k special values must sum to 1, since the k population proportions sum to 1. The special values of interest are denoted

$$p_{10}, p_{20}, \ldots, p_{k0}.$$

The symbol p_{10} is read "pea sub one naught" or "pea sub one zero;" it is never read "pea sub ten." Similarly, p_{20} is "pea sub two naught" or "pea sub two zero," and p_{k0} is "pea sub kay naught" or "pea sub kay zero." This is analogous to the earlier notation; a subscript 0 on a symbol denotes the special value of interest for the quantity denoted by that symbol.

Example 10.5 introduced a study of peas by Gregor Mendel. Mendel's theory suggested that the four types of peas—round and yellow, round and green, wrinkled and yellow, and wrinkled and green—should occur in the ratio 9:3:3:1. With a small amount of mathematics, this theory can be fit into the current framework. Make the following assignments:

- p_1 denotes the probability of obtaining a round and yellow pea.
- p_2 denotes the probability of obtaining a round and green pea.
- p_3 denotes the probability of obtaining a wrinkled and yellow pea.
- p_4 denotes the probability of obtaining a wrinkled and green pea.

These probabilities are unknown, but of course their sum must be one. According to Mendel's theory,

$$p_1 = 9p_4, \quad p_2 = 3p_4, \quad \text{and} \quad p_3 = 3p_4.$$

Substituting these identities into $p_1 + p_2 + p_3 + p_4 = 1$ yields

$$9p_4 + 3p_4 + 3p_4 + p_4 = 1,$$

or $16p_4 = 1$, or $p_4 = 1/16$. Thus, if Mendel's theory is correct, $p_1, p_2, p_3,$ and p_4 are 9/16, 3/16, 3/16, and 1/16, respectively. These four numbers are taken to be the special values of interest. More precisely,

$$p_{10} = \tfrac{9}{16}, \quad p_{20} = \tfrac{3}{16}, \quad p_{30} = \tfrac{3}{16}, \quad \text{and} \quad p_{40} = \tfrac{1}{16}.$$

The general form of the null hypothesis is

$$H_0: p_1 = p_{10},\ p_2 = p_{20}, \ldots, p_k = p_{k0}.$$

For Mendel's experiment this null hypothesis is

$$H_0: p_1 = \tfrac{9}{16},\ p_2 = \tfrac{3}{16},\ p_3 = \tfrac{3}{16},\ p_4 = \tfrac{1}{16}.$$

In words, the null hypothesis states that Mendel's theory is correct.

The mathematical theory allows for only one choice for the alternative,

$$H_1: \text{not } H_0.$$

In words, the alternative states that at least one of the true proportions is different from its hypothesized value. (Actually, if one is different, at least two must be different, since the collection of true proportions and the collection of their hypothesized values each sum to 1.)

The second step in a hypothesis test is the specification of the test statistic and the determination of its sampling distribution under the assumption that the null hypothesis is correct. As in step 1, the test of Chapter 6 for a dichotomous response provides guidance to the correct answer. In Chapter 6 the data were summarized by x, the observed count of the number of successes in the sample, and by n, the sample size. For the current problem, the data are summarized by the collection of observed counts $O_1, O_2, \ldots, O_k$, whose sum is the sample size, n. The sampling distribution for X is the binomial distribution; it can be shown that the sampling distribution for the collection of observed counts is a generalization of the binomial distribution called the multinomial distribution. Use of the multinomial distribution to obtain the exact P-value is very tedious, so researchers usually settle for an approximate P-value. The formula for the test statistic for the approximate method is quite messy, but it is easy to follow the derivation and to perform the computation by creating a table with seven columns. This table is constructed in Table 10.4 for the general problem

TABLE 10.4 Computing the Chi-squared Test Statistic for the General Case

Response	p_0	O	E	$O - E$	$(O - E)^2$	$(O - E)^2/E$
1	p_{10}	O_1	E_1			
2	p_{20}	O_2	E_2			
$\vdots$	$\vdots$	$\vdots$	$\vdots$			
k	p_{k0}	O_k	E_k			
Total	1	n	n			χ^2

TABLE 10.5 Computing the Chi-squared Test Statistic for Mendel's Pea Study

Response	p_0	O	E	O − E	$(O - E)^2$	$(O - E)^2/E$
Round and yellow	$\frac{9}{16}$	315	312.8	2.2	4.84	0.015
Round and green	$\frac{3}{16}$	108	104.2	3.8	14.44	0.139
Wrinkled and yellow	$\frac{3}{16}$	101	104.2	−3.2	10.24	0.098
Wrinkled and green	$\frac{1}{16}$	32	34.8	−2.8	7.84	0.245
Total	1	556	556	0		$\chi^2 = 0.497$

and Table 10.5 for the analysis of Mendel's data. The first three columns of the table are easy to obtain:

- The first column lists the category name or number of the response.
- The second column lists the special values of interest.
- The third column lists the observed counts.

The remaining entries in these tables will be discussed below.

Recall that in Chapter 6 an approximate P-value was obtained by considering the standardized version of X with the unknown parameters p and q replaced by their special values of interest, p_0 and q_0. The result was

$$Z = \frac{X - np_0}{\sqrt{np_0 q_0}}. \tag{10.1}$$

The numerator in Equation 10.1 compares, by subtraction, the number of successes, X, with its expected value under the assumption that the null hypothesis is true, np_0. In words, the numerator compares the actual number of successes, X, with the number of successes one would expect if the null hypothesis is true, np_0. The denominator in Equation 10.1 can be viewed as a correction factor that adjusts for the fact that the larger the value of n, the greater is the likely chance variation in the value of the numerator. These ideas extend to the problem of the current chapter; namely, each observed count is standardized in a similar way.

Recall that O_1 denotes the number of times the first category of the response occurs in the sample. For Mendel's data, O_1 is the number of round and yellow peas, and its observed value is 315. This random variable has a binomial distribution with sample size n and probability of success p_1. Thus, its expected value is np_1. If the null hypothesis is true, the unknown p_1 can be replaced by the known p_{10}, yielding

$$E(O_1) = np_{10}.$$

For notational simplicity, let

$$E_1 = E(O_1) = np_{10}.$$

In words, E_1 is the null expected value of O_1; it is called the expected count of the first category. For example, for Mendel's data

$$E_1 = np_{10} = 556\left(\tfrac{9}{16}\right) = 312.8.$$

The above reasoning for the first category can be extended to each of the categories. For example, the null expected value of the number of observations in the second category is denoted by E_2 and is equal to np_{20}. In general, each category has a null expected count, which is the product of its special value of interest and the sample size, n. The expected counts are listed in the fourth column of Table 10.4; the entries are obtained by multiplying each entry in the second column (the special values of interest) by the total of the third column (n). As a check of the arithmetic, note that the sum of the entries in the fourth column must equal, except for possible round-off error, the sample size, n. You should verify the values of E given in the fourth column of Table 10.5.

The numerator of Equation 10.1, the standardized test statistic for the problem of Chapter 6, compares the observed count with its expected value by subtraction. Analogously, for the current problem one computes the observed value of the difference $O - E$ for each category. At this point, the analogy between the current solution and the solution of Chapter 6 weakens. Each difference is squared, yielding the random variables

$$(O - E)^2.$$

Next, each of these squared differences is rescaled by dividing by E. The result is the random variable

$$\frac{(O - E)^2}{E}$$

for each category. The sum of these random variables taken over all categories is the test statistic and is denoted by the symbol X^2, which is the uppercase Greek letter chi squared (unfortunately, it looks like an "ex" squared). The observed results of these operations are presented in columns 5–7 of Table 10.5 for the analysis of Mendel's pea data. You should verify that the values in this table are correct. Note that the observed numerical value of the test statistic X^2 is denoted by χ^2, the lowercase Greek letter chi squared.

The above method for computing X^2 can be expressed with the formula

$$X^2 = \sum \frac{(O - E)^2}{E},$$

where the sum is taken over all of the categories. The next fact indicates why the symbol X^2 is used to denote the test statistic.

If the null hypothesis is true, the sampling distribution of the test statistic X^2 can be well approximated by the chi-squared curve with the number of degrees of freedom equal to the number of categories minus one,

$k - 1$. The general guideline for using this approximation is that every E should be greater than or equal to five.

The third step in a hypothesis test is the statement of the rule of evidence. For the chi-squared goodness of fit test the rule is

> The larger the value of the test statistic X^2, the stronger is the evidence in support of the alternative.

Intuitively, the greater the discrepancy between what is observed (O) and what is expected to be observed if the null hypothesis is true (E), the stronger is the evidence in support of the alternative. It then follows that the larger the value of $(O - E)^2$, the stronger is the evidence in support of the alternative. Finally, after rescaling, the larger the value of

$$\frac{(O - E)^2}{E},$$

the stronger the evidence in support of the alternative. It now follows from summing over all categories that the larger the value of the test statistic, X^2, the stronger is the evidence in support of the alternative.

Finally, the fourth step of a hypothesis test is the computation of the P-value. Recall that the P-value is the probability of obtaining data of the same or stronger evidence in support of the alternative hypothesis than what was actually obtained in the study. If this general definition of the P-value is combined with the above results for steps 2 and 3, it follows that the approximate P-value is equal to the area to the right of the value of the test statistic under the chi-squared curve with $k - 1$ degrees of freedom. For the analysis of Mendel's pea data, for example, the approximate P-value is equal to the area to the right of the observed value of the test statistic, $\chi^2 = 0.497$, under the chi-squared curve with three degrees of freedom. Using Table A.5 in conjunction with the invisible column argument presented earlier in this chapter yields

$$0.90 < P < 0.95.$$

Thus, the data are not close to being statistically significant. (With the help of Minitab, the area to the right of 0.497 under the chi-squared curve with three degrees of freedom is found to equal 0.9196.) Based on these data, there is little reason to doubt the null hypothesis that Mendel's theory is correct.

It is important to assess the practical importance of the results of a study. Such assessment is aided by comparing the relative frequency of each category with its special value of interest, as presented in Table 10.6 and Figure 10.10 for Mendel's pea data. The numbers in this table show that the absolute difference between the relative frequencies of the categories (the point estimates) and the hypothesized probabilities is at most 0.6 percentage points. This is a remarkably close agreement! At the scale of Figure 10.10, the difference is barely discernable.

TABLE 10.6 *Comparison of the Point Estimates and Hypothesized Values of the Probabilites for Mendel's Pea Study*

Category	Estimated Probability	Hypothesized Probability	Estimated Minus Hypothesized
1. Round and yellow	0.567	0.562	0.005
2. Round and green	0.194	0.188	0.006
3. Wrinkled and yellow	0.182	0.188	−0.006
4. Wrinkled and green	0.058	0.062	−0.004

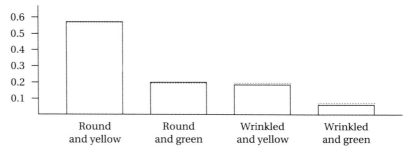

FIGURE 10.10 *Bar charts of relative frequency distribution (solid rectangles) and hypothesized probabilities (dashed rectangles) of the types of peas growns by Mendel.*

The hypothesis-testing procedure of this section is called the chi-squared goodness of fit test. It measures how well the hypothesized proportions fit the observed data, in much the same way that clothes may or may not fit a person well.

10.3.1 Summary

The foregoing derivation of the chi-squared goodness of fit test is rather lengthy. The procedure can be summarized as follows:

1. Select the k special values of interest, $p_{10}, p_{20}, \ldots, p_{k0}$. The null hypothesis is

$$H_0: p_1 = p_{10}, p_2 = p_{20}, \ldots, p_k = p_{k0},$$

and the alternative is

$$H_1: \text{not } H_0.$$

2. Select a random sample of size n from the population and compute the observed counts in each category, $O_1, O_2, \ldots, O_k$.

3. Construct a table like Table 10.4. Enter the k special values of interest and the observed counts. Compute the column of expected counts by multiplying each entry in the p_0 column by n. Complete the remainder of the table and obtain the value of the test statistic.

4. The approximate P-value equals the area to the right of the test statistic under the chi-squared curve with degrees of freedom equal to the number of categories minus one. The general guideline for using this approximation is that the expected count of each category equals or exceeds five.

5. Assess the practical importance of the data by comparing the relative frequency of each category with its special value of interest.

This section ends with three more examples of the use of the chi-squared goodness of fit test.

EXAMPLE 10.8: Minitab's Electronic Die

Consider a study designed to investigate whether or not a particular die is balanced. There are six possible outcomes when a die is cast: 1, 2, 3, 4, 5, and 6. Let $p_1, p_2, p_3, p_4, p_5,$ and p_6 denote the probabilities of these six outcomes, respectively. If the die is fair, each of the outcomes is equally likely. Thus, the six special values of interest for a balanced die are each equal to $\frac{1}{6}$. More precisely, let the hypotheses be

$$H_0: p_1 = \tfrac{1}{6},\ p_2 = \tfrac{1}{6},\ p_3 = \tfrac{1}{6},\ p_4 = \tfrac{1}{6},\ p_5 = \tfrac{1}{6},\ p_6 = \tfrac{1}{6}.$$

$H_1:$ not H_0.

The alternative hypothesis, not H_0, means that at least one of the true probabilities does not equal $\frac{1}{6}$; that is, the die is not fair.

You may toss a die a large number of times to investigate whether or not it is fair. Alternatively, Minitab has a command that supposedly generates any number of outcomes of the tosses of a fair die. The chi-squared goodness of fit test can be used to test the claim that the Minitab program in fact generates the outcomes of tossing a balanced die. Minitab generated $n = 600$ tosses of its (supposedly fair) die. The results of the computer generation and subsequent computations are given in Table 10.7. For example, 4 was the most frequent outcome, and it occurred 116 times. In addition, the value of the test statistic is 6.10. The approximate P-value equals the area to the right of 6.10 under the chi-squared curve with $k - 1 = 6 - 1 = 5$ degrees of freedom. Reading from Table A.5,

$$0.10 < P < 0.50.$$

The exact area to the right of 6.10 is 0.2963. The data are not statistically significant; thus, there is no compelling reason to doubt that the Minitab command works as advertised.

10.3 The Chi-squared Goodness of Fit Test

TABLE 10.7 Computing the Test Statistic for the Die Example

Outcome	p_0	O	E	O − E	$(O − E)^2$	$(O − E)^2/E$
1	$\frac{1}{6}$	94	100	−6	36	0.36
2	$\frac{1}{6}$	102	100	2	4	0.04
3	$\frac{1}{6}$	107	100	7	49	0.49
4	$\frac{1}{6}$	116	100	16	256	2.56
5	$\frac{1}{6}$	97	100	−3	9	0.09
6	$\frac{1}{6}$	84	100	−16	256	2.56
Total	1	600	600	0		$\chi^2 = 6.10$

TABLE 10.8 Comparison of the Point Estimates and Hypothesized Values of the Probabilities for the Die Example

Outcome	Sample Proportion	Hypothesized Probability	Difference $(\hat{p} − p_0)$
1	0.157	0.167	−0.010
2	0.170	0.167	0.003
3	0.178	0.167	0.011
4	0.193	0.167	0.026
5	0.162	0.167	−0.005
6	0.140	0.167	−0.027

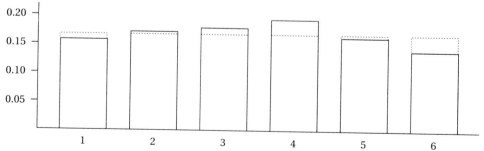

FIGURE 10.11 Bar charts of the relative frequency distribution (solid rectangles) and hypothesized probabilities (dashed rectangles) for the die example.

The practical importance of the results is assessed by comparing the sample proportions with the hypothesized special values of interest as illustrated in Table 10.8 and Figure 10.11. The sample proportions are all close to one-sixth; thus, the results do not seem to have practical importance, but you may reasonably disagree with this conclusion. ▲

10.3.2 Canada Geese

A common assumption in practice is that a sample has been obtained by taking a random sample from a population. If the composition of the population is known for some variable, perhaps from census data, the assumption of a random sample can be examined via the chi-squared goodness of fit test.

The example of this section is adapted from the handbook for Minitab [2], which in turn refers one to [3] for more details.

The Missouri Department of Conservation wanted to study the migration habits of Canada geese by studying one flock. The scientists were especially interested in comparing and contrasting the behaviors of four types of geese: adult males, adult females, yearling males, and yearling females. The researchers believed there was a known normal migration route for the flock and wanted to learn whether these four types of geese had the same propensity to stray from the normal route.

It is not practical to monitor the flights of individual geese; in fact, any attempt to follow a goose may change its flight pattern. Thus, the researchers faced the problem of how to measure the number and type of geese following or not following the normal route. One approach would be to obtain data from hunters on each goose shot, including its type and the location of the kill. This method of data collection has two obvious problems:

1. It may be difficult to contact all hunters and inform them of the need to submit data.
2. Suppose that most of the geese shot outside the normal route are, say, yearling males. This does not indicate that yearling males have a greater propensity to stray and be shot; perhaps yearling males outnumber the other types in the flock.

The scientists decided instead to capture and band a large number of geese. The results of the banding are given in Table 10.9. Information on the bands asked all hunters who shot a banded goose to send the band to a central office and specify where the goose was shot. Consider the first banded goose shot outside the normal migratory route. Make the following definitions:

- p_1 denotes the probability that the first banded goose shot is an adult male.
- p_2 denotes the probability that the first banded goose shot is an adult female.

TABLE 10.9 Canada Geese Study

Type	Number Banded	Proportion Banded
Adult males	4,144	0.226
Adult females	3,597	0.197
Yearling males	5,034	0.275
Yearling females	5,531	0.302
Total	18,306	1.000

- p_3 denotes the probability that the first banded goose shot is a yearling male.
- p_4 denotes the probability that the first banded goose shot is a yearling female.

The values of p_1, p_2, p_3, and p_4 are unknown. They would be known, however, if it were valid to assume that each type of goose had the same propensity to stray and be shot. If this assumption were true, then

$$p_1 = \frac{4,144}{18,306} = 0.226,$$

because 4,144 of the 18,306 banded geese were adult males. Similarly, the other entries in the third column of Table 10.9 give the probabilities of the various types being shot *on the assumption that all types have the same propensity to stray and be shot.*

The researchers did not want simply to adopt this assumption; they wanted to test it. In the notation of this chapter, they wanted to test

H_0: $p_1 = 0.226$, $p_2 = 0.197$, $p_3 = 0.275$, $p_4 = 0.302$.
H_1: not H_0.

Of course, data on more than one goose are needed to have any chance of learning something interesting. A total of 112 banded geese were shot outside the normal migration route. If the null hypothesis is true, each of these kills has probability 0.226 of being an adult male, 0.197 of being an adult female, and so on. (Actually, killing banded geese is sampling without replacement from the population of banded geese, because a goose cannot be killed twice. But the sample size, 112, is less than 1 percent of the population size, 18,306, making it valid to approximate random sampling without replacement by random sampling with replacement.) The chi-squared goodness of fit test can be used to measure the evidence in the sample of 112 geese. The data and the computation of the test statistic are given in Table 10.10. Note in particular that the value of the test statistic is 4.60. The approximate P-value is equal to the area to the right of 4.60 under the

TABLE 10.10 Computing the Test Statistic for the Canada Geese Study

Type	H_0 Probability	O	E	O − E	$(O - E)^2$	$(O - E)^2/E$
Adult males	0.226	17	25.3	−8.3	68.89	2.72
Adult females	0.197	21	22.1	−1.1	1.21	0.06
Yearling males	0.275	38	30.8	7.2	51.84	1.68
Yearling females	0.302	36	33.8	2.2	4.84	0.14
Total	1.000	112	112	0		$\chi^2 = 4.60$

chi-squared curve with $k - 1 = 4 - 1 = 3$ degrees of freedom. Reading from Table A.5, $0.10 < \mathbf{P} < 0.50$. (The exact area to the right of 4.60 is 0.2035.) Thus, the results are not statistically significant; there is no compelling reason to discard the null hypothesis that the four types of geese have the same propensity to stray and be shot. Practical importance is explored by comparing the sample proportions with the hypothesized probabilities, as illustrated in Table 10.11 and Figure 10.12. The female sample proportions are very close to the hypothesized

TABLE 10.11 Comparison of the Point Estimates and Hypothesized Values of the Probabilities for the Canada Geese Study

Type	Sample Proportion	Hypothesized Probability	Difference $(\hat{p} - p_0)$
Adult males	0.152	0.226	−0.074
Adult females	0.188	0.197	−0.009
Yearling males	0.339	0.275	0.064
Yearling females	0.321	0.302	0.019

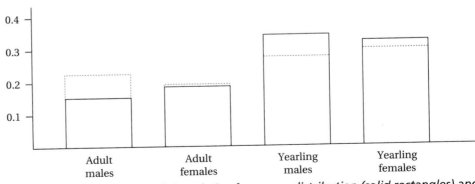

FIGURE 10.12 Bar charts of the relative frequency distribution (solid rectangles) and hypothesized probabilities (dashed rectangles) for the Canada Geese study.

probabilities. Adult males are somewhat underrepresented and yearling males are somewhat overrepresented in the sample. These differences appear to be of practical importance. Thus, more data would be desirable.

10.3.3 The Daily Game

The Wisconsin daily lottery game was introduced in Example 10.3 (p. 319). Recall that the data consist of 1,410 numbers that were selected by the state. Consider the first number selected by the state. Define the following:

- p_1 denotes the probability that the first number selected equals 1.
- p_2 denotes the probability that the first number selected equals 2.
- ⋮
- p_{36} denotes the probability that the first number selected equals 36.

If the state's selection method is fair, each number is equally likely to be selected and each of the probabilities is $\frac{1}{36}$. Thus, one might be tempted to use the 1,410 observations to test the null hypothesis,

$$H_0: p_1 = \tfrac{1}{36}, \; p_2 = \tfrac{1}{36}, \; \ldots, \; p_{36} = \tfrac{1}{36},$$

versus the alternative that the numbers are not equally likely by using the chi-squared goodness of fit test. I performed this analysis and found that the expected count for each outcome is $1{,}410/36 = 39.17$. In words, if the null hypothesis is true, the expected number of times each ball would be selected is 39.17. I further obtained the value of the test statistic, $\chi^2 = 30.664$. With the help of Minitab, I found that the area under the chi-squared curve with $k - 1 = 36 - 1 = 35$ degrees of freedom to the right of 30.664 is equal to 0.6775. Thus, the approximate P-value is equal to 0.6775.

Before reading further, try to determine what is wrong with the above analysis.

The problem with the above analysis is that the successive numbers selected by the state do not satisfy the conditions for multinomial trials. In particular, the trials are not independent. Each day six balls are selected from a collection of 36 without replacement. Frequently in this book it has been stated that an analyst can pretend that random sampling without replacement is actually random sampling with replacement provided the number of items sampled is 5 percent or less of the number of items from which the sample is taken. In the current example, the fraction is 6 out of 36, or 16.7 percent, so the general guideline is violated. As the analysis below indicates, ignoring the method of sampling leads to a serious error.

There are two ways to proceed at this point. First, a mathematical statistician could attempt to find a hypothesis test which is valid for the dependence inherent in the generation of the winning numbers. A second approach is described below.

It can be shown (and it is perfectly reasonable) that, even with dependence, the expected number of occurrences of each ball is $E = 39.17$, the number that was obtained with the assumption of independence. In addition, the argument given earlier can be extended to establish that

$$X^2 = \sum \frac{(O-E)^2}{E}$$

is still a *reasonable* way to measure the evidence against the null hypothesis in the data. (It is not necessarily the best way to measure evidence for dependence, but it is reasonable.) Thus, the P-value is the probability, computed under the dependence stricture, that the value of X^2 equals or exceeds 30.664, the value actually obtained in the study. This P-value is very difficult to compute mathematically, but, with a computer, it is easily approximated with a simulation experiment. Each run of the simulation experiment consists of 235 selections of 6 numbers at random, without replacement, from the first 36 integers. Each run yields 1,410 numbers, and these numbers are used to compute the value of X^2. A simulation experiment with 300 runs yielded 119 runs with a value of the test statistic that was greater than or equal to 30.664. Thus, the simulation estimate of the P-value is $119/300 = 0.397$. Note the following:

- This simulation estimate is not likely to be exactly correct. A 95 percent confidence interval for the actual P-value is

$$0.397 \pm 1.96\sqrt{\frac{0.397(0.603)}{300}} = 0.397 \pm 0.055 = [0.342, 0.452].$$

- The simulation experiment clearly shows that the evidence in the data against the null hypothesis of fairness is quite weak.
- The simulation experiment clearly shows that the approximate P-value obtained by using the (inappropriate) chi-squared curve with 35 degrees of freedom (0.6775) is, in fact, a very poor approximation. *Do not use the chi-squared curve to obtain an approximate P-value unless the conditions for the chi-squared goodness of fit test are reasonable.*

10.3.4 Degrees of Freedom

The approximate P-value for the chi-squared goodness of fit test is obtained by using the chi-squared curve with $k-1$ degrees of freedom. The expression *degrees of freedom* will appear in Chapters 11, 12, 13, 15, and 16 in a variety of statistical contexts. This section presents a brief explanation of why the quantity is called degrees of freedom and why its value is $k-1$ for the test of this chapter. Related explanations will be provided with each subsequent use of the expression *degrees of freedom* throughout this book.

The goodness of fit test compares the observed frequencies of the k categories $O_1, O_2, \ldots, O_k$ with their null expected values $E_1, E_2, \ldots, E_k$. In words, the n subjects in the sample are summarized by the k observed frequencies of the categories, and these k values are the fuel for the hypothesis test. These k

observed frequencies are restricted by one condition: they must sum to the sample size n. In symbols,

$$O_1 + O_2 + \cdots + O_k = n.$$

Thus, once the values of $O_1, O_2, \ldots, O_{k-1}$ are known, the value of O_k is determined because the sum of all observed frequencies is n. In other words, only $k-1$ of the k observed frequencies are *free to vary*. Thus, statisticians say that the k observed frequencies have $k-1$ *degrees of freedom*.

For example, consider Mendel's pea data, displayed in Table 10.5 (p. 336). The four observed frequencies are constrained to total 556. Thus, once it is known that the first three frequencies are equal to 315, 108, and 101, it follows that the fourth observed frequency must equal 32 so that the total of the frequencies equals 556. As a result, these frequencies are said to have three degrees of freedom.

EXERCISES 10.3

1. (Hypothetical) A scientist looks at the results of the Canada Geese study in Section 10.3.2 and states

    I have always believed there is no difference in migration habits between adult females and yearling females. Therefore, those two categories should be collapsed into one and the analysis should be repeated.

    Humor the scientist and do the suggested analysis. Does this bother you? Why?

2. (Hypothetical) A scientist looks at the results of the Canada Geese study in Section 10.3.2 and states

    I am not interested in female geese. Throw away the female data. This leaves a dichotomous response, and the method of Chapter 6 gives a value of $Z = 1.99$ for the test statistic. Thus, I conclude that yearling males are significantly more likely to stray and be shot than are adult males.

    What do you think?

3. Exercise 3 on p. 321 presented data on the distribution of flavors in a box of dog biscuits. Recall that the results were

    bacon 19, cheese 17, liver 20, meat 11, milk 21, poultry 12, and vegetable 20.

    Test the null hypothesis that the seven flavors are equally likely versus the alternative that they are not. Briefly summarize your findings.

4. Refer to Example 10.8. I further checked Minitab's electronic die by generating 6,000 tosses. The number 6 occurred 1,041 times, 2 occurred 1,019 times, 1 and 4 occurred 999 times each, 3 occurred 985 times, and 5 occurred 957 times. Analyze these data. Comment on the practical importance of these data. Do you think Minitab should be tested further?

5. Refer to Exercise 3. A second box of dog biscuits yielded the following frequencies of flavors:

    bacon 10, cheese 9, liver 17, meat 14, milk 13, poultry 15, and vegetable 9.

    Test the null hypothesis that the seven flavors are equally likely versus the alternative that they are not. Briefly summarize your findings.

6. Exercise 4 on p. 321 presented data on the distribution of flavors in a box of Trix cereal. Recall that the results were

    grape 420, lemon 530, lime 610, orange 585, and raspberry 470.

    Test the null hypothesis that the five flavors are equally likely versus the alternative that they are not. Briefly summarize your findings.

7. Refer to Exercises 3 and 5. If the data from the two boxes of biscuits are combined, the following frequencies of flavors are obtained:

>bacon 29, cheese 26, liver 37, meat 25, milk 34, poultry 27, and vegetable 29.

Test the null hypothesis that the seven flavors are equally likely versus the alternative that they are not. Briefly summarize your findings.

REFERENCES

1. A. Agresti, *Analysis of Ordinal Categorical Data* (New York: Wiley, 1984).
2. T. A. Ryan, B. L. Joiner, and B. F. Ryan, *Minitab Handbook*, 2d ed., rev. printing (Boston: PWS-KENT, 1985).
3. F. Mosteller, ed., *Statistics by Example* (Reading, Mass.: Addison-Wesley, 1973).

Tests of Homogeneity and Independence

11.1 CONTROLLED STUDIES
11.2 TWO OR MORE POPULATIONS
11.3 ONE POPULATION, TWO RESPONSES
11.4 THE 2 X 2 TABLE REVISITED*

Chapter 11

Chapters 1 through 3, 7, and 8 considered studies that yield data that can be summarized in a 2×2 contingency table. The studies could be

- Controlled studies with subjects assigned to treatments by randomization, with no extension of conclusions to a population (Chapters 1 through 3)
- Controlled or observational studies on independent random samples of subjects from two populations (Chapter 7)
- Observational studies with two responses on subjects selected at random from a population (Chapter 8)

Further, the subjects could be distinct individuals or a sequence of trials. In all cases, the treatments or populations can be compared by applying Fisher's test.

In the present chapter, each type of study mentioned above is generalized in either or both of two directions:

1. The response has (or responses have) more than two possible categories.
2. More than two treatments or populations are compared.

As in the earlier chapters, all of the types of studies listed above, with either or both of these generalizations, are analyzed with a single hypothesis test, called the chi-squared test of homogeneity or the chi-squared test of independence. The ideas and methods will be introduced through a number of examples. Inference will be restricted to point estimation and hypothesis testing. There are no widely used methods of confidence interval estimation for the problems considered in this chapter.

The topics of Chapters 4 and 9, blocking and controlling for factors, will not be extended to the situations studied in this chapter.

11.1 CONTROLLED STUDIES

The ideas of this section are introduced with the following example of a controlled study with a four-category response and three treatments.

EXAMPLE 11.1: Duchenne's Muscular Dystrophy

Duchenne's muscular dystrophy (DMD) is an inexorably progressive disease that results in severe disability by the age of 10 to 12 years and in death in early adulthood [1]. A randomized, double-blind, six-month trial was performed on 97 patients to investigate the efficacy of prednisone in the treatment of DMD. The following three treatments were compared:

- Placebo
- Low dose of prednisone (0.75 mg/kg body weight per day)
- High dose of prednisone (1.50 mg/kg body weight per day)

Both doses of prednisone were found to have several beneficial effects, including improvement on several measures of muscle strength, leg function, and pulmonary function. There was unconvincing evidence that the high dose was better than the low dose. Unfortunately, prednisone treatment was found to be associated with a number of undesirable side effects, including weight gain. This example explores how weight gain depends on treatment.

The response is the subject's six-month weight gain, from the beginning of the study to the end. The researchers decided to classify weight gain into one of the four categories: less than 5 percent, 5 percent to 10 percent, 11 percent to 20 percent, and greater than 20 percent. Of course, three treatments are to be compared: the placebo, and the low and high doses of prednisone.

The data are displayed in Table 11.1, which is a 3×4 contingency table of observed counts. Each row represents a different treatment, and each column

TABLE 11.1 Table of Observed Counts for the Prednisone Duchenne's Muscular Dystrophy Study

Treatment	Weight Gain (Percent)				Total
	<5	5–10	11–20	>20	
Placebo	18	10	5	2	35
Low dose	3	3	13	11	30
High dose	3	5	14	10	32
Total	24	18	32	23	97

TABLE 11.2 *Table of Row Proportions for the Prednisone Duchenne's Muscular Dystrophy Study*

Treatment	Weight Gain (Percent)				
	<5	5–10	11–20	>20	Total
Placebo	0.514	0.286	0.143	0.057	1.000
Low dose	0.100	0.100	0.433	0.367	1.000
High dose	0.094	0.156	0.438	0.312	1.000

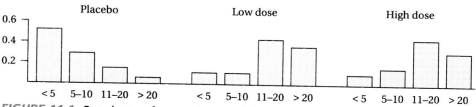

FIGURE 11.1 *Bar charts of percentage weight gain by treatment for the Prednisone Duchenne's Muscular Dystrophy study*

represents a different category of the response. For example, the entry 18 in the first row and first column indicates that 18 subjects received the placebo and experienced a weight gain of less than 5 percent. Whenever the response categories are ordered, as in this example, the columns of the contingency table should preserve the ordering. For descriptive purposes, each entry in the table of observed counts is divided by its row total. The resultant table of row proportions is given in Table 11.2 and is depicted in contiguous bar charts in Figure 11.1. The bar charts show that the placebo yielded a very different distribution of weight gain than either dose of prednisone. In addition, the distributions of weight gain are very similar for the two doses of prednisone. This study, which will be abbreviated as the P-DMD study, will be used to introduce a method of hypothesis testing. ▲

The first step of a hypothesis test is the selection of the hypotheses. It will not be assumed that the 97 subjects were selected at random from a population of subjects. (Hypotheses for a population model will be discussed later in this chapter.) The null hypothesis is that for each subject the treatments have the same effect on the response. In other words, if the null hypothesis is true, each subject's weight gain reflects characteristics of that particular subject and is not influenced by the treatment received. If given a different treatment, the subject would have experienced the same weight gain. The test of this section is called the chi-squared test of homogeneity. The null hypothesis is that the treatments have the same effect on a given subject—that is, the treatments have homogeneous effects.

There is only one choice for the alternative—that for at least one subject the treatments do not have the same effect on the response. This lack of choice in specifying the alternative forces a researcher either to perform a number of hypothesis tests on the same set of data to explore different possibilities (illustrated below), or to use methods beyond the level of this book (for example, log-linear models).

The second step in the hypothesis test is the determination of the test statistic and its null sampling distribution. Define r and c:

- r is the number of treatments being compared.
- c is the number of categories for the response variable.

For the P-DMD study, $r = 3$ and $c = 4$. When the data are presented in a contingency table of observed counts, the table has r rows, one for each treatment, and c columns, one for each response category. Thus, the symbols r and c are easy to remember: they are the number of rows and columns, respectively, in the contingency table of observed counts.

The number of cells in an $r \times c$ contingency table is equal to the product of r and c. For example, the 2×2 tables studied in Part I of this text had $2 \times 2 = 4$ cells, and the 3×4 table of the P-DMD study has $3 \times 4 = 12$ cells. Each cell has an observed frequency, denoted by O, as in Chapter 10. Also, as in Chapter 10, each cell has an expected frequency computed on the assumption that the null hypothesis is true, denoted by E. There is a simple rule for determining the expected frequency of any cell in the $r \times c$ contingency table:

> The expected frequency of a cell equals the product of the row and column totals for that cell divided by the total sample size.

First, use of this rule is illustrated, and then a motivation for it is given.

Consider the data for the P-DMD study in Table 11.1. The observed frequency in the upper left cell, subjects assigned to the placebo who had a six-month weight gain of less than 5 percent, is 18. The row total for that cell is 35, its column total is 24, and the number of subjects is 97. Thus, the expected frequency for the cell is

$$E = \frac{35 \times 24}{97} = 8.66.$$

The observed frequency in the cell, 18, is over twice as large as the frequency, 8.66, that would be expected if the null hypothesis were true. As in Chapter 10, this discrepancy between O and E casts doubt on the validity of the null hypothesis.

As another example, the cell corresponding to high dose and weight gain of less than 5 percent has an observed frequency of 3, a row total of 32, and a column total of 24. Thus, its expected frequency is

$$E = \frac{24 \times 32}{97} = 7.92.$$

As for the previously examined cell, the difference here between $O = 3$ and $E = 7.92$ casts doubt on the assumption that the null hypothesis is true.

The rule for computing the expected frequency can be motivated quite easily. Consider the upper left cell in Table 11.1 again. Its row total, 35, equals the number of subjects placed on the placebo. A natural question is

> What proportion of those 35 subjects are expected to give the lowest response (weight gain of less than 5 percent) if the null hypothesis is true?

The key step in answering this question is to realize that if the null hypothesis is true (there is no difference between treatments), one expects that each treatment will have the same proportion of subjects giving the lowest response. For the entire study, 24 out of 97 subjects gave the lowest response. Thus, if the null hypothesis is true, $\frac{24}{97}$ of the subjects of each treatment are expected to give the lowest response. Therefore, the expected frequency for the upper left cell is $\frac{24}{97}$ times 35:

$$E = \frac{24}{97} \times 35.$$

Note that 24 is the column total for the upper left cell and 97 is the total number of subjects. Thus, this last equation can be written

$$E = \frac{\text{Column total}}{\text{Total sample size}} \times \text{Row total} = \frac{\text{Row total} \times \text{Column total}}{\text{Total sample size}},$$

which is simply the rule given earlier. Thus, the rule for computing the expected frequency has been shown to be reasonable for the upper left cell. A similar argument can be applied to any cell in the table.

As in Chapter 10, the discrepancy between the observed and expected frequency in a cell is summarized by the quantity

$$\frac{(O - E)^2}{E}.$$

The test statistic, denoted X^2, is the sum of these quantities over all the cells

$$X^2 = \sum \frac{(O - E)^2}{E}.$$

It can be shown that if the null hypothesis is true, the sampling distribution of the test statistic X^2 can be well approximated by a chi-squared curve with the number of degrees of freedom equal to

$$(r - 1) \times (c - 1).$$

The general guideline for using this approximation is that all of the Es must be greater than or equal to 1 and that at least 75 percent of the Es must be greater than or equal to 5.

The rule of evidence is

The larger the value of X^2, the stronger is the evidence in support of the alternative.

The rationale behind this rule is identical to the rationale given for the rule of evidence of the goodness of fit test, on p. 338.

Finally, combining the above results, the approximate P-value is equal to the area to the right of the observed value of the test statistic under the chi-squared curve with

$$(r - 1) \times (c - 1)$$

degrees of freedom.

The following example illustrates the computation of the approximate P-value for the P-DMD study.

EXAMPLE 11.2: Hypothesis Test for the P-DMD Study

The computation of the value of the test statistic for the P-DMD study is formidable. The value of the expected frequency E and the summary

$$\frac{(O - E)^2}{E}$$

must be computed for each of 12 cells. A computer, however, can perform these computations quickly. Table 11.3 presents the Minitab output for the P-DMD study. Note the following features of Table 11.3:

1. The contingency table of observed frequencies is presented, including row and column totals.

2. The expected frequency of each cell appears directly below its observed frequency. For example, the upper left cell has an observed frequency of 18 with an expected frequency of 8.66 immediately below it. The lower left cell has an observed frequency of 3 and an expected frequency of 7.92. These values of E agree with the numbers computed earlier by hand.

3. Below the contingency table are three lines labeled with χ^2 and ending with the number 32.466. These lines list the 12 values of

$$\frac{(O - E)^2}{E}$$

for the 12 cells of the table, written in the same order as the cells. For example, the cell in the upper right corner of the table has

$$\frac{(O - E)^2}{E} = 4.781.$$

TABLE 11.3 Minitab Output for the Duchenne's Muscular Dystrophy Study

	Percentage of Weight Gain*				
Treatment	<5	5–10	11–20	>20	Total
Placebo	18	10	5	2	35
	8.66	6.49	11.55	8.30	
Low dose	3	3	13	11	30
	7.42	5.57	9.90	7.11	
High dose	3	5	14	10	32
	7.92	5.94	10.56	7.59	
Total	24	18	32	23	97
$\chi^2 =$	10.074+	1.892+	3.712+	4.781+	
	2.635+	1.184+	0.973+	2.124+	
	3.054+	0.148+	1.123+	0.767=	32.466
df = 6					

*Expected counts are printed below observed counts

The last entry in these three lines, 32.466, is the value of the test statistic for the P-DMD study.

4. The number of degrees of freedom is given at the bottom of the table. For the P-DMD study its value is 6, which agrees with the rule:

$$(r - 1) \times (c - 1) = (3 - 1)(4 - 1) = 2(3) = 6.$$

Minitab does not provide the approximate P-value. For the current study, the approximate P-value is the area to the right of 32.466 under the chi-squared curve with six degrees of freedom. According to Table A.5,

$$P < 0.01.$$

As mentioned in Chapter 10, a computer can yield areas under a chi-squared curve. In particular, the area under the chi-squared curve with six degrees of freedom to the right of 32.466 is calculated to be 0.0000. Thus, a more precise approximation for **P** for the P-DMD study is 0.0000. It seems absurd, then, to believe the null hypothesis in the face of these data. ▲

The test of homogeneity indicates that it is unreasonable to believe that the three treatments for the P-DMD study have the same effect on the weight gain of subjects. There is, however, an extremely frustrating characteristic of this test: the test does not reveal which treatments are different and which are not. More precisely, for the P-DMD study, it seems clear from an inspection of the bar charts in Figure 11.1 that the placebo is significantly different from either dose of prednisone and that the doses are not significantly different from each other. Although this conclusion seems clear, it does not follow from the test of homogeneity.

If the test of homogeneity is statistically significant, much insight can be gained by subsequently conducting a number of pairwise comparisons of treatments. I recommend proceeding somewhat cautiously:

> If the number of treatments is six or fewer—a condition that is almost always met in practice—and the test of homogeneity is statistically significant, it is appropriate to make any number of additional pairwise comparisons of treatments. But do not judge any pair of treatments to be significantly different unless the P-value for comparing them is less than 0.01.

This analysis strategy is illustrated in the following example.

EXAMPLE 11.3: *Further Analysis of the P-DMD Study*

Three pairwise comparisons of treatments are possible for the P-DMD study. The results of the analysis (details not given) are as follows:

Treatments Compared	χ^2	Approximate P-value
Placebo, low dose	24.027	0.0000
Placebo, high dose	21.887	0.0001
Low dose, high dose	0.521	0.9143

The values of χ^2 and the approximate P-values in this display were obtained using Minitab. For example, the entries in the row "Placebo, low dose" were obtained by performing a test of homogeneity on the 2×4 table with one row for placebo and one row for low dose, and the columns corresponding to the four categories of response. In other words, this table is obtained by deleting the high dose row from the original data, presented in Table 11.1. (Note: If you perform these computations by hand, remember that deleting a row from the original table changes the column totals.)

It is clear from these P-values that the placebo is significantly different from both the low dose and the high dose of prednisone, but the two doses of prednisone are not significantly different. This pairwise analysis provides a useful insight beyond that provided by the test of homogeneity. The test of homogeneity only concludes that the null hypothesis is not true—that is, that the three treatments are not identical. The test of homogeneity does not indicate which treatments differ and which do not. The pairwise analysis can help a researcher decide which treatments differ. ▲

11.2 TWO OR MORE POPULATIONS

The ideas of this section are introduced with the following example of an observational study with three populations and a four-category response.

EXAMPLE 11.4

Question 14 of the 1982–1984 Wisconsin Driver Surveys read

> For statistical purposes, would you tell us how often, if at all, you drink alcoholic beverages?

The possible responses were "several times a week," "several times a month," "less often," and "not at all." Table 11.4 presents the frequencies of the various responses for the years 1982, 1983, and 1984. Table 11.5 and Figure 11.2 present the row proportions for these data. The hypothesis-testing analysis below

TABLE 11.4 Responses to Question 14—How Often Do You Drink Alcoholic Beverages?—by Year

Year	Response*				Total
	1	2	3	4	
1982	188	420	640	270	1,518
1983	142	262	407	207	1,018
1984	353	632	974	506	2,465
Total	683	1,314	2,021	983	5,001

*Key to responses: 1 = several times a week, 2 = several times a month, 3 = less often, and 4 = not at all.

TABLE 11.5 Row Proportions of Responses to Question 14—How Often Do You Drink Alcoholic Beverages?—by Year

Year	Response*				Total	Sample Size
	1	2	3	4		
1982	0.124	0.277	0.422	0.178	1.000	1,518
1983	0.139	0.257	0.400	0.203	1.000	1,018
1984	0.143	0.256	0.395	0.205	1.000	2,465

*Key to responses: 1 = several times a week, 2 = several times a month, 3 = less often, and 4 = not at all.

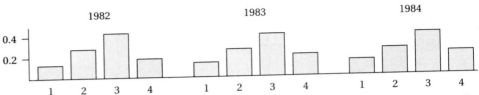

FIGURE 11.2 *Self-reported drinking frequency in the Wisconsin Driver Survey. Response 1 is "several times a week," response 2 is "several times a month," response 3 is "less often," and response 4 is "not at all."*

indicates that the changes in the distribution of sample responses for the three years are not statistically significant. Figure 11.2 shows that the interyear changes were small. ▲

In order to proceed beyond a simple description of the data, a population model is needed. It is more efficient to generalize the ideas beyond this driver survey example. Suppose the researcher wants to compare some number of populations. Let r represent the number of populations. The response variable has c categories, which may be ordered or unordered. Chapter 7 considered this model for the case of $r = 2$ populations and $c = 2$ response categories. The example of this section has $r = 3$ and $c = 4$. If the subjects in the P-DMD study of the previous section are assumed to be a random sample from a superpopulation of DMD sufferers, then for the population model there are $r = 3$ populations, generated by the three treatments considered, and $c = 4$. (Note: These are the same values of r and c as for the original—no population—analysis.)

Each population has its own population box of cards. On each subject's card is his or her value of the response variable. An examination of the entire box would yield the population distribution for the response variable. For example, among the population of 1982 Wisconsin licensed drivers there is a distribution of responses to question 14. This distribution consists of four proportions:

- $p_1^{(1982)}$ = the proportion of all 1982 Wisconsin licensed drivers who would have responded "several times a week" to question 14.
- $p_2^{(1982)}$ = the proportion of all 1982 Wisconsin licensed drivers who would have responded "several times a month" to question 14.
- $p_3^{(1982)}$ = the proportion of all 1982 Wisconsin licensed drivers who would have responded "less often" to question 14.
- $p_4^{(1982)}$ = the proportion of all 1982 Wisconsin licensed drivers who would have responded "not at all" to question 14.

Without a census, the distribution for each population is unknown. If the researcher is willing to assume that the subjects selected from the various populations are independent random samples, a hypothesis test can be performed to compare the distributions of the response across the populations.

The null hypothesis is that each of the r populations has exactly the same population distribution for the response variable. There is only one choice for the alternative. It is that the null hypothesis is not true; that is, the r population distributions are not identical. As in Part I of the text, the remaining steps of the hypothesis test are identical to the steps for the test based on the random assignment of subjects to treatments. In short, the chi-squared test of homogeneity introduced in the previous section for the P-DMD study is appropriate for the model of this section, too. The details of the test will be given for the survey of this section.

Table 11.6 presents the Minitab output for the responses to question 14 for the years 1982–1984. The value of the test statistic is 9.848, and there are six degrees of freedom. From Table A.5,

$$0.10 < P < 0.50.$$

In fact, with the help of Minitab, the approximate P-value is found to be 0.1310. The differences in responses between years are not statistically significant. A pairwise comparison of years is inappropriate because the test of homogeneity is not statistically significant. In other words, since the test concludes that the researcher should continue to believe there is no difference between years, one should not conduct additional tests that look for a difference.

TABLE 11.6 Minitab Output for the Response to Question 14 by Year

Year	Drinking Frequency*				Total
	Several Times a Week	Several Times a Month	Less Often	Not at All	
1982	188 207.32	420 398.85	640 613.45	270 298.38	1,518
1983	142 139.03	262 267.48	407 411.39	207 200.10	1,018
1984	353 336.65	632 647.67	974 996.15	506 484.52	2,465
Total	683	1,314	2,021	983	5,001
$\chi^2 =$	1.800+ 0.063+ 0.794+	1.121+ 0.112+ 0.379+	1.149+ 0.047+ 0.493+	2.699+ 0.238+ 0.952=	9.848

df = 6

*Expected counts are printed below observed counts.

11.3 ONE POPULATION, TWO RESPONSES

This section extends the methods of Chapter 8 to the case in which either or both responses have more than two categories.

EXAMPLE 11.5

Question 10 on the Wisconsin Driver Survey read

> Do you believe that drinking black coffee or taking a cold shower will help sober up a drunk?

The possible responses were "yes," "sometimes," "no," and "don't know." The researchers at the Wisconsin Department of Transportation believed that "no" was the correct response; this belief will not be challenged in this example. The researchers wanted to see how the response to question 10 was related to the age of the subject. To this end, each subject was asked to self-report age, and the number reported was transformed into a variable with six categories:

14–20, 21–25, 26–35, 36–45, 46–55, and 56–85 years.

One can follow the development presented in Chapter 8 and define the conditional probabilities of the responses to question 10 given age and of ages given the response to question 10. The notion of independence can be defined and studied. The researchers' goal in the current study, however, was to investigate how the response to question 10 depended on age. The most efficient way to do that is to imagine the population of all drivers to be divided into six smaller populations, one for each age group. Next, pretend that the random sample from the original population of all drivers comprises independent random samples from the six smaller populations. As discussed in Chapter 7 (p. 232) and in Chapter 8 (p. 277), it is mathematically valid to pretend this. Thus, the method of the previous section can be used to test the null hypothesis that the distribution of responses is the same in the six age-group populations. Since the response is dichotomous, the null hypothesisis is actually that the probability of a success (correct answer) is the same in each population (if the probability of success is the same in each population, the probability of failure must also be constant across populations, and the distributions are therefore identical).

The data from the 1983 survey are presented in a 6×2 contingency table, which is included in the Minitab output of Table 11.7. Table 11.8 and Figure 11.3 are simple and efficient modifications of the table of row proportions and the bar chart for a dichotomous variable measured on several populations. The bar chart indicates that the success rate decreases with age and that the two largest declines occur between the two youngest and the two oldest age groups. As given in Table 11.7, the value of the test statistic is 35.210; the approximate P-value is the area to the right of 35.210 under the chi-squared curve with five degrees of

TABLE 11.7 Minitab Output for the Response to Question 10

	Response*		
Age	Correct	Incorrect	Total
14–20	108 87.26	21 41.74	129
21–25	96 89.97	37 43.03	133
26–35	150 148.82	70 71.18	220
36–45	152 156.26	79 74.74	231
46–55	72 77.12	42 36.88	114
56–85	43 61.56	48 29.44	91
Total	621	297	918
$\chi^2 =$	4.927+ 0.404+ 0.009+ 0.116+ 0.340+ 5.595+	10.302+ 0.845+ 0.019+ 0.243+ 0.710+ 11.699=	35.210

df = 5

*Expected counts are printed below observed counts.

freedom. Reading from Table A.5, **P** < 0.01. With the help of Minitab, the area to the right of 35.210 is found to equal 0.0000. The data are highly statistically significant, and one should conclude that the six populations do not have the same proportions of successes.

It is natural to wonder which pairs of age groups differ significantly. Table 11.9 presents the approximate P-value for each of the 15 possible pairwise comparisons, with the P-values listed from smallest to largest. The P-values for the first five pairwise comparisons in the table are less than one in one thousand. These extremely small P-values indicate that the researcher should feel quite comfortable in concluding the corresponding pairwise differences are real. The next two P-values are considerably less than 0.01. Again, this indicates that the researcher can be quite comfortable concluding that these two differences are real. The next two P-values fall between 0.01 and 0.05. Finally, the last six P-values

TABLE 11.8 Response to Question 10 as a Function of Age, 1983 Survey

Age	Proportion Correct	Sample Size
14–20	0.837	129
21–25	0.722	133
26–35	0.682	220
36–45	0.658	231
46–55	0.632	114
56–85	0.472	91

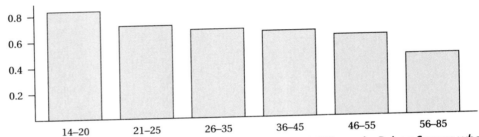

FIGURE 11.3 Proportion of respondents in the 1983 Wisconsin Driver Survey who responded no to the question, "Do you believe that drinking black coffee or taking a cold shower will help sober up a drunk?" plotted versus age.

TABLE 11.9 Approximate P-values for All Pairwise Comparisons of Responses to Question 10 as a Function of Age

Age Groups	P-value	Age Groups	P-value	Age Groups	P-value
14–20, 56–85	0.0000	14–20, 26–35	0.0014	21–25, 36–45	0.2085
21–25, 56–85	0.0002	36–45, 56–85	0.0022	26–35, 46–55	0.3566
14–20, 46–55	0.0003	46–55, 56–85	0.0226	21–25, 26–35	0.4285
14–20, 36–45	0.0003	14–20, 21–25	0.0245	26–35, 36–45	0.5909
26–35, 56–85	0.0005	21–25, 46–55	0.1296	36–45, 46–55	0.6286

are larger than 0.10. Following the cautious analysis strategy outlined earlier in this chapter, one concludes that the first seven pairwise comparisons are clearly significant and the other eight are not clearly significant. In fact, there is no good reason to believe the last six are significant. A brief examination of the first seven pairs shows that this analysis can be summarized quite easily:

The youngest drivers, aged from 14 to 20, had significantly higher success rates than drivers in each of the four age groups above 25. All groups of drivers aged 45 or younger had significantly higher success rates than the oldest drivers. No other pairwise differences were clearly significant. ▲

For the type of study discussed in this section (that is, when at least one response can take on more than two values), the chi-squared test of homogeneity is frequently called the chi-squared test of independence since the null hypothesis can be viewed as being that the response variables are statistically independent.

The rest of this section is optional. It is intended to provide insight into why there are $(r - 1) \times (c - 1)$ degrees of freedom for the hypothesis test of this chapter. You may want to review Section 10.3.4.

The test of homogeneity utilizes the $r \times c$ observed frequencies that appear in the $r \times c$ contingency table. These observed frequencies must sum to the total sample size, n. Thus, following the argument of Chapter 10, one might conjecture that these observed frequencies have $rc - 1$ degrees of freedom. In fact, this is exactly what the famous mathematician Karl Pearson conjectured many years ago. Pearson was very famous as the developer of the chi-squared goodness of fit test, which was very useful in the infant field of genetics, so his conjectures were taken seriously by his contemporaries. Eventually, however, Sir Ronald A. Fisher showed that Pearson's conjecture was incorrect.

Fisher argued that the row totals and the column totals in the contingency table give no information on whether or not there is a difference between treatments or populations. (You may recall that a similar argument was made in Chapter 2 for Fisher's test for a 2×2 table obtained from a completely randomized design.) Thus, the observed frequencies are constrained to add to the row and column totals, which are viewed as fixed. For example, consider the following artificial 3×4 contingency table:

	Response				
Treatment	1	2	3	4	Total
1	25	25	25		100
2	20	30	40		100
3					200
Total	100	100	100	100	400

This table has 12 cells, but the specification of the row totals, the column totals, and the six entries in the cells above, completely determine the remaining observed frequencies. More precisely, the last entry in the first row must equal 25 to make the first row frequencies sum to 100. The last entry in the second row must equal 10 to make the second row frequencies sum to 100. At this point the remaining unknown entries all lie in the bottom (third) row. But these values are all determined by the fact that the column entries must sum to the column totals. For example, the first entry in row 3 must be 55 to make the first column frequencies sum to their total, 100.

In general, for an arbitrary $r \times c$ table the entries in the last column and last row are determined by the remaining observed frequencies (and, of course, the row and column totals). Dropping the last row and column leaves $r - 1$ rows and $c - 1$ columns, in which the observed frequencies are free to vary: that is, the observed frequencies have

$$(r - 1) \times (c - 1)$$

degrees of freedom.

11.4 THE 2 × 2 TABLE REVISITED*

Nothing in the mathematical development of the chi-squared test of homogeneity requires either r or c to be larger than 2. In other words, the test of homogeneity is an alternative to Fisher's test for the analysis of a 2×2 table. This optional section investigates the difference between these two tests. For either test, the practical importance of the data should be investigated by examining the values of $\hat{p}_1$, $\hat{p}_2$, and $\hat{p}_1 - \hat{p}_2$. The tests will be compared by their formulas for computing the approximate P-value.

For the test of homogeneity on a 2×2 table, the approximate P-value is the area to the right of χ^2 under the chi-squared curve with

$$(r - 1)(c - 1) = (2 - 1)(2 - 1) = 1$$

degree of freedom. The reader who enjoys algebra can show that for a 2×2 table, the observed value of

$$\chi^2 = \sum \frac{(O - E)^2}{E}$$

can be written as

$$\chi^2 = \frac{n(ad - bc)^2}{n_1 n_2 m_1 m_2}.$$

(For the upper left cell, make the substitutions

$$O = a \quad \text{and} \quad E = \frac{n_1 m_1}{n}.$$

Similar formulas are available for the other three cells. After these substitutions, the rest is just algebra.)

Recall that in order to obtain an approximate P-value for Fisher's test, the observed value

$$z = \frac{\sqrt{n - 1}(ad - bc)}{\sqrt{n_1 n_2 m_1 m_2}}$$

of the random variable Z must be computed. The sampling distribution of Z is approximated by the standard normal curve. As stated in Chapter 10 (p. 332), this approximation is equivalent to approximating the sampling distribution of Z^2 by the chi-squared curve with one degree of freedom. Squaring each side of the equation for z gives

$$z^2 = \frac{(n-1)(ad-bc)^2}{n_1 n_2 m_1 m_2}.$$

Thus, the formulas for z^2 and χ^2 differ only in that the former has $n-1$ in the numerator and the latter has n. Unless n is very small, the difference between z^2 and χ^2 is negligible. For example, for the analysis of the previous section comparing drivers aged 21–25 with those aged 56–85,

$$z^2 = (3.77)^2 = 14.21,$$

and χ^2 equals 14.26 (computations not shown). The two test statistics, z (or z^2) and χ^2, give somewhat different approximate P-values, and neither is universally better than the other. Finally, if n is very small, the general guideline for using the standard normal curve or chi-squared approximations is not satisfied.

The computations of the test statistics for the 15 pairwise comparisons in Table 11.9 were performed by Minitab. Minitab uses the test statistic χ^2 rather than z or z^2.

EXERCISES FOR CHAPTER 11

1. Exercises 3 and 5 on p. 347 in Chapter 10 presented data from two boxes of dog biscuits. The data can be presented in the 2 × 7 contingency table at the bottom of this page. The test of homogeneity yields $\chi^2 = 7.167$ (the smallest value of E is 10.5). Use this value of the test statistic to obtain a P-value. Carefully state the null and alternative hypotheses, in words, and summarize your findings.

2. The following passage is taken from a study of the sexual behavior of college women (at an unspecified university in the Northeast) [2]:

   To compare sexual practices in college women before and after the start of the current epidemics of Chlamydia trachomatis, genital herpes virus, and human immunodeficiency virus type 1 infec-

| | | | | Flavor | | | | |
Box	Bacon	Cheese	Liver	Meat	Milk	Poultry	Vegetable	Total
1	19	17	20	11	21	12	20	120
2	10	9	17	14	13	15	9	87
Total	29	26	37	25	34	27	29	207

tion, we surveyed 486 college women who consulted gynecologists at a student health service in 1975, 161 in 1986, and 132 in 1989 at the same university. There were no statistically significant differences in age, age at menarche, or reason for visiting the gynecologist. The percentages of women in this population who were sexually active were the same in all three years (88 percent in 1975, 87 percent in 1986 and 87 percent in 1989).

The women who were sexually active were also asked to report on the frequency of condom use by their partner(s); the results were as follows:

Year	Always or Almost Always	Seldom or Never	Total
1975	49	370	419
1986	30	99	129
1989	46	66	112
Total	125	535	660

(a) Construct the table of row proportions. Draw a bar chart of the sample proportions of the response "always or almost always" for the three years. Describe what you have learned.

(b) Do you think it is reasonable to use this study to infer characteristics of all college women at the unknown university? All college women in the United States?

(c) Compute the value of E for the cell 1975, "always or almost always." Compute

$$\frac{(O-E)^2}{E}$$

for the same cell.

3. In the 1986 Wisconsin Driver Survey, for some unknown reason, the responses to question 14 (frequency of drinking alcohol) were changed from

"several times a week," "several times a month," "less often," and "not at all"

to

"every day," "more than once a week," "a few times a month," and "rarely or never."

The observed counts for these categories are, respectively, 52, 304, 605, and 687. Compare the 1986 results with the data from the earlier years (Table 11.4). (Hint: Clearly, the researchers made a major error. A very imperfect comparison of 1986 with the earlier years can be made by collapsing the four-category response variables into a three-category response variable according to the following display:

1982–1984	1986	New Response
Several times a week	Every day More than once a week	Heavy
Several times a month	A few times a month	Moderate
Less often, Not at all	Rarely or never	Light

Create the four bar charts for this new response and comment on your findings. No formal test is needed.)

4. Refer to Exercise 2.

(a) On the assumption of independent random samples, conduct a test of the null hypothesis that the proportion of those whose partner(s) used condoms "always or almost always" was the same in each of the three years studied. (Hint: $\chi^2 = 51.625$.) Based on your answer, is it appropriate to make pairwise comparisons of the years?

(b) On the assumption of independent random samples, conduct a test of the null hypothesis that the proportion who would answer "always or almost always" was the same in 1975 and 1986.

(c) On the assumption of independent random samples, conduct a test of the null hypothesis that the proportion who would answer "always or almost always" was the same in 1975 and 1989.

(d) On the assumption of independent random samples, conduct a test of the null hypothesis that the proportion who would answer "always or almost always" was the same in 1986 and 1989.

Exercises 5 through 8 pertain to a study on risk factors for HIV infection [3]. The abstract of the study stated, in part,

> To identify risk factors for HIV infection in intravenous drug users, we undertook a study of the seroprevalence of HIV antibody in 452 persons enrolled in a methadone-treatment program in the Bronx, New York.
>
> The presence of HIV antibody was associated with the number of injections per month, the percentage of injections with used needles, the average number of injections with cocaine per month, and the percentage of injections with needles that were shared with strangers or acquaintances....
>
> The number of heterosexual sex partners who used intravenous drugs was associated with HIV infection in women and was the only risk factor for users who had not injected drugs after 1982.

Table 11.10 is taken from the report; use it to complete Exercises 5–8.

5. Complete the following for the variables "average number of injections per month" and "HIV antibody status."

    (a) Draw a bar chart of the sample proportions of HIV antibody positive responses for the three levels of the injections variable. Briefly describe what you find.

    (b) Present the data in a 3 × 2 table.

    (c) Compute E for the cell corresponding to subjects who make more than 100 injections monthly and are HIV antibody positive.

    (d) Compute the value of

    $$\frac{(O - E)^2}{E}$$

    for the same cell.

    (e) Do you think it is reasonable to assume that the subjects were a random sample from all IV drug users in New York City?

TABLE 11.10 *HIV Antibody Status of 452 Subjects, according to Reported Injection Behavior*

Value of Variable	No. HIV-Positive/ No. Tested	% HIV-Positive
Average number of injections per month		
0	9/75	12.0
1–100	98/242	40.5
>100	71/132	53.8
Duration of intravenous drug use (month)		
0	9/75	12.0
1–50	81/218	37.2
>50	88/158	55.7
Percentage of total injections in shooting galleries		
0	84/282	29.8
1–25	54/107	50.5
>25	37/56	66.1
Percentage of total injections with used needles		
0	55/183	30.1
1–25	31/91	34.1
>25	92/176	52.3
Percent of total injections with needles shared with strangers or acquaintances		
0	56/196	28.6
1–25	31/79	39.2
>25	85/162	52.5

For the remainder of this exercise, assume that the data are a random sample from a population that represents the process of the infection of IV drug users in New York City with HIV.

(f) Conduct the chi-squared test of homogeneity. (Hint: $\chi^2 = 35.063$.) Describe what you have learned. Based on your answer, is it appropriate to make pairwise comparisons of the levels of the injection variable?

(g) Perform three Fisher's tests for the pairwise comparisons of the three injection groups. Describe what you have learned.

6. Complete the following for the variables "percentage of total injections with used needles" and "HIV antibody status."

    (a) Draw a bar chart of the sample proportions of HIV antibody positive responses for the three levels of the injection variable. Briefly describe what you find.

    (b) Present the data in a 3 × 2 table.

    (c) Note that the 75 persons who reported no intravenous drug use (see the average number of injections per month and the duration of IV drug use) are included in the 0 percent category for the current analysis. Do you think the researcher should have done that, or should those 75 persons have been dropped from the analysis?

    (d) Refer to your answer in part (b) and the information in part (c). Construct the 3 × 2 table for the data after deleting the 75 persons who do not inject drugs intravenously. Construct the table of row proportions and a bar chart of the proportions who are HIV antibody positive. Describe what you find and comment on the effect of dropping the 75 subjects. Which strategy do you think is more reasonable?

7. Complete the following for the variables "duration of intravenous drug use" and "HIV antibody status."

    (a) Draw a bar chart of the sample proportions of HIV antibody positive responses for the three levels of the duration variable. Briefly describe what you find.

    (b) Present the data in a 3 × 2 table.

    (c) Compute the value of E for the cell for those who respond "more than 50 months" and are HIV antibody negative.

    (d) Compute the value of
    $$\frac{(O-E)^2}{E}$$
    for the same cell.

    For the remainder of this exercise, assume that the data are a random sample from a population that represents the process of the infection of IV drug users in New York City with HIV.

    (e) Conduct the chi-squared test of homogeneity. (Hint: $\chi^2 = 41.590$.) Describe what you have learned. Based on your answer, is it appropriate to perform pairwise comparisons of the levels of the duration variable?

    (f) Perform three Fisher's tests for the pairwise comparisons of the three duration groups. Describe what you have learned.

8. Refer to Exercise 6. Label the three levels of injection with used needles "none," "low," and "high" (in the obvious way). This exercise examines the analysis after the 75 persons who do not use IV drugs have been deleted from the data set.

    (a) The chi-squared test of homogeneity on the 3 × 2 table gives $\chi^2 = 8.407$; compute the approximate P-value. Do you believe it is appropriate to perform pairwise comparisons of the levels of the injection variable?

    (b) The chi-squared test of homogeneity on the 2 × 2 table comparing "none" to "low" gives $\chi^2 = 1.514$; use Table A.2 to compute the approximate P-value.

    (c) The chi-squared test of homogeneity on the 2 × 2 table comparing "none" to "high" gives $\chi^2 = 2.511$; use Table A.2 to compute the approximate P-value.

    (d) The chi-squared test of homogeneity on the 2 × 2 table comparing "low" to "high" gives $\chi^2 = 8.003$; use Table A.2 to compute the approximate P-value.

Exercises 9 through 12 pertain to the information in Table 11.11. Table 11.11 appeared in a study of the effectiveness of zidovudine (formerly AZT) in preventing the development of AIDS or advanced AIDS-related complex (ARC) among subjects who had tested positive for the HIV antibody [4]. The design was an RBD with subjects divided into two blocks by baseline CD4+ cell count. Subject follow-up varied from 19 to 107 weeks with an arithmetic average (mean) of 55 weeks. There are two choices for a dichotomous response, "AIDS" or

TABLE 11.11 *Frequency of Clinical Progression of Disease to AIDS or Advanced AIDS-Related Complex, according to CD4+ Cell Count (cells/mm³) at Entry into the Study*

			Study Group	
CD4+	Characteristic	Placebo	500 mg Zidovudine	1,500 mg Zidovudine
<200	No. of subjects	56	55	51
	AIDS	12	7	3
	AIDS or ARC	13	10	6
200–499	No. of subjects	372	398	406
	AIDS	21	4	11
	AIDS or ARC	25	7	13

"not AIDS," and "AIDS or ARC" or "neither AIDS nor ARC." Read each exercise carefully to determine which response is appropriate. There were three treatments in the study; placebo, 500 mg of zidovudine, and 1,500 mg of zidovudine.

9. For the subjects with a CD4+ count below 200 at baseline and the response "AIDS" or "not AIDS," the value of the test statistic for the test of homogeneity is 5.548. What is the approximate P-value? Do you think it is appropriate to make pairwise comparisons of the treatment levels?

10. For the subjects with a CD4+ count below 200 at baseline and the response "AIDS or ARC" or "neither AIDS nor ARC," the value of the test statistic for the test of homogeneity is 2.385. What is the approximate P-value? Do you think it is appropriate to make pairwise comparisons of the treatment levels?

11. For the subjects with a CD4+ count between 200 and 499 at baseline and the response "AIDS" or "not AIDS," the value of the test statistic for the test of homogeneity is 14.210. What is the approximate P-value? Do you think it is appropriate to make pairwise comparisons of the treatment levels?

12. For the subjects with a CD4+ count between 200 and 499 at baseline and the response "AIDS or ARC" or "neither AIDS nor ARC," the value of the test statistic for the test of homogeneity is 13.520. What is the approximate P-value? Do you think it is appropriate to make pairwise comparisons of the treatment levels?

13. Question 1 on the Wisconsin Driver Survey read

    How serious a problem do you think drunk driving is in Wisconsin?

    A summary of results from the 1984 survey is given in Table 11.12. Analyze these data using both descriptive and inferential techniques.

14. Question 11(c) on the Wisconsin Driver Survey asked subjects to evaluate the effectiveness of more education in achieving the goal of "keeping people from driving after (having) too much to drink." A summary of results from the 1984 survey is given in Table 11.13. Analyze the data using both descriptive and inferential techniques.

15. Question 11(d) on the Wisconsin Driver Survey asked subjects to evaluate the effectiveness of mandatory jail terms in achieving the goal of "keeping people from driving after (having) too much to drink." A summary of results from the 1983 and 1984 surveys is given in Table 11.14. Analyze the data using both descriptive and inferential techniques.

16. Question 11(d) on the Wisconsin Driver Survey asked for a rating of the effectiveness of mandatory jail terms in "keeping people from driving after (having) too much to drink." The output from a Minitab

TABLE 11.12 Minitab Output for Exercise 13 (Expected Counts Are Printed below Observed Counts)

	Attitude				
Gender	Not Very Serious	Somewhat Serious	Very Serious	Extremely Serious	Total
Female	5	152	516	524	1,197
	25.67	197.64	495.56	478.12	
Male	48	256	507	463	1,274
	27.33	210.36	527.44	508.88	
Total	53	408	1,023	987	2,471
$\chi^2 =$	16.648+	10.541+	0.843+	4.402+	
	15.642+	9.904+	0.792+	4.136 =	62.907
df = 3					

TABLE 11.13 Minitab Output for Exercise 14 (Expected Counts Are Printed below Observed Counts)

	Attitude				
Gender	Not Very Important	Somewhat Important	Fairly Important	Most Important	Total
Female	102	203	286	435	1,026
	110.98	205.01	266.07	443.94	
Male	127	220	263	481	1,091
	118.02	217.99	282.93	472.06	
Total	229	423	549	916	2,117
$\chi^2 =$	0.727+	0.020+	1.493+	0.180+	
	0.684+	0.018+	1.404+	0.169 =	4.695
df = 3					

TABLE 11.14 Minitab Output for Exercise 15 (Expected Counts Are Printed below Observed Counts)

	Attitude				
Year	Not Very Important	Somewhat Important	Fairly Important	Most Important	Total
1983	123	174	226	456	979
	102.12	179.00	258.55	439.33	
1984	221	429	645	1,024	2,319
	241.88	424.00	612.45	1,040.67	
Total	344	603	871	1,480	3,298
$\chi^2 =$	4.271+	0.140+	4.099+	0.632+	
	1.803+	0.059+	1.730+	0.267 =	13.001
df = 3					

TABLE 11.15 Minitab Output for Exercise 16 (Expected Counts and Row Proportions Are Printed below Observed Counts)

Drinking Frequency	Attitude				Total
	Not Very Important	Somewhat Important	Fairly Important	Most Important	
Not at all	30	55	116	243	444
	41.58	81.16	123.73	197.53	
	0.068	0.124	0.261	0.547	
Less often	62	167	254	408	891
	83.43	162.87	248.30	396.40	
	0.070	0.187	0.285	0.458	
Several times per month	68	126	161	225	580
	54.31	106.02	161.63	258.04	
	0.117	0.217	0.278	0.388	
Several times per week	49	60	91	117	317
	29.68	57.95	88.34	141.03	
	0.155	0.189	0.287	0.369	
Total	209	408	622	993	2,232
$\chi^2 =$	3.223+	8.433+	0.483+	10.466+	
	5.505+	0.105+	0.131+	0.340+	
	3.451+	3.765+	0.002+	4.230+	
	12.571+	0.073+	0.080+	4.095 =	56.951
df = 9					

analysis of the 1984 data is given in Table 11.15. (Row proportions have been added to the output below the expected counts.)

(a) Draw bar charts of the responses to the question by drinking frequency. Describe what they reveal.

(b) Find the approximate P-value for the test of homogeneity. Do you think it is appropriate to make a pairwise comparison of the drinking frequency groups?

(c) It is possible to perform six pairwise comparisons of the drinking frequency groups. Each pairwise comparison will yield a 2 × 4 table. Here is a summary of the six tests:

Groups	1, 2	1, 3	1, 4	2, 3	2, 4	3, 4
χ^2	12.677	33.110	31.592	14.664	22.632	3.220

Find the approximate P-value for each of the six tests. Briefly describe what you learn.

17. Does knowledge of the legal limit for blood alcohol concentration depend on drinking habits for Wisconsin drivers? Answer this question with the help of the Minitab output in Table 11.16. Your answer should contain all components of your answer to the previous exercise that are appropriate.

18. Recall that question 9 on the Wisconsin Driver Survey asked the respondent

   Which do you think is more intoxicating?

The possible responses were "can of beer," "glass of wine," "mixed drink," or "about the same" (the correct answer). Use the Minitab output of the 1984 data given in Table 11.17 to answer the question, Does the ability to identify the correct answer to question 9 depend on drinking habits? Your answer should contain all appropriate components of your answers to the previous two exercises.

Chapter 11 / Tests of Homogeneity and Independence

TABLE 11.16 Minitab Output for Exercise 17 (Expected Counts Are Printed below Observed Counts)

Drinking Frequency	Correct	Incorrect	Total
Not at all	290 303.52	213 199.48	503
Less often	552 584.72	417 384.28	969
Several times per month	401 380.16	229 249.84	630
Several times per week	236 210.60	113 138.40	349
Total	1,479	972	2,451
$\chi^2 =$	0.603+ 1.831+ 1.143+ 3.064+	0.917+ 2.786+ 1.738+ 4.663 =	16.745

df = 3

TABLE 11.17 Minitab Output for Exercise 18 (Expected Counts Are Printed below Observed Counts)

Drinking Frequency	Correct	Incorrect	Total
Not at all	258 286.91	242 213.09	500
Less often	556 554.89	411 412.11	967
Several times per month	369 359.79	258 267.21	627
Several times per week	220 201.41	131 149.59	351
Total	1,403	1,042	2,445
$\chi^2 =$	2.913+ 0.002+ 0.236+ 1.715+	3.923+ 0.003+ 0.318+ 2.310 =	11.420

df = 3

REFERENCES

1. J. R. Mendell et al., "Randomized, Double Blind Six-Month Trial of Prednisone in Duchenne's Muscular Dystrophy," *New England Journal of Medicine*, Vol. 320, June 15, 1989, pp. 1592–1597.
2. B. A. DeBuono et al., "Sexual Behavior of College Women in 1975, 1986, and 1989," *New England Journal of Medicine*, Vol. 322, March 22, 1990, pp. 821–825.
3. E. E. Schoenbaum et al., "Risk Factors for Human Immunodeficiency Virus Infection in Intravenous Drug Users," *New England Journal of Medicine*, Vol. 321, September 28, 1989, pp. 874–878.
4. P. A. Volberding et al., "Zidovudine in Asymptomatic Human Immunodeficiency Virus Infection," *New England Journal of Medicine*, Vol. 322, April 5, 1990, pp. 941–949.

Part III

NUMERICAL RESPONSES

Chapter 12 **Describing One Numerical Response**

Chapter 13 **Correlation and Regression**

Chapter 14 **Time Series**

Chapter 15 **Inference for One Numerical Population**

Chapter 16 **Numerical Data from Two Sources**

Describing One Numerical Response

12.1 INTRODUCTION
12.2 PLOTS AND HISTOGRAMS
12.3 MEASURES OF CENTER
12.4 MEASURES OF SPREAD
12.5 ASSORTED TOPICS*

Chapter 12

12.1 INTRODUCTION

A variable is **numerical** if its possible values are numbers. One way to obtain the value of a numerical variable is by *counting*. If a subject is a person, examples of variables obtained by counting include number of times married, number of children, number of siblings, and number of statistics courses completed.

A second way to obtain the value of a numerical variable is by assigning numbers to the categories of an ordered categorical variable. The first question on the 1982 Wisconsin Driver Survey is

> How serious a problem do you think drunk driving is in Wisconsin at this time?

The possible responses are "not serious at all," "not very serious," "somewhat serious," and "extremely serious." These four responses are ordered, and one may assign numbers to them, such as 1, 2, 3, and 4, respectively. The order of the numbers assigned must agree with the order of the categories. It is common to have successive numbers differ by a fixed amount, usually one.

A third way to obtain a numerical variable is by *measuring*. If a subject is a person, examples of variables obtained by measuring include height, weight, and time to complete a task.

For a measurement variable, a certain amount of imprecision is inevitable because of limitations of measuring devices. For example, suppose a woman's height is measured and reported to be 67 inches. This is not her exact height, but merely her height measured to the nearest inch. A more precise measuring device might find her height to be 66.8 inches, or another measuring device might yield 66.82 inches. (You may decide whether "exact height" has any practical meaning.) This text does not present a careful mathematical development of the effect of imprecision on measurement data. Simply use common sense to select a measuring device sufficiently precise for the problem at hand. For example, to report a two-year-old child's height to the nearest yard is too imprecise, since virtually all children will be 1 yard tall. On the other hand, for measuring the distance from our sun to its planets, 1 million miles is a common unit of measurement.

A fourth and final way to obtain a numerical variable is to create a mathematical function of one or more numerical variables. For example, IQ (intelligence quotient) was originally defined to be a person's "mental age" divided by chronological age; the Dow Jones Industrial Average is a summary of the prices of a collection of stocks; the difference or sum of two numerical variables creates a new variable.

Chapters 12 through 14 present methods for displaying and describing numerical data. No assumptions are made about how the subjects are selected for inclusion in the study, for example, by random sampling. The interpretation of the results, of course, depends a great deal on how the subjects are obtained. Chapters 15 and 16 introduce inference methods for a numerical response.

As discussed earlier in this text, data can arise in a number of different ways. The following three dichotomies will be especially useful in this text.

- Subjects may be either distinct individuals or trials.
- Data may be obtained either from one source or from more than one source.
- There may be either one or two variables per subject.

(Obviously, there could be more than two variables per subject, but that topic will be left to a more advanced text.) These three dichotomies can be combined to yield eight different possibilities, for example, two variables measured on distinct individuals from one source. The four possibilities with one variable per subject are studied in this chapter. Chapter 13 presents methods for data consisting of two variables per subject, and Chapter 14 introduces methods appropriate for one variable obtained from sequential trials.

If there is one variable per subject, it is called the response, as in Parts I and II of the text. If there are two variables, usually one is called the response, or dependent variable, and the other the predictor, or independent variable; see Chapter 13 for more details.

12.2 PLOTS AND HISTOGRAMS

This section begins with an example of data from individuals—students enrolled in one of my statistics classes. Each student's height was self-reported

TABLE 12.1 Heights, in Inches, of 70 Students

59	60	61	61	62	62	62	62	63	63
64	64	64	64	65	65	65	65	66	66
66	66	66	66	66	66	66	67	67	67
67	67	67	68	68	68	68	68	68	68
69	69	69	69	69	69	69	69	70	70
70	70	71	71	71	71	71	72	72	72
72	73	73	73	73	73	74	74	76	76

TABLE 12.2 Distributions of Heights, in Inches, of 70 Students

Value	Frequency	Value	Frequency	Value	Frequency
59	1	65	4	71	5
60	1	66	9	72	4
61	2	67	6	73	5
62	4	68	7	74	2
63	2	69	8	76	2
64	4	70	4		

to the nearest inch. A sorted list of the 70 reported heights is presented in Table 12.1. When the order in which the data are collected is irrelevant, as in this example, no information is lost by sorting the values, and it is usually helpful for later work to do so.

The **distribution** is a list of the distinct values in a data set and the frequency of occurrence of each distinct value. The distribution of the heights of the 70 students is presented in Table 12.2. A quick inspection of the distribution reveals that the smallest reported height is 59 inches and the largest is 76 inches. The most frequently reported height is 66 inches, which was reported by nine persons. Two or more subjects who give the same value for the response are said to be **tied;** there are many ties in the height data. When a distribution contains many ties, it is called **granular.**

A good way to study a distribution is to draw a picture of it. This section considers three popular pictures, dot plots, histograms, and stem and leaf plots.

12.2.1 The Dot Plot

Figure 12.1 presents the dot plot of the distribution of heights of 70 students. Each observation in a data set contributes a dot to the dot plot. For example, there is a single dot above 59 because one person in the study reported a height of 59 inches, and there are four dots above 62 inches because four persons reported that height. An examination of the dot plot reveals the following features:

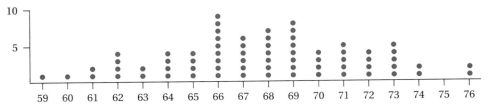

FIGURE 12.1 Dot plot of heights, in inches, of 70 students.

- There are **peaks** at 62, 66, 69, 71, and 73 inches.
- There is a **gap** at 75 inches: there are no observations at that value and there is at least one observation both above and below it.
- The dot plot is approximately symmetric about 67.5 inches.

The range of values between two peaks is called a **valley.** The dot plot of the height data has valleys extending from 63 inches to 65 inches and from 67 inches to 68 inches. It also has valleys at 70 inches and at 72 inches.

It is often instructive to compare and contrast a dot plot with a standard, the ideal bell-shaped dot plot presented in Figure 12.2. The ideal bell-shaped dot plot has a single peak, is symmetric, and has no gaps. It has a central value (or two) that occurs more frequently than any other value. Moving away from the center in either direction, the frequency of occurrence declines, and the decline is the same in each direction.

For any dot plot with a single peak, the portion of the dot plot to the right of the peak is called the **right tail** of the distribution and the portion to the left of the peak is called the **left tail.**

A researcher typically examines a dot plot with the goal of understanding or explaining any departure from the bell-shaped standard. The achievement of this goal often depends critically on the researcher's understanding of the

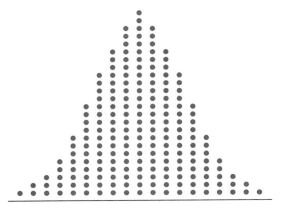

FIGURE 12.2 Ideal bell-shaped dot plot.

phenomenon being studied. This procedure is illustrated for a number of distributions, beginning with the heights of the 70 students in my class.

The dot plot of heights has several peaks. The presence of two or more peaks can have a number of explanations, three of which are discussed below.

First, if the data come from two or more importantly different sources, each source might contribute one or more peaks to the dot plot. In such cases it is usually fruitful to (try to) separate the data set into component pieces. For example, the class of 70 students consisted of 40 women and 30 men. Since men are typically taller than women, it is natural to split the original data set into two parts, one for women and one for men. Figure 12.3 presents dot plots of the distributions of women's and men's heights. These dot plots reveal several features of the data. First, as expected, the men in the class, as a group, are noticeably taller than the women in the class. The dot plot of female heights has peaks at 62 inches and 66 inches, has a gap at 70 inches, and is not close to being symmetric. The dot plot of male heights has peaks at 69 inches and 73 inches, has gaps at 67 inches and 75 inches, and is approximately symmetric about 71 inches. The gaps in the female and male dot plots are not particularly exciting; see the Rolling Stones data below for an example in which gaps are important.

A second explanation for the presence of two or more peaks is simply that it need not signify anything important in the data. For example, if one of the women who reported 62 inches is replaced by one who reports 63 inches, the peak at 62 inches would disappear. Similarly, the dot plot of male heights could be described as having a single broad peak extending from 69 inches to 73 inches.

A third explanation for a peak is that it reflects a known characteristic of the phenomenon that yields the data, as illustrated in the following example.

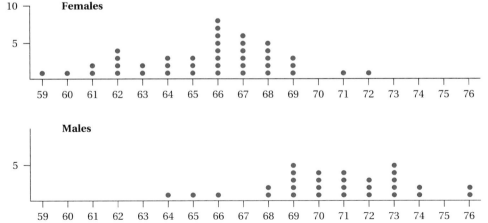

FIGURE 12.3 Dot plots of heights, in inches, of 40 female students and 30 male students.

The subjects are the 171 World Series games played between the years 1960 and 1987, inclusive. Figure 12.4 is the dot plot of the sum of the number of runs scored by both teams [1]. This dot plot has six valleys—at 2, 4, 6, 10, 13, and 17–18 runs. In fact, there are many more occurrences of odd than even numbers for the response; the count is 101 to 70. Here are two reasons that contribute to this difference as well as to the valleys at 2, 4, 6, and 10:

1. Ten games were tied at the end of nine innings. In each of these games, the eventual total score was an odd number. If the games had been allowed to end as a tie, the count in favor of odd totals would be reduced from 101-70 to 91-80.

2. If the home team bats in the bottom of the ninth inning, the game ends immediately if they take the lead. Since a victory by one run implies an odd total for the number of runs scored, an odd total appears to be more likely than an even total. In fact, five games ended while the home team was batting in the ninth, all with a victory by one run and a subsequent odd total.

The valley at 13 runs is not a result of these facts; it is likely due to chance variation and therefore is unimportant.

Sometimes an analyst cannot explain multiple peaks, as illustrated by the following example. The subjects are again the 171 World Series games played between 1960 and 1987. Figure 12.5 is a dot plot of the response obtained by taking the home team's score and subtracting the visiting team's score, *both measured at the end of nine innings*. Thus, this response is larger than 0 if the home team wins in regulation play, it is smaller than 0 if the visiting team wins in regulation play, and it equals 0 if the game requires extra innings. The dot

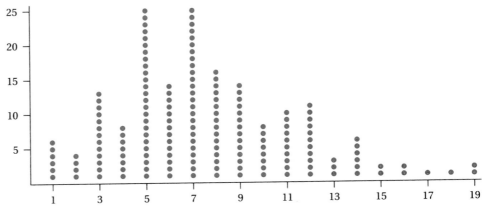

FIGURE 12.4 *Dot plot of the total number of runs scored in 171 World Series games, 1960–1987.*

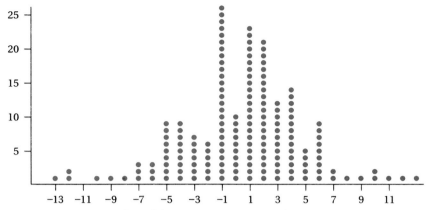

FIGURE 12.5 *Dot plot of home team's score minus visiting team's score after nine innings in 171 World Series games, 1960–1987.*

plot contains several peaks and valleys, the most striking of which is the deep valley at 0. Many games end after nine innings with one team ahead by one run, but relatively few require extra innings. I do not know how to explain this pattern.

12.2.2 Histograms

Recall that the probability histogram was introduced in Chapter 3 as a picture of a sampling distribution. Similarly, a histogram provides a picture of the distribution of a set of data.

Table 12.3 presents the lengths, in seconds, of 50 songs by the Rolling Stones. The songs are from a number of albums of mine and the lengths are taken from the record labels. The distribution is not very granular, resulting in a dot plot (Figure 12.6) that is difficult to interpret. A histogram sacrifices some of the detail of a dot plot to obtain a picture of the distribution that is easier to understand.

Although the dot plot of the Rolling Stones data is difficult to interpret, it does reveal an important feature of the distribution. Two values in the data set, 327 and 341, are isolated from the other 48 points. An **outlier** of a distribution is any value that is markedly different from all other values in the distribution except,

TABLE 12.3 Lengths, in Seconds, of 50 Rolling Stones Songs

108	108	128	135	136	143	149	152	152	155
158	160	165	167	167	169	170	172	172	172
178	183	186	188	188	190	190	190	192	194
200	200	202	204	205	209	218	219	220	225
228	230	232	239	252	259	270	287	327	341

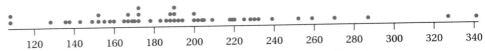

FIGURE 12.6 Dot plot of the lengths, in seconds, of 50 Rolling Stones songs.

possibly, other outliers. The word *markedly* is key: different analysts will attach different operational meanings to the word. Thus, labeling a point an outlier is a subjective endeavor. In general, each case that leads to an outlier should be examined carefully. Is the outlier the result of an error in recording the value? An experiment gone awry? Or simply a case that is unusual? For the Rolling Stones data, the two extreme values are recorded accurately; as a fan of theirs, I do not consider these songs ("Wild Horses" and "Let It Bleed") to be failed experiments; others may disagree. In my opinion, the two songs are simply unusual; it took a bit longer for the Stones to finish their message. (Not to be confused with repetition of a message ad nauseum; listen to their "Going Home" for 11 minutes and 35 seconds!)

There are three versions of the histogram, namely,

- The frequency histogram
- The relative frequency histogram
- The density scale histogram

The first step in obtaining a histogram is the construction of a relative frequency table. The first column of a relative frequency table lists the **class intervals,** and the second column gives the number of observations in each class interval. For example, in Table 12.4 the first class interval is 100–150 seconds, and, referring to the Rolling Stones data in Table 12.3, one sees seven observations with a response value between 100 seconds and 150 seconds. Thus, the number 7 appears in the first row of the frequency column. A difficulty arises when one attempts to

TABLE 12.4 Relative Frequency Table of the Lengths, in Seconds, of 50 Rolling Stones Songs

Class Interval*	Frequency	Relative Frequency
100–150	7	0.14
150–200	23	0.46
200–250	14	0.28
250–300	4	0.08
300–350	2	0.04
Total	50	1.00

*Each class interval includes its left endpoint but not its right.

determine the frequency of the second class interval, 150–200. Should the two observations in the data set equal to 200 seconds be counted in the second class interval, 150–200, or the third class interval, 200–250? This dilemma is solved by adopting an **endpoint convention.** The endpoint convention used throughout this text is that each class interval includes its left, but not its right, endpoint. Thus, the second class interval includes 150 but not 200, and the third includes 200 but not 250, and so on. Therefore, the two observations at 200 are counted in the third class interval, 200–250. You should use the data in Table 12.3 to verify the entries in the second column of Table 12.4.

The third column of a relative frequency table contains the relative frequencies of the class intervals. For example, the relative frequency of the class interval 100–150 is its frequency, 7, divided by the sample size, 50, yielding $7/50 = 0.14$, as presented in the first row of Table 12.4. You should verify the remaining entries in the relative frequency column of this table.

Figures 12.7 through 12.9 present the frequency, relative frequency, and density scale histograms for the Rolling Stones data. The three histograms have identical shapes and differ only in the vertical scale.

- For a frequency histogram, the height of the rectangle above a class interval equals the frequency of the class interval.
- For a relative frequency histogram, the height of the rectangle above a class interval equals the relative frequency of the class interval.
- For a density scale histogram, the height of the rectangle above a class interval equals the relative frequency of the class interval divided by the width of the class interval.

A consequence of the last method of computing the height is that

> For a density scale histogram, the *area* of a rectangle above a class interval is equal to the relative frequency of the interval.

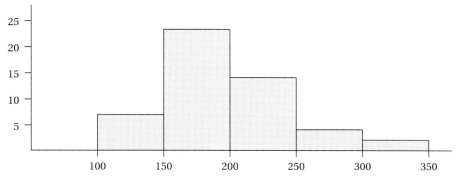

FIGURE 12.7 Frequency histogram of the lengths, in seconds, of 50 Rolling Stones songs.

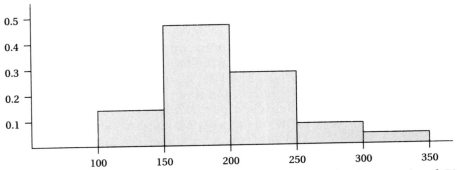

FIGURE 12.8 *Relative frequency histogram of the lengths, in seconds, of 50 Rolling Stones songs.*

(You might be reminded of an analogous property of probability histograms introduced in Chapter 3.) To prove this claim, let w and r denote the width and relative frequency, respectively, of a class interval, and recall that the area of a rectangle is equal to its base times its height. Thus, the area of a rectangle in a density scale histogram is

$$\text{Area} = \text{Base} \times \text{Height} = w \times \frac{r}{w} = r,$$

its relative frequency.

The ideas introduced earlier for examining a dot plot can be applied to examining a histogram. The histogram of the Rolling Stones data has a single peak and no gaps. It is noticeably asymmetrical, with its right tail longer than its left. Whenever the right tail is longer than the left, the histogram is called **skewed to the right.** If the left tail is longer than the right, then the histogram is called **skewed to the left.** Finally, note that the histogram does not reveal the two outliers.

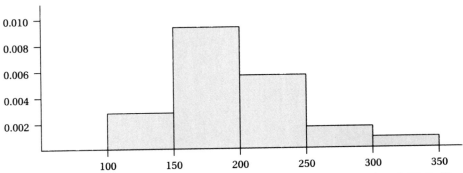

FIGURE 12.9 *Density scale histogram of the lengths, in seconds, of 50 Rolling Stones songs.*

Obviously, there is no reason to construct all three histograms. The following facts can help you decide which histogram to draw. (These facts are illustrated with later examples.)

- It is sometimes useful to have area correspond to relative frequency. In such situations, the density scale histogram should be used.
- If not all class intervals have the same width, the frequency and relative frequency histograms give misleading pictures of the distribution, but the density scale histogram does not. This feature is illustrated in Section 12.2.4.
- A researcher who wants to compare two distributions is always safe using density scale histograms. If the two histograms use the same class intervals, relative frequency histograms are acceptable. Finally, if the two histograms have the same class intervals and the sample sizes are not too different, frequency histograms work fine.
- The frequency histogram is the easiest and the density scale histogram is the most difficult to construct by hand. With a computer, the relative difficulties of the three histograms depend on the available software. Ease of construction should not outweigh the above considerations.

Table 12.5 presents a relative frequency table for the Rolling Stones data for a new choice of class intervals. (You are encouraged to verify the entries in this

TABLE 12.5 *A Second Relative Frequency Distribution of the Lengths, in Seconds, of 50 Rolling Stones Songs*

Class Interval	Frequency	Relative Frequency
100–120	2	0.04
120–140	3	0.06
140–160	6	0.12
160–180	10	0.20
180–200	9	0.18
200–220	8	0.16
220–240	6	0.12
240–260	2	0.04
260–280	1	0.02
280–300	1	0.02
300–320	0	0.00
320–340	1	0.02
340–360	1	0.02
Total	50	1.00

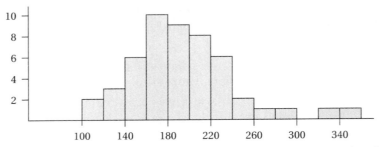

FIGURE 12.10 *Frequency histogram of the lengths, in seconds, of 50 Rolling Stones songs.*

table.) Figure 12.10 presents the frequency histogram for the Rolling Stones data for the class intervals of Table 12.5. Like the earlier histogram for the Rolling Stones data, the current histogram has a single peak and is skewed to the right. In addition, the current histogram has a gap which alerts the analyst that the two largest observations might be labeled outliers.

A researcher has a tremendous amount of freedom in choosing the class intervals for a relative frequency table, but the following rules must be obeyed:

1. The lower endpoint of the first class interval is less than or equal to every number in the data set.
2. The upper endpoint of the last class interval is larger than every number in the data set.
3. The upper endpoint of one class interval equals the lower endpoint of the next class interval.
4. Each class interval is the same width (that is, the upper endpoint minus the lower endpoint is the same number for each class interval).

The last rule can be relaxed; refer to Section 12.2.4 for details.

It is useful to contrast the dot plot and histogram. First, a dot plot reproduces the distribution exactly, whereas a histogram groups values into class intervals. Second, although there is a unique dot plot for a distribution, there are many choices of the class intervals that yield a histogram, and, unfortunately, different choices can lead to strikingly different histograms. See Exercise 27 at the end of Section 12.3 for an extreme example of this phenomenon. Finally, if the distribution has few ties, a histogram is usually better than a dot plot for determining the shape of the distribution, as illustrated by the Rolling Stones data.

The above presentation of histograms for the Rolling Stones data suggests that histograms are useful only when values have been collapsed into categories. That is not true; for very granular data a histogram, without collapsing, is an alternative to a dot plot as a picture of the data. This idea is presented in this book only for count data (data that take on integer values). For example, Figure 12.11 is a frequency histogram of the total number of runs scored in 171 World

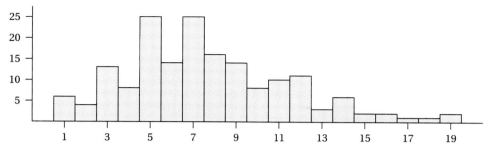

FIGURE 12.11 *Frequency histogram of the total number of runs scored in 171 World Series games, 1960–1987.*

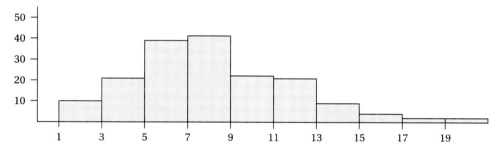

FIGURE 12.12 *A second frequency histogram of the total number of runs scored in 171 World Series games, 1960–1987.*

Series games, a distribution that was displayed with a dot plot in Figure 12.4. I find the histogram to be aesthetically more pleasing than the dot plot. Note the following two features of a histogram obtained without collapsing values:

1. The rectangle for each value is centered at the value; there is no need for an endpoint convention.
2. The width of each rectangle is 1. Therefore, the height of each rectangle in a density scale histogram equals the relative frequency divided by 1, that is, the relative frequency. Thus, the relative frequency and density scale histograms are identical.

It is instructive to collapse these baseball data into class intervals:

1–3, 3–5, 5–7, ..., 19–21.

The resulting frequency histogram is presented in Figure 12.12. This histogram is much easier to describe than the uncollapsed one: it has a single peak and is skewed to the right. It could be criticized, however, for hiding a striking feature—the many valleys—of the distribution. Regardless of which picture you prefer, these distinctive pictures illustrate the following general principle:

Before presenting data with a histogram, always examine the dot plot to determine whether the grouping used for the histogram hides or distorts any important features of the data.

12.2.3 Stem and Leaf Plots

A stem and leaf plot combines features of the dot plot and a frequency histogram:

- Like a dot plot, a stem and leaf plot displays exact data values.
- Like a frequency histogram, a stem and leaf plot allows for grouping values for an improved picture.

This section concentrates exclusively on data that assume integer values; see Exercises 23 and 24 for generalizations of this method.

Each observation in a data set is divided into two parts: its stem and its leaf. The stem is the number of tens in the observation, and the leaf is the number of units. For example, an observation of 70 is equal to 7 tens plus 0 units; thus, its stem is 7 and its leaf is 0. Similarly, 101 has a stem of 10 and a leaf of 1.

Figure 12.13 presents a stem and leaf plot of the 30 daily high temperatures in Madison, Wisconsin, in June 1988. To read this plot, note that the numbers to the left of the vertical line segment are the stems and the numbers to the right are leaves. Each of the 30 observations in the data set is represented by its leaf in the row of its stem. For example, a high of 85 is represented by a 5 (its leaf) in row 8 (its stem). Thus, the stem and leaf plot provides a sorted list of the 30 observations, namely,

70	72	73	75	75	75	76	78	80	82
83	83	84	84	84	85	86	88	88	89
89	90	91	91	92	93	96	97	100	101

In addition, if rotated 90 degrees counterclockwise, the stem and leaf plot becomes a frequency histogram with class intervals 70–80, 80–90, 90–100, and 100–110, provided you can imagine that a string of numbers is a rectangle.

This stem and leaf plot has a single peak and is slightly skewed to the right. The plot, however, could be criticized for having too little detail; that is, the class intervals are too wide. This criticism can be answered by splitting each stem

7	02355568
8	0233444568899
9	0112367
10	01

FIGURE 12.13 Stem and leaf plot of the June 1988 daily high temperatures in Madison, Wisconsin; 7 | 0 represents a high of 70 degrees.

		7	0
		7	23
		7	555
		7	6
7	023	7	8
7	55568	8	0
8	0233444	8	233
8	568899	8	4445
9	01123	8	6
9	67	8	8899
10	01	9	011
		9	23
		9	
		9	67
		9	
		9	01

FIGURE 12.14 *Two more stem and leaf plots of the June 1988 daily high temperatures in Madison, Wisconsin; 7 | 0 represents a high of 70 degrees.*

into two stems, with leaves 0–4 placed next to the first occurrence of a stem and leaves 5–9 placed next to the second occurrence of a stem. Alternatively, each original stem can be split into five stems with leaves 0 and 1 placed with the first occurrence of a stem, 2 and 3 with the second occurrence of a stem, and so on. These two different methods of splitting are illustrated in Figure 12.14. The first of these new plots has a single peak and is skewed to the right. The second has multiple peaks, valleys, and gaps. In my opinion, these features are unimportant; splitting each original stem into five stems gives too much detail. If each original stem is split into ten stems the result is the dot plot. Splitting a stem into some number of stems other than 2 or 5 is not allowed, because an unequal number of leaf values would be associated with each stem, causing a misleading picture (see Section 12.2.4).

12.2.4 Unequal Class Intervals

The data examined in this section indicate that a more informative histogram might be obtained if the analyst drops the restriction that the class intervals have the same width. Further, it is shown that for unequal class intervals a frequency or relative frequency histogram is misleading, but a density scale histogram is not misleading.

The large salaries paid to the men who play major league baseball is a frequent topic of conversation among sports fans and reporters. A serious discussion of salaries should consider what proportion of men who become professional baseball players actually reach the major leagues and how long a typical major league career lasts. This section explores the second topic, the length of major league careers.

The *Baseball Encyclopedia* lists career statistics for every man who ever has played in the major leagues [1]. The *Baseball Encyclopedia* divides the men into two groups, players and pitchers. I selected a random sample of 200 men from the population of all players and recorded the total number of games that each played during his major league career. The 200 numbers are presented in Table 12.6. Before collecting data, I suspected that the typical career is short. The actual data, however, amazed me—over one-half of the men sampled (103 of 200) played in fewer than 100 games during their careers. Figure 12.15 presents a density scale histogram of the distribution of number of games played. The histogram indicates that the distribution is strongly skewed to the right with as many as eight values, those above 1,500 games, that could be labeled large outliers. An undesirable feature of this histogram is that it collapses over one-half of the observations into a single class interval (0–100). This weakness can be partially alleviated by forming smaller class intervals, as in Figure 12.16.

Of these two histograms, the second histogram is better for values below 100 games; it provides more insight into the distribution of these short careers. On the other hand, the second histogram provides too much detail for the longer careers. Its narrow class intervals result in numerous peaks and valleys that are clearly the result of chance variation.

TABLE 12.6 Total Number of Career Games for a Random Sample of 200 Baseball Players

1	1	1	1	1	1	1	1	1	1
1	2	2	2	2	2	2	2	2	2
2	2	3	3	4	4	4	4	5	5
5	5	5	5	6	6	6	6	6	7
7	7	8	8	8	9	9	10	10	11
12	12	13	14	15	16	16	17	17	18
20	21	22	22	23	23	24	25	26	26
28	28	29	31	33	33	34	35	35	37
38	39	40	41	41	43	45	46	47	47
50	51	54	55	55	56	70	72	73	82
85	94	99	105	108	111	115	119	120	120
124	128	132	156	168	172	197	208	211	215
222	224	230	233	252	257	261	275	280	286
289	290	292	297	314	319	337	338	356	376
378	383	386	391	401	409	437	447	468	470
498	545	557	569	582	602	608	629	641	657
667	668	682	689	695	704	744	756	760	769
790	811	822	876	893	936	946	959	960	1,023
1,057	1,071	1,107	1,112	1,115	1,245	1,290	1,317	1,374	1,421
1,446	1,454	1,766	1,774	2,071	2,158	2,475	2,616	2,951	3,308

12.3 MEASURES OF CENTER

The dot plot, histogram, and stem and leaf plot provide pictures of a distribution. Often it is useful to summarize a distribution by computing one or more numbers. Most numerical summaries fall into either of two categories, measures of center or measures of spread. The former are examined in this section and the latter in the following section. The two most common measures of the center of a data set are the sample mean and sample median.

The **sample mean** of a data set is its arithmetic average; that is, the sum of the observations divided by the sample size. For example, below is the sorted list of the 30 daily high temperatures in Madison, Wisconsin, in June 1988.

70	72	73	75	75	75	76	78	80	82
83	83	84	84	84	85	86	88	88	89
89	90	91	91	92	93	96	97	100	101

The sum of these 30 values is

$$70 + 72 + 73 + \cdots + 101 = 2{,}550.$$

Thus, the sample mean is $2{,}550/30 = 85.0$ degrees.

It is convenient to introduce some notation. Let n denote the number of observations in the data set; $n = 30$ for the above example. The responses are denoted by

$$x_1, x_2, \ldots, x_n.$$

The sample mean is denoted by $\bar{x}$, which is read "ex bar," and can be written as

$$\bar{x} = \frac{x_1 + x_2 + \cdots + x_n}{n} = \frac{\sum x}{n}.$$

The sample mean of a set of data is equal to the balance point of the dot plot of the distribution. (Recall from Chapter 3 that the mean of a sampling distribution has a similar property.) If the dot plot is symmetric, the sample mean equals the point of symmetry.

The **sample median** of a set of data is equal to the center value in the sorted list of the observations. The sample median is denoted by $\tilde{x}$, which is read "ex tilde." If the sample size, n, is an odd number, the sample median is observation $(n + 1)/2$ in the sorted list. If the sample size is an even number, then observations $n/2$ and $(n/2) + 1$ are in the center of the list and the median is defined to be the arithmetic average of these two values.

The sample size is an even number, 30, for the June 1988 daily high-temperature data. Observations in positions $n/2 = 30/2 = 15$ and 16 are in the center of the list. In the sorted list of the data, the 15th observation is 84 and the 16th is 85; thus,

$$\tilde{x} = \frac{84 + 85}{2} = 84.5 \text{ degrees.}$$

Below is a comparison of the sample mean and sample median for each of the distributions introduced in the previous section:

- The total of the heights of 70 students listed in Table 12.1 (p. 381) can be shown to equal 4,743 inches; thus, the sample mean $\bar{x} = 4{,}743/70 = 67.8$ inches. Observations $n/2 = 70/2 = 35$ and 36 in the sorted list both equal 68 inches; thus, the sample median $\tilde{x} = 68$ inches.
- The heights of the 40 females displayed in Figure 12.3 have a sample mean of 65.6 inches and a sample median of 66 inches. The heights of the 30 males displayed in the same figure have a sample mean of 70.7 inches and a sample median of 71 inches.
- The total number of runs in 171 World Series games displayed in Figure 12.4 has a sample mean of 7.6 runs and a sample median of 7 runs.
- The difference between the scores of the home and visiting teams at the end of nine innings in 171 World Series games displayed in Figure 12.5 has a sample mean of 0.4 runs and a sample median of 1 run.
- The lengths of 50 Rolling Stones songs listed in Table 12.3 have a sample mean of 193.7 seconds and a sample median of 189 seconds.
- The number of career games played by 200 major league baseball players listed in Table 12.6 has a sample mean of 354.0 games and a sample median of 83.5 games.

In the last of these examples—the number of career games—the two measures of center differ markedly, but in the other examples they have similar values. These examples illustrate some of the general comments discussed below.

If a dot plot is symmetric, the sample mean and sample median are equal to the point of symmetry and therefore are equal to each other. Real data sets, however, are rarely perfectly symmetric, so it is important to realize that if a distribution is approximately symmetric, the sample mean and the median are similar. This fact is illustrated above by the heights of the 70 students, the heights of the 30 male students, and the June high temperatures.

In a distribution skewed to the right, the sample mean is larger than the sample median. This makes sense: the longer right tail pulls the center of gravity (mean) to the right of the middle value (median). This phenomenon is illustrated by the total number of runs in World Series games, the lengths of Rolling Stones songs, the June high temperatures, and, very dramatically, the number of games in baseball players' careers. By similar reasoning, the sample mean of a distribution skewed to the left is smaller than its sample median. With 20-20 hindsight I can see weak skewness to the left in the distributions of the heights of females, the heights of the males, and the differences of runs scored, thus explaining why the sample means are smaller than the respective medians for these data sets.

It is instructive to contrast the data from the Rolling Stones with the data on the lengths of baseball careers. The former distribution is slightly skewed to the right, so the sample mean is somewhat larger than the sample median. By contrast, the data on length of career is strongly skewed to the right, with as many as eight observations that could be labeled outliers. Thus the center of gravity—the sample mean—of the career games data is pulled dramatically to the right of the sample median.

When the mean and the median of a distribution differ radically, it may be tempting to state that one measure of center is superior to the other. The preferred measure of center, however, usually depends on the scientific goals that motivate the study. Nevertheless, it is worth noting the following characteristic of the distribution of lengths of baseball careers: the sample mean is larger than the career number of games for 69 percent (138 of 200) of the men. Many people find it peculiar for a center to exceed 69 percent of the observations.

EXERCISES for Sections 12.2 and 12.3

1. Brian Peterson performed a balanced completely randomized design (CRD) with twenty trials. The response was the time, measured to the nearest second, required by Brian to run 1 mile. Wearing combat boots, his sorted times were

321	323	329	330	331
332	337	337	343	347

    Wearing jungle boots, Brian's sorted times were

301	315	316	317	321	
321	323	327	327	327	

    (a) Construct dot plots for the responses on each treatment.
    (b) Compute the sample mean and median for each treatment.
    (c) Write a few sentences that summarize what these pictures and numbers reveal.

2. This study was performed by Phil Coan and Jean Schoeni. The Pipetman is an aid for accurately measuring and transferring liquid volumes. A disposable tip is placed on this hand-held device for each measurement. Tips are of two sizes, 100 μl and 1,000 μl. Tips must be placed on a rack for sterilization and prevention of contamination during use. Each rack holds 100 tips. Treatment 1 was filling a Pipetman tip rack of 1,000-μl tips, and treatment 2 was filling a rack of 100-μl tips. The response is the time, measured to the nearest second, required to fill a rack. The metasubject believed that his fine motor control was deficient and that he should have been able to fill racks of 1,000-μl tips faster than racks of 100-μl tips. A randomized pairs design (RPD) was conducted to investigate this issue. The sorted observations on the first treatment are

92	92	94	95	96	97	99	99	102	102
103	104	104	105	106	106	107	108	108	110
111	114	114	116	116	116	116	117	120	120
121	123	124	126	126	127	129	129	136	137

The sorted observations on the second treatment are

110 111 111 113 114 114 115 116 116 116
116 119 119 119 121 121 121 122 123 124
124 126 126 127 128 129 130 130 131 131
132 134 135 136 138 139 140 140 144 165

The sorted differences (second treatment minus first) are

2   2   3   4   5   5   5   6   6
7   7   7   8   9 10 11 11 11 11
12 14 17 17 17 17 18 19 19 19
19 19 20 22 22 24 28 32 33 36

(a) Construct dot plots for the responses for each treatment; for the differences.

(b) Construct density scale histograms for the responses for each treatment using 90–100, 100–110, ... as the class intervals.

(c) Construct density scale histograms for the differences using 0–8, 8–16, ... as the class intervals.

(d) Based on the pictures you have obtained, is it obvious whether the sample mean or sample median is larger for the first treatment? For the second treatment? For the differences? Given that the sample means are 111.68, 125.65, and 13.98, respectively, how do they compare with the sample medians?

(e) Write a few sentences that summarize what you have learned. Is the metasubject's suspicion supported?

3. Jennifer Erickson and Michelle Hansen performed a CRD to compare Jennifer's hitting ability with aluminum and wooden baseball bats. The response was the distance, measured to the nearest foot, that the ball traveled when hit. The sorted responses with an aluminum bat are

105 110 115 120 123 125 130 135 139 141
142 145 147 148 149 150 151 155 156 159
160 163 167 167 172 174 174 177 182 186
189 192 200 202 209 214 231 241 260 350

The sorted responses with a wooden bat are

86   114 117 119 121 122 123 127 127 132
134 137 139 141 142 144 147 149 151 152
153 153 154 155 156 157 159 162 163 164
166 167 167 168 171 176 177 181 189 200

(a) Construct stem and leaf plots for the responses on each treatment.

(b) An analyst looks at the plot for the data on the aluminum bat and states, "Clearly the sample mean is larger than the sample median." How does she know this? Given that the sample mean equals 168.88, is she correct? Recompute the sample mean after deleting the largest observation and comment.

(c) The total of the 40 observations on the wooden bat is 5,962 feet. Compute the sample mean and the sample median.

(d) Write a few sentences that summarize what you have learned.

(e) Occasionally Jennifer missed the ball or tipped it foul. On those trials, the card denoting the treatment was returned to the unused pile of cards, which was then reshuffled. Do you think this was a good strategy? Give reasons for your answer.

4. In the dart game "301" the object is to be the first player to score exactly 301 points. In each round a player throws three darts and the total score of the three darts is added to the player's previous score. If the new total is greater than 301, the player's score reverts to the previous total. Thus, for example, if a player reaches a total of 300, then he or she needs exactly one point to win; any larger score will be ignored. Doug H. performed a CRD to compare his personal darts to bar darts. The response was the number of rounds Doug required to score 301. The sorted responses with Doug's darts are

12 13 14 14 15 15 17 18 18 19
19 19 20 20 21 21 22 23 25 27

The sorted responses with the bar darts are

13 15 16 16 17 17 17 18 19 21
21 22 23 25 26 26 27 27 28 30

(a) Construct dot plots for the responses for each treatment.

(b) Construct stem and leaf plots for the responses for each treatment.

(c) Repeat part (b) after dividing each stem into five stems.

(d) Compute the sample mean and sample median for each treatment.

(e) Write a few sentences that summarize what you have learned.

5. Kymn Fischer is a member of the women's varsity crew at the University of Wisconsin-Madison. When she cannot practice on a lake, she works out on a rowing simulation device called an ergometer. There are two settings on the "erg" that she uses: the small gear with the vent closed and the large gear with the vent open. (The large gear with the vent closed would be too easy and the small gear with the vent open too difficult.) Kymn performed a CRD with 10 trials to compare the two settings. Her response is the time, measured to the nearest second, required to row the equivalent of 2,000 meters. With the small gear and the vent closed her times were

493 489 492 490 493

With the large gear and the vent open her times were

485 483 488 479 486

(a) Construct dot plots for the responses on each treatment.

(b) Compute the sample mean and the sample median for each treatment.

(c) Write a few sentences that summarize what you have learned.

(d) Kymn's coxswain told her that people who are muscularly very strong perform better with the small gear than with the large gear, whereas those who are aerobically very fit show the reverse tendency. If the coxswain is correct, which type do you think Kymn is?

6. Peter Olson wanted to compare his accuracy throwing heavy and light darts. He performed a CRD with response equal to the distance, rounded up to the nearest 2 cm, from the center of a target. Below is the distribution of his responses with the heavy darts:

Resp.	Freq.	Resp.	Freq.	Resp.	Freq.
2	1	10	7	18	1
4	2	12	3	22	1
6	8	14	9	24	1
8	2	16	4		

Below is the distribution of his responses with the light darts:

Resp.	Freq.	Resp.	Freq.	Resp.	Freq.
2	1	8	9	14	2
4	8	10	8	16	1
6	5	12	3	18	2

(a) Construct dot plots for the responses on each treatment.

(b) Compute the sample mean and the sample median for each treatment.

(c) Write a few sentences that summarize what you have learned.

7. Amy Zeik recalls her high school teacher's telling her that she could type faster if she did not look at the keys. Amy decided to test this advice with an RPD of 40 trials. Each trial consisted of selecting an alphabetic sentence—a sentence that includes every letter of the alphabet—of 70 strokes and typing it (repeatedly) as many times as possible in 1 minute. Each trial used a different alphabetic sentence. The first treatment was touch-typing without looking at the keys, and the second treatment was typing with no restriction. The response was the number of key strokes completed during the 1-minute test period. Below are the sorted responses with the first treatment:

111 118 120 120 122 127 128 129 132 135
137 139 140 140 142 147 150 150 160 164

Below are the sorted responses with the second treatment:

131 132 135 135 142 146 150 154 154 157
159 161 161 163 164 165 165 166 170 177

(a) Construct stem and leaf plots for the responses on each treatment.

(b) Repeat part (a) after dividing each stem into two stems.

(c) Compute the sample mean and the sample median for each treatment. (Hint: The total number of strokes with the first treatment is 2,711; with the second treatment it is 3,087.)

(d) Write a few sentences that summarize what you have learned.

8. Deborah Martin White's son Scotty is convinced that his Snowbie sled is slower than his friend Sam's Sno-Racer. An RPD was conducted to investigate this issue. A trial consisted of a slide down a local hill. The response is the time, measured to the nearest 10th of a second, that Scotty required to complete a slide. In the first treatment Scotty rode his Snowbie, and in the second treatment Scotty rode Sam's Sno-Racer. Below are the results of the study:

Trial	Treatment	Time
1	2	11.3
2	1	12.0
3	2	11.3
4	1	11.1
5	2	10.1
6	1	8.4
7	2	9.0
8	1	12.1
9	1	8.9
10	2	10.7
11	2	10.6
12	1	9.8
13	2	10.1
14	1	8.8
15	2	9.9
16	1	10.5
17	1	12.2
18	2	12.1

(a) Construct dot plots for the responses on each treatment.

(b) For each pair of trials, compute the time on the Sno-Racer minus the time on the Snowbie. Construct the dot plot of these nine differences.

(c) Compute the sample mean and the sample median of the times on the Sno-Racer, the times on the Snowbie, and the differences.

(d) Write a few sentences that summarize what you have learned.

9. Refer to Exercise 7. Unfortunately, Amy made some mistakes in every trial. A different response is the number of strokes minus five times the number of errors. (The idea is that because five strokes is usually counted as a word, this new response subtracts one word for each error.) Below are the sorted responses on the first treatment:

| 70 | 70 | 70 | 80 | 86 | 87 | 87 | 88 | 89 | 97 |
| 99 | 99 | 108 | 110 | 110 | 112 | 112 | 112 | 115 | 120 |

Below are the sorted responses on the second treatment:

| 105 | 116 | 122 | 130 | 132 | 134 | 135 | 136 | 139 | 139 |
| 141 | 142 | 144 | 147 | 153 | 155 | 156 | 160 | 160 | 161 |

(a) Construct stem and leaf plots for the responses on each treatment.

(b) Repeat part (a) after dividing each stem into two stems.

(c) Compute the sample mean and the sample median for each treatment. (Hint: The total of the responses with the first treatment is 1,921; with the second treatment it is 2,807.)

(d) Write a few sentences that summarize what you have learned.

10. Peter Zarov performed an RPD with 50 trials. The response was the time, measured to the nearest 0.5 second, required by Peter to swim two lengths of a 25-yard pool. The first treatment was swimming the backstroke and the second treatment was swimming the breaststroke. The responses he obtained are at the top of p. 405.

(a) Construct dot plots for the responses with each treatment.

(b) Construct the dot plot of the differences.

(c) The sample mean and the sample median are 40.18 and 40.00, respectively, for the backstroke, 38.74 and 38.50, respectively, for the breaststroke, and 1.44 and 1.50, respectively, for the differences. Using these numbers and your pictures, write a few sentences that summarize what you have learned.

Pair:	1	2	3	4	5	6	7	8	9
Backstroke	40.0	39.5	39.5	41.0	39.0	38.0	38.5	38.5	39.0
Breaststroke	37.0	37.5	37.5	37.0	38.0	38.0	38.5	40.5	39.0
Difference	3.0	2.0	2.0	4.0	1.0	0.0	0.0	−2.0	0.0

Pair:	10	11	12	13	14	15	16	17	18
Backstroke	39.5	40.5	40.0	40.0	40.5	45.0	40.0	39.5	40.5
Breaststroke	39.0	39.5	38.0	39.0	38.5	38.0	38.5	43.0	39.0
Difference	0.5	1.0	2.0	1.0	2.0	7.0	1.5	−3.5	1.5

Pair:	19	20	21	22	23	24	25
Backstroke	41.0	41.5	41.0	40.5	41.5	40.0	40.5
Breaststroke	39.5	40.0	39.0	38.0	39.0	39.5	38.0
Difference	1.5	1.5	2.0	2.5	2.5	0.5	2.5

11. Christine McDonald and Steve Allen performed an RPD entitled "Skill versus Luck: The Darts Study." In the first treatment Christine aimed and threw, one at a time, three plastic-tipped darts at a standard dart board. In the second treatment Christine held all three darts in her hand and threw them, simultaneously, at the target. The response was the total score of the three darts. The sorted responses with the first treatment were

```
 2  2  6  7  9 11 12 18 18 19
21 21 22 24 25 25 26 27 29 30
31 33 34 36 38 38 40 40 41 45
51 53 56 57 58 70 70 70 81 124
```

The sorted responses with the second treatment were

```
 0  0  0  3  3  6  7 10 13 15
15 16 17 17 17 21 23 24 25 27
27 28 31 32 35 36 36 37 42 43
49 53 55 56 59 60 61 65 68 85
```

(a) Construct density scale histograms for the responses with each treatment using 0–20, 20–40, ... as the class intervals.

(b) Construct stem and leaf plots for the responses on each treatment.

(c) An analyst looks at these pictures and comments, "It is obvious that the sample mean is smaller than the sample median for each treatment." Is the analyst correct? Explain your answer, but no computations are needed.

(d) Write a few sentences that summarize what you have learned.

12. Terry Cefalu-Haney, John Sopha, and Steven Unbehaun performed an RPD to compare heavy and light darts. The first treatment was throwing three heavy darts and the second treatment was throwing three light darts. The response was the total score for the three darts. The sorted responses with the first treatment were

```
 8 17 18 20 23 26 27 27 27 29
29 30 34 34 36 39 39 39 40 42
43 44 47 47 48 50 52 53 56 58
58 61 66 66 68 68 73 77 93 103
```

The sorted responses with the second treatment were

```
 5 14 18 19 21 24 24 24 25 26
26 27 29 30 30 33 34 34 35 37
38 46 46 47 50 57 57 63 64 68
69 73 73 73 85 85 86 109 110 114
```

The sorted responses of the differences (treatment 1 minus treatment 2) were

−90	−62	−61	−55	−38	−38	−37	−34	−34	−30
−29	−24	−21	−19	−14	−13	−9	−8	−6	−3
−3	−2	1	4	6	9	10	10	11	15
17	20	27	31	45	47	59	63	69	73

(a) Construct density scale histograms for the responses with each treatment using 0–20, 20–40, . . . as the class intervals.

(b) Construct a density scale histogram for the differences, with the first class interval beginning at −100 and with each interval having a width of 20.

(c) With the heavy darts the total of the 40 scores is 1,815; the total with the light darts is 1,928. Use this information to compute the sample mean and the sample median for the heavy darts, for the light darts, and for the differences.

(d) Write a few sentences that summarize what you have learned.

13. Sara Lamers performed a balanced CRD of 80 trials at a local driving range. Her two treatments were hitting a golf ball with a 3-wood and hitting a golf ball with a 3-iron. Her response was the distance the ball traveled, in yards, before coming to rest. Below are her sorted responses with the 3-wood:

22	32	38	56	58	77	81	93	99	101
101	101	104	107	107	108	109	109	110	111
113	114	115	116	118	122	122	127	127	128
128	128	129	131	131	137	139	139	140	147

Below are her sorted responses with the 3-iron:

27	52	53	57	58	59	68	68	68	82
84	88	92	92	92	92	97	97	98	99
100	101	105	107	107	107	108	109	110	116
118	127	132	132	136	136	137	138	139	139

(a) Construct a density scale histogram of the data for the 3-wood using the class intervals 20–40, 40–60, . . . , 140–160.

(b) Repeat (a) with the data for the 3-iron, using the same class intervals.

(c) The sample means for the two treatments are 106.87 and 98.18; match the sample mean with the treatment. Compute the sample medians.

(d) Write a few sentences that summarize what you have learned.

14. Reggie Holt performed a balanced CRD of 30 trials. Each trial consisted of a game of darts, where a game is defined as throwing 12 darts. His two treatments were throwing darts from a distance of 10 feet and throwing darts from a distance of 12 feet. Reggie's response was his total score in a game, with larger numbers better. Below are Reggie's sorted scores from 10 feet:

181	184	189	197	198	198	200	200
205	205	206	210	215	215	220	

Below are Reggie's sorted scores from 12 feet:

163	164	168	174	175	186	191	196
196	197	200	200	201	203	206	

(a) Construct a dot plot of the scores from 10 feet.

(b) Construct a dot plot of the scores from 12 feet.

(c) The total of Reggie's scores from 10 feet is 3,023; the total from 12 feet is 2,820. Use this information to compute the sample mean and median for each distance.

(d) Write a few sentences that summarize what you have learned.

15. Mei Lan Chan, Sin Fai Cheung, and Todd Willer performed a balanced CRD of 30 trials. Each trial consisted of accelerating an automobile from 40 miles per hour to 65 miles per hour. Their two treatments were driving the car with all windows open (first) and with all windows closed (second). The response is the time, measured to the nearest 0.01 second, required to reach 65 miles per hour. Below are the sorted times with the windows open:

7.34	7.56	7.67	7.67	7.77	7.89	7.95	7.96
8.01	8.07	8.12	8.17	8.22	8.34	8.34	

Below are the sorted times with the windows closed:

6.67	6.78	6.79	6.86	6.88	6.92	6.94	7.05
7.11	7.12	7.23	7.34	7.36	7.56	7.86	

(a) Construct a dot plot of the times with the windows open.

(b) Construct a dot plot of the times with the windows closed.

(c) Compute the sample median for each treatment. The sample means equal 7.939 and 7.098; match these means with their treatments.

(d) Write a few sentences that summarize what you have learned.

16. Stir-fry is a popular meal for college students at Madison. Aimee Becker, Kevin Rathke, and David Schwartz performed an RPD to compare chopping carrots with chopping celery for stir-fry. Their primary question was whether the slices were uniform in thickness. The response was the thickness of each piece measured to the nearest millimeter. Below is the distribution of the thicknesses of 50 carrot slices:

Resp.	Freq.	Resp.	Freq.	Resp.	Freq.
9	1	15	3	21	1
10	3	16	3	22	2
11	7	17	4	23	1
12	8	18	2	24	1
13	8	19	1	27	1
14	3	20	1		

Below is the distribution of the thicknesses of 50 celery slices:

Resp.	Freq.	Resp.	Freq.	Resp.	Freq.
10	1	15	7	20	2
11	2	16	6	21	1
12	2	17	3	22	1
13	2	18	10	24	1
14	10	19	2		

(a) Construct dot plots for the responses on each treatment.

(b) Write a few sentences that summarize what you have learned.

17. Stacy Markowitz, Lauren Trais, and Jay and Jeff Hengel performed a CRD on 40 persons. Subjects were individually shown for 10 seconds a 5-inch by 10-inch scenic drawing with 13 faces hidden in it. The response was the number of faces located by the subject. Subjects receiving the first treatment were told, "You are going to see a scenic drawing with hidden faces in it for 10 seconds. Count the number of hidden faces you see." Subjects receiving the second treatment were told, "You are going to see a scenic drawing with hidden things in it for 10 seconds. Count the number of hidden things you see." The first treatment is called the prior knowledge condition, and the second is called the no prior knowledge condition. The sorted responses for the prior knowledge condition were

2 3 3 4 4 4 4 5 5 5
6 6 6 6 6 7 7 8 9 10

The sorted responses for the no prior knowledge condition were

0 1 1 1 1 2 2 2 2 2
2 3 3 3 3 4 4 5 5 6

(a) Construct dot plots for the responses with each treatment.

(b) Compute the sample mean and median for the prior knowledge condition and for the no prior knowledge condition.

(c) Write a few sentences that summarize what you have learned.

18. Dale Mischke performed an RPD to compare bowling with light and heavy bowling balls. A trial consisted of a game of bowling, and the response was Dale's score. Below are the results of the study:

Game	Ball	Score	Game	Ball	Score
1	Light	184	8	Heavy	148
2	Heavy	164	9	Light	205
3	Light	166	10	Heavy	149
4	Heavy	191	11	Heavy	177
5	Light	164	12	Light	165
6	Heavy	160	13	Light	169
7	Light	157	14	Heavy	158

(a) Construct dot plots for the responses with each treatment.

(b) Construct stem and leaf plots for the responses with each treatment.

(c) For subject equal to a pair of trials, construct the dot plot for the score with the light ball minus the score with the heavy ball.

(d) Compute the sample mean and sample median for each treatment.

(e) Write a few sentences that summarize what these pictures and numbers reveal.

19. Alisa Evans performed an RPD to compare bowling with one hand (the usual method!) and bowling two handed ("granny style"). A trial consisted of a game of bowling, and the response was Alisa's score. Below are the results of the study:

Game	Hands	Score	Game	Hands	Score
1	One	97	6	One	110
2	Two	85	7	One	123
3	Two	91	8	Two	96
4	One	108	9	One	125
5	Two	95	10	Two	94

(a) Construct dot plots for the responses on each treatment.

(b) For subject equal to a pair of trials, construct the dot plot for the score using one hand minus the score using two hands.

(c) Compute the sample mean and median for each treatment and for the differences.

(d) Write a few sentences that summarize what these pictures and numbers reveal.

20. Bascom Hill is a long, steep hill (by Wisconsin standards!) in the center of the university campus in Madison. Damion Clayton wondered whether smoking a cigarette affected his climbing of Bascom Hill. He performed an RPD with response equal to the time, measured to the nearest second, he needed to walk from the bottom to the top of the hill. The first treatment consisted of walking while smoking a cigarette, and the second consisted of walking while not smoking a cigarette. Below are the results of the study:

Trial	Treatment	Time
1	2	163
2	1	173
3	1	175
4	2	160
5	1	169
6	2	157
7	2	165
8	1	175
9	1	180
10	2	167
11	2	159
12	1	184
13	2	170
14	1	182
15	2	155
16	1	186
17	1	190
18	2	153
19	2	164
20	1	188

(a) Construct dot plots for the responses for each treatment.

(b) Construct stem and leaf plots for the responses for each treatment.

(c) For subject equal to a pair of trials, construct the dot plot for the score on treatment 1 minus the score on treatment 2.

(d) Compute the sample mean and the sample median for each treatment.

(e) Write a few sentences that summarize what you have learned.

21. Martha Ehlinger and Lisa Holle performed an RPD to study Martha's juggling skill. The first treatment was juggling three tennis balls, and the second treatment was juggling three large apples. The response was the time, measured to the nearest second, during which the three items were in "a regular cycle of juggling." The sorted responses with the tennis balls were

1	1	2	2	2	2	2	3	3
4	4	4	4	4	5	5	5	6
6	6	6	6	6	6	6	6	7
7	8	10	11	12	12	13	13	14	17

The sorted responses with the large apples were

```
1  1  1  1  1   2   2   2   2   2
2  3  3  3  3   3   4   4   4   5
5  5  5  5  5   5   6   6   7   8
8  9  9  9  9  10  11  12  13  13
```

The sorted responses of the differences (balls minus apples) were

```
−11  −9  −8  −8  −8  −5  −4  −4  −3  −3
 −3  −3  −2  −1  −1   0   0   0   0   1
  1   1   1   2   3   3   3   3   4   4
  5   5   5   6   7   9   9  10  13  14
```

(a) Construct dot plots for the responses with each treatment.

(b) Construct the dot plot of the differences.

(c) Compute the sample medians of these three sets of data. The total of the responses with the tennis ball is 244; for the apples the total is 209. What is the total for the differences? Use this information to compute the sample means for these three sets of data.

(d) Write a few sentences that summarize what you have learned.

22. Becky Blachowiak, Rachel Fleming, and Rachel Lewis performed a CRD with 40 trials. In each trial Becky shot five free throws, and the response was the number she made. The first treatment was shooting a women's basketball, and the second treatment was shooting a men's basketball. (A women's ball is somewhat smaller.) With the women's ball Becky made one basket in each of 4 trials, two baskets in 8 trials, three baskets in 5 trials, and four baskets in 3 trials. With the men's basketball Becky made no baskets in 4 trials, one basket in 10 trials, two baskets in 5 trials, and three baskets in 1 trial.

(a) Construct dot plots for the responses with each treatment.

(b) Compute the sample mean for each treatment.

(c) Write a few sentences that summarize what you have learned.

23. Amy Glassman and Jim MacDougall measured the length, to the nearest 0.01 second, of 40 notes played on a piano without using the damper pedal. Below is the distribution of their responses:

Resp.	Freq.	Resp.	Freq.	Resp.	Freq.
17	1	25	8	28	2
19	7	26	1	30	3
21	2	27	1	31	1
22	14				

They also played 40 notes with the damper pedal depressed. Below are their sorted values, again in units of 0.01 second.

```
  813    838    839    847    849
  851    892    914    925    936
  942    953    957    967    989
  993  1,015  1,016  1,026  1,031
1,042  1,043  1,054  1,078  1,088
1,102  1,103  1,109  1,112  1,112
1,121  1,134  1,139  1,142  1,143
1,182  1,197  1,202  1,206  1,209
```

(a) Construct a dot plot for the responses obtained without the damper pedal. The sample mean equals 23.25; what is the sample median? Write a few sentences that summarize what you have learned.

(b) Consider the data with the damper pedal depressed. A stem and leaf plot would require 40 stems (81–120), which is too many for a sample of size 40. (Why?) An alternative approach is to *truncate*, or round down, each observation by changing its unit to 0. For example, 813 becomes 810 and 838 becomes 830. A stem and leaf plot can then be formed with the stem equaling the number of hundreds and the leaf equaling the number of tens. For example, 830 has a stem of 8 and a leaf of 3. Use this technique to obtain a stem and leaf plot for the data obtained with the damper pedal depressed.

(c) Refer to part (b). Divide each stem into two stems.

(d) Write a few sentences that summarize what you have learned.

24. This exercise demonstrates how to construct a stem and leaf plot for data with a fractional component. Below are the lengths, in minutes, of 20 Buddy Holly songs from the album *Buddy Holly Lives*:

2.27 2.58 2.02 2.20 2.42 2.18 2.10 3.43 2.22 2.83
2.27 2.23 2.30 1.88 2.87 2.85 2.17 2.40 2.12 2.13

Define the stems to equal the number of 10ths of a minute and the leaves to equal the number of 100ths of a minute. For example, 2.27 has a stem of 22 and a leaf of 7, and 1.88 has a stem of 18 and a leaf of 8. Use these definitions of stems and leaves to construct a stem and leaf plot of these data. Describe what your picture reveals.

25. Select a trial and two treatments of interest to you that yield a numerical response. (Exercises 1–23 of this section are examples of selections by my students.) Perform a balanced completely randomized design to study the effects of the treatments on the response. The number of trials should reflect the difficulty of the trials. For instance, if the trial is easy to perform, like hitting a golf ball, you should perform at least 50 trials. If, however, trials are time consuming, like bowling a game or running a mile, 10 trials will suffice.

(a) Use dot plots, histograms, or stem and leaf plots to present your data. Write a few sentences that discuss what your pictures reveal.

(b) Compute the sample mean and sample median for the data from each treatment. Discuss what these numbers reveal.

Save your data for later use.

26. Select a trial and two treatments of interest to you that yield a numerical response. (Exercises 1–23 of this section are examples of selections by my students.) Perform a randomized pairs design to study the effects of the treatments on the response. The number of trials should reflect the difficulty of the trials, as discussed in the previous exercise.

(a) Use dot plots, histograms, or stem and leaf plots to present your data. Write a few sentences that discuss what your pictures reveal.

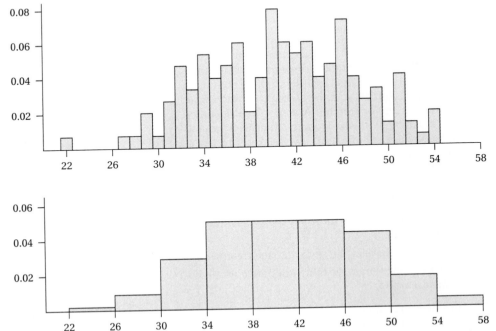

FIGURE 12.19 Density scale histograms of 153 cash prizes for matching four numbers in Lotto America.

(b) Compute the sample mean and sample median for the data from each treatment and for the differences. Discuss what these numbers reveal. Save your data for later use.

27. Consider the following artificial data:

0	1	2	3	4	15	16	17	18	19
20	21	22	23	24	35	36	37	38	39

Construct a frequency histogram using the class intervals 0–10, 10–20, 20–30, and 30–40. Construct a frequency histogram using the class intervals 0–5, 5–10, ..., 35–40. Compare the histograms.

28. Lotto America is a multistate semiweekly lottery. A player selects six numbers from the first 54 positive integers and wins if four, five, or six of the selections match the six numbers selected by the lottery commission. Figure 12.19 presents two density scale histograms of the prizes, in dollars, for matching four numbers for a total of $n = 153$ drawings. (The numbers were provided by the Multi-State Lottery Association for the dates February 8, 1989, through July 25, 1990, inclusive.) All prizes are awarded in whole dollar amounts. Use these two histograms to answer the following questions:

(a) Describe the shape of each histogram.
(b) Are there any outliers in the data set? Which histogram is more useful for answering this question? Why?

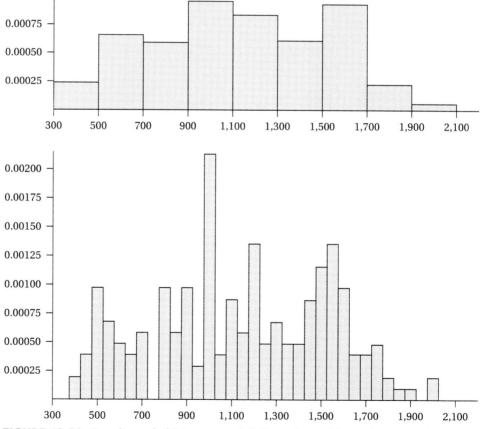

FIGURE 12.20 *Density scale histograms of dollar values of 208 prizes on "The Price Is Right" television game show.*

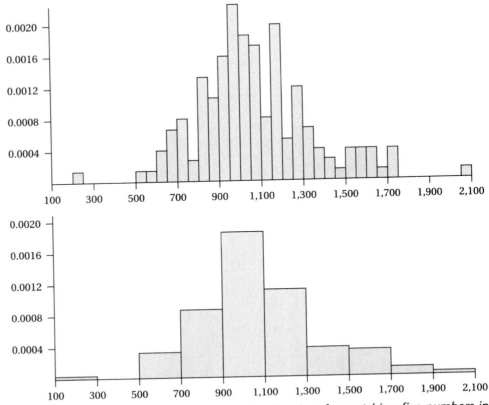

FIGURE 12.21 Density scale histograms of cash prizes for matching five numbers in Lotto America.

(c) What is the most frequent dollar value prize in this data set?

(d) Use the bottom picture to obtain a quick approximation to the proportion of cash prizes between $34 and $45, inclusive.

29. Figure 12.20 on p. 411 presents two density scale histograms for the dollar value of 208 prizes given on the television game show "The Price Is Right." Write a few sentences that summarize what the two histograms reveal, including a comparison of the utility of each histogram.

30. Refer to Exercise 28. Figure 12.21 presents two density scale histograms of the prizes for matching five numbers for a total of $n = 152$ drawings. (One of the numbers provided by the Multi-State Lottery Association was $29; this value, which I believe to be a mistake, has been deleted. This fact accounts for the sample size being one smaller for this exercise.) Use these two histograms to answer the following questions:

(a) Describe the shape of each histogram.

(b) Are there any outliers in the data set? Which histogram is more useful for answering this question? Why?

(c) What is the most frequent dollar-value prize in this data set?

(d) Use the bottom picture to obtain a quick approximation to the proportion of cash prizes between 900 and 1,099 dollars, inclusive.

(e) Which of the numbers 1,029.5 and 1,068.5 is the sample mean, and which is the sample median? Explain your answer.

31. Below is a set of 28 observations:

−3	−3	−3	−3	−2	−2	−2
−2	−1	−1	−1	−1	0	0
0	0	1	1	1	1	2
2	2	2	3	3	3	3

(a) Construct a dot plot of these data.

(b) A stem and leaf plot (arbitrarily and, quite frankly, ridiculously) defines one-half of the zeros in a data set to be +0 and the other half to be −0. A resulting stem and leaf plot for these data is

−0	33332222
−0	111100
0	001111
0	22223333

Compare this plot to the dot plot and comment.

32. Table 12.7 presents a sorted list of the sales price, in thousands of dollars, of 225 property transfers in Madison, Wisconsin, during February and March, 1991.

(a) Construct a density scale histogram for these data with class intervals 0–50, 50–100, 100–150, 150–200, 200–250, 250–300, 300–350, 350–400, 400–450, 450–500, and 500–550.

(b) Construct a density scale histogram for these data with class intervals 0–50, 50–60, 60–70, 70–80, 80–90, 90–100, 100–120, 120–150, 150–200, 200–350, and 350–550.

(c) Which histogram is more informative? Why?

TABLE 12.7 Sales price, in thousands of dollars, of 225 property transfers in Madison, Wisconsin, during February and March, 1991

17.3	18.9	19.9	19.9	20.0	20.0	20.0	21.0	22.0	25.0
25.5	25.7	25.9	26.9	28.4	30.3	31.4	31.7	32.0	34.5
34.5	34.9	35.0	35.0	36.2	39.0	39.5	42.9	43.0	44.6
45.0	45.9	46.0	47.7	49.0	49.0	49.8	49.9	50.0	50.5
51.0	52.0	52.0	53.5	53.5	53.6	53.9	54.0	54.9	55.0
55.0	55.0	55.5	55.5	55.9	56.0	57.0	58.0	58.0	58.5
59.0	59.4	59.9	59.9	60.0	60.0	60.0	60.0	60.5	61.0
61.0	61.0	62.0	62.2	63.6	63.9	64.0	64.5	64.9	65.0
65.0	65.0	65.0	66.0	67.0	67.0	67.5	68.0	68.2	68.5
68.6	69.0	69.5	70.0	70.0	70.0	70.4	71.0	71.9	72.0
72.0	72.0	72.3	72.9	73.0	73.0	73.7	73.8	74.0	74.0
74.5	74.9	75.0	75.5	75.7	75.9	76.0	76.6	77.0	77.5
78.0	78.5	78.5	78.9	79.5	79.9	80.0	80.0	80.0	80.5
81.3	81.5	81.9	82.0	82.0	82.0	83.9	84.0	84.4	84.6
85.9	86.0	86.0	87.0	87.0	87.0	88.4	88.5	89.0	89.9
90.0	90.3	91.9	92.0	92.0	92.3	92.5	92.9	93.0	93.5
94.3	94.5	94.5	94.5	95.0	95.0	95.0	96.1	96.4	98.0
103.0	103.4	104.0	104.0	104.0	104.9	105.0	105.4	107.0	109.0
109.9	110.0	112.9	116.8	119.5	121.0	121.0	121.0	122.8	123.0
123.7	124.9	125.0	125.0	125.0	125.0	125.5	129.3	129.5	129.8
129.9	131.7	132.0	138.0	138.8	145.0	146.0	147.5	150.0	155.0
159.9	160.0	162.9	165.0	171.8	178.3	184.2	186.2	210.0	235.7
240.0	280.0	318.8	320.3	514.0					

(d) Find the sample median. Given that the sum of the 225 sales prices is equal to 19,313, find the sample mean.

(e) Explain why the sample mean is larger than the sample median.

(f) Delete the largest value from the data set and recompute the sample median and sample mean. Comment.

12.4 MEASURES OF SPREAD

Figure 12.22 presents dot plots of the daily high temperatures in Madison, Wisconsin, in January and June 1988. Even a quick examination of this figure shows that the highs were more variable in January than in June. In other words, the January highs exhibit greater spread. This section examines three ways to measure the spread in a distribution: the sample range, interquartile range, and standard deviation. Statisticians have found spread to be an inherently more difficult concept than center; as discussed in this section, each measure of spread has some desirable properties, but each also has some serious shortcomings.

The **sample range** is denoted by R and is equal to the largest observation minus the smallest observation. For example, for the January highs,

$$R = 46 - (-4) = 50 \text{ degrees},$$

and for the June highs,

$$R = 101 - 70 = 31 \text{ degrees}.$$

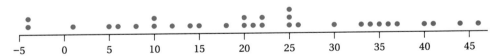

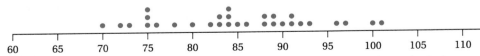

FIGURE 12.22 Dot plots of daily high temperatures in Madison, Wisconsin, in January and June, 1988.

The main advantages of the range are that it is easy to compute and interpret. The main disadvantages of the range are that it ignores the dispersion of all but the two extreme observations and it is influenced tremendously by even one wild observation (see Exercise 19).

The sample interquartile range keeps the advantages of the range while lessening its weaknesses. In words, the interquartile range equals the range of the middle half of the data. Thus, it is easy to compute and interpret. Since the interquartile range concentrates on the middle half of the data, it is not influenced by a small number of wild observations, but it does ignore one-half of the data.

Statisticians have a number of different definitions of the middle half of the data, and these lead to a variety of ways to compute the interquartile range. The method used in this book is particularly easy for hand computation. You should be aware, however, that a computer package may give a somewhat different answer.

Below is a sorted list of the Madison daily high temperatures for January, 1988:

−4	−4	1	5	6	8	10	10	12	14
15	18	20	20	21	22	22	25	25	25
26	30	33	34	35	36	37	40	41	44
46									

The median is the 16th observation in the sorted list; namely, 22 degrees. The first 15 values in the sorted list are called the lower half and the last 15 values the upper half of the data. Note that if the sample size is an odd number, the center observation is not included in either the lower or the upper half of the data set. The first quartile, denoted Q_1, is defined as the median of the lower half of the data. The third quartile, denoted Q_3, is defined as the median of the upper half of the data. You should verify that for the January data,

$$Q_1 = 10 \text{ degrees} \quad \text{and} \quad Q_3 = 34 \text{ degrees}.$$

The **interquartile range** is denoted by IQR and is defined to equal the third quartile minus the first quartile:

$$\text{IQR} = Q_3 - Q_1.$$

For the January data,

$$\text{IQR} = Q_3 - Q_1 = 34 - 10 = 24 \text{ degrees}.$$

The sorted list of June highs was presented on p. 399. The first 15 values in the sorted list form the lower half and the last 15 values form the upper half of the data. Note that if the sample size is an even number, then each observation belongs to either the lower or upper half of the data. You should verify that for the June data

$$Q_1 = 78, \quad Q_3 = 91, \quad \text{and} \quad IQR = 13 \text{ degrees.}$$

Approximately one-half—the center half—of the data fall between the first and third quartiles. The interquartile range measures the distance between the top and bottom values—the range—of the middle half of the data. The interquartile range of the January data is almost twice as large as the interquartile range of the June data, because of the much greater spread in the middle half of the January data.

It is helpful to remember the following simple interpretation of the sample quartiles:

- About one-quarter of the data are smaller than the first sample quartile.
- About one-quarter of the data are between the first and second sample quartiles.
- About one-quarter of the data are between the second and third sample quartiles.
- About one-quarter of the data are larger than the third sample quartile.
- About one-half of the data, the middle half, are between the first and third sample quartiles.

Before the sample standard deviation is introduced, it is useful to digress to a new visual presentation of a distribution, called the **box plot.**

12.4.1 The Box Plot

Table 12.8 presents a number of summary statistics of the daily high temperatures in Madison for each of the 12 months of 1988. The third column of the table lists the smallest (minimum) high temperature for each month, and the seventh column lists the largest (maximum) high. The headings on the other columns are familiar. In addition, the information in this table, except for the sample sizes, can be displayed visually by drawing 12 box plots, as shown in Figure 12.23. A box plot consists of

- A number line for reference
- A rectangle, called the box
- A vertical line segment within the box
- Two horizontal line segments, called whiskers, emanating from the midpoints of the left and right ends of the box

The left end of the box is located at the first quartile, and the right end of the box is located at the third quartile. Thus, the width of the box equals the interquartile

TABLE 12.8 Selected Summary Statistics of the Daily High Temperatures in Madison, Wisconsin, for the 12 Months of 1988

Month	n	Min	Q_1	$\tilde{x}$	Q_3	Max	IQR	R
January	31	−4	10.0	22.0	34.0	46	24.0	50
February	29	2	17.0	24.0	35.5	51	18.5	49
March	31	25	38.0	45.0	53.0	59	15.0	34
April	30	41	51.0	58.0	68.0	78	17.0	37
May	31	58	72.0	78.0	81.0	89	9.0	31
June	30	70	78.0	84.5	91.0	101	13.0	31
July	31	75	83.0	87.0	93.0	100	10.0	25
August	31	71	79.0	84.0	95.0	102	16.0	31
September	30	57	71.0	77.0	83.0	89	12.0	32
October	31	39	48.0	54.0	63.0	74	15.0	35
November	30	29	40.0	46.5	52.0	62	12.0	33
December	31	14	22.0	32.0	43.0	53	21.0	39

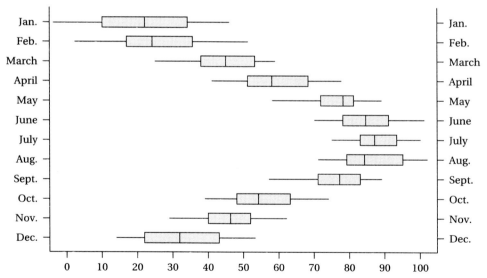

FIGURE 12.23 Box plots of daily high temperatures in Madison, Wisconsin, for the 12 months of 1988.

range. The height of the box has no meaning; simply choose a height that gives an aesthetically pleasing picture. The vertical line segment in the box is located at the sample median. Finally, the left whisker extends to the minimum value in the data set, and the right whisker extends to the maximum value.

With practice, it is easy to extract information from a box plot. For example, the vertical line segment divides the January box into congruent rectangles; thus, the distances from the median to the quartiles are identical. In a rough sense this suggests that the middle half of the January data is symmetric. By contrast, for the February data the median is much closer to the first quartile than the third, suggesting that the middle half of the data is skewed to the right. The lengths of the whiskers give information on how far the tails of the distribution extend. The September data have a much longer lower tail than upper tail, showing that the cold days in September (the lowest quarter of the data) were much more variable than the hot days (the upper quarter). A long whisker suggests that there might be outliers in the data. Further inspection of the September data shows that there were two especially cold days with highs of 57 degrees each, compared to the next lowest value of 64 degrees.

The real strength of a box plot, however, is that it helps a researcher obtain a quick comparison of several distributions. Figure 12.23 allows a researcher to detect easily the monthly patterns in

- The median high temperature, revealing the strong seasonality
- The IQR of highs, revealing larger values (above 20°) in December and January, smaller values (below 14°) from May to July, and moderate values the remainder of the year with the exception of smaller values in September and November.

12.4.2 The Sample Standard Deviation

Recall that the population standard deviation was defined in Chapter 3 and was shown to be a measure of the spread in a sampling distribution.

There are four basic ideas underlying the formula for the **sample standard deviation.** The first idea is that the spread in a data set can be measured by computing how far the values lie from the center, where the center is measured by the sample mean. In particular, to each observation, x, in a data set is associated the deviation

$$x - \bar{x}.$$

Consider the high temperatures for June presented in Figure 12.22. It was stated earlier that the sample mean of these values is 85 degrees. Thus, for each value x in this data set, the deviation of x is $x - 85$. For example, the observation $x = 73$ has a deviation of $73 - 85 = -12$. Geometrically, this means that the observation $x = 73$ lies 12 units (degrees) to the left of the mean. Similarly, the observation $x = 93$ has a deviation of $93 - 85 = 8$ degrees. In words, the observation $x = 93$

lies 8 units to the right of the mean. These deviations are presented graphically in Figure 12.24. Below are the 30 deviations for the June data; these numbers were obtained by subtracting 85 ($\bar{x}$) from each of the June observations.

−15	−13	−12	−10	−10	−10	−9	−7	−5	−3
−2	−1	−1	−1	−1	0	1	2	3	4
4	5	6	6	7	8	11	12	15	16

These deviations sum to 0. In fact, the sum of the deviations of any set of data equals 0. Thus, if an analyst knows all deviations but one, the remaining deviation is determined. Statisticians summarize this fact by saying that only $n - 1$ of the n deviations are free to vary, or, more succinctly, that the deviations have $n - 1$ degrees of freedom. (You may have noticed that this use of "degrees of freedom" is analogous to the use in Chapters 10 and 11.)

The sample standard deviation provides a summary of the deviations. The second basic idea alluded to earlier is that deviations of, for example, +5 and −5 should contribute the same amount to the value of the sample standard deviation. The rationale behind this idea is that the observations leading to these deviations are five units from the center and the direction from the center (left for negative deviations or right for positive deviations) is not important. Of course, there is nothing special about the number 5 in this argument; similarly, +2 and −2 contribute the same amount to the value of the standard deviation—but a different amount than 5 contributes. (In mathematical jargon, the standard deviation is a function of the absolute value of the deviations.)

The third basic idea is to *square* each deviation; note that this accords with the second idea since the squares of +5 and −5 are equal, the squares of +2 and −2 are equal, and so on. Symbolically, a squared deviation is given by $(x - \bar{x})^2$.

The sample variance is denoted by s^2 and is defined to equal the sum of the squared deviations divided by the number of degrees of freedom. In symbols,

$$s^2 = \frac{\sum (x - \bar{x})^2}{n - 1}. \tag{12.1}$$

For example, it can be shown that the sample variances for the January and June highs are 170.24 and 69.38, respectively.

The value of the sample variance is difficult to interpret because it is measured in squared units. For example, the high temperatures are measured in degrees and the deviations are measured in degrees, but the squared deviations are

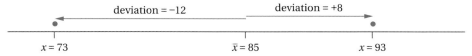

FIGURE 12.24 Two deviations for the June high temperatures.

measured in squared degrees. The fourth basic idea is to transform back to the original units by taking the square root of the sample variance to obtain the sample standard deviation. The sample variance is denoted by s^2, so naturally the sample standard deviation is denoted by s. In symbols,

$$s = \sqrt{s^2} = \sqrt{\frac{\sum (x - \bar{x})^2}{n - 1}}. \tag{12.2}$$

For the January and June highs,

$$s = \sqrt{s^2} = \sqrt{170.24} = 13.05 \text{ degrees} \quad \text{and} \quad s = \sqrt{69.38} = 8.33 \text{ degrees}.$$

It is very tedious to compute the sample variance and sample standard deviation by hand using Equations 12.1 and 12.2. In practice, a researcher typically uses a calculator or computer to obtain these values. In the remainder of this chapter and this book, the values of the sample variance or sample standard deviation will be given so that we can focus on interpreting the values.

The sample standard deviation has two important characteristics:

- Statisticians have developed a number of formulas for inferential procedures for the mean of one or more populations. The evaluation of each of these formulas requires the value of the sample standard deviation.
- If the data have a bell-shaped distribution, the sample mean and sample standard deviation provide a good approximation to the entire set of data.

Chapters 15 and 16 will demonstrate the importance of the sample standard deviation in inference. The second characteristic is illustrated by the following example.

EXAMPLE 12.1

Each day the *Wisconsin State Journal*, a morning newspaper, publishes its forecast, denoted by f, of the day's high temperature in Madison, Wisconsin. Let h denote the actual high temperature in Madison for the day. Both f and h are reported as integers in degrees Fahrenheit. The forecast error, denoted by x, is given by the equation

$$x = f - h.$$

For example, if on a particular day the forecasted high is 63 and the actual high is 60, the forecast error for that day is

$$x = f - h = 63 - 60 = 3 \text{ degrees}.$$

Table 12.9 is the distribution of 400 forecast errors for the period from August 16, 1987, to September 30, 1988. Figure 12.25 is a density scale histogram of these 400 forecast errors. The histogram reveals two small outliers, -17 and -19, and no large outliers. In addition, the distribution is approximately bell-

TABLE 12.9 Distribution of 400 High-Temperature Forecast Errors

Value	Frequency	Value	Frequency	Value	Frequency	Value	Frequency
−19	1	−6	11	1	40	8	7
−17	1	−5	23	2	29	9	3
−12	1	−4	27	3	28	10	2
−10	5	−3	34	4	15	11	1
−9	2	−2	44	5	10		
−8	6	−1	41	6	7	Total	400
−7	11	0	47	7	4		

shaped. The sample mean and sample standard deviation of the 400 forecast errors are

$$\bar{x} = -0.712 \quad \text{and} \quad s = 4.024.$$

Suppose a researcher is interested in some characteristic of the 400 observations; for example, the proportion of forecast errors between −2 and +2. The exact answer is available from Table 12.9. The values −2, −1, 0, 1, and 2 have frequencies 44, 41, 47, 40, and 29, respectively. Thus, the number of observations between −2 and +2 is

$$44 + 41 + 47 + 40 + 29 = 201.$$

Equivalently, the proportion of forecast errors between −2 and +2 is

$$\frac{201}{400} = 0.5025.$$

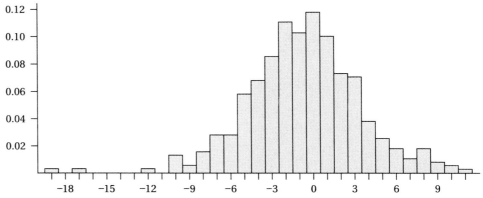

FIGURE 12.25 Density scale histogram of 400 high-temperature forecast errors in Madison, Wisconsin.

This proportion also can be obtained by computing the area under the density scale histogram between −2.5 and +2.5. (Note that the rectangle centered at 2 extends up to 2.5, halfway between 2 and the next larger possible value of the response, 3. Similarly, the rectangle centered at −2 extends down to −2.5, halfway between −2 and the next smaller possible value of the response, −3.)

Because the proportion can be obtained by computing an area and the density scale histogram is approximately bell-shaped, the standard normal curve can be used to obtain an approximate proportion. This is analogous to using the standard normal curve to obtain an approximate P-value or a confidence interval, as illustrated numerous times earlier in this text. Proceeding as before, the boundary values −2.5 and +2.5 are standardized by subtracting the sample mean and dividing by the sample standard deviation:

$$-2.5 \to \frac{-2.5 - (-0.712)}{4.024} = -0.44.$$

$$2.5 \to \frac{2.5 - (-0.712)}{4.024} = 0.80.$$

Thus, the original question,

What proportion of forecast errors fall between −2 and +2?

is equivalent to

What proportion of standardized values of forecast errors fall between −0.44 and 0.80?

An approximate answer to this last question is obtained by finding the area under the standard normal curve between −0.44 and 0.80. In the method developed in Chapter 3, this area is obtained by looking up these two numbers in Table A.2 and subtracting the smaller number found from the larger number found. In the current case, the area is 0.6700 − 0.2119 = 0.4581. This approximate answer differs from the exact proportion by 0.4581 − 0.5025 = −0.0444. Exercise 22 shows that the error in the approximation is largely attributable to the two outliers. Changing the values −2 and +2 to −2.5 and +2.5 before standardizing is referred to as a **continuity correction.** ▲

The previous example can be summarized as follows:

Exact answers to questions about a data set can be obtained only by examining the entire data set. If the sample size is large, this determination can be tedious; if the data are not accessible, this determination can be impossible. If the data have a bell-shaped distribution, approximate answers can be obtained quite easily by using only the sample mean, sample standard deviation, and a table of standard normal curve areas.

An abbreviated version of this result is referred to as the **empirical rule for interpreting the value of s:**

If the distribution of the data is bell-shaped, then

1. Approximately 68 percent of the observations are between the numbers
$$\bar{x} - s \quad \text{and} \quad \bar{x} + s.$$

2. Approximately 95 percent of the observations are between the numbers
$$\bar{x} - 2s \quad \text{and} \quad \bar{x} + 2s.$$

3. Approximately 99.7 percent of the observations are between the numbers
$$\bar{x} - 3s \quad \text{and} \quad \bar{x} + 3s.$$

The empirical rule will be illustrated for the temperature forecast data. According to the empirical rule, about 68 percent of the 400 observations should be between
$$\bar{x} - s = -0.712 - 4.024 = -4.736$$
and
$$\bar{x} + s = -0.712 + 4.024 = +3.312.$$

The actual count is 290, which is 72.5 percent of the 400 observations. Note that a continuity correction is not used with the empirical rule.

The empirical rule has hidden uses. For example, suppose you read that a random sample of 500 college women has a mean height of 65 inches, a standard deviation of 9 inches, and approximately a bell-shaped distribution. The value of the mean looks plausible, but the value for s must be incorrect. If it were correct, the empirical rule would state that approximately 68 percent of the women have heights between
$$\bar{x} - s = 65 - 9 = 56 \quad \text{and} \quad \bar{x} + s = 65 + 9 = 74 \text{ inches.}$$

Thus the remaining 32 percent of the women are either shorter than 56 inches or taller than 74 inches. Heights of women are not that variable! As another example, suppose the report had stated that the sample mean is 65 inches and the sample standard deviation is 0.5 inch. This would imply too little variability in the data. The assumption of a random sample in these two scenarios is crucial. The value $s = 9$ is possible if a researcher deliberately oversamples extremely short and extremely tall women; the value $s = 0.5$ is possible if the researcher only measures women who appear to be approximately 65 inches tall.

The empirical rule also may be used to rule out a bell-shaped distribution. Suppose a report states that the mean number of cats in a sample (randomness is not important for this example) of households is 2.0 with a sample standard deviation of 3.0. (These are hypothetical numbers!) It can be concluded, by contradiction, that the distribution of cats per household is not bell-shaped. To see this, assume that the distribution is bell-shaped. Then about 68 percent of the households would have between

$$\bar{x} - s = 2 - 3 = -1 \quad \text{and} \quad \bar{x} + s = 2 + 3 = 5 \text{ cats.}$$

About one-half (by the symmetry of the bell-shaped curve) of the remaining households would have fewer than -1 cats. Clearly that is not possible. Note that this argument is not restricted to the variable *number of cats*; it applies to any variable that must be nonnegative. For nonnegative data, unless the sample mean is at least twice the value of the sample standard deviation, an approximately bell-shaped curve is not possible.

The empirical rule states that approximately 99.7 percent of the observations fall between $\bar{x} - 3s$ and $\bar{x} + 3s$. If the sample size, n, is smaller than 1,000, approximately all of the observations fall between these limits. This approximation can be used to obtain a quick guess of the value of the sample standard deviation, as demonstrated in the following example.

EXAMPLE 12.2

Consider the forecast error data presented in this section. The observations fall in the interval $[-19, 11]$. The length of this interval is 30. By the empirical rule approximately all of the observations fall in the interval $[\bar{x} - 3s, \bar{x} + 3s]$, which has a length of $6s$. Thus, $6s \approx 30$ or, after algebra, $s \approx \frac{30}{6} = 5$. It was stated earlier that $s = 4.024$; the approximation is poor because of the two outliers. If the two outliers are deleted, the distance between the smallest observation, $-12°$, and the largest, $11°$, is $23°$. It can be shown that the sample standard deviation of the remaining 398 observations is $s = 3.826$. By the above argument, $6s \approx 23$ or, after algebra, $s \approx \frac{23}{6} = 3.833$. This is an excellent approximation to $s = 3.826$. ▲

It is important to avoid becoming overzealous in the use of the standard normal curve and the empirical rule approximations. Remember that they should be used only if the data have approximately a bell-shaped distribution. For example, Figure 12.17 (p. 397) is a density scale histogram of the number of career games for a random sample of 200 major league baseball players. This histogram is not bell-shaped; it is strongly skewed to the right. It can be shown that the sample mean of the 200 values is 354.0 and the sample standard deviation is 560.4. The list of the ordered values in Table 12.6 (p. 394) can be used to verify the entries in the second column of Table 12.10. According to the

TABLE 12.10 Inappropriate Application of the Empirical Rule to the Number of Career Games for a Random Sample of 200 Baseball Players

Interval	Number of Observations	Proportion of Observations	Empirical Rule Approximation
$\bar{x} \pm s = [-206.4, 914.4]$	175	0.875	0.680
$\bar{x} \pm 2s = [-766.8, 1474.8]$	192	0.960	0.950
$\bar{x} \pm 3s = [-1327.2, 2035.2]$	194	0.970	0.997

This equation for the sample mean can be rewritten to yield

$$\sum x = n\bar{x}. \tag{12.3}$$

In words,

> The total (of the observations) is equal to the product of the sample size and the sample mean.

For instance, suppose it is known that a sample of size $n = 10$ has a sample mean $\bar{x} = 13$; it follows that the total of the observations is

$$\sum x = 10(13) = 130.$$

The utility of this result is illustrated in the following example.

EXAMPLE 12.3

Consider again the high temperatures in Madison, Wisconsin, in 1988. The histograms for spring and autumn (not shown) are remarkably similar, suggesting that it might be reasonable to combine these seasons (for the purpose of examining high temperatures only!). The 91 spring observations have a sample mean of 59.19, and the 90 autumn observations have a sample mean of 59.34.

An application of Equation 12.3 gives

$$\text{Total of spring highs} = 91(59.19) = 5{,}386.$$

(The highs were measured to the nearest degree; thus, each observation and the total must be an integer.) Another application of this equation gives

$$\text{Total of autumn highs} = 90(59.34) = 5{,}341.$$

Thus, the total of the highs for all 181 days is $5{,}386 + 5{,}341 = 10{,}727$, and the sample mean of the 181 days is $10{,}727/181 = 59.27$. The method used to answer the question of this example is presented, in general terms, below. ▲

Suppose an analyst has two samples of data. The sample size of the first is denoted by n_1 and its sample mean by $\bar{x}_1$; the sample size of the second is denoted by n_2 and its sample mean by $\bar{x}_2$. If the two samples are combined into a single sample, the mean of the combined sample, denoted by $\bar{x}_c$ (c is for *combined*), can be computed with the following formula:

$$\bar{x}_c = \frac{n_1 \bar{x}_1 + n_2 \bar{x}_2}{n_1 + n_2}. \tag{12.4}$$

If interested, you may prove the validity of Equation 12.4. It will now be shown that this equation gives the correct answer for the high-temperature example above. In the previous example

$$n_1 = 91, \quad \bar{x}_1 = 59.19, \quad n_2 = 90, \quad \text{and} \quad \bar{x}_2 = 59.34.$$

Substituting these values into Equation 12.4 gives the mean of the combined sample,

$$\bar{x}_c = \frac{91(59.19) + 90(59.34)}{91 + 90} = \frac{5{,}386 + 5{,}341}{181} = 59.27,$$

agreeing with the earlier answer.

Equation 12.4 can be extended to the problem of combining three or more samples; the details will not be given. Equation 12.4, however, *cannot* be extended to sample medians. The median of the combined sample cannot be obtained by combining, in any manner, the medians of the original two samples. This fact is illustrated in Exercise 3 at the end of this section.

12.5.3 Change of Units

Here are the ordered times, in seconds, of ten songs by Samantha Phillips (from the album *The Indescribable Wow*):

$$148 \quad 166 \quad 169 \quad 175 \quad 187 \quad 189 \quad 209 \quad 216 \quad 217 \quad 255$$

If interested, you may verify the following values for selected summary statistics for these data:

$$\bar{x} = 193.1, \; s = 31.47, \; \tilde{x} = 188, \; Q_1 = 169, \; Q_3 = 216, \text{ and IQR} = 47,$$

all measured in seconds. Suppose, however, that the songs are measured in minutes instead of seconds. In other words, suppose that each of the 10 numbers above is divided by 60 (to convert from seconds to minutes). The resulting ten observations are

$$2.47 \quad 2.77 \quad 2.82 \quad 2.92 \quad 3.12 \quad 3.15 \quad 3.48 \quad 3.60 \quad 3.62 \quad 4.25$$

A useful mathematical fact is that any of the earlier (six) summary statistics for the data measured in minutes can be obtained by taking the corresponding summary statistic for the data measured in seconds and dividing by 60. For example,

$$\text{Sample mean for data in minutes} = \frac{\text{Sample mean for data in seconds}}{60}.$$

For the Samantha Phillips data, this gives

$$\text{Sample mean for data in minutes} = \frac{193.1}{60} = 3.22.$$

This is a special case of the following fact:

Let

$$x_1, x_2, \ldots, x_n$$

be any set of data. Define a new set of data

$$y_1, y_2, \ldots, y_n$$

by the rule

$$y_1 = c + dx_1, \; y_2 = c + dx_2, \; \ldots, \; y_n = c + dx_n, \qquad (12.5)$$

where c is any number and d is any *positive* number. Let s_x, $Q_1(x)$, $Q_3(x)$, and IQR(x) denote the sample standard deviation, sample first and third quartiles, and sample interquartile range, respectively, of $x_1, x_2, \ldots, x_n$. Define s_y, $Q_1(y)$, $Q_3(y)$, and IQR(y) in the analogous way. The following equations are true:

$$\bar{y} = c + d\bar{x}. \qquad (12.6)$$

$$\tilde{y} = c + d\tilde{x}. \qquad (12.7)$$

$$Q_1(y) = c + dQ_1(x). \qquad (12.8)$$

$$Q_3(y) = c + dQ_3(x). \qquad (12.9)$$

$$s_y = ds_x. \qquad (12.10)$$

$$\text{IQR}(y) = d\,\text{IQR}(x). \qquad (12.11)$$

This result looks a bit formidable, so it is illustrated with the Samantha Phillips data. First, one must determine the rule relating the x's and y's. The x's are measured in seconds and the y's are measured in minutes, so the rule is

$$y_1 = \frac{x_1}{60}, \; y_2 = \frac{x_2}{60}, \; \ldots, \; y_n = \frac{x_n}{60}.$$

This fits the form of Equation 12.5 with the identifications

$$c = 0 \quad \text{and} \quad d = \frac{1}{60}.$$

Thus, the equations yield the following:

$$\bar{y} = c + d\bar{x} = 0 + \frac{193.1}{60} = 3.22.$$

$$\tilde{y} = c + d\tilde{x} = 0 + \frac{188}{60} = 3.13.$$

$$Q_1(y) = c + dQ_1(x) = 0 + \frac{169}{60} = 2.82.$$

$$Q_3(y) = c + dQ_3(x) = 0 + \frac{216}{60} = 3.60.$$

$$s_y = ds_x = \frac{31.47}{60} = 0.5245.$$

$$\text{IQR}(y) = d\text{IQR}(x) = \frac{47}{60} = 0.78.$$

As another example, suppose a researcher has a number of temperatures for which the sample mean and sample standard deviation are 20 and 10 degrees Celsius, respectively. Next, suppose the researcher wants to know the sample mean and standard deviation in degrees Fahrenheit. The mean and standard deviation are known for degrees Celsius, so the x's represent the measurements in degrees Celsius and the y's the measurements in degrees Fahrenheit. The relationship between x and y is

$$y = 32 + 1.8x;$$

thus, $c = 32$ and $d = 1.8$ in Equations 12.5. Substituting these values in Equations 12.6 and 12.10 gives

$$\bar{y} = c + d\bar{x} = 32 + 1.8(20) = 32 + 36 = 68$$

and

$$s_y = ds_x = 1.8(10) = 18.$$

12.5.4 Shortcut Formulas

If the sample size is large relative to the number of distinct values in the data set, there is a useful shortcut method for computing the sample mean and sample standard deviation. The formulas are demonstrated for data from the 1982 Wisconsin Driver Survey.

Question 3 on the survey read

If a person drives while drunk, what is his/her chance of being stopped by the police?

The relative frequency distribution for the response to this question for 1982 is given in Table 12.15. This ordered categorical variable can be made numerical by assigning the values 1, 2, 3, 4, and 5 to the ordered categories "very low," "low," "about 50:50," "high," and "very high," respectively. With these assignments,

TABLE 12.15   Responses to Question 3 (1982 Data)

Response	Frequency	Relative Frequency
Very low	160	0.105
Low	407	0.266
About 50:50	699	0.457
high	172	0.112
Very high	92	0.060
Total	1,530	1.000

the values of the sample mean and sample standard deviation can be obtained from the computations given in Table 12.16. The first column of the table lists the distinct values of the response (x), the second column lists the relative frequencies (r) of each response, the third column lists the row-by-row products of the entries in the first two columns (xr), and the fourth column lists the row-by-row products of the entries in the first and third columns ($x^2 r$). It can be shown that

$$\bar{x} = \sum xr \quad \text{and} \quad s^2 = \frac{n}{n-1}\sum x^2 r - \frac{n}{n-1}\bar{x}^2. \tag{12.12}$$

(If interested, you may verify algebraically the validity of these equations.) Equations 12.12 and the computations in Table 12.16 can be combined to obtain the following values for the sample mean and sample variance:

$$\bar{x} = \sum xr = 2.756$$

and

$$s^2 = \frac{n}{n-1}\sum x^2 r - \frac{n}{n-1}\bar{x}^2 = 8.580 - 7.601 = 0.979.$$

The sample standard deviation can now be obtained easily:

$$s = \sqrt{s^2} = \sqrt{0.979} = 0.989.$$

A consequence of summarizing the distribution of responses with the mean is that a response of 1, "very low," and a response of 3, "about 50:50," are interpreted as equivalent to two responses of 2, "low." This may or may not be desirable, depending on the application. For example, consider the outcome of a serious illness as a variable of interest. Suppose the ordered categorical outcomes are recovery in two months or less, recovery in more than two months, and death. Clearly, assigning the values 1, 2, and 3 and then averaging would not be appropriate!

TABLE 12.16 Computations Needed to Obtain $\bar{x}$ and s

Value (x)	Relative Frequency (r)	xr	$x^2 r$
1	0.105	0.105	0.105
2	0.266	0.532	1.064
3	0.457	1.371	4.113
4	0.112	0.448	1.792
5	0.060	0.300	1.500
Total	1.000	2.756	8.574

EXERCISES 12.5

1. Construct a dot plot of the following data set of $n = 26$ observations:

   1 1 1 1 1 1 2 2 2 2 2 3
   3 3 3 4 4 4 5 5 5 6 6 7

   What is the shape of your plot? Based on the shape, which do you believe is larger, the sample mean or the sample median? Compute the sample mean and the sample median and comment.

2. Table 12.6 (p. 394) lists the number of career games for a random sample of 200 baseball players. The sample mean of these 200 numbers is 354.0. A second random sample is selected from the population of baseball pitchers. For the sample of pitchers the mean number of games played is 83.65.

   (a) Compute the sample mean number of games played by the 300 men sampled.
   (b) Is it reasonable to compare pitchers and players on games played?
   (c) Is it reasonable to assume that the 300 men studied are a random sample from the population of all men who have played in the major leagues?

3. Consider the following three data sets:

   $A: 1, 3, 6;$ $B: 0, 5, 7;$ and $C: 4, 5, 6.$

   The sample median of A is 3, and the sample medians of both B and C are 5.

   (a) Find the sample median of the data set obtained by combining sets A and B.
   (b) Find the sample median of the data set obtained by combining sets A and C.
   (c) In view of the previous two answers, discuss why a result similar to Equation 12.4 cannot be obtained for the sample median.

4. The lengths of sixteen songs by Dire Straits have the following summary statistics (in seconds):

   $\bar{x} = 321.1,$ $\tilde{x} = 299.5,$ and $s = 86.0.$

   Find the values these summary statistics would have if the song lengths were measured in minutes instead of seconds.

5. Comment on the following reasoning:

   The mean of 400 same-day high-temperature forecast errors is -0.712 degrees Fahrenheit. If the errors are converted to degrees Celsius, the mean is

   $$\frac{5}{9}(-0.712 - 32) = -18.173.$$

6. Compute the sample mean and sample standard deviation for the sample of $n = 1{,}000$ observations described in the following table:

Value	Frequency
1	143
2	192
3	322
4	207
5	136
Total	1,000

REFERENCE

1. J. L. Reichler, *The Baseball Encyclopedia* (New York: Macmillan, 1988).

Correlation and Regression

13.1 **INTRODUCTION**

13.2 **THE SCATTERPLOT AND THE CORRELATION COEFFICIENT**

13.3 **THE REGRESSION LINE**

13.4 **RESIDUALS, OUTLIERS, AND ISOLATED CASES**

13.5 **DATA FROM TWO SOURCES***

Chapter 13

13.1 INTRODUCTION

This chapter introduces methods for describing data sets that consist of two numerical variables per subject. The main topics are the scatterplot, the correlation coefficient, the principle of least squares, and the regression line. Before these topics are introduced, however, some of the elementary properties of straight lines that are used in this chapter are presented.

Mathematicians study **deterministic,** or **functional,** relationships between two variables. More precisely, let y and x denote the variables. The variable y is said to be deterministically, or functionally, related to x if the value of x determines the exact value of y. The functional relationship is usually expressed as an equation, as in the following examples:

$$y = 5 + 4x$$
$$y = 10 - 2x$$
$$y = 3(10)^x$$
$$y = 3 + 2x - x^2$$

It is often insightful to draw a picture of a function. This is achieved by plotting on a coordinate system all ordered pairs of numbers (x, y) that satisfy the functional equation. The plot of these points is called the **graph** of the function.

The variable y is said to be a **linear function** of x if

$$y = a + bx, \tag{13.1}$$

for some numbers a and b.

The name *linear* comes from the fact that the graph of a linear function is a straight line. (Conversely, if the graph of a function is a straight line, it is a linear function; that is, its equation has the form given in Equation 13.1.) Of the four functions listed above, the first two are linear ($a = 5$ and $b = 4$ in the first; $a = 10$ and $b = -2$ in the second) and the others are not. Only linear functions are considered in the remainder of this section.

The value of y corresponding to a given value of x can be obtained either by using the functional equation or by examining a graph of the function, as illustrated in the following familiar example. It is well known that temperature measured in degrees Celsius can be converted easily to degrees Fahrenheit. The temperature in Celsius is multiplied by 1.8 and then 32 is added; the result is the temperature in degrees Fahrenheit. This rule can be expressed as an equation:

$$y = 32 + 1.8x, \tag{13.2}$$

where x denotes temperature in degrees Celsius and y denotes temperature in degrees Fahrenheit. This equation can be used to obtain the value of y for any given value of x. For example, if the temperature is 10 degrees Celsius ($x = 10$), the temperature in degrees Fahrenheit is

$$y = 32 + 1.8(10) = 32 + 18 = 50.$$

The graph of Equation 13.2 is given in Figure 13.1. This figure also illustrates how to determine graphically that $x = 10$ yields $y = 50$ by applying the following two-step algorithm:

1. The given value of x is located on the horizontal (x) axis. A vertical line segment is drawn from the value of x to the line.

2. From the point of intersection on the line, a horizontal line segment is drawn to the vertical (y) axis. The coordinate on the vertical axis equals the value of y corresponding to the given value of x.

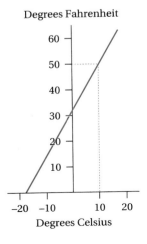

FIGURE 13.1 Graph of $y = 32 + 1.8x$ and the determination that $x = 10$ yields $y = 50$.

If $x = 0$ is substituted into Equation 13.1, the result is
$$y = a + b(0) = a.$$
Thus, the values $x = 0$ and $y = a$ satisfy the functional relationship, implying that the point $(0, a)$ is on the graph of the function. Since this point lies on the y-axis, because $x = 0$, it is called the **y-intercept** of the line.

The number b in Equation 13.1 is called the **slope** of the line; it measures the change in y corresponding to a unit change in x. Consider two values of x that differ by one unit and call the values $(c + 1)$ and c. If $x = c$, then $y = a + bc$. If $x = c + 1$, then
$$y = a + b(c + 1) = a + bc + b.$$
Thus, the result of increasing x by one unit is that y changes by b units. (If b is positive, y increases; if b is negative, y decreases.)

Straight lines are particularly important in mathematics for two reasons:

1. For many real problems, the relationship between y and x is given by a straight line.
2. Straight-line relationships are much easier to study than curved relationships.

EXERCISES 13.1

1. The variables x and y are related by the equation $y = 3 + 5x$. Use this equation to find the value of y that corresponds to $x = 0$; $x = -1$; $x = 2$.

2. The variables x and y are related by the equation $y = 10 - 2x$. Use this equation to find the value of y that corresponds to $x = 3$; $x = 4$; $x = 5$.

3. Refer to Exercise 1.
    (a) Draw the graph of the equation $y = 3 + 5x$.
    (b) Use the graph to find the value of y that corresponds to $x = 0$; $x = -1$; $x = 2$.

4. Refer to Exercise 2.
    (a) Draw the graph of the equation $y = 10 - 2x$.
    (b) Use the graph to find the value of y that corresponds to $x = 3$; $x = 4$; $x = 5$.

13.2 THE SCATTERPLOT AND THE CORRELATION COEFFICIENT

As mentioned in the previous section, mathematicians study deterministic relationships between variables. By contrast, statisticians study relationships that are weaker than deterministic relationships but are still of interest. Fortunately, many of the mathematician's techniques for deterministic relationships can be used by the statistician.

The ideas and methods of this chapter are introduced through a number of examples. For the first example, called the Batting Averages study, the subjects are the 124 baseball players who had 200 or more official at bats during both the 1985 and 1986 American League seasons. Each player's two variables are his batting averages (number of hits divided by official number of at bats) for the two seasons [1]. These data will be examined a number of different ways. The data for either year can be examined individually using the methods of the previous chapter. In addition, this chapter presents methods for studying the relationship between the two batting averages. First, the scatterplot is a visual presentation of the data that helps the researcher see the relationship between the two variables. In particular, the researcher looks for a general pattern to the data and identifies any unusual cases. The correlation coefficient is a number that summarizes the linear relationship between the variables. Finally, in the next section, the least squares regression line is introduced. This line enables the researcher to use the value of one variable to predict the value of the other variable.

The variable the analyst wants to predict is called the **response,** or **dependent variable,** and the variable to be used in making the prediction is called the **predictor,** or **independent variable.** Let Y denote the response variable and X the predictor. For the current example, because of the chronological order of the variables, it is natural to use the 1985 batting average to predict the 1986 batting average. Therefore, Y is a player's 1986 batting average and X is his 1985 batting average. Sample values of X and Y are denoted by lowercase letters, x and y. Each case in the data set provides two numbers, x and y, which are written as an ordered pair, (x, y). For example, Wade Boggs batted $x = 0.368$ in 1985 and $y = 0.357$ in 1986; he gives the ordered pair $(0.368, 0.357)$. Table 13.1 presents the values of x and y for the 124 players in the data set. The name corresponding to each player number in this table is given in the following list (you may skip ahead to the next paragraph without reading these names):

1 Tony Armas; 2 Harold Baines; 3 Dusty Baker; 4 Steve Balboni; 5 Jesse Barfield; 6 Marty Barrett; 7 Don Baylor; 8 George Bell; 9 Juan Beniquez; 10 Tony Bernazard; 11 Bruce Bochte; 12 Wade Boggs; 13 Bob Boone; 14 Phil Bradley; 15 George Brett; 16 Tom Brookens; 17 Tom Brunansky; 18 Bill Buckner; 19 Steve Buechele; 20 Randy Bush; 21 Brett Butler; 22 Joe Carter; 23 Dave Collins; 24 Cecil Cooper; 25 Julio Cruz; 26 Alvin Davis; 27 Mike Davis; 28 Doug DeCinces; 29 Rick Dempsey; 30 Brian Downing; 31 Mike Easler; 32 Darrell Evans; 33 Dwight Evans; 34 Tony Fernandez; 35 Carlton Fisk; 36 Scott Fletcher; 37 Julio Franco; 38 Gary Gaetti; 39 Greg Gagne; 40 Jim Gantner; 41 Damaso Garcia; 42 Rich Gedman; 43 Kirk Gibson; 44 Bobby Grich; 45 Alfredo Griffin; 46 Ozzie Guillen; 47 Toby Harrah; 48 Ron Hassey; 49 Billy Hatcher; 50 Dave Henderson; 51 Rickey Henderson; 52 Larry Herndon; 53 Donnie Hill; 54 Kent Hrbek; 55 Tim Hulett; 56 Garth Iorg; 57 Reggie Jackson; 58 Brook Jacoby; 59 Cliff Johnson; 60 Ruppert Jones; 61 Bob Kearney; 62 Dave Kingman; 63 Ron Kittle; 64 Lee Lacy; 65 Carney Lansford; 66 Rudy Law; 67 Chet Lemon; 68 Fred Lynn; 69 Steve

TABLE 13.1   1985 and 1986 Batting Averages for 124 American League Players

No.	1985	1986	No.	1985	1986	No.	1985	1986	No.	1985	1986
1	0.265	0.264	32	0.248	0.241	63	0.230	0.218	94	0.286	0.252
2	0.309	0.296	33	0.263	0.259	64	0.293	0.287	95	0.282	0.282
3	0.268	0.240	34	0.289	0.310	65	0.277	0.284	96	0.251	0.210
4	0.243	0.229	35	0.238	0.221	66	0.259	0.261	97	0.300	0.233
5	0.289	0.289	36	0.256	0.300	67	0.265	0.251	98	0.219	0.249
6	0.266	0.286	37	0.288	0.306	68	0.263	0.287	99	0.262	0.272
7	0.231	0.238	38	0.246	0.287	69	0.264	0.227	100	0.283	0.228
8	0.275	0.309	39	0.225	0.250	70	0.218	0.254	101	0.228	0.237
9	0.304	0.300	40	0.254	0.274	71	0.324	0.352	102	0.280	0.264
10	0.274	0.301	41	0.282	0.281	72	0.239	0.266	103	0.258	0.246
11	0.295	0.256	42	0.295	0.258	73	0.259	0.252	104	0.257	0.287
12	0.368	0.357	43	0.287	0.268	74	0.297	0.281	105	0.245	0.212
13	0.248	0.222	44	0.242	0.268	75	0.232	0.260	106	0.275	0.326
14	0.300	0.310	45	0.270	0.285	76	0.259	0.253	107	0.251	0.204
15	0.335	0.290	46	0.273	0.250	77	0.222	0.203	108	0.215	0.187
16	0.237	0.270	47	0.270	0.218	78	0.295	0.259	109	0.236	0.229
17	0.242	0.256	48	0.296	0.323	79	0.233	0.252	110	0.313	0.265
18	0.299	0.267	49	0.282	0.278	80	0.297	0.305	111	0.258	0.277
19	0.219	0.243	50	0.241	0.265	81	0.267	0.290	112	0.275	0.251
20	0.239	0.269	51	0.314	0.263	82	0.290	0.283	113	0.258	0.277
21	0.311	0.278	52	0.244	0.247	83	0.267	0.277	114	0.287	0.316
22	0.262	0.302	53	0.285	0.283	84	0.259	0.231	115	0.280	0.269
23	0.251	0.270	54	0.278	0.267	85	0.239	0.238	116	0.249	0.272
24	0.293	0.258	55	0.268	0.231	86	0.273	0.257	117	0.245	0.268
25	0.197	0.215	56	0.313	0.260	87	0.249	0.276	118	0.285	0.251
26	0.287	0.271	57	0.252	0.241	88	0.257	0.258	119	0.189	0.219
27	0.287	0.268	58	0.274	0.288	89	0.275	0.265	120	0.244	0.237
28	0.244	0.256	59	0.260	0.250	90	0.288	0.328	121	0.278	0.269
29	0.254	0.208	60	0.231	0.229	91	0.276	0.276	122	0.275	0.262
30	0.263	0.267	61	0.243	0.240	92	0.306	0.176	123	0.273	0.252
31	0.262	0.302	62	0.238	0.210	93	0.291	0.324	124	0.277	0.312

Lyons; 70 Rick Manning; 71 Don Mattingly; 72 Oddibe McDowell; 73 Hal McRae; 74 Paul Molitor; 75 Charlie Moore; 76 Lloyd Moseby; 77 Darryl Motley; 78 Rance Mulliniks; 79 Dwayne Murphy; 80 Eddie Murray; 81 Pete O'Brien; 82 Ben Oglivie; 83 Jorge Orta; 84 Spike Owen; 85 Mike Pagliarulo; 86 Lance Parrish; 87 Larry Parrish; 88 Gary Pettis; 89 Jim Presley; 90 Kirby Puckett; 91 Willie Randolph; 92 Floyd Rayford; 93 Jim Rice; 94 Ernest Riles; 95 Cal Ripken; 96 Ed Romero; 97 Mark Salas; 98 Dick Schofield; 99 Larry Sheets; 100 John Shelby; 101 Pat Sheridan; 102 Don Slaught; 103 Roy Smalley; 104 Lonnie Smith; 105 Jim Sundberg; 106 Pat Tabler; 107 Mickey Tettleton; 108 Gorman Thomas; 109 Andre Thornton; 110 Wayne Tolleson; 111 Alan Trammel; 112 Willie Upshaw;

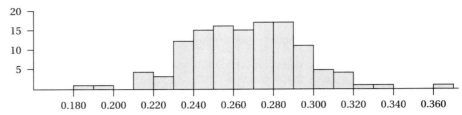

FIGURE 13.2 *Frequency histogram of 1985 American League batting averages.*

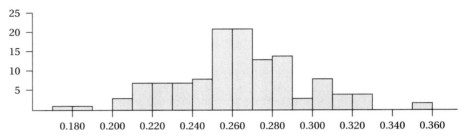

FIGURE 13.3 *Frequency histogram of 1986 American League batting averages.*

113 Greg Walker; 114 Gary Ward; 115 Lou Whitaker; 116 Frank White; 117 Ernie Whitt; 118 Alan Wiggins; 119 Rob Wilfong; 120 Curt Wilkerson; 121 Willie Wilson; 122 Dave Winfield; 123 Mike Young; 124 Robin Yount.

The analysis usually begins with an examination of the individual variables. Figures 13.2 and 13.3 present frequency histograms of the 1985 and 1986 batting averages for the 124 players. Each distribution is roughly bell-shaped with two small values that are outliers and either one (1985) or two (1986) large values that are outliers.

For the predictor variable X,

- $\bar{x}$ denotes the sample mean.
- $\tilde{x}$ denotes the sample median.
- s_X denotes the sample standard deviation.

Similarly, for the response variable Y,

- $\bar{y}$ denotes the sample mean.
- $\tilde{y}$ denotes the sample median.
- s_Y denotes the sample standard deviation.
- SSTO denotes the total sum of squares,

$$\text{SSTO} = \sum (y - \bar{y})^2.$$

Note that

$$s_Y = \sqrt{\frac{\text{SSTO}}{n-1}}.$$

TABLE 13.2 *Summary Statistics for American League Batting Averages by Year*

Year	n	Mean	Standard Deviation	Median	Q_1	Q_3	IQR	Range
1985	124	0.2664	0.0280	0.2655	0.2455	0.2865	0.0410	0.179
1986	124	0.2636	0.0320	0.2645	0.2445	0.2830	0.0385	0.181

Table 13.2 presents an assortment of summary statistics for each year. Each year the two measures of center, the sample mean and median, have similar values. This is not surprising given the shape of the histograms. The mean of the batting averages dropped almost 0.003 (3 points) from 1985 to 1986, and the median dropped 0.001 (1 point). The sample ranges are nearly identical for the two seasons. The values of the interquartile ranges indicate that the middle half of the batters were more similar (less dispersed) in 1986 than in 1985. Finally, the standard deviation for 1986 is about 14 percent larger than the value for 1985.

It is impossible to study the dependence of Y on X unless both of the variables exhibit some variation in the data set. Thus, the methods of this chapter apply only to data sets for which

$$s_X > 0 \quad \text{and} \quad s_Y > 0.$$

This condition is met by the batting averages data since, according to Table 13.2, $s_X = 0.028$ and $s_Y = 0.032$.

13.2.1 The Scatterplot

A **scatterplot** consists of a Cartesian coordinate system with its vertical axis labeled Y and its horizontal axis labeled X and each case plotted on it. Figure 13.4 is the scatterplot of the data for the 124 baseball players. The O in the upper right-hand corner of the scatterplot has a y-coordinate of 0.357 and an x-coordinate of 0.368. (Such precise values cannot be read from the scatterplot; these numbers come from the list of the data in Table 13.1.) You might recall that these values, and hence this O, correspond to Wade Boggs. In addition to the letter O, a scatterplot may contain numbers or plus signs. A 2 indicates that two cases give the same values for the two variables. For example, Figure 13.4 contains a 2 at the coordinates (0.262, 0.302) corresponding to players 22 (Joe Carter) and 31 (Mike Easler), each of whom hit 0.262 in 1985 and 0.302 in 1986. (Recall that batting averages are given in Table 13.1 and that player names are given on p. 442.) A 3 in a scatterplot denotes three cases, and so on. If more than nine cases fall at the same point, a + is placed there. A scatterplot may use a symbol other than O to denote cases, but the letter O is particularly good at distinguishing between similar cases. For example, players 51, 56, and 110, had similar batting averages in 1985 (0.314, 0.313, and 0.313, respectively) and 1986 (0.263, 0.260, 0.265, respectively); the result is that their Os overlap but are clearly distinguishable. Of course, a researcher may decide that such precision is

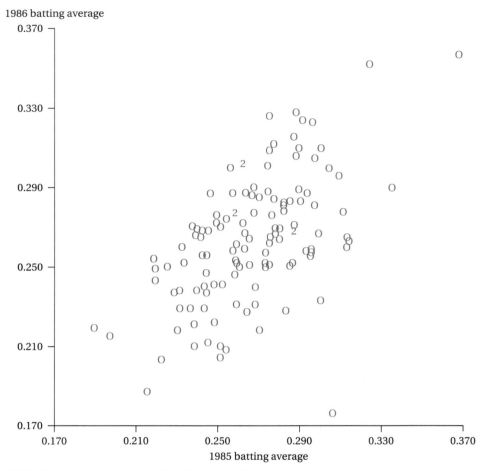

FIGURE 13.4 *Scatterplot of 1986 versus 1985 American League batting averages.*

not needed; in such instances these three cases could be displayed as a 3 in the scatterplot without a substantial loss of information.

It is stating the obvious, but note that the scatterplot reveals that the relationship between the 1985 and 1986 batting averages is not deterministic. More precisely, many players with the same 1985 batting averages have different, sometimes dramatically different, 1986 batting averages. For one of many possible examples, players 14 and 97 each batted 0.300 in 1985, but batted 0.310 and 0.233 in 1986, respectively.

After it is obtained, the scatterplot should be scanned for **isolated cases,** which are cases whose positions in the scatterplot are isolated from all other cases, except, possibly, other isolated cases. For the baseball data, three cases appear to be isolated: $x = 0.368$ and $y = 0.357$ (Wade Boggs), $x = 0.324$ and $y = 0.352$ (Don Mattingly), and $x = 0.306$ and $y = 0.176$ (Floyd Rayford).

Next, the analyst should look for a general pattern in the scatterplot by scanning it visually from left to right, that is, from smaller to larger X values. For the baseball data, as X increases, Y tends to increase. In words, players with lower 1985 batting averages tend to have lower 1986 batting averages, and players with higher 1985 batting averages tend to have higher 1986 batting averages. Such a relationship is certainly not surprising since a player's batting average presumably reflects his skill at hitting. The relationship between Y and X for the batting average data is called **positive, direct,** or **increasing**. Alternatively, if smaller values of X tend to produce *larger* values of Y, and larger values of X tend to produce *smaller* values of Y, the relationship is called **negative, indirect,** or **decreasing**. Figure 13.5 shows two scatterplots of increasing relationships, two scatterplots of decreasing relationships, and one scatterplot of a relationship that is neither increasing nor decreasing. This figure also contains two scatterplots of **linear** (straight line) relationships and three of **curved** relationships. It is not always easy to decide whether a scatterplot reflects a linear or a curved relationship; a **residual plot,** defined later, can help one decide. The scatterplot in Figure 13.4 suggests that it is not unreasonable to believe that the relationship between 1986 and 1985 batting averages is linear.

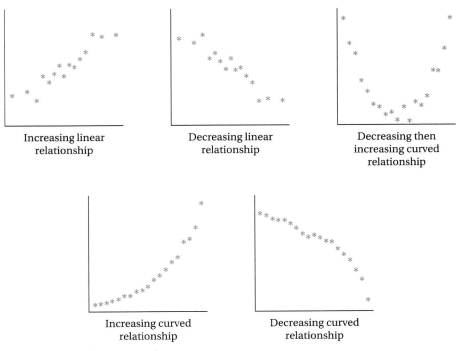

FIGURE 13.5 *Five scatterplots.*

In order to understand scatterplots better, it is helpful to consider some data sets smaller than that of the Batting Averages study. Researchers measured the heart rate and body weight of 40 spiders [2]. In addition, each spider was classified into one of four types: small hunter, tarantula, large hunter, or web weaver. Figure 13.6 presents scatterplots of heart rate versus body weight for

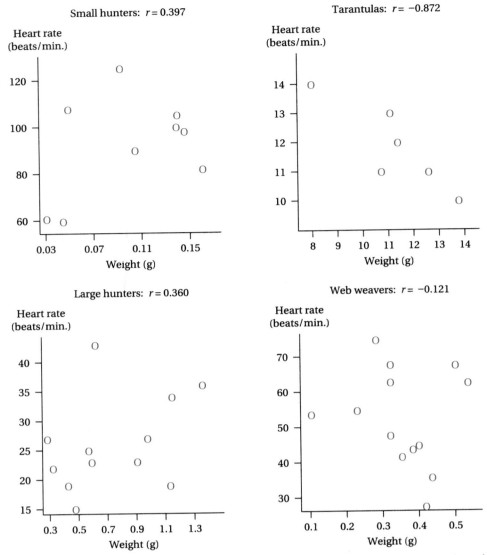

FIGURE 13.6 Scatterplots and correlation coefficients of heart rate (beats per minute) versus body weight (grams) for four types of spiders.

each of the four types of spiders. (Table 13.3 presents the weights and heart rates of the 40 spiders.) Note the following features revealed by these scatterplots. First, across scatterplots:

- Comparing the vertical coordinates of the cases in the plots, one finds that small hunters have, as a group, by far the highest heart rates, followed by web weavers and then large hunters. Finally, every tarantula has a lower heart rate than every other spider of a different type.
- Making a similar comparison of the horizontal coordinates, one finds that tarantulas are by far the heaviest, followed by large hunters, then web weavers, and finally small hunters.

Next, examining each scatterplot separately:

- The two small hunters with the lowest heart rates and the large hunter with the highest heart rate seem to be isolated cases for their scatterplots. No other cases appear to be isolated, though you may reasonably disagree. (For example, an argument could be made that the lightest tarantula is an isolated case.)
- The relationship between heart rate and weight is strong and negative for the tarantulas, weak and negative for the web weavers, and moderate and positive for the small hunters and large hunters. (The adjectives *strong*, *weak*, and *moderate* will become clearer after the correlation coefficient is defined below.)

TABLE 13.3 Body Weights and Heart Rates of 40 Spiders

Small Hunters		Tarantulas		Large Hunters		Web Weavers	
Weight*	Rate†	Weight*	Rate†	Weight*	Rate†	Weight*	Rate†
0.045	60	10.75	11	0.980	27	0.422	27
0.031	61	11.10	13	0.623	43	0.387	44
0.105	90	8.01	14	0.483	15	0.324	48
0.093	125	13.80	10	0.431	19	0.234	55
0.139	100	12.60	11	0.324	22	0.439	36
0.050	108	11.40	12	0.289	27	0.357	42
0.161	82			1.135	19	0.325	68
0.146	98			0.906	23	0.106	54
0.140	105			0.591	23	0.325	63
				0.570	25	0.287	75
				1.152	34	0.404	45
				1.363	36	0.540	63
						0.506	68

*Weights are in grams.
†Heart rates are in beats per minute.

- If the two isolated cases are deleted from the scatterplot for small hunters, the relationship between heart rate and weight changes from moderate and positive to fairly strong and negative. But is there any reason to delete them? Perhaps: these two spiders belong to the taxonomic group *Misumenops asperatus*, and the other seven do not. Perhaps the small hunters designation is too broad.
- If the datum of the one isolated large hunter is deleted, the relationship between heart rate and weight remains positive but becomes stronger. Unlike the previous item, here there seems to be no reason to justify deleting this case. This spider is one of two *Dolomedes tenebrosus* in the data set, and the other one is similar to the other large hunters—it is the case with $x = 0.980$ and $y = 27$.

13.2.2 The Correlation Coefficient

The examination of the spider scatterplots at the end of the last section suggested a need for a numerical summary of the relationship between two variables. The most popular and useful such summary is Pearson's product moment correlation coefficient, or correlation coefficient for short. I recommend against calling it simply the correlation because the word *correlation* is used in so many statistical contexts that it is confusing to use it as a specific technical term.

It is important to remember that the correlation coefficient is a measure of the *linear*, or straight-line, relationship between two variables. As illustrated later in this section, the correlation coefficient can provide a misleading summary of a curved relationship. This section begins with a careful consideration of how to measure the strength of evidence in support of a direct or indirect relationship in a set of data. These ideas lead to the definition of the correlation coefficient. The ideas are introduced with the data from the Batting Averages study.

Figure 13.7 contains a redrawing of the scatterplot of the batting averages data with two lines drawn in for reference: a vertical line through the point

$$X = \bar{x} = 0.2664,$$

and a horizontal line through the point

$$Y = \bar{y} = 0.2636.$$

These two lines divide the scatterplot into four quadrants, labeled I, II, III, and IV in the figure. Note that

- Cases in the first quadrant satisfy

$$x > \bar{x} \quad \text{and} \quad y > \bar{y}.$$

These players had batting averages above the mean each year.
- Cases in the second quadrant satisfy

$$x < \bar{x} \quad \text{and} \quad y > \bar{y}.$$

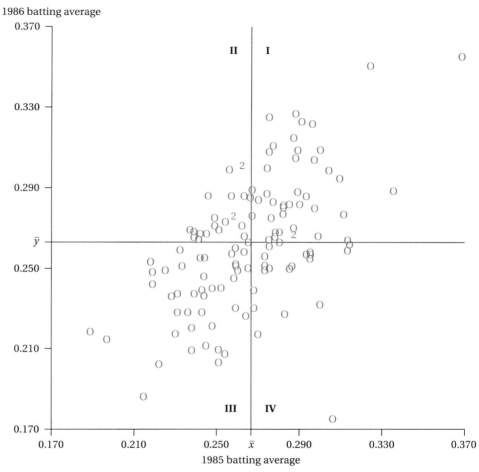

FIGURE 13.7 *Scatterplot of 1986 versus 1985 American League batting averages with four quadrants.*

These players had batting averages below the mean in 1985 but above the mean in 1986.

- Cases in the third quadrant satisfy

$$x < \bar{x} \quad \text{and} \quad y < \bar{y}.$$

These players had batting averages below the mean each year.

- Cases in the fourth quadrant satisfy

$$x > \bar{x} \quad \text{and} \quad y < \bar{y}.$$

These players had batting averages above the mean in 1985 but below the mean in 1986.

Notice that cases in the first and third quadrants give evidence of a direct relationship between Y and X; in the first, values of X above its mean occur with values of Y above its mean, whereas in the third, values of X below its mean occur with values of Y below its mean. In other words, larger values of X are paired with larger values of Y, and smaller values of X are paired with smaller values of Y. Similarly, cases in the second and fourth quadrants give evidence of an indirect relationship between Y and X; in the second, values of X below its mean occur with values of Y above its mean, and in the fourth, values of X above its mean occur with values of Y below its mean. In other words, smaller values of X are paired with larger values of Y, and larger values of X are paired with smaller values of Y. One way to summarize this evidence is by counting the number of cases that fall in each quadrant. The following display gives the results of this counting.

II	I
22	42
III	IV
41	19

Eighty-three cases fall in quadrant I or III and provide evidence of a direct relationship between Y and X, compared with only 41 that fall in quadrant II or IV and provide evidence of an indirect relationship. In summary, this simple counting suggests that the batting averages data set contains more evidence in support of a direct relationship than in support of an indirect relationship between Y and X.

Although this counting method can be helpful in assessing the direction of the relationship between Y and X, it is not sensitive enough to measure the strength of the relationship. To see this, examine the scatterplots with quadrants in Figure 13.8. Note the following features of these two scatterplots:

- Each scatterplot has 10 cases. In addition, the 10 values of X are the same in each data set and the 10 values of Y are the same in each data set. Thus, the only difference between these plots is the manner in which the values of X and Y are paired.
- Each scatterplot has five cases in quadrant I and five cases in quadrant III. Based on counting, each scatterplot provides the same evidence in support of a direct relationship between Y and X.
- The strength of the relationship is very different in the plots. In data set B all values of Y and X fall exactly on a straight line—the strongest possible linear relationship. Set A has a weaker linear relationship.

Each ordered pair (x, y) of values in the data set yields an ordered pair of standardized values (x', y'), where

$$x' = \frac{x - \bar{x}}{s_X} \quad \text{and} \quad y' = \frac{y - \bar{y}}{s_Y}.$$

13.2 The Scatterplot and the Correlation Coefficient

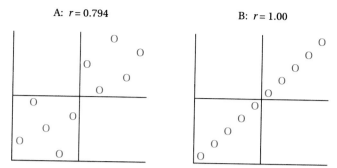

FIGURE 13.8 Scatterplots and correlation coefficients of two phony data sets.

For example, in the Batting Averages study Wade Boggs had values $x = 0.368$ and $y = 0.357$, yielding standardized values

$$x' = \frac{x - \bar{x}}{s_X} = \frac{0.368 - 0.26644}{0.02802} = 3.625$$

and

$$y' = \frac{y - \bar{y}}{s_Y} = \frac{0.357 - 0.26375}{0.03204} = 2.910.$$

Note that a standardized value is positive if, and only if, the original value is larger than its mean.

Consider the product

$$z' = x' \times y'.$$

For any case in the first or third quadrants of the scatterplot, z' is positive. (In the first quadrant x' and y' are both positive; in the third quadrant both are negative; in either event, their product is positive.) For any case in the second or fourth quadrants of the scatterplot, z' is negative. (In each quadrant one of x' and y' is positive and the other is negative; hence, their product is negative.) Combining these remarks concerning the value of z' with the earlier properties of the quadrants gives the following rule:

A case in a data set that yields a positive value of z' provides evidence of a positive relationship between Y and X; one that yields a negative value of z' provides evidence of a negative relationship between Y and X.

The values of z' are summarized by **Pearson's product moment correlation coefficient**, or **correlation coefficient** for short, which is denoted by r and is defined by the formula

$$r = \frac{\sum z'}{n - 1}.$$

For the batting averages data, the value of the correlation coefficient is $r = 0.554$. For other examples, Figure 13.8 presents the values of the correlation coefficient for the phony data, and Figure 13.6 presents the values of the correlation coefficient for the spider data.

It is very tedious to compute r by hand. The use of a statistical computer package or a calculator with an "r button" is recommended. Note, however, that as discussed below it is always absolutely necessary to examine the scatterplot of the data before interpreting the value of r.

Figure 13.9 presents twelve scatterplots and their values of the correlation coefficient, r. The correlation coefficient has many useful and important properties; some are listed below, and others are revealed later.

1. If the correlation coefficient is greater than 0, Y and X are said to have a **positive linear relationship;** if it is less than 0, the variables are said to have a **negative linear relationship;** if it equals 0, the variables are said to have **no linear relationship,** or to be **uncorrelated.**

2. The correlation coefficient is not appropriate for summarizing a curved relationship between Y and X. This fact is illustrated by the bottom right scatterplot in Figure 13.9, in which there is a perfect (deterministic) curved relationship between Y and X and yet the correlation coefficient equals 0. Therefore, it is *always* necessary to examine a scatterplot of the data to determine whether computation of the correlation coefficient is appropriate.

3. The correlation coefficient is always between -1 and $+1$. It equals $+1$ if, and only if, all data points lie on a straight line with positive slope; it equals -1 if, and only if, all data points lie on a straight line with negative slope.

4. The farther the value of the correlation coefficient is from 0, in either direction, the stronger is the linear relationship. This fact can be seen by examining the scatterplots in Figure 13.9 and will also be made more precise in two ways later in this chapter.

5. The value of the correlation coefficient does not depend on the units of measurement chosen by the experimenter. More precisely, if X is replaced by $aX + b$ and Y is replaced by $cY + d$, where $a, b, c,$ and d are any numbers and a and c are bigger than 0, the correlation coefficient of the new variables is equal to the correlation coefficient of X and Y. (The numbers a and c must be positive to prevent reversal of the direction of the relationship; a related result can be obtained if a or c is negative, but it will not be needed in this text.) This result is true because the correlation coefficient is defined in terms of the standardized values of X and Y and these do not change when the units change. For example, changing from miles to inches, pounds to kilograms, degrees Celsius to degrees Fahrenheit, or seconds to hours does not change the correlation coefficient.

13.2 The Scatterplot and the Correlation Coefficient

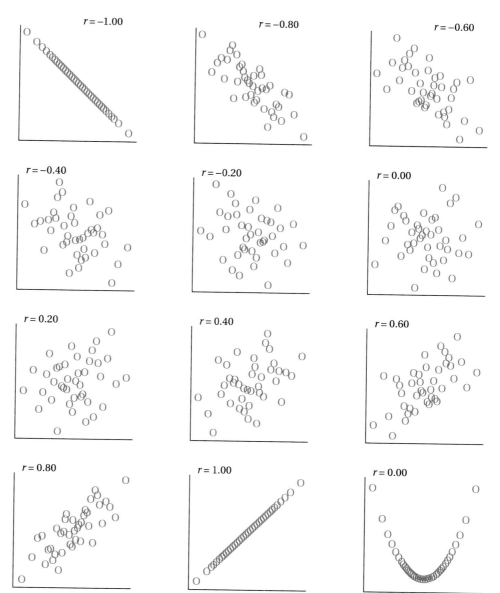

FIGURE 13.9 Twelve scatterplots and their correlation coefficients.

6. The correlation coefficient is symmetric in X and Y. In other words, if the researcher changes perspective and relabels the predictor and response, the correlation coefficient will not change. In particular, if there is no natural assignment of the labels predictor and response to the two numerical variables, the value of the correlation coefficient is not affected by which assignment is chosen.

EXERCISES 13.2

1. The feature "You Be the Critic" appears weekly in the *Capital Times*, a Madison, Wisconsin, newspaper. For example, one week Table 13.4 appeared with the following explanation [3]:

    Tabulations are based on an exit poll of more than 14,000 filmgoers. Viewers rate the movie between 1 and 4 stars. TOP MARKS is the percent who gave the film either 3 and $\frac{1}{2}$ or 4 stars. It indicates the potential of your liking the film.

    (a) Discuss this explanation.
    (b) Construct a dot plot of the top marks. Comment.
    (c) Construct a stem and leaf plot of the top marks. Comment.

TABLE 13.4 Display from "You Be the Critic"

Movie	Top Marks	Number of Stars	
		Public	Critics
1 Dances with Wolves	93	3.9	3.4
2 Home Alone	92	3.8	2.5
3 Ghost	92	3.8	3.0
4 Misery	86	3.6	3.0
5 Kindergarten Cop	84	3.6	2.8
6 Awakenings	84	3.7	3.1
7 Three Men and Little Lady	84	3.6	2.0
8 Rescuers Down Under	84	3.5	3.2
9 Rocky V	77	3.4	2.3
10 Edward Scissorhands	72	3.5	2.7
11 Child's Play 2	69	3.4	1.5
12 The Rookie	67	3.2	2.2
13 Predator 2	59	3.1	2.3
14 Look Who's Talking Too	58	3.3	1.2
15 Mermaids	57	3.3	2.3
16 The Godfather III	52	3.1	3.3
17 Almost an Angel	46	2.8	1.7
18 The Russia House	40	3.1	3.0
19 Havana	27	2.7	3.2
20 Bonfire of the Vanities	15	2.0	1.7

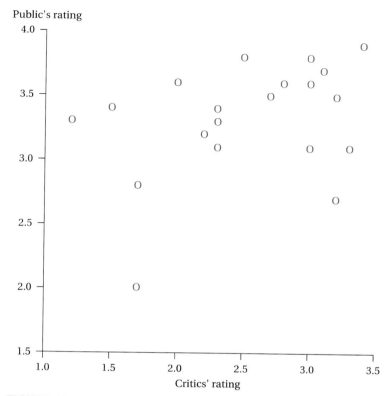

FIGURE 13.10 Scatterplot of the public's versus the critics' ratings of twenty movies.

(d) Construct dot plots of the number of stars for the public's ratings and the critics' ratings. Comment.

(e) The scatterplot of the public's rating versus the critics' rating is given in Figure 13.10.

- Write a few sentences that describe what the scatterplot reveals.
- One of the following numbers is the correlation coefficient for these data; which one is it?

$$-0.83, -0.36, 0.00, 0.36, 0.83$$

2. Eighty college students are given a 20-question trivia quiz and then a second (different) 20-question trivia quiz. Figure 13.11 presents the scatterplot of the score (the number correct) on the second quiz versus the score on the first quiz.

(a) Write a few sentences to describe the scatterplot.

(b) One of the following numbers is the correlation coefficient for these data; which one is it?

$$-0.52, -0.12, 0.12, 0.52$$

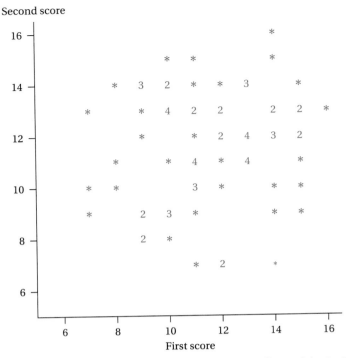

FIGURE 13.11 Scatterplot of second versus first trivia test scores for 80 college students.

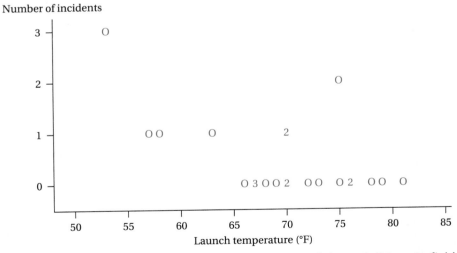

FIGURE 13.12 Scatterplot of the number of incidents of thermal distress to field joint O-rings versus launch temperature for 23 space shuttle flights before the Challenger accident.

3. On January 28, 1986, shortly after lift-off, one of the rocket boosters on the space shuttle *Challenger* exploded, resulting in the deaths of the seven crew members. A presidential commission concluded that the disaster was caused by the failure of an O-ring in a field joint on the rocket booster. The commission further concluded that the failure was due to a faulty design that made the O-ring unacceptably sensitive to a number of factors, including temperature. The O-rings had been damaged on several of the 24 previous shuttle program flights. Figure 13.12 is the scatterplot of the number of incidents of thermal distress to field joint O-rings versus the launch temperature for 23 shuttle flights before the *Challenger* disaster. (The hardware from the fourth shuttle flight was lost at sea.) (See [4] and [5] for additional details.)

   (a) Write a few sentences that describe what the scatterplot reveals.

   (b) Based on the scatterplot (and not hindsight) criticize the decision to launch the *Challenger* when the temperature was 31 degrees.

   (c) Unfortunately, the night before the *Challenger* launch, when managers discussed the effect of temperature on field joint O-rings, they decided that the launches that yielded $y = 0$ were irrelevant. Look at the seven cases in Figure 13.12 that have $y > 0$; is there a convincing relationship between Y and X?

4. Columns is a video game available on the Sega Genesis system. In the flashing version of the game, the player attempts to capture a flashing block in the bottom row of the screen as quickly as possible. The game may be played at levels 2 through 9. The level denotes the number of complete rows of blocks on the screen at the beginning of the game. For example, at level 2 the player must break through the second row of blocks in order to reach the flashing block in the first row. At level 9, the player must break through eight rows of blocks in order to reach the flashing block. I performed a CRD with five games (trials) at each of the eight levels. The response Y is the time needed to capture the flashing block, and the predictor X is the level. Before I collected any data, it seemed obvious, based on my prior experience playing the game, that increasing the level makes the game more difficult. Thus, I anticipated finding an increasing relationship between Y and X. I further anticipated three sources of variation that would cloud the relationship between Y and X. First, although a game at level 5, for example, always begins with 5 rows of blocks, the game's assignment of the colors to the blocks can affect the difficulty of my task. Second, the colors of new blocks (which either help me capture blocks or get in the way) given to me by the computer can affect the difficulty of the game. Finally, my decisions on where to put the new blocks presumably vary from game to game. I believed that the higher the level, the greater would be the variability in the responses. In summary, I suspected that as the level increased, both the response and its variability would increase.

   Figure 13.13 presents the scatterplot of the data from the CRD. Use this plot to answer the following:

   (a) Do you think there are any isolated cases? If yes, which one(s)?

   (b) One of the numbers below is the correlation coefficient for these data; which one is it?

   $$-1.00, -0.524, 0.026, 0.315, 0.716$$

   (c) Table 13.5 presents summary statistics for the responses for each of the eight levels. Discuss how these numbers support or fail to support my conjectures that the response and its variability should increase as the level increases.

5. The study of spider body weight and heart rate discussed earlier in this chapter included a fifth type of spider, primitive hunters and weavers. Figure 13.14 presents the scatterplot of heart rate versus body weight for eight spiders of this type in the study. Use the scatterplot to answer the following:

   (a) Do you think there are any isolated cases? If yes, which one(s)?

   (b) One of the following numbers is the correlation coefficient for these data; which one is it?

   $$-0.726, -0.413, 0.055, 0.487, 0.823$$

   (c) Compare and contrast the primitive hunters and weavers with the four types of spiders presented in Figure 13.6 (p. 448).

6. In a campcraft class Susan Robords learned that there is a formula for computing air temperature from cricket chirps. She decided to investigate this issue

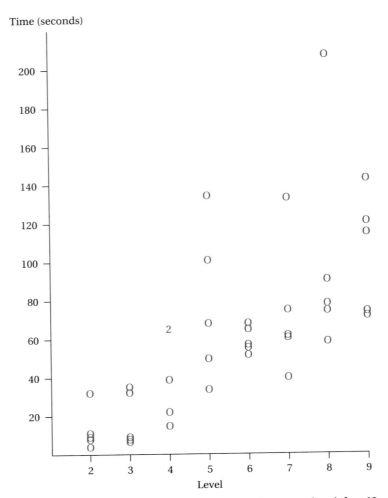

FIGURE 13.13 Scatterplot of time to finish versus level for 40 plays of the Columns video game.

TABLE 13.5 Selected Summary Statistics of Response by Level for 40 Games of Columns

Level	2	3	4	5	6	7	8	9
Mean	13.8	19.2	42.2	78.4	60.0	75.2	103.0	106.2
Median	10.0	10.0	40.0	69.0	58.0	63.0	80.0	116.0
Standard Deviation	11.0	14.0	23.4	40.3	7.4	35.2	59.8	30.8
IQR	15.5	26.0	46.5	75.5	14.0	53.5	82.5	58.5

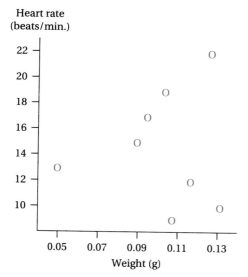

FIGURE 13.14 Scatterplot of heart rate versus body weight for eight spiders classified as primitive hunters and weavers.

empirically and collected data on temperature and number of chirps per minute on 11 occasions. (The number of chirps per minute was obtained by listening to the crickets for between two and five minutes.) This is one of many examples in which each variable could be viewed as the predictor or as the response. First, from a strictly cause-effect perspective, temperature can influence cricket behavior, but cricket behavior cannot influence the temperature. This argument suggests that Y should equal the number of chirps and X should equal air temperature. On the other hand, a person lacking a thermometer might want to count the number of chirps in order to predict the temperature. This perspective suggests that Y should equal temperature and X should equal the number of chirps. Figure 13.15 presents scatterplots of Susan's data from each perspective.

(a) What do these scatterplots reveal?

(b) The correlation coefficient for chirps versus temperature is one of the following numbers; which one is it?

$$-0.25, 0.13, 0.46, 0.82$$

(c) What is the correlation coefficient for temperature versus chirps?

7. A student read in her biology book that fish activity increases with water temperature, and she decided to investigate this issue empirically. On nine successive days she measured fish activity and water temperature in her aquarium. (Her definition of fish activity is involved and not of interest here; it suffices to know that larger values of her measure of fish activity denote

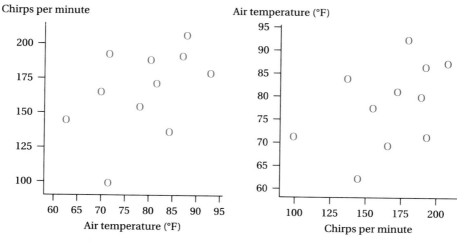

FIGURE 13.15 Scatterplots of air temperature and number of chirps per minute for 11 Madison crickets.

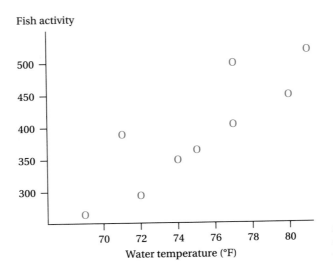

FIGURE 13.16 Scatterplot of fish activity versus water temperature.

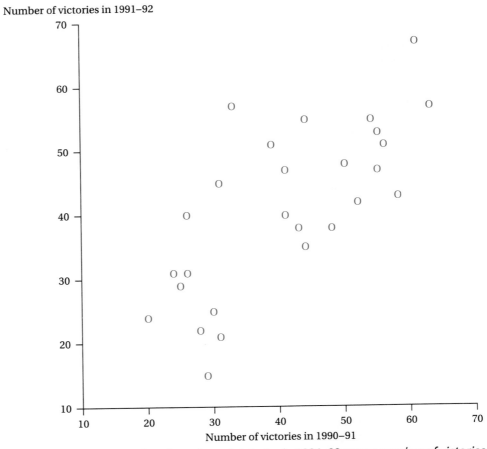

FIGURE 13.17 Scatterplot of number of victories in 1991–92 versus number of victories in 1990–91 for 27 National Basketball Association teams.

more activity.) Figure 13.16 presents the scatterplot of her data.

(a) What does the scatterplot reveal?
(b) One of the following numbers is the correlation coefficient for these data; which one is it?

$$-0.20, 0.03, 0.52, 0.86$$

8. Figure 13.17 presents a scatterplot of the number of victories in 1991–92 (Y) versus the number of victories in 1990–91 (X) for the 27 National Basketball Association (NBA) teams. (Each team played 82 games each season.)

(a) Write a few sentences that summarize what this plot reveals.
(b) One of the following numbers is the correlation coefficient for these data; which one is it?

$$0.000, 0.052, 0.302, 0.724, 1.000$$

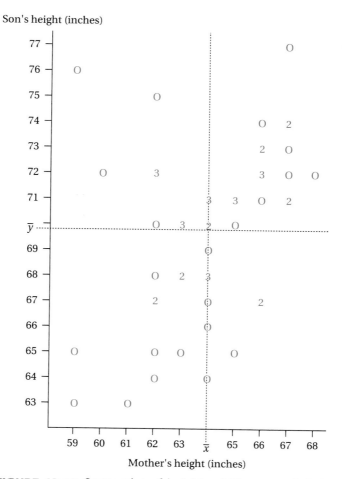

FIGURE 13.18 Scatterplot of heights of 55 sons and their mothers.

9. Figure 13.18 presents a scatterplot of the heights of 55 adult sons and their mothers. The data were obtained from 55 men in my statistics class who self-reported (as opposed to actually measuring) heights to the nearest inch.

   (a) Construct a dot plot of the sons' heights and a dot plot of the mothers' heights. Write a few sentences that explain what these dot plots reveal.
   (b) Does the scatterplot contain any isolated cases? If yes, which one(s)?
   (c) The scatterplot includes lines that delineate the four quadrants. Count the number of observations in each quadrant. (Note: The 12 cases above $x = 64$ lie on the boundary and should not be counted in any quadrant.) What do these counts reveal?
   (d) One of the following numbers is the correlation coefficient; which one is it?

   $-0.41, 0.02, 0.41, 0.82$

10. Figure 13.19 presents a scatterplot of the number of victories versus the team earned run average (ERA) for the 14 American League baseball teams during the 162-game 1987 season.

    (a) Write a few sentences to describe what the scatterplot reveals. (Note: It is generally believed that the larger the value of the ERA, the weaker the pitching staff.)
    (b) One of the following numbers is the correlation coefficient; which one is it?

    $-1.00, -0.76, -0.38, 0.00, 0.38, 0.76, 1.00$

11. A researcher reads the third property of the correlation coefficient (p. 454) and asks

    What is the value of the correlation coefficient if all cases lie on a horizontal straight line (slope equals 0)?

    Answer this question. (Hint: Remember the restriction that s_X and s_Y must both be greater than 0.)

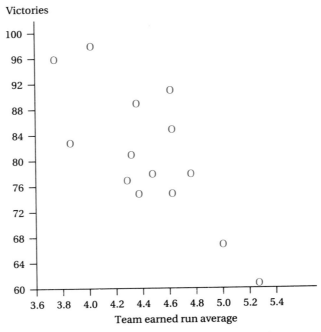

FIGURE 13.19 Scatterplot of number of victories versus team ERA for 1987 American League teams.

13.3 THE REGRESSION LINE

13.3.1 Comparing Two Prediction Lines

One use of the scatterplot is to examine whether there is a linear or nonlinear relationship between Y and X. The correlation coefficient gives a numerical summary of the strength and direction of a linear relationship. With so much attention being paid to linear relationships, it seems natural to describe the data with a line. This section investigates the question of finding the best line to describe a set of data. Consider the phrase from the last sentence, "finding the best line." The word *best* is being used in a precise technical sense. In order to understand what a statistician means by "the best line," it is instructive to compare two lines to see which is better.

Exercise 6 on p. 459 introduced a study of crickets by Susan Robords. Susan was interested in using the number of chirps per minute by a cricket to predict the air temperature. Susan had read that a formula exists for converting the number of chirps per minute into a prediction of the air temperature. In words, the formula takes one-quarter of the number of chirps per minute and adds 37.5 to the product. The resultant sum is the prediction of the temperature. It is convenient to convert this rule to symbols. Let x denote the number of chirps per minute and let $\dot{y}$ (which is read "y dot") denote the predicted temperature obtained from this rule. The rule is

$$\dot{y} = 37.5 + 0.25x \tag{13.3}$$

For comparison, another rule is

$$\hat{y} = 56.2 + 0.136x \tag{13.4}$$

This second rule multiplies the number of chirps per minute by 0.136 and adds 56.2 to the product. The resultant sum is the predicted temperature and is denoted by $\hat{y}$. (The motivation for choosing this particular second rule is made clear later in this section.)

Note that for each of these prediction rules, the predicted value is a linear function of the number of chirps per minute. More precisely, $\dot{y}$ is a linear function of x with slope equal to 0.25 and y-intercept equal to 37.5, and $\hat{y}$ is a linear function of x with slope equal to 0.136 and y-intercept equal to 56.2. Thus, if one draws the graph of $\dot{y}$ versus x or the graph of $\hat{y}$ versus x, the result is a straight line. In view of this fact, these prediction rules can be referred to as prediction lines. These graphs are presented in Figure 13.20; these pictures also include the scatterplot of Susan's data for reference.

It is natural to wonder how well these prediction lines perform, both comparatively (which line is better at describing the data?) and absolutely (does the better line give a useful description of the data?). First, the lines are compared by

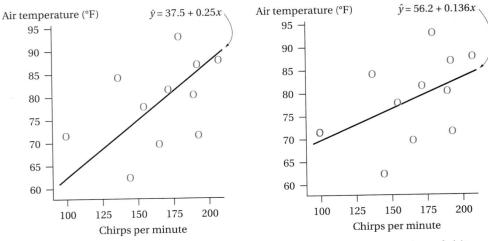

FIGURE 13.20 Two prediction lines for air temperature versus mean number of chirps per minute for 11 Madison crickets.

seeing how they perform on Susan's data. The comparison can be made either with a table or a picture. Because this is a new idea, they will be compared both ways. Table 13.6 presents the results of the computations that are necessary for the comparison of the two lines. The first column of this table presents the case numbers for the 11 cases. The second and third columns present the number of

TABLE 13.6 Comparison of the Prediction Lines $\dot{y} = 37.5 + 0.25x$ and $\hat{y} = 56.2 + 0.136x$ for Susan Robords's Cricket Data*

1	2	3	4	5	6	7	8	9
Case	x	y	$\dot{y}$	$y - \dot{y}$	$(y - \dot{y})^2$	$\hat{y}$	$y - \hat{y}$	$(y - \hat{y})^2$
1	145.0	62.6	73.75	−11.15	124.32	75.92	−13.32	177.42
2	172.0	81.5	80.50	1.00	1.00	79.59	1.91	3.64
3	155.0	77.9	76.25	1.65	2.72	77.28	0.62	0.38
4	137.0	84.2	71.75	12.45	155.00	74.83	9.37	87.76
5	179.5	92.8	82.37	10.43	108.68	80.61	12.19	148.55
6	192.0	86.9	85.50	1.40	1.96	82.31	4.59	21.05
7	207.0	87.8	89.25	−1.45	2.10	84.35	3.45	11.89
8	165.5	69.8	78.87	−9.07	82.36	78.71	−8.91	79.35
9	193.0	71.6	85.75	−14.15	200.22	82.45	−10.85	117.68
10	100.0	71.6	62.50	9.10	82.81	69.80	1.80	3.24
11	189.0	80.4	84.75	−4.35	18.92	81.90	−1.50	2.26
Total					SSE($\dot{y}$) = 780.10			SSE($\hat{y}$) = 653.23

*These numbers are rounded off from a computer analysis; hence, for example, the numbers in column 9 are not exactly equal to the squares of the numbers in column 8.

chirps per minute, x, and the actual temperature, y, for the 11 cases in Susan's data set. The fourth column presents the values of $\dot{y}$ obtained from repeated application of Equation 13.3. For example, for the first case, $x = 145.0$ chirps per minute. Thus, the symbol x in the rule for $\dot{y}$,

$$\dot{y} = 37.5 + 0.25x,$$

is replaced by 145.0, yielding

$$\dot{y} = 37.5 + 0.25(145.0) = 73.75.$$

This value appears in the fourth column and first row of Table 13.6. Similarly, the seventh column of the table presents the values of $\hat{y}$ obtained from repeated application of Equation 13.4. Again, the first case has $x = 145.0$ chirps per minute, yielding

$$\hat{y} = 56.2 + 0.136(145.0) = 75.92,$$

which appears in the seventh column and first row of Table 13.6.

By comparison, for the first case the actual temperature was 62.6 degrees. In an absolute sense, neither prediction rule did very well, but in a comparative sense, the first prediction rule did better because $\dot{y}$ is closer than $\hat{y}$ to the actual y.

The remaining entries in the fourth and seventh columns can be obtained similarly, but the computations are very tedious. Verify enough of these values to convince yourself that you understand the procedure involved, and then proceed below.

The remaining columns of Table 13.6 enable the analyst to make a careful and precise comparison of the two prediction rules. The fifth column compares the actual temperature, y, with the value predicted by the first rule, $\dot{y}$. As usual, the comparison is made by subtracting; in this case the predicted value is subtracted from the actual value. The result is called the error in the observation y relative to the prediction $\dot{y}$. For example, for the first case, $y = 62.6$ and $\dot{y} = 73.75$, yielding

$$y - \dot{y} = 62.6 - 73.75 = -11.15.$$

The sign of this error is negative, which indicates that the actual observation is smaller than the prediction. In words, the observation is too small, or, if you prefer, the prediction is too large. The magnitude of this error, 11.15 degrees, indicates the size of the error. Clearly, it is best to have this magnitude equal 0, but, failing that, it is better to have it close to 0.

(Historical note: It may seem peculiar that statisticians refer to the error *in the observation*. Clearly, the observation represents reality and the error *should* be assigned to the prediction. In fact, historically this was done. Originally, the error was defined as "the predicted value minus the actual value" which is the reverse of the definition above and is naturally called the error in the prediction. The reason for the current usage is derived from the following equation. One can write

$$y = \text{predicted value} + (y - \text{predicted value}).$$

In words, the actual observation is equal to the predicted value plus the amount by which the observation differs from the predicted value. This second term can be viewed as the effect of all of the factors that contribute to the actual value but are not included in the model that yields the predicted value. The contribution of these unused factors is sometimes called noise, or error, with the term *error* not implying any mistake.)

The remaining entries in the fifth column aid the analyst in assessing the performance of the first prediction rule. For example, the rule seemed to perform quite well for cases 2, 3, 6, and 7, because an error of plus or minus 2 degrees is negligible. On the other hand, the rule performed very badly for cases 1, 4, 5, 8, 9, and 10, because a prediction that misses the actual temperature by nine degrees or more—in either direction—is poor. Finally, the prediction rule performed reasonably well for the last case because an error of roughly -4 degrees is not too bad.

Next, in many applications it seems reasonable to view an error of, say, -10 as equally serious as an error of $+10$. In other words, the seriousness of the error is determined by its magnitude, not its sign. (You may remember that this argument was used in Chapter 12 as part of the motivation for the standard deviation.) The sixth column of Table 13.6 contains the squares of the entries in the fifth column; that is, the squared errors. Note that all squared errors must be greater than or equal to 0 (why?) and that good predictions yield squared errors that are smaller than those obtained from poor predictions. Thus, smaller squared errors are preferred. These squared errors are summarized by computing their sum, which is denoted by SSE($\dot{y}$). The symbol SSE($\dot{y}$) is read "the sum of squared errors (or error sum of squares) for the predictions obtained by using $\dot{y}$."

The eighth and ninth columns of Table 13.6 do for the second prediction rule what the fifth and sixth columns do for the first prediction rule. In particular, SSE($\hat{y}$) is the sum of squared errors for the predictions obtained from $\hat{y}$.

13.3.2 The Least Squares Line

The **principle of least squares** is that the prediction rule whose error sum of squares is smaller is the better rule. According to this principle, the second prediction rule, $\hat{y}$, is better than the first prediction rule, $\dot{y}$, because it has a smaller sum of squared errors, 653.23 compared with 780.10.

Historical and technical note: The principle of least squares has a long and illustrious history in science, statistics, and mathematics. It has applications far beyond those illustrated in this chapter. It was originally adopted because it made computations relatively easy, and later it was found to lead to valid answers to many important scientific questions. In this chapter, the principle will be used to define the "best line" mentioned earlier. It is important to realize, however, that although the principle often yields valid answers, there are many data sets for which following the principle's directive can yield misleading answers. This chapter presents a number of easily recognizable indicators that will alert you to situations for which the best line may be misleading. With the tremendous advances in computing in recent years, researchers have developed alternative methods that frequently provide better answers than those obtained from following the principle of least squares. These methods, however, are beyond the scope of this book.

There is a strong connection between the computations in Table 13.6 and the scatterplots and lines in Figure 13.20. Figure 13.21 makes the connection easier to see by focusing only on the first case in the data set.

Consider the second prediction rule, which is given by the formula

$$\hat{y} = 56.2 + 0.136x.$$

The graph of this rule is presented in the right picture in Figure 13.21. The first case is represented by the O at coordinates $x = 145$ and $y = 62.6$. Note in particular that the height (vertical coordinate) of this O corresponds to the value of y. The predicted value of the temperature, $\hat{y} = 75.92$, is equal to the height of the prediction line above the value $x = 145$. The O for the first case is 13.32 degrees below the line; this vertical distance equals the magnitude of the error in the observation. In short, for cases that lie below the line, the value of y is smaller than the predicted value; thus, the observation is too small, or the prediction is too large. For cases that lie above the line, the value of y is larger than the predicted value and the observation is too large, or the prediction is too small. Finally, the vertical distance from the case to the line equals the magnitude of the error, so cases with errors close to 0 lie close to the line and cases with errors far from 0 lie far from the line. An examination of Figure 13.20 indicates that $\hat{y}$ is a better prediction rule than $\dot{y}$ because it is closer to the data, where "closer" is in terms of the sum of the squared vertical distances from the data to the line. This visual conclusion agrees with the more precise comparison of SSE($\hat{y}$) and SSE($\dot{y}$).

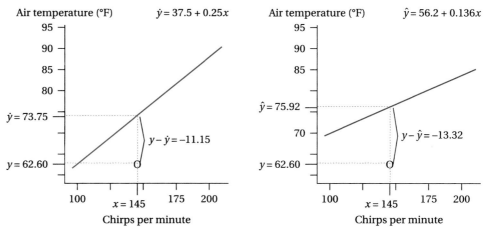

FIGURE 13.21 *Graphical presentation of the prediction error for the case $x = 145$ and $y = 62.6$ for the two prediction rules for the Cricket study.*

The above example for the cricket data shows that the principle of least squares can be used to compare two prediction rules. The rule with the smaller sum of squared errors is better, according to the principle. Some elementary arguments from calculus can be used to prove that a prediction line exists whose sum of squared errors is smaller than that of any other prediction line. This line is, therefore, the best line according to the principle of least squares. Fortunately, this best line can be obtained easily from the values $\bar{x}, s_X, \bar{y}, s_Y$, and r. This best line (we will henceforth suppress reference to the fact that the line is best according to the principle of least squares and simply call it best) has slope denoted by b_1 and y-intercept denoted by b_0; these values are obtained by evaluating the following equations:

$$b_1 = r \frac{s_Y}{s_X} \quad (13.5)$$

and

$$b_0 = \bar{y} - b_1 \bar{x}. \quad (13.6)$$

The predictions from this best line are denoted by $\hat{y}$. Thus, the equation for the best line is

$$\hat{y} = b_0 + b_1 x. \quad (13.7)$$

For the cricket data, for example, it can be shown that

$$\bar{x} = 166.8, \quad s_X = 31.0, \quad \bar{y} = 78.83, \quad s_Y = 9.11, \quad \text{and} \quad r = 0.461.$$

Substituting these values into Equations 13.5 and 13.6 yield

$$b_1 = 0.461 \frac{9.11}{31.0} = 0.1355 \quad \text{and} \quad b_0 = 78.83 - 0.1355(166.8) = 56.23.$$

Thus, the equation of the best prediction line is

$$\hat{y} = 56.23 + 0.1355x.$$

This last equation agrees, except for round-off error, with the definition of $\hat{y}$ given in Equation 13.4 (p. 465). (This fulfills the earlier promise of explaining the motivation for the second prediction rule for the cricket study—it is the best prediction line. The earlier equation for $\hat{y}$ was taken from a statistics computer package that does not use the rounded-off values of $\bar{x}, s_X$, and so on, which I used above; the computer values are better, but the difference is inconsequential.) Usually the best prediction line is called the **least squares line** or more simply the **regression line.** (The reason for the adjective *regression* is discussed below.)

Recall that for the batting averages data,

$$\bar{x} = 0.2664, \quad s_X = 0.028, \quad \bar{y} = 0.2636, \quad s_Y = 0.032, \quad \text{and} \quad r = 0.554.$$

Substituting these values into Equations 13.5 and 13.6 yields

$$b_1 = \frac{0.554(0.032)}{0.028} = 0.633 \quad \text{and} \quad b_0 = 0.2636 - 0.633(0.2664) = 0.095.$$

Thus, the equation of the regression line is

$$\hat{y} = 0.095 + 0.633x.$$

Figure 13.22 is a scatterplot of the batting averages data with the regression line superimposed.

The entire formula for the regression line is of interest because it allows one to compute a predicted value of Y for a given value of X. The slope of the equation is also of independent interest. For example, in the cricket study the slope, b_1, equals 0.1355. This means that for every unit increase in X, the number of cricket chirps per minute, the predicted temperature increases by 0.1355 degrees Fahrenheit. Alternatively, since $0.1355(7.4) = 1$, for every increase of 7.4 chirps per

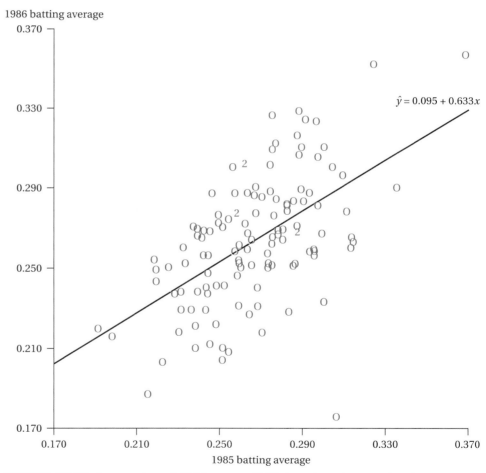

FIGURE 13.22 Scatterplot of 1986 versus 1985 American League batting averages with regression line.

minute the predicted temperature increases by 1 degree. This type of information usually is considered to be of interest. As another example, in the batting averages study the slope equals 0.633. For a 10-point (0.010) increase in 1985 batting average, the predicted 1986 batting average increases by 6.33 points ($0.633 \times 0.010 = 0.00633$). This result could be mildly surprising to a baseball fan who might suspect that a 10-point increase in 1985 batting average would yield a 10-point increase in predicted 1986 batting average. This issue is discussed further below.

An easy mistake for a researcher to make is automatically to assign meaning to the y-intercept, b_0, of the regression line. In the cricket study, for example, the y-intercept equals 56.2. Literally, this means that the height of the regression line is 56.2 at $x = 0$. Does this mean that if the number of cricket chirps per minute equals 0, the predicted temperature is 56.2 degrees Fahrenheit? Most certainly not! The study provided data on cricket chirp rates between 100 and 207 per minute; thus, the study provided evidence of a linear relationship between Y and X only for values of X in the range from 100 to 207. It can be very misleading to extend the conclusions of the study to values of X outside the range of data. In fact, in the cricket study the same source that provided the first prediction equation stated that crickets do not chirp when the temperature falls below 55 degrees; thus, a lack of chirping gives very little information about the temperature (especially with Madison's climate). Similarly, for the batting averages data, the y-intercept of 0.095 does not imply that one should predict that a player who hit 0.000 in 1985 would hit 0.095 in 1986. With the possible exception of pitchers, there has never been a player who had 200 at bats in a season and hit 0.000. If $X = 0$ is included in the range of the data, however, it is valid to interpret the y-intercept, b_0, as the predicted value of Y given $X = 0$.

Based on the argument of the previous paragraph, one might be tempted to say, "If the value $X = 0$ is not included in the data set, the y-intercept, b_0, is not important." This statement is not true. The value of b_0 is important because it appears in the regression equation and is needed to obtain predicted values. Unlike the slope, however, the y-intercept is not of independent interest unless $X = 0$ is included in the range of the data.

In view of the possible misinterpretation of the y-intercept as discussed above, some researchers prefer the following alternate presentation of the equation of the regression line:

$$\hat{y} = \bar{y} + r\frac{s_Y}{s_X}(x - \bar{x}). \tag{13.8}$$

For example, for the cricket study,

$\bar{x} = 166.8,$ $\quad s_X = 31.0,$ $\quad \bar{y} = 78.83,$ $\quad s_Y = 9.11,$ $\quad$ and $\quad r = 0.461.$

Substituting these values into Equation 13.8 yields

$$\hat{y} = 78.83 + 0.461\frac{9.11}{31.0}(x - 166.8) = 78.83 + 0.1355(x - 166.8).$$

It looks different, but this is the same equation obtained earlier, namely
$$\hat{y} = 56.23 + 0.1355x.$$
This new form of the regression line has the advantage that there is a simple interpretation of each number in it. For example, for the cricket study, 0.1355 is the slope, 166.8 is the sample mean of the values of X, and 78.83 is both the sample mean of the values of Y and the predicted value of Y when X is equal to its sample mean. Note that this last interpretation makes sense: since $\bar{x}$ is the center of the values of X, it must be included in the range of the data.

13.3.3 The Regression Effect

There is a third popular way to write the equation of the regression line:
$$\frac{\hat{y} - \bar{y}}{s_Y} = r\left(\frac{x - \bar{x}}{s_X}\right). \tag{13.9}$$

This form of the equation is not employed to compute predicted values; rather, it provides an important insight into the regression line when $r \geq 0$. (Related remarks are possible for $r < 0$, but they are more complicated.) This insight is first illustrated for some simple examples from the Batting Averages study and then is discussed in general. Recall that $r = 0.554$ for the batting averages data. Suppose that a particular player's 1985 batting average was one (1985) standard deviation above the (1985) mean. For this player, the right side of Equation 13.9 would equal $r(1) = r = 0.554$. The left side of this equation equals this player's predicted 1986 batting average standardized by the mean and standard deviation of the 1986 values. In particular, note that his predicted 1986 batting average is only 0.554 standard deviation above the mean of the 1986 batting averages. I say "only" because this player was one full standard deviation above the mean in 1985. Part of this player's skill seems to have disappeared! Similarly, if a player's 1985 batting average was two standard deviations above the mean, the player's predicted 1986 batting average would be only $0.554(2) = 1.108$ standard deviations above the 1986 mean. Next, consider players who were below the mean in 1985. A player whose 1985 batting average was one standard deviation below the mean would have a predicted 1986 batting average that is only $r = 0.554$ standard deviation below the mean. (Thus, part of his inferiority has disappeared!)

This argument can be generalized. For ease of exposition, assume that larger values of Y and X are preferred to smaller values. If $r = 1$, the standardized value of the prediction $\hat{y}$ is equal to the standardized value of x. All of the advantage (for x greater than the mean) or disadvantage (for x less than the mean) in x is passed on to $\hat{y}$. If, however, $0 < r < 1$, only a proportion of the advantage or disadvantage is passed on to $\hat{y}$, and the proportion equals the correlation coefficient r. (This is another property of the correlation coefficient, as promised earlier.) Note that if $r = 0$, then $\hat{y} = \bar{y}$ for all values of x; in words, if the variables are uncorrelated, the advantage or disadvantage in x is irrelevant for predicting y.

This loss of advantage for $r < 1$ is sometimes called the **regression effect**. The term is inspired by noting that the predicted value of the response is closer to its mean, in terms of standardized values, than is the predictor. The response is regressing to its mean or, more colorfully, to mediocrity.

Presumably, few readers of this book have been professional baseball players, but all have taken tests. The above argument can be extended to this situation, too. Exercise 2 on p. 457 examined a study in which 80 college students were given two trivia quizzes of 20 questions each. Figure 13.11 (p. 458) presents a scatterplot of the score (the number correct) on the second quiz versus the score on the first quiz. The distribution of scores on the first quiz is very similar to the distribution of scores on the second quiz (details are not given). Thus, there is little evidence that the two quizzes had different levels of difficulty. Further evidence of the similarity in difficulty of the tests is obtained by computing the usual summary statistics. It can be shown that

$$\bar{x} = 11.587, \quad s_X = 2.226, \quad \bar{y} = 11.563, \quad s_Y = 2.146,$$
$$\text{and} \quad r = 0.121.$$

Note that the sample means are nearly identical and the sample standard deviations have very similar values. The equation of the regression line is

$$\hat{y} = \bar{y} + r\frac{s_Y}{s_X}(x - \bar{x}) = 11.563 + 0.121\frac{2.146}{2.226}(x - 11.587)$$
$$= 11.563 + 0.1167(x - 11.587).$$

Eight persons scored 15 on the first quiz, well above the mean. The predicted score on the second quiz for these persons is

$$\hat{y} = 11.563 + 0.1167(15 - 11.587) = 11.961,$$

or approximately 12. This answer typically surprises my students. As noted above, there is virtually no evidence that the tests have different degrees of difficulty; thus, it seems strange to predict such a large drop in score for students who did so well on the first quiz. Perhaps, a student will suggest, the regression line, although best according to the principle of least squares, is flawed in some way. An examination of the scatterplot of the quiz scores indicates, however, that the eight students who scored 15 on the first quiz scored 14, 13, 13, 12, 12, 11, 10, and 9 on the second quiz. The mean of these eight scores on the second quiz is 11.75. Thus, instead of underpredicting, the regression line slightly overpredicted these scores, although the difference is small.

Similarly, the nine persons who scored 9 on the first quiz have a predicted second quiz score of

$$\hat{y} = 11.563 + 0.1167(9 - 11.587) = 11.261.$$

Actually, these nine persons scored 14, 14, 14, 13, 12, 9, 9, 8, and 8 correct on the second quiz. The mean of these nine second quiz scores is 11.22, which is very close to the regression prediction for each person.

The above results on the trivia quizzes seem counterintuitive to many of my students. The following argument makes these results more palatable to them. My students are comfortable with the idea that if a person takes two different quizzes on the same subject (in this case trivia), the test scores will not necessarily coincide. For example, Sally might score 15 on the first quiz and 13 on the second quiz without anything being amiss. Each of a student's two scores can be viewed as being the sum of two parts, his or her true ability and chance variation. (In this context, chance variation includes the sum effect of all factors other than true ability that contribute to the score.) If the quizzes are of roughly the same level of difficulty and if the student's knowledge neither grows nor diminishes between exams, the true ability portion might be viewed as being constant. Thus, a change in a student's score from the first to the second quiz would be the result of a change in the value of the chance variation from the first to the second quiz. In everyday language, chance variation is often called luck, and I will make that identification here for ease of exposition. Consider *a person* who obtained a very high score, say 19, on the first quiz. There are three possibilities: the person's true ability is less than 19 and she had good luck on the quiz; her true ability is 19 and she had no luck on the quiz; or her true ability is greater than 19 and she had bad luck on the quiz. Now consider *all persons* who scored 19 on the first quiz. Because 19 is such a good score, it seems intuitively correct to believe that more of these persons had good luck than bad luck; that is, that more of them have a true value below 19 rather than above 19. Since there is no reason to believe that the preponderance of good luck for these persons should persist to the second quiz, it seems reasonable to predict that the scores will decrease. (This argument can be made mathematically rigorous, but it is beyond the scope of this text.) If $r = 1$, there is no luck in the scores and no regression to the mean. If $r < 1$, luck plays some role and there is some regression toward the mean. The smaller the value of r is (remembering that $r \geq 0$), the stronger is the influence of luck and the greater is the regression effect.

13.3.4 The Coefficient of Determination

Section 13.4 addresses the issue of how well the regression line performs in an absolute sense. In this section an easier question is considered, namely, how well does the regression line perform in a relative sense?

The regression line has the smallest sum of squared errors among the collection of all prediction lines. In other words, it gives the best predictions of the response values Y that can be obtained by using the values of X in a straight-line equation. But just because something is the best, it is not necessarily any good! The relative value of the regression line is assessed by comparing it to the best prediction rule that *ignores* the value of X. The idea is that if the predictions obtained by using X are little or no better than predictions obtained by ignoring X, there is not much reason to use X.

But how well can a researcher make predictions without using X? "Without using X" means that regardless of the value of X, the rule must yield the same predicted value of Y. Thus, the prediction rule must always yield the value c, where c is some constant. But which constant c should be used? Using the constant prediction c, the error in an observation y is $y - c$, the squared error is $(y - c)^2$, and the sum of squared errors is

$$\sum (y - c)^2.$$

According to the principle of least squares, the value of c that makes this sum as small as possible is the best constant prediction rule. Algebra can be used to show that the best choice is $c = \bar{y}$; thus, the sample mean of Y is the best constant prediction rule. The sum of squared errors for the prediction $\bar{y}$ is

$$\sum (y - \bar{y})^2,$$

which has previously been denoted as SSTO, the total sum of squares of the values of Y.

The following display presents the values of SSE and SSTO for three familiar studies.

Study	SSE	SSTO
Batting Averages	0.08757	0.12635
Chirping Crickets	653.19	829.54
Trivia Quizzes	358.38	363.69

Note that for each study, SSE is smaller than SSTO. This pattern is not surprising because of the following argument. SSTO is the sum of squared errors for the constant prediction $\bar{y}$. In addition, $\bar{y}$ can be viewed as $\bar{y} + 0x$; that is, it is a linear prediction with slope equal to 0 and y-intercept equal to $\bar{y}$.

- If $r = 0$, then Equation 13.8 implies that $\hat{y} = \bar{y}$; that is, the regression line is the constant prediction rule and SSTO equals SSE.
- If $r \neq 0$, however, the regression line has a nonzero slope and does not coincide with $\bar{y}$. The regression line has the property that its sum of squared errors, SSE, is smaller than the sum of squared errors for any other linear prediction rule. In particular, SSE is smaller than SSTO, the sum of squared errors for $\bar{y}$.

According to the principle of least squares, $\bar{y}$ is the best prediction rule that ignores X, and $\hat{y}$ is the best prediction rule that uses X in a linear manner. SSTO and SSE measure the sum of squared errors for these two prediction rules, respectively. Thus, the difference between SSTO and SSE is a measure of how much of the total squared error of Y can be "removed," "explained," or "accounted for" by using the linear relationship between Y and X. This difference, SSTO − SSE, is not a suitable summary statistic, however, for measuring the utility of using X because its value depends on the units of measurement of X and Y. For the units of the crickets study, for example,

$$\text{SSTO} - \text{SSE} = 829.54 - 653.19 = 176.35.$$

If temperature is measured in degrees Celsius, however, it can be shown that SSTO = 256.03 and SSE = 201.60, yielding

$$\text{SSTO} - \text{SSE} = 256.03 - 201.60 = 54.43.$$

It can be shown that this disagreement disappears if the difference is divided by SSTO, yielding

$$R^2 = \frac{\text{SSTO} - \text{SSE}}{\text{SSTO}}. \tag{13.10}$$

For the crickets data with temperature measured in degrees Fahrenheit,

$$R^2 = \frac{829.54 - 653.19}{829.54} = 0.213,$$

and with temperature measured in degrees Celsius,

$$R^2 = \frac{256.03 - 201.60}{256.03} = 0.213.$$

These two values agree, as claimed above.

The number R^2 is the **coefficient of determination**. As another example of its computation, for the batting averages data,

$$R^2 = \frac{0.12635 - 0.08757}{0.12635} = 0.307.$$

The value of R^2 is often reported as a percentage; thus, R^2 is 30.7 percent for the batting averages data. The coefficient of determination measures the proportion (or percentage) of the total squared error in the best constant prediction that can be *removed* by using X to obtain the best linear prediction. Researchers are fond of various picturesque ways to interpret the value of R^2; for example, the use of the word "removed" above. Here are some other interpretations illustrated with the batting averages data. (The words in brackets improve the accuracy of the statements but are often omitted. The words in parentheses are often substituted for the words immediately preceding them.)

> The 1985 batting averages account for (are responsible for, explain) [via a linear relationship] 30.7 percent of the [squared] variation in the 1986 batting averages.

Or, if you prefer to interchange the subject and predicate,

> Of the [squared] variation in the 1986 batting averages, 30.7 percent is explained by (due to, accounted for by) [a linear relationship with] the 1985 batting averages.

It can be shown that the coefficient of determination is equal to the square of the correlation coefficient: $R^2 = r^2$. This identity can be illustrated with the batting averages data. It was stated earlier that $r = 0.554$; thus,

$$r^2 = (0.554)^2 = 0.307,$$

which equals R^2 computed above.

The identity $R^2 = r^2$ provides another interpretation of the correlation coefficient; namely, its square equals the proportion of the squared variation in Y that can be explained by a linear relationship with X. In addition, since we already know that the value of r does not depend on the units of the variables (since it is defined in terms of standardized values), this identity shows, as claimed earlier when the cricket data were considered, that the value of R^2 does not depend on the units of the variables.

EXERCISES 13.3

1. Refer to Exercise 1 on p. 456, a comparison of critics' and the public's ratings of movies. In the table at the bottom of the page are selected summary statistics for these data. In addition, the correlation coefficient equals 0.360.

    (a) Write a few sentences describing the difference between the distributions of the critics' ratings and the public's ratings.
    (b) i. Obtain the equation of the regression line.
       ii. Find the value of $\hat{y}$ for the movie *Ghost;* for the movie *Havana.*

2. Refer to Exercise 2 on p. 457, a comparison of scores on two trivia tests for 80 university students. In the table at the top of the next page are selected summary statistics for these data. In addition, the correlation coefficient equals 0.121.

    (a) Use Figure 13.11 (p. 458) to construct dot plots of the scores on the first quiz and the scores on the second quiz. Comment on what these dot plots reveal.

    (b) This study was used in the text to illustrate the regression effect. This item demonstrates that the regression effect also happens when one moves backwards in time. Let Y denote the score on the first quiz and let X denote the score on the second quiz.

       i. What is the equation of the regression line for using the score on the second quiz to predict the score on the first quiz?
       ii. What is the regression line prediction of the score on the first quiz for a person who scored 15 on the second quiz? Compare this predicted value with the actual scores on the first quiz for these subjects. Comment.
       iii. What is the regression line prediction of the score on the first quiz for a person who scored 9 on the second quiz? Compare this predicted value with the actual scores on the first quiz for these subjects. Comment.

Source	n	Mean	Standard Deviation	Median	Q_1	Q_3	IQR	Range
Public	20	3.32	0.448	3.40	3.10	3.60	0.50	1.9
Critics	20	2.52	0.653	2.60	2.10	3.05	0.95	2.2

Test	n	Mean	Standard Deviation	Median	Q_1	Q_3	IQR	Range
First	80	11.587	2.226	11	10	13	3	9
Second	80	11.563	2.146	12	10	13	3	9

3. Exercise 7 on p. 461 presents a study of the effect of water temperature (X) on fish activity (Y). For these data,

$$\bar{x} = 75.11, \ s_X = 4.04, \ \bar{y} = 393.33,$$

$$s_Y = 86.3, \text{ and } r = 0.860.$$

(a) Obtain the equation of the regression line for using water temperature to predict fish activity. One case had $x = 74$ and $y = 350$; compute $\hat{y}$ for this case and comment.

(b) Obtain the equation of the regression line for using fish activity to predict water temperature. Obtain the prediction of water temperature for the case given in part (a) and comment.

4. Figure 13.6 (p. 448) and Figure 13.14 (p. 461) present scatterplots of heart rate versus weight for five types of spiders. The following summary statistics are needed to answer the questions below.

Type	$\bar{x}$	s_X	$\bar{y}$	s_Y	r
Small hunters	0.1011	0.0491	92.11	21.51	0.397
Tarantulas	11.277	1.955	11.833	1.472	−0.872
Large hunters	0.737	0.357	26.08	8.03	0.360
Web weavers	0.3582	0.1140	52.92	14.13	−0.121
Primitive hunters and weavers	0.1029	0.0259	14.63	4.50	0.055

(a) Obtain the equation of the regression line for the small hunters. One of these spiders had $x = 0.161$ and $y = 82$; compute $\hat{y}$ for this spider and comment.

(b) Obtain the equation of the regression line for the tarantulas. One of these spiders had $x = 10.75$ and $y = 11$; compute $\hat{y}$ for this spider and comment.

(c) Obtain the equation of the regression line for the large hunters. One of these spiders had $x = 1.135$ and $y = 19$; compute $\hat{y}$ for this spider and comment.

(d) Obtain the equation of the regression line for the web weavers. One of these spiders had $x = 0.422$ and $y = 27$; compute $\hat{y}$ for this spider and comment.

(e) Obtain the equation of the regression line for the primitive hunters and weavers. One of these spiders had $x = 0.108$ and $y = 9$; compute $\hat{y}$ for this spider and comment.

5. Exercise 9 on p. 463 presents a study of the relationship between the heights of adult sons (Y) and the heights of their mothers (X). For these data,

$$\bar{x} = 64.00, \ s_X = 2.219, \ \bar{y} = 69.78,$$

$$s_Y = 3.287, \text{ and } r = 0.411.$$

(a) Obtain the equation of the regression line for using the mother's height to predict the son's height. For one case, the mother is 66 inches tall and the son is 71 inches tall; compute $\hat{y}$ for this case and comment.

(b) Obtain the equation of the regression line for using the son's height to predict the mother's height. Obtain the prediction of the mother's height for the case given in part (a) and comment.

6. Exercise 10 on p. 464 presents a study of the relationship between the number of victories (Y) and the team ERA (X) for 14 American League baseball teams in 1987. For these data,

$$\bar{x} = 4.456, \ s_X = 0.420, \ \bar{y} = 81,$$

$$s_Y = 10.38, \text{ and } r = -0.758.$$

(a) Obtain the equation of the regression line.

(b) The Milwaukee Brewers had a team ERA of 4.62 and won 91 games. Compute the predicted number of victories for the Brewers and comment.

TABLE 13.7 Number of Victories for Each NBA Team during the 1990–91 and 1991–92 Regular Seasons

Team	Number of Victories 1990–91	Number of Victories 1991–92	Team	Number of Victories 1990–91	Number of Victories 1991–92
Portland	63	57	Seattle	41	47
Chicago	61	67	New York	39	51
L.A. Lakers	58	43	Cleveland	33	57
Boston	56	51	L.A. Clippers	31	45
Phoenix	55	53	Orlando	31	21
San Antonio	55	47	Washington	30	25
Utah	54	55	Minnesota	29	15
Houston	52	42	Dallas	28	22
Detroit	50	48	Charlotte	26	31
Milwaukee	48	31	New Jersey	26	40
Golden State	44	55	Sacramento	25	29
Philadelphia	44	35	Miami	24	38
Atlanta	43	38	Denver	20	24
Indiana	41	40			

7. The purpose of this exercise is to give you some practice computing predicted values, errors, and sums of squares and in applying the principle of least squares. An artificial data set consists of five cases: (1,3), (1,1), (2,2), (2,3), and (3,4). One prediction line is given by the equation $\hat{y} = x + 1$, and a second prediction line is given by the equation $\hat{y} = 2x - 1$. According to the principle of least squares, which of these lines is better?

8. This exercise looks at the data introduced in Exercise 8 on p. 463. Table 13.7 contains the number of victories for each of the 27 National Basketball Association (NBA) teams during the 82-game 1990–91 (X) and 1991–92 (Y) regular seasons. The correlation coefficient for these data is 0.724; additional summary statistics are provided in the table at the bottom of the page.

   (a) Write down the equation of the regression line. Use this line to predict the number of victories obtained in 1991–92 by Chicago; by the Los Angeles Lakers.

   (b) An analyst states

   > Because $\bar{x} = 41$ and $\bar{y} = 41$, I can conclude that the average ability of the teams remained unchanged from 1990–91 to 1991–92.

   What do you think?

   (c) An analyst examines Table 13.7 and states

   > Of the 13 teams that had a winning season (42 or more victories) in 1990–91, only three (Chicago, Utah, and Golden State) won more games in 1991–92. By contrast, of the 12 teams that had a losing season (40 or fewer victories) in 1990–91, eight (all but Orlando, Washington, Minnesota, and Dallas) won more games in 1991–92.

Season	n	Mean	Standard Deviation	Median	Q_1	Q_3	IQR	Range
1990–91	27	41	12.96	41	29.5	53.0	23.5	43
1991–92	27	41	13.08	42	31.0	51.0	20.0	52

The analyst provides an explanation and a prediction:

- These data show that teams quickly become complacent with success, but increase their efforts in adversity.
- Since the strong teams declined and the weak teams improved, eventually all teams will have the same ability.

What do you think?

9. (Hypothetical) A self-proclaimed mathematical wizard claims to be able to improve children's math test scores after only one hour of training. The wizard gives every student in grade three a standardized test of math skills. He then takes the one-quarter of the students with the lowest scores and gives them one hour of his special training. The next day each child in the training program retakes the standardized test. Almost always, the wizard reports, the majority of his students improve their test scores. Comment.

10. Baseball experts have discovered the following phenomenon. Teams that win very few games one year but have a good record during spring training the next year (spring training is practice before the season begins) usually experience an increase (over the previous season) in games won during the coming season. Comment.

13.4 RESIDUALS, OUTLIERS, AND ISOLATED CASES

13.4.1 Residuals and Outliers

Each case in a data set has a response y, a predictor x, and a predicted response $\hat{y}$. The difference between the response and the predicted response was defined earlier and called the error in the observation relative to the prediction $\hat{y}$. This difference usually is called the **residual** and is denoted by e,

$$e = y - \hat{y}.$$

Note that every case has its own residual. If a residual is close to 0, whether positive or negative, the predicted response and actual response are close to each other and the prediction is good for that case. On the other hand, if a residual is far from 0, whether positive or negative, the predicted response and actual response are far from each other and the prediction is poor for that case. The distinction between close and far is not an absolute; it depends on the particular application. Several examples below will make this point clear.

An analyst usually constructs a dot plot or histogram of the residuals. A frequency histogram of the residuals for the batting averages data, shown in Figure 13.23, is approximately bell-shaped with one small outlier for Floyd Rayford. Rayford batted 0.306 in 1985; the predicted value of his 1986 batting average is

$$\hat{y} = 0.2636 + 0.6331(0.306 - 0.2664) = 0.289.$$

Unfortunately for Mr. Rayford and his team, the Baltimore Orioles, his batting average dropped to 0.176 in 1986, giving a residual of

$$e = y - \hat{y} = 0.176 - 0.289 = -0.113.$$

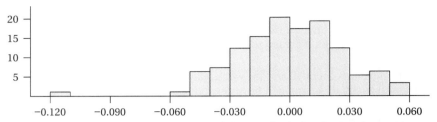

FIGURE 13.23 *Frequency histogram of the residuals for the batting averages data.*

A case is called an **outlier** if its residual is labeled an outlier. For the batting averages data, Floyd Rayford is the only outlier.

Taking another example from the batting averages data, Wade Boggs's residual is

$$e = 0.357 - 0.328 = 0.029.$$

(If interested, you can verify that for Boggs $\hat{y} = 0.328$.) A quick glance at Figure 13.23 shows that 0.029 is not an unusual value for a residual; it is certainly not an outlier. Recall, however, that Boggs was earlier classified as an isolated case. This shows that outlier and isolated case are different concepts.

Figure 13.18 (p. 463) is a scatterplot of the heights of adult sons versus the heights of their mothers. The 55 subjects were students in my statistics class; see Exercise 9 on p. 463 for more details of this study. Figure 13.24 presents a histogram of the residuals for these data. The most isolated case in the scatterplot, (59,76), yields the most extreme residual,

$$e = y - \hat{y} = 76 - 66.73 = 9.27.$$

(If interested, you can use the information in Exercise 9 to verify the value of $\hat{y}$.) I label this case an outlier.

A good researcher attempts to determine what causes an outlier. In the batting averages data, for example, Floyd Rayford is simply an unusual case. It is rare (though not completely unheard of) for a player to bat so well one year and so badly the next. (Incidentally, the next season Rayford had 8 hits in 50 at bats for a 0.160 batting average; he never played in the majors again.) When I discuss the Height study in my class, I ask my students to conjecture why the son in the outlying case is so tall. Almost immediately someone will state, "Prob-

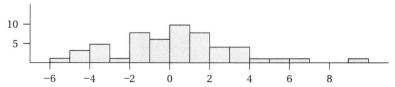

FIGURE 13.24 *Frequency histogram of residuals for the Height study.*

ably his father was tall." This is a natural conjecture. A person receives genes from both parents; thus, a model that employs both parents' heights to predict the son's height is more reasonable than a model that uses only one parent. In fact, in this case the father was 75 inches tall. Another possible reason for this outlier is that for height, short is dominant over tall. Thus, the mother in this outlying case apparently had an unexpressed tall gene. These considerations suggest that a model that employed both parents' heights and the idea of genetic dominance would be better than the model employed in this chapter. Unfortunately, these more complicated models are beyond the scope of this book.

Figure 13.19 (p. 464) presents a scatterplot of the number of victories versus team earned run average (ERA) for the 14 American League baseball teams in 1987. Figure 13.25 presents a dot plot of the residuals obtained from an analysis of these data. The most extreme residual, which I would not label an outlier, is for the Milwaukee Brewers. With a team ERA of 4.62, their predicted number of victories is

$$\hat{y} = 81 - 18.733(4.62 - 4.456) = 81 - 3.07 = 77.93.$$

(See Exercise 10 on p. 464 for the information needed to obtain the equation of the regression line.) The Brewers' actual number of victories, 91, gives the residual

$$e = y - \hat{y} = 91 - 77.93 = 13.07.$$

Below are two possible explanations why my favorite team overachieved in 1987:

1. The team ERA does not distinguish between the pitching performance in tight games, in which one run can be crucial, and one-sided games, in which several runs can be irrelevant. One indication that the Brewers' pitchers were especially effective in tight games is their total of 45 saves, second best in the American League in 1987.

2. The Brewers offense had the second highest total of runs scored in 1987. The strong offense no doubt compensated for many weak pitching performances.

The regression line provides the analyst with predicted values of the response; the square of the correlation coefficient (the coefficient of determination) measures how good the predictions are relative to the sample mean, and the distribution of residuals indicates how good the predictions are in an absolute sense. For example, for the batting averages data one sees by counting (details not shown) that only 37 of the 124 predictions (29.8 percent) are within 10 points (0.010) of

FIGURE 13.25 *Dot plot of residuals for the Victories-ERA study.*

the actual 1986 batting average and only 71 (57.3 percent) are within 20 points of the actual 1986 batting average. It would seem, then, that the predictions are not particularly accurate.

A residual is a generalization of the deviation defined on p. 418. The analyst must compute the regression line in order to obtain the residuals. A line is determined by two values, its slope and intercept, and each of these computations places a restriction on the residuals. The first restriction is that the residuals must sum to 0; hence their mean is 0. The second restriction is not important now but will arise later. Because they have two restrictions, the residuals have $n-2$ degrees of freedom. In order to obtain the standard deviation of the residuals, the sum of their squared deviations is needed. This sum of squares is

$$\sum (e - \bar{e})^2 = \sum (e - 0)^2 = \sum e^2 = \sum (y - \hat{y})^2.$$

Thus, the sum of the squared deviations of the residuals is simply the sum of squared errors, SSE.

The variance of the residuals is denoted by s^2, with no subscript, and is equal to the sum of their squared deviations divided by their number of degrees of freedom,

$$s^2 = \frac{\text{SSE}}{n-2}.$$

The standard deviation of the residuals, denoted s, is the positive square root of this variance.

For the batting averages data, the standard deviation of the residuals is $s = 0.0268$, which is just under 27 points. For the heights of sons and mothers, the standard deviation of the residuals is $s = 3.024$ inches. For the relationship between victories and team ERA, the standard deviation of the residuals is $s = 7.04$ victories.

If the residuals have a bell-shaped distribution, the empirical rule introduced in Chapter 12 can be used to obtain approximate answers to questions about their distribution. For the batting averages data, for example, the empirical rule states that about 68 percent of the residuals should fall between

$$\bar{e} - s = 0 - 0.0268 = -0.0268 \quad \text{and} \quad \bar{e} + s = 0 + 0.0268 = 0.0268.$$

Actually, 88 of 124 (71.0 percent) of the residuals lie within these bounds. For the height data, 38 of 55 residuals (69.1 percent) are between -3.024 inches and $+3.024$ inches (plus or minus one standard deviation).

If the residuals do not have a bell-shaped distribution, the researcher may prefer a measure of spread other than the standard deviation, such as the interquartile range.

13.4.2 The Residual Plot

It is useful to construct a scatterplot of the residuals versus the values of the independent variable X. As mentioned previously, correlation measures the strength of the linear relationship between Y and X and therefore should not be used if the relationship is curved. Unfortunately, curvature in a scatterplot is not always as obvious as suggested by the examples in Figures 13.5 (p. 447) and 13.9 (p. 455). The curvature in the scatterplot in Figure 13.26 is more subtle. Suppose the analyst does not notice the curvature in these data and obtains the regression line. Figure 13.27 is a **residual plot** of these data—a scatterplot of the residuals versus X. It is impossible to miss the curvature in this scatterplot! The residual plot makes it easier for an analyst to assess the linearity of the relationship.

(Technical note: It is easier to see the curvature in the residual plot because in the scatterplot of Y versus X the curvature is a perturbation of a positive linear relationship between the variables. This difficulty is removed in the residual plot; it can be shown mathematically that there is never a linear relationship between e and X. The second restriction on the residuals states that

$$\sum (xe) = 0;$$

it can be shown that this fact along with the fact that the mean of the residuals is 0 implies that the residuals and the values of X are uncorrelated; hence there is no linear relationship between them.)

Scatterplots of the residuals versus X for the batting averages, heights, and ERA data, which are not shown here, give no indication of curvature.

13.4.3 Isolated Cases

For the batting averages data, $r = 0.554$, $R^2 = 0.307$, and $s = 0.0268$. If the single outlier, Floyd Rayford, is dropped from the data set, these summaries become $r = 0.609$, $R^2 = 0.371$, and $s = 0.0248$. Dropping just one case from 124—less

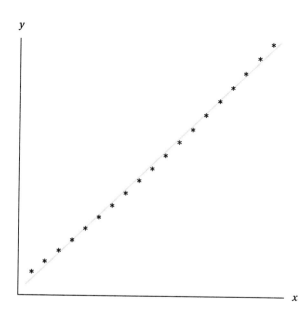

FIGURE 13.26 Scatterplot of a subtle curved relationship between Y and X.

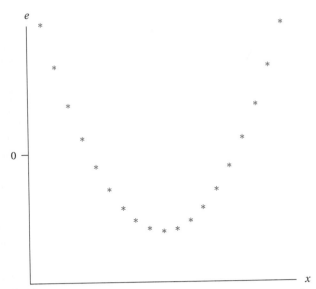

FIGURE 13.27 *Scatterplot of the residuals versus X for the subtle curved relationship of Figure 13.26.*

than 1 percent of the data—results in a 10 percent increase in the correlation coefficient, a 21 percent increase in the coefficient of determination, and a 7 percent decrease in the standard deviation of the residuals! Is it any wonder that researchers who need to find what they consider to be good answers are tempted to delete outliers? Earlier in this section it was shown that Rayford was an isolated case and an outlier. By contrast, Wade Boggs is an isolated case, but not an outlier. If Boggs is dropped from the original data set (including Rayford), the above summaries become $r = 0.513$, $R^2 = 0.263$, and $s = 0.0268$. Compared with the entire data set, dropping Boggs has decreased the correlation coefficient by 7 percent, decreased the coefficient of determination by 14 percent, and left the standard deviation of the residuals unchanged. Unlike Rayford, the large effect Boggs has on the analysis cannot be attributed to his having an unusual residual (recall that his residual equals 0.029). Instead it is due to the fact that his value of X is so extreme. It can be shown that extreme values of X potentially have more effect on the regression analysis than moderate values of X. (See the discussion of Figure 13.28 given below.)

For the Height study, recall that the most isolated case is the ordered pair (59,76). Not only is this case an outlier because of its residual, but it has an extreme value of X. The deletion of this single case decreases the standard deviation of the residuals by 9 percent, from 3.024 to 2.732, increases the correlation coefficient by 30 percent, from 0.411 to 0.536, and increases the coefficient of determination by 70 percent, from 0.169 to 0.287!

13.4 Residuals, Outliers, and Isolated Cases

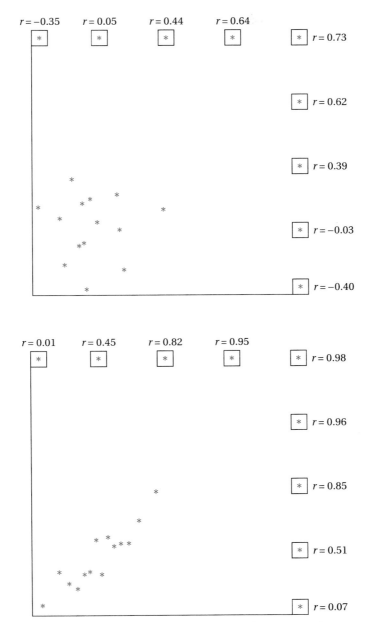

FIGURE 13.28 Examples of the effect of an isolated case on the value of the correlation coefficient. The 14 cases clustered in the upper figure have $r = -0.02$; those in the lower figure have $r = 0.91$. If one of the cases indicated by boxes is added to the data, r changes as indicated.

Figure 13.28 provides additional examples of the effect of one isolated case on the value of the correlation coefficient. In the top scatterplot in this figure, the 14 clustered points are nearly uncorrelated ($r = -0.02$). As illustrated in the figure, the addition of just one of the isolated cases (the boxed stars) can change r dramatically. In the bottom scatterplot the clustered cases have a strong positive relationship ($r = 0.91$). The addition of just one isolated case can alter r substantially in this plot, too.

This section has presented a number of examples that have illustrated the degree to which the correlation coefficient r can be influenced by as few as one isolated case. Recall that the correlation coefficient is included in the equation of the regression line; thus, it is not surprising to find that as few as one isolated case can have a tremendous impact on the regression line and, subsequently, the predictions $\hat{y}$. The details will not be given.

A related phenomenon can occur when data from two sources are combined into one scatterplot. Figure 13.29, for example, combines data for the small hunter and the web weaver spiders, which were presented earlier in this chapter.

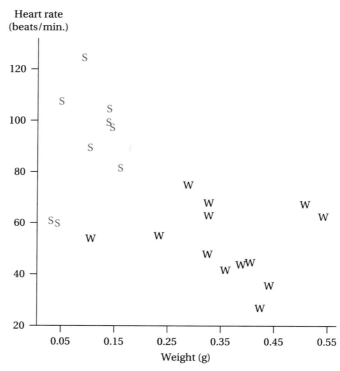

FIGURE 13.29 *Scatterplots of heart rate versus body weight for small hunters (S) and web weavers (W). For the small hunters only, $r = 0.397$; for the web weavers only, $r = -0.121$; for the combined data, $r = -0.606$.*

For the small hunters, the correlation coefficient for heart rate and weight is 0.397. For the web weavers, the correlation coefficient is −0.121. Thus, these two data sets tell somewhat different stories; in the first set, the linear relationship is positive and accounts for about 16 percent of the squared variation, and in the second set, the linear relationship is very weak and negative, and accounts for only about 1.5 percent of the squared variation. If the data sets are combined, however, the correlation coefficient is −0.606; the linear relationship is fairly strong and negative, accounting for almost 37 percent of the squared variation. The problem in practice is not so much that devious researchers combine data from different sources; rather, more often this happens inadvertently.

EXERCISES 13.4

1. Refer to the 12 scatterplots in Figure 13.9 (p. 455). Identify the two of these 12 plots that could possibly be a residual plot (that is, a plot of e versus x) and explain why they are the only possibilities.

2. Can a case be an outlier but not isolated? If your answer is yes, sketch a scatterplot with a case that is an outlier but not isolated. If your answer is no, explain why not.

3. Refer to Exercise 1 on p. 456, a comparison of critics' and the public's rating of movies. Figure 13.30 is a dot plot of the residuals, and Figure 13.31 is a scatterplot of the residuals versus X for these data.

   (a) Write a few sentences to describe the dot plot and the scatterplot.

   (b) Which movie has the smallest residual? The second smallest residual?

   (c) If the movies corresponding to the two smallest residuals are dropped from the study, the correlation coefficient increases from 0.360 to 0.414. Comment.

4. For each of the following, sketch a scatterplot that satisfies the stated conditions. (Hint: Refer to Figures 13.28 and 13.29.)

   (a) The scatterplot consists of two distinct groups of cases. For each group the relationship is linear, direct, and strong, but for the entire scatterplot the correlation coefficient is approximately 0.

   (b) The scatterplot consists of two distinct groups of cases. For one group the relationship is linear, direct, and strong, and for the other group the relationship is linear, indirect, and strong, but for the entire scatterplot the correlation coefficient is approximately 0.

   (c) The scatterplot consists of two distinct groups of cases. For each group the relationship is linear and the correlation coefficient is approximately 0, but for the entire scatterplot the relationship is linear, direct, and strong.

   (d) The scatterplot consists of two distinct groups of cases. For each group the relationship is linear and the correlation coefficient is approximately 0, but for the entire scatterplot the relationship is linear, indirect, and strong.

5. Glenda M. worked for an organization that provides people with training for the Law School Admission Test (LSAT). At the end of training, each client is given a practice version of the LSAT. Glenda was

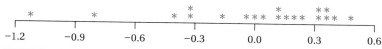

FIGURE 13.30 Dot plot of residuals for the Movie Rating study.

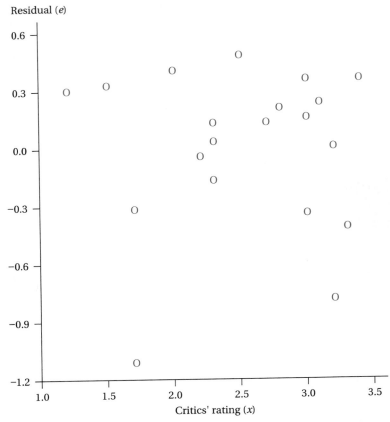

FIGURE 13.31 Scatterplot of residuals versus critics' rating for 20 movies.

interested in whether there was a relationship between a person's score on the practice exam, Y, and the person's undergraduate grade point average (based on a four-point scale), X. Glenda collected data on 34 of the organization's clients. A scatterplot suggests that the relationship between Y and X is approximately linear. A computer analysis yields $R^2 = 0.311$ and

$$\hat{y} = 117 + 12.3x.$$

(a) What is the value of the correlation coefficient, r?

(b) Find the predicted test score for a client whose undergraduate GPA was $x = 3.6$.

(c) Refer to your answer to part (b). Given that the client's actual test score was $y = 153$, find the value of the residual $e = y - \hat{y}$.

6. In the National Hockey League a team is awarded two points in the standings for each victory and one point for each tie. Table 13.8 presents data from the 1991–92 NHL regular season. A regression analysis was performed to investigate the relationship between points (Y) and the difference between goals scored and goals allowed in games (X). A scatterplot suggests that the relationship between Y and X is approximately linear. A computer analysis yields $R^2 = 0.932$ and

TABLE 13.8 *Data from the 1991–92 National Hockey League Regular Season*

Team	Points	Goals Scored	Goals Allowed	Difference (Scored − Allowed)
Boston	84	270	275	−5
Buffalo	74	289	299	−10
Calgary	74	296	305	−9
Chicago	87	257	236	21
Detroit	98	320	256	64
Edmonton	82	295	297	−2
Hartford	65	247	283	−36
Los Angeles	84	287	296	−9
Minnesota	70	246	278	−32
Montreal	93	267	207	60
New Jersey	87	289	259	30
New York Islanders	79	291	299	−8
New York Rangers	103	314	245	69
Philadelphia	75	252	273	−21
Pittsburgh	87	342	301	41
Quebec	52	255	318	−63
St. Louis	83	279	266	13
San Jose	39	219	359	−140
Toronto	67	234	294	−60
Vancouver	96	285	250	35
Washington	98	330	275	55
Winnipeg	81	251	244	7

$$\hat{y} = 79.9 + 0.3x.$$

(a) What is the correlation coefficient, r?

(b) Find the predicted number of points for Quebec; for Chicago.

(c) Refer to your answer to part (b). Find the residual for Quebec; for Chicago.

7. Each team in the National Basketball Association played 82 games during the 1991–92 regular season. Table 13.9 presents for each team the number of victories, the mean number of points scored per game, the mean number of points allowed per game, and the difference of these two means. A regression analysis was performed to investigate the relationship between the number of victories (Y) and the difference between the mean number of points scored and the mean number of points allowed per game (X). A scatterplot suggests that the relationship between Y and X is approximately linear. A computer analysis yields $R^2 = 0.917$ and

$$\hat{y} = 41 + 2.50x.$$

(a) What is the correlation coefficient, r?

(b) Find the predicted number of victories for Philadelphia; for Chicago.

(c) Refer to your answer to part (b). Find the residual for Philadelphia; for Chicago.

8. Each team in the National Basketball Association played 82 games during the 1990–91 regular season. Table 13.10 presents for each team the number of victories, the mean number of points scored per game, the mean number of points allowed per game, and the difference of these two means. A regression analysis was performed to investigate the relationship between the number of victories (Y) and the difference between the mean

TABLE 13.9 Data from the 1991–92 National Basketball Association Regular Season

Team	Victories	Points Scored per Game	Points Allowed per Game	Difference (Scored – Allowed)
Atlanta	38	106.2	107.7	−1.5
Boston	51	106.6	103.0	3.6
Charlotte	31	109.5	113.4	−3.9
Chicago	67	109.9	99.5	10.4
Cleveland	57	108.9	103.4	5.5
Dallas	22	97.6	105.3	−7.7
Denver	24	99.7	107.6	−7.9
Detroit	48	98.9	96.9	2.0
Golden State	55	118.7	114.8	3.9
Houston	42	102.0	103.7	−1.7
Indiana	40	112.2	110.3	1.9
L.A. Clippers	45	102.9	101.9	1.0
L.A. Lakers	43	100.4	101.5	−1.1
Miami	38	105.0	109.2	−4.2
Milwaukee	31	105.0	106.7	−1.7
Minnesota	15	100.5	107.5	−7.0
New Jersey	40	105.4	107.1	−1.7
New York	51	101.6	97.7	3.9
Orlando	21	101.6	108.5	−6.9
Philadelphia	35	101.9	103.2	−1.3
Phoenix	53	112.1	106.2	5.9
Portland	57	111.4	104.1	7.3
Sacramento	29	104.3	110.3	−6.0
San Antonio	47	104.0	100.6	3.4
Seattle	47	106.5	104.7	1.8
Utah	55	108.3	101.9	6.4
Washington	25	102.4	106.8	−4.4

number of points scored and the mean number of points allowed per game (X). A scatterplot suggests that the relationship between Y and X is approximately linear. A computer analysis yields $R^2 = 0.966$ and

$$\hat{y} = 41 + 2.48x.$$

(a) What is the correlation coefficient, r?
(b) Find the predicted number of victories for Philadelphia; for Chicago.
(c) Refer to your answer to part (b). Find the residual for Philadelphia; for Chicago.

9. The previous two exercises stated that the regression lines for the 1990–91 and 1991–92 seasons are

$\hat{y} = 41 + 2.48x$ and $\hat{y} = 41 + 2.50x$,

respectively. Compare these answers and comment.

10. Each team in the American League played 162 games during the 1987 season. Table 13.11 presents for each team the number of wins, the total number of runs scored during the season, the total number of runs allowed during the season, and the difference of these last two numbers. A regression analysis was performed to investigate the relationship between the number of wins (Y) and the difference between the number of runs scored and the number of runs allowed during the season (X). A scatterplot suggests that the relationship between Y and X is

TABLE 13.10 Data from the 1990–91 National Basketball Association Regular Season

Team	Victories	Points Scored per Game	Points Allowed per Game	Difference (Scored − Allowed)
Atlanta	43	109.8	109.0	0.8
Boston	56	111.5	105.7	5.7
Charlotte	26	102.8	108.0	−5.2
Chicago	61	110.0	101.0	9.0
Cleveland	33	101.7	104.2	−2.5
Dallas	28	99.9	104.5	−4.6
Denver	20	119.9	130.8	−10.9
Detroit	50	100.1	96.8	3.3
Golden State	44	116.6	115.0	1.6
Houston	52	106.7	103.2	3.5
Indiana	41	111.7	112.1	−0.4
L.A. Clippers	31	103.5	107.0	−3.5
L.A. Lakers	58	106.3	99.6	6.7
Miami	24	101.8	107.8	−6.0
Milwaukee	48	106.4	104.0	2.4
Minnesota	29	99.6	103.5	−3.9
New Jersey	26	102.9	107.5	−4.6
New York	39	103.1	103.3	−0.2
Orlando	31	105.9	109.9	−4.0
Philadelphia	44	105.4	105.6	−0.2
Phoenix	55	114.0	107.5	6.5
Portland	63	114.7	106.0	8.7
Sacramento	25	96.7	103.5	−6.8
San Antonio	55	107.1	102.6	4.5
Seattle	41	106.6	105.4	1.2
Utah	54	104.0	100.7	3.3
Washington	30	101.4	106.4	−5.0

TABLE 13.11 Data from the 1987 American League Baseball Season

Team	Wins	Runs Scored	Runs Allowed	Difference (Scored − Allowed)
Baltimore	67	729	880	−151
Boston	78	842	825	17
California	75	770	803	−33
Chicago	77	748	746	2
Cleveland	61	742	957	−215
Detroit	98	896	735	161
Kansas City	83	715	691	24
Milwaukee	91	862	817	45
Minnesota	85	786	806	−20
New York	89	788	758	30
Oakland	81	806	789	17
Seattle	78	760	801	−41
Texas	75	823	849	−26
Toronto	96	845	655	190

approximately linear. A computer analysis yields $R^2 = 0.879$ and
$$\hat{y} = 81 + 0.0941x.$$
(a) What is the correlation coefficient, r?

(b) Find the predicted number of victories for Minnesota; for Toronto.

(c) Refer to your answer to part (b). Find the residual for Minnesota; for Toronto.

13.5 DATA FROM TWO SOURCES*

Sometimes a researcher wants to compare similar data from two sources. Figure 13.29 (p. 488) provided an example that showed that combining the data from the two sources can be misleading. This optional section gives two examples of a valid way to compare data from two sources.

My house is heated with steam fueled by a natural gas boiler. In April 1990 the boiler cracked and I replaced it with a new high-efficiency model. I am interested in how much energy I save with the new boiler and have collected some data. This example will examine data I collected from 1989 to 1991.

The monthly bill from the local utility, Madison Gas and Electric, reports the amount of gas used, measured in therms, and the total number of degree days during the billing period. The weather bureau labels the mean of each day's high and low temperatures, in Fahrenheit, the "mean" temperature for the day. If this "mean" is below 65, then 65 minus the "mean" is a measure of the need for heating for that day, called the number of degree days. It seems natural to suspect that the use of gas in therms during a billing period depends upon the number of degree days. Thus, the number of therms is the dependent variable, Y, and the total number of degree days is the independent variable, X. The subjects are different billing periods, and the two sources are data collected on the old and new boilers.

One approach is to apply the methods introduced earlier in this chapter to the data from each source. For example, scatterplots should be drawn to obtain pictures of the data. Figure 13.32 presents both scatterplots drawn on the same coordinate system. In this figure, each case for the old boiler is plotted as an O and each case for the new boiler is plotted as an N. (This method will not work if numbers appear in the scatterplot.) Examining only the data from the old boiler (the Os), one sees that there are no isolated values and that there is a strong linear relationship between the variables. The same is true for the new boiler.

If the data from the two boilers are compared, it appears that for similar values of X the cases on the old boiler, O, are plotted higher than the cases on the new boiler, N. Thus, the old boiler appears to have required more gas to operate. Further insight can be gained by finding the regression line for each source. For the old boiler,
$$\hat{y} = 4.3 + 0.258x,$$
and for the new boiler,
$$\hat{y} = -4.0 + 0.230x.$$

Time Series

14.1 THE TIME SERIES PLOT
14.2 AUTOCORRELATION AND SMOOTHING

Chapter 14

Chapter 12 developed methods for presenting and summarizing data sets that consist of one numerical variable per subject. If the subjects are a sequence of trials observed in temporal order, the resulting data are called a **time series.**

The initial question to ask for a time series is whether the time order affects the response. If the analyst knows that the time order *does not* affect the response, the time order can be ignored and the data can be presented and summarized using the methods of Chapter 12. In addition, the methods of Chapter 15 can be used to perform inference, if appropriate. On the other hand, if the analyst knows that the time order *does* affect the response, the time order should not be ignored. For such data, the methods of Chapter 12 will fail to reveal important features of the data and the methods of Chapter 15 will give answers that are either very inefficient or incorrect.

The **time series plot** is the standard way to present time series data. The **autocorrelation** of a time series measures the strength of a particular type of time effect. In many cases, **smoothing** a time series makes it easier to see a time trend. This chapter provides a brief introduction to the time series plot, autocorrelation, and smoothing. Our modest goal is to gain some insight into whether a time trend is present in the data. If the analyst decides that there is no time trend in a time series, the methods in Chapter 12 and Chapter 15 may be used to study the data. If, however, the analyst decides that there is a time trend in the time series, more advanced methods are needed. Cryer [1] provides an excellent introduction to these more advanced methods of time series analysis.

14.1 THE TIME SERIES PLOT

EXAMPLE 14.1

Figure 14.1 is a time series plot of the high temperature in Madison, Wisconsin, on 27 alternate Sundays—January 2, January 16, January 30, ..., and December 31, in 1988. The plot contains 27 stars, one for each response and date combination. The vertical coordinate of a star denotes its response value (high temperature), and the horizontal coordinate denotes the trial (date). Two features of the data are readily apparent from a quick examination of Figure 14.1:

1. There is a strong pattern in the data. High temperatures vary substantially from one season to the next.
2. There is a great deal of variation superimposed upon the pattern. In other words, the transition from season to season is not a simple monotonic process. For example, the fourth day in the data set, February 13, was much colder than either of its neighboring points, January 30 or February 27.

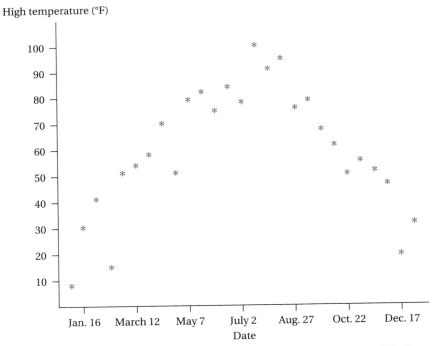

FIGURE 14.1 *Time series plot of biweekly high temperatures in Madison, Wisconsin, in 1988.*

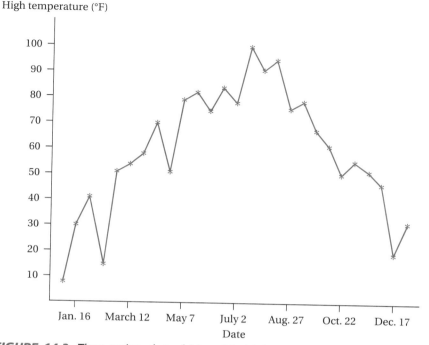

FIGURE 14.2 *Time series plot of biweekly high temperatures in Madison, Wisconsin, in 1988.*

These features can be emphasized by drawing connecting line segments between successive values, as illustrated in Figure 14.2.

Although it is not generally recommended, it is possible to compute the correlation coefficient, introduced in Chapter 13, of the response and the trial number. For these high-temperature data, $r = 0.175$. Obviously, this is an inappropriate measure because the time series plot clearly reveals a curved relationship between time (trial number) and response.

Figure 14.3 presents a dot plot of the 27 high temperatures. The main feature of the data revealed in the dot plot is the large variation in high temperatures in Madison. Although it is not literally *incorrect* to draw a dot plot of these data, the dot plot hides a main feature of the data, namely, their seasonality. ▲

EXAMPLE 14.2

Figure 14.4 is a time series plot of data obtained during August, 1988. Each day's response is $f - h$, where h denotes the high temperature in Madison and f denotes the forecasted high that appeared in that morning's edition of the

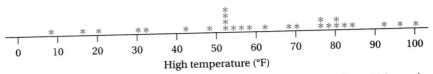

FIGURE 14.3 Dot plot of biweekly high temperatures in Madison, Wisconsin, in 1988.

Wisconsin State Journal. In words, the response equals the error in the forecast. For example, the forecast high for August 1 was 92 degrees and the actual high was 95 degrees, giving a forecast error of

$$f - h = 92 - 95 = -3 \text{ degrees.}$$

Two features are readily apparent from an inspection of the time series plot:

1. The value 10 for the response is quite different from the other values; it might be labeled an outlier.
2. Unlike the data of the previous example, these data provide no evidence of a strong pattern.

Although there is no obvious strong pattern in the series, I detect a downward trend in the responses over time. To investigate this further, the correlation coefficient can be computed and found to equal -0.047, reflecting a very weak

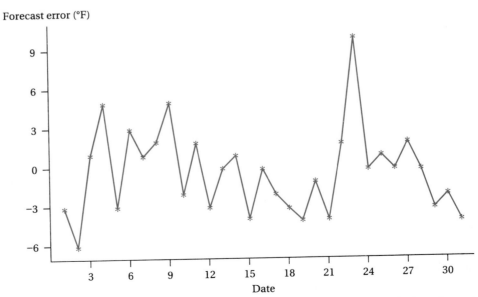

FIGURE 14.4 Time series plot of high-temperature forecast errors during August 1988 in Madison, Wisconsin.

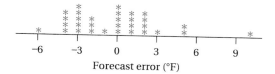

FIGURE 14.5 Dot plot of high-temperature forecast errors during August 1988 in Madison, Wisconsin.

downward linear trend. As discussed in the previous chapter, even one isolated case can have a large effect on the value of the correlation coefficient. If the isolated case, the response of 10 on August 23, is replaced by a less extreme value, 5, the resulting correlation coefficient decreases to -0.098. Thus, for this data set replacing the isolated value by a more typical value has only a minor effect on the value of the correlation coefficient.

Figure 14.5 is a dot plot of these data. Again, the response of 10 stands out as unusual. ▲

EXAMPLE 14.3

Figures 14.6 and 14.7 are a time series plot and a frequency histogram of the number of victories achieved by the National League pitcher with the most victories for each of the 51 seasons from 1937 to 1987 [2]. It is easier to examine the histogram first. There is one outlier, a low value of 14, that occurred in 1981. This low value is undoubtedly due to the player strike in 1981, which resulted in the cancellation of about one-third of the games. (Note that 14 is one-third smaller than the most common value in the data set, 21.)

The histogram is noticeably asymmetric and has an absolute peak at 21. In fact, one of the values 21, 22, or 23 occurred in 30 out of the 51 years. There is a minor peak at 27, one gap because of the strike value, 14, and no meaningful valleys. It is instructive to examine the values of several summary statistics, both

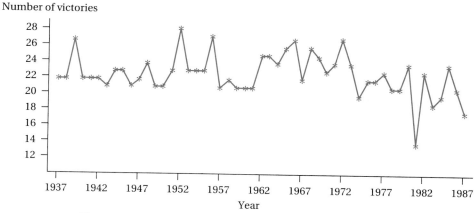

FIGURE 14.6 Time series plot of maximum number of victories for a National League pitcher, 1937–1987.

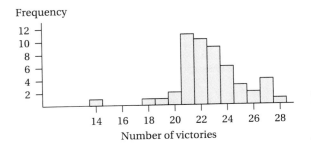

FIGURE 14.7 Frequency histogram of maximum number of victories for a National League pitcher, 1937–1987.

for the entire data set and for the data set with the value 14 deleted. The results are given in Table 14.1. In agreement with the general discussion of Chapter 12, the deletion of a single outlier greatly affects the standard deviation and the range; the former is reduced by 12 percent, from 2.503 to 2.198, and the latter is reduced by 29 percent, from 14 to 10. The quartiles do not change, and the mean increases by less than 1 percent, from 22.67 to 22.84. Curiously, the median increases more than the mean does, by 2 percent, from 22 to 22.5; this is due to the granularity of the data (see Section 12.5).

The time series plot indicates that 7 of the 10 highest responses (25 or more victories) occurred during the period from 1962 to 1972 and 6 occurred between 1962 and 1969. Possible explanations for this phenomenon include the following:

- It could be due to chance variation.
- As a reaction to Roger Maris's 61 home runs in 1961, baseball rules were changed to favor pitchers. In addition, in 1962 the National League added two expansion teams. The two expansion teams, the Mets and the Colt 45s (later Astros), were very bad, giving pitchers on other teams more games against weak opposition. By 1969 hitting statistics had fallen considerably, and the rules were modified to help batters. Perhaps these rule changes influenced the maximum number of victories by a pitcher.
- A number of excellent power pitchers were active during this time, including Don Drysdale, Sandy Koufax, Juan Marichal, Tom Seaver, and Steve Carlton. Koufax obtained three of the overall seven highest responses in just a four-year period, from 1963 to 1966, before arm trouble ended his career.

TABLE 14.1 Summary Statistics for Maximum Number of Victories for a National League Pitcher, 1937–1987

Values Deleted	Mean	Median	Standard Deviation	First Quartile	Third Quartile	Range
None	22.67	22.0	2.503	21	24	14
14	22.84	22.5	2.198	21	24	10

None of the 10 highest values occurred after 1972. Again, because this is an observational study, there is no obvious reason for this phenomenon. Perhaps it is due to the lack of a dominant pitcher, the increased importance of relief pitchers (which reduces the number of decisions—wins plus losses—for starting pitchers), or the increased reliance on a five-man rotation (which leads to fewer starts per pitcher).

The correlation coefficient between the responses and trial number (year) is -0.200. If, however, the value of 14 in 1981 is dropped from the series, the correlation coefficient changes to -0.129, a weakening of the evidence of a negative trend. On the other hand, if the strike year, 1981, is omitted, the correlation coefficient between the responses and trial number data for the post-1961 data is -0.687, a fairly stong negative trend. ▲

It is easy to see the underlying pattern in the high-temperature data in Figure 14.2. In addition, the pattern is not surprising to anyone who is familiar with the seasonal weather variations in the midwestern United States. In fact, patterns that are expected based on other knowledge of the data-generating process often are easy to spot. It is somewhat more difficult to spot a pattern in the baseball data in Figure 14.6. I see only the increase in level between 1962 and 1969 and the downward trend since 1972. Finally, I am unable to identify visually any significant pattern in the temperature forecast data presented in Figure 14.4.

The next section presents a method for quantitatively measuring the strength of a particular kind of pattern and a method for smoothing a time series to make its underlying pattern more apparent.

14.2 AUTOCORRELATION AND SMOOTHING

14.2.1 Autocorrelation

Recall that correlation was introduced in Chapter 13 as a measure of the strength of a linear relationship between two variables. The autocorrelation of a time series is, as the name suggests, its correlation with itself, in a sense that is made precise in the following example.

EXAMPLE 14.4

Consider the high-temperature data in Figure 14.2, a time series with 27 observations. Each observation in the series, except the first one, has a previous value. For example, the second observation, 30° on January 16, has a previous value of 8° on January 2. Figure 14.8 is a scatterplot of the observations versus their respective previous values. For example, the case plotted as an A instead of

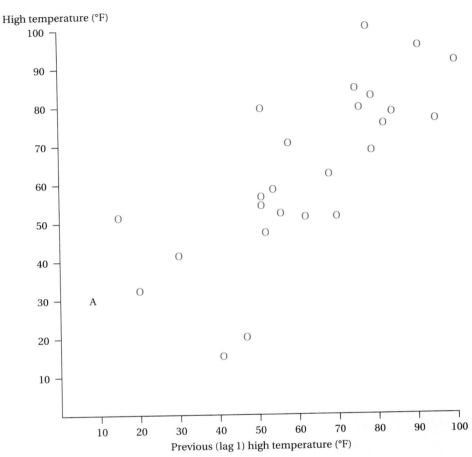

FIGURE 14.8 *Scatterplot of high temperature versus previous high temperature.*

an O is for January 16; its *y*-coordinate is the high on January 16, 30°, and its *x*-coordinate is the previous high, 8° on January 2. The scatterplot in Figure 14.8 reveals a strong positive linear relationship between the observations and their previous values. This linear relationship can be summarized by computing the correlation coefficient. For the current data, the correlation coefficient equals 0.775. Thus,

$$R^2 = (0.775)^2 = 0.601,$$

which implies that 60.1 percent of the variation in the 26 high temperatures (all but the first observation) can be explained by a linear relationship with the previous high. This large value of the coefficient of determination is a result of the strong seasonality in the data. ▲

More generally, **lagging** is the association of observations with previous values in a time series. The **lag** is the interval of time between consecutive observations in a time series. (In this book, only series with constant lags will be considered.) For example, in the above high-temperature study, the lag is two weeks; for the earlier examples of forecast errors and maximum victories, the lags are one day and one year, respectively. The **lag length** is the number of time periods from a current value to its associated previous value. For the example in the preceding paragraph, the lag length is one because each value is associated with its immediate predecessor. The correlation in the previous paragraph, which has numerical value 0.775, is called the **autocorrelation of lag 1;** "auto" is for self-correlation, and "lag 1" refers to the lag length. The autocorrelations of lags 2, 3, and so on, are defined analogously.

EXAMPLE 14.5

Figure 14.9 is a scatterplot of the maximum number of victories by a National League pitcher versus the lag one (immediately previous) value. Recall that this series has one outlier, the value 14 for the strike year, 1981. Note that this

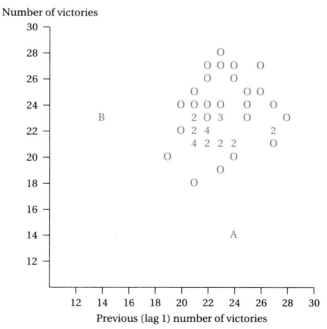

FIGURE 14.9 *Scatterplot of maximum number of victories versus previous maximum number of victories, 1937–1987.*

outlier results in two isolated cases in Figure 14.9, the case marked *A* when the outlier is a current value and the case marked *B* when the outlier is a lag 1 value. These two cases should be deleted before the lag 1 autocorrelation is computed; the result is 0.212. (If the two cases are not deleted, the autocorrelation drops to 0.116.) Note that

$$R^2 = (0.212)^2 = 0.045;$$

thus, only 4.5 percent of the variation in the values in the series can be attributed to a linear relationship with the previous value. For this example, the autocorrelation suggests a weak relationship between each value and its immediate predecessor. ▲

EXAMPLE 14.6

Figure 14.10 is a scatterplot of the same-day high-temperature forecast errors for August 1988 versus their previous values. The autocorrelation of lag 1 equals 0.108, reflecting the extremely weak linear relationship. This value is not influenced by the outlier of 10 in the series; if it is replaced by the next largest value, 5, the lag 1 autocorrelation changes only slightly, to 0.091. ▲

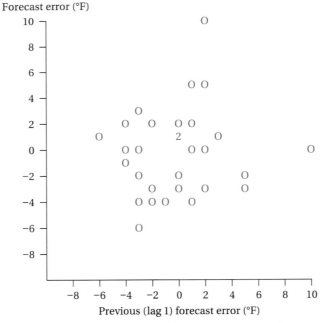

FIGURE 14.10 *Scatterplot of high-temperature forecast error versus previous high-temperature forecast error.*

14.2.2 Smoothing a Time Series

A time series plot can be **smoothed** to make its underlying pattern, if there is one, more visible. A single or repeated application of the **running median of three smoother** is examined in the remainder of this section.

 The running median of three smoother replaces each response in a time series by the median of three numbers: the response itself and the responses immediately before and after it in time. The first and last responses are not changed by the smoother.

EXAMPLE 14.7

Consider the high-temperature data in Figure 14.2 (p. 503). The result of applying the running median of three smoother to this series is given in Figure 14.11. The computations leading from a series to a smoothed series are tedious and are usually performed by a computer. It is instructive, however, to inspect a few of the computations to discover what the smoother does. The first five responses in the original time series are 8, 30, 41, 15, and 51. After the running

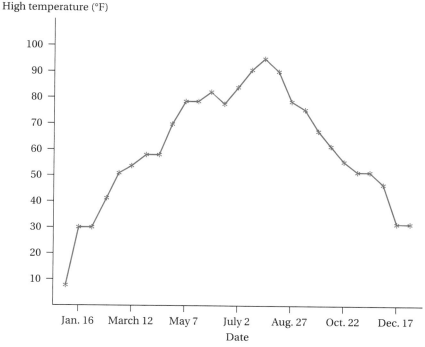

FIGURE 14.11 Biweekly high-temperature series after one application of the running median of three smoother.

median of three smoother is applied to the time series, the first response remains 8 because the smoother does not change the first and last responses. The second response in the original series is 30, and its immediate neighbors in the original series include a smaller number, 8, and a larger number, 41. The median of these three numbers is 30; thus, the smoother does not change the value of the second response. In general, if a response in the original series has one immediate neighbor that is less than or equal to it, and one immediate neighbor that is greater than or equal to it, the response is not changed by the smoother. By contrast, the third response in the original series, 41, is larger than both its immediate neighbors, 30 and 15. In the time series plot of the original series the relationship between these three numbers contributes a jagged up-and-down pattern to the series. The smoother softens, or smooths, this pattern by replacing 41 by 30, the closer value of its immediate neighbors. Similarly, the fourth response, 15, is smaller than both its immediate neighbors, 41 and 51. In the time series plot of the original series the relationship between these three numbers contributes a jagged down-and-up pattern to the series. The smoother softens this pattern by replacing 15 by 41, the closer value of its immediate neighbors.

Compared with the original time series, the smoothed series, Figure 14.11, more clearly shows the seasonal pattern for high temperatures in Madison.

A time series can be smoothed a second time, a third time, and so on. Simply use the output time series from one smoothing as the input for the next smoothing. For the current example, the result of a second application of the smoother, given in Figure 14.12, is a small, but clear, improvement (in terms of smoothness) over the once-smoothed series. Eventually, further smoothing ceases to change the series. For example, Figure 14.12, the result of the second smoothing, cannot be smoothed further. This process, smoothing until no further change is possible, is called the repeated running median of three smoother. ▲

EXAMPLE 14.8

Figure 14.13 presents the twice-smoothed time series of the maximum number of victories by a National League pitcher (see Figure 14.6). Further smoothing will not change the series. The smoothing provides a clearer picture of the series; it is fairly constant from 1937 to 1961, and an increase in level occurs in 1962, which persists for a few years and is followed by a long, gradual decline. ▲

EXAMPLE 14.9

Figure 14.14 presents the temperature forecast error series after two smoothings. The original series had eight of 31 values outside the range from -3 to $+3$; the smoothing has removed all of these values except for the last observation, -4, which, of course, is never changed by the smoother. Thus, the smoothed series gives an overly optimistic picture of the accuracy of the forecasts. This is a good example of the following rule:

Do not smooth a series that has no apparent pattern.

14.2 Autocorrelation and Smoothing • **513**

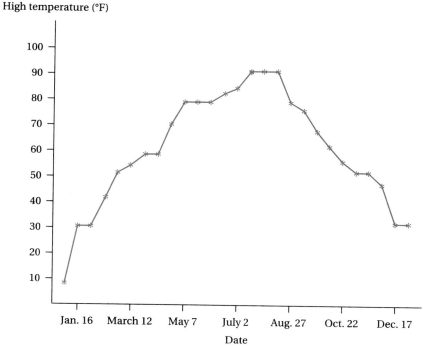

FIGURE 14.12 *Biweekly high-temperature series after two applications of the running median of three smoother.*

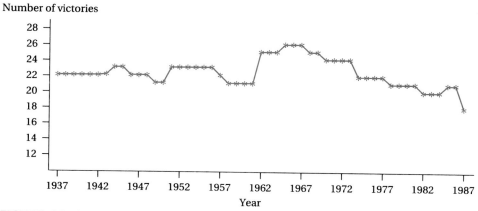

FIGURE 14.13 *Maximum number of victories series after two applications of the running median of three smoother.*

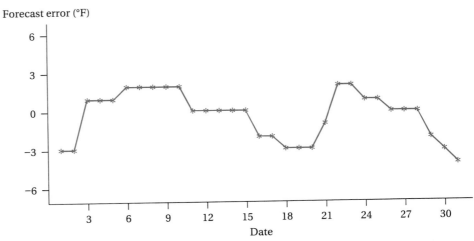

FIGURE 14.14 *Temperature forecast error series after two applications of the running median of three smoother.*

(If you think that Figure 14.14 reveals a hidden pattern in the time series, consider the following. I took the 31 forecast errors and arranged them in random order. I smoothed the resultant time series three times and obtained the time series presented in Figure 14.15. The pattern in this plot is meaningless because there was no pattern in the original series.) ▲

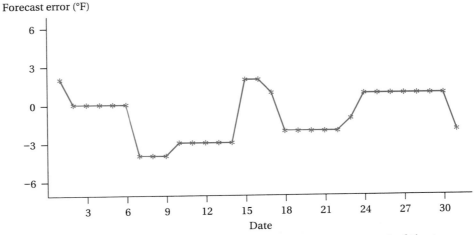

FIGURE 14.15 *Result of three smoothings of a random arrangement of the temperature forecast error series.*

EXERCISES FOR CHAPTER 14

1. Hank Aaron holds the major league baseball record for the most career home runs. Figures 14.16, 14.17, and 14.18 present a time series plot of his annual home run total, the running median of three smoothed version of the time series, and a scatterplot of the number of home runs versus the lag 1 number of home runs [2].

   (a) Examine the time series and describe what it reveals.

   (b) Examine the smoothed series and describe what it reveals.

   (c) Compare the utility of the previous two series.

   (d) One of the following numbers is the lag 1 autocorrelation; which one?

   $-0.823, -0.405, 0.003, 0.405, 0.823$

   (e) If the last two years of Aaron's career are dropped from the time series, the two lowest points in Figure 14.18 are deleted and the lag 1 autocorrelation becomes 0.015. Comment.

2. The Milwaukee Brewers baseball team has existed since 1970. Here are its winning percentages in chronological order for 1970 through 1987 [2]:

   0.401, 0.429, 0.417, 0.457, 0.469, 0.420, 0.410, 0.414, 0.574, 0.590, 0.531, 0.569, 0.586, 0.537, 0.416, 0.441, 0.478, 0.562

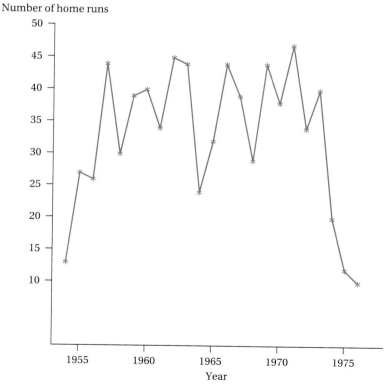

FIGURE 14.16 Time series plot of annual number of home runs hit by Hank Aaron during his major league career.

FIGURE 14.17 Hank Aaron's annual home run series after one application of the running median of three smoother.

(a) Construct a time series plot of these values. What does it reveal?

(b) Construct the running median of three smoothed version of the series. What does it reveal? Compare it with the original time series.

3. The methods of this chapter can be used to look for a time trend in a CRD on a sequence of trials. For example, consider the CRD to examine the effect of distance on score in darts, as described in Exercise 14 on p. 406. It can be shown that the sample mean of the 15 scores from the shorter distance is 201.5 and the sample mean of the 15 scores from the longer distance is 188.0. The raw data are presented in Table 14.2. The entry in the "Deviation" column of this table equals the score minus 201.5 for games from the shorter distance and equals the score minus 188.0 for games from the longer distance.

(a) Construct the time series of the score versus the game number. Comment.

(b) Construct the time series of the deviation versus the game number. Comment.

(c) Compare your two time series. Which do you think is better for detecting a time trend in the data? Why?

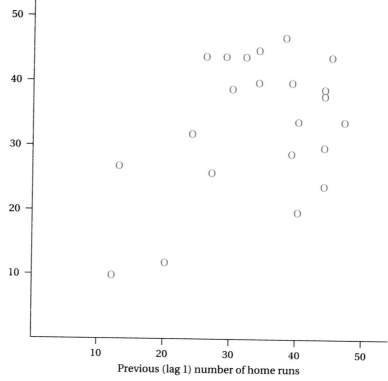

FIGURE 14.18 Scatterplot of Hank Aaron's home run total versus previous home run total.

TABLE 14.2 Scores of 30 Dart Games

Game	Distance	Score	Deviation	Game	Distance	Score	Deviation
1	10 ft.	200	−1.5	16	10 ft.	220	18.5
2	12 ft.	175	−13.0	17	12 ft.	206	18.0
3	12 ft.	174	−14.0	18	10 ft.	205	3.5
4	10 ft.	210	8.5	19	10 ft.	200	−1.5
5	12 ft.	186	−2.0	20	12 ft.	196	8.0
6	10 ft.	205	3.5	21	10 ft.	198	−3.5
7	12 ft.	191	3.0	22	12 ft.	196	8.0
8	10 ft.	206	4.5	23	10 ft.	198	−3.5
9	12 ft.	201	13.0	24	10 ft.	197	−4.5
10	12 ft.	200	12.0	25	12 ft.	164	−24.0
11	10 ft.	215	13.5	26	10 ft.	184	−17.5
12	10 ft.	215	13.5	27	10 ft.	181	−20.5
13	12 ft.	197	9.0	28	12 ft.	168	−20.0
14	12 ft.	203	15.0	29	12 ft.	163	−25.0
15	12 ft.	200	12.0	30	10 ft.	189	−12.5

TABLE 14.3 1990 Bus Ridership by Month in Madison, Wisconsin

Month	Number of Riders	Cumulative Number of Riders
Jan.	777,681	777,681
Feb.	853,738	1,631,419
March	872,574	2,503,993
April	772,592	3,276,585
May	757,757	4,034,342
June	584,372	4,618,714
July	573,488	5,192,202
Aug.	626,308	5,818,510
Sep.	797,166	6,615,676
Oct.	925,222	7,540,898
Nov.	857,917	8,398,815
Dec.	697,670	9,096,485

4. Table 14.3 presents the number of riders on the Madison bus system, per month, for 1990 [3].

   (a) Construct a time series plot of the number of riders versus month. Describe the pattern in the data.

   (b) Construct a time series plot of the cumulative number of riders versus month. Describe the pattern in the data.

   (c) Compare your two time series. Which do you prefer?

REFERENCES

1. Jonathan D. Cryer, *Time Series Analysis* (Boston: Duxbury Press, 1986).
2. J. L. Reichler, ed., *The Baseball Encyclopedia* (New York: Macmillan, 1988).
3. *Wisconsin State Journal*, March 10, 1991.

Inference for One Numerical Population

15.1 INTRODUCTION
15.2 INFERENCE FOR THE POPULATION MEDIAN
15.3 INFERENCE FOR THE POPULATION MEAN
15.4 INFERENCE FOR TIME SERIES DATA*
15.5 PREDICTIONS*

Chapter 15

15.1 INTRODUCTION

Chapters 12, 13, and 14 presented methods for displaying and describing numerical data. This chapter focuses on statistical inference for numerical data consisting of one variable measured on subjects from one source. In particular, methods are developed that enable a researcher to use the data from a sample of subjects to infer characteristics of a larger population of subjects. All inference procedures assume that the subjects under study are selected at random from the population of interest.

The broad goal of this chapter, to use sample data to infer population characteristics, is familiar. Chapters 6 and 10 considered this problem for dichotomous and multicategory variables, respectively. You may want to review the earlier material. In particular, the following ideas should be familiar.

First, a population can represent a well-defined collection of distinct individuals, or it can represent a mathematical model of a process that generates data. In either case, a population is identified as a sampling distribution, or probability distribution.

Second, population inference methods in this text are based on the assumption that the subjects comprise a random sample from a population. For each type of population, the phrase *random sample* has a unique meaning.

- For a population corresponding to distinct individuals, a random sample can be obtained as follows:
    (a) Construct a population box containing one card per population member.

(b) Select n cards at random, with replacement, from the population box. The subjects whose cards are selected constitute the sample.

This method of sampling is important mathematically because, among other reasons, it enables the analyst to use the multiplication rule to compute probabilities of complicated events. In practice, it is common for a researcher to select subjects at random without, rather than with, replacement. As before, if the sample size, n, is less than 5 percent of the population size, N, the formulas of this chapter apply to sampling without replacement. If the sample size is greater than 5 percent of the population size, the formulas of this chapter can be improved (details will not be given).

- For a population that models a process, a random sample is obtained whenever the analyst is willing to assume the following:

  (a) **Stationarity:** There is no change in the process generating the data; the sampling distribution is the same for all trials.

  (b) **Independence:** The value of the response on any trial is not influenced by the outcomes of any collection of other trials.

The assumption of independence is important mathematically because it allows the analyst to use the multiplication rule to compute probabilities. Stationarity is important because if the population changes, more complicated analysis methods are needed. (Time series inference methods may be appropriate if either or both of these assumptions are violated.)

Each type of population presents a challenge to the researcher. For a collection of distinct individuals, the challenge lies in obtaining a list of the population members and actually selecting a sample of size n at random. (The 1990 U.S. census, with a suspected undercount, demonstrates that it is difficult to obtain a complete list of a population.) For a process, the analyst must critically examine the assumptions of stationarity and independence, a topic considered in Section 15.4. In particular, the prediction of an outcome of a future trial should be interpreted cautiously; any such prediction implicitly depends on the conditions of stationarity and independence continuing to hold in the future. That is, at best, a risky assumption.

For a numerical response, the population is again identified as a probability distribution, but it is necessary to distinguish numerical variables obtained by counting from those obtained by measuring. The following artificial example of a count response is used for illustration throughout this chapter because of its arithmetic simplicity. It is called the cat population.

EXAMPLE 15.1: The Cat Population

A community consists of 100,000 households. Suppose that 10,000 households have no cats, 50,000 households have one cat, 30,000 households have two cats, and the remaining 10,000 households have three cats. Let X denote the number

TABLE 15.1 Sampling Distribution of X, the Number of Cats in a Randomly Selected Household

x	P(X = x)	xP(X = x)	(x − μ)	(x − μ)²P(X = x)
0	0.1	0.0	−1.4	0.196
1	0.5	0.5	−0.4	0.080
2	0.3	0.6	0.6	0.108
3	0.1	0.3	1.6	0.256
Total	1.0	μ = 1.4		σ² = 0.640

of cats in a randomly selected household. The first two columns of Table 15.1 give the sampling distribution of X. (For example, the probability of selecting a household with two cats is 30,000 divided by 100,000, or 0.3, agreeing with the tabulated value of $P(X = 2)$.) Therefore, the first two columns of the table also give the population. The remaining entries in the table show, following the formulas given in Chapter 3, that the population mean, μ, is equal to 1.4 and that the population variance, σ^2, is equal to 0.64. (For the inference methods of this chapter, it is not necessary for you to remember how to compute the population mean and variance.)

For a numerical response it turns out to be particularly helpful to draw a picture, namely the probability histogram, of the population. The probability histogram for the cat population is presented in Figure 15.1. Recall that a probability histogram can be used to compute probabilities. For example, for the cat population the probability of the value 1 can be obtained by computing the area of the rectangle centered at 1. This rectangle has a base of 1 and a height of 0.5; thus, its area is $1(0.5) = 0.5$; this area agrees with the value of $P(X = 1)$ given in Table 15.1. ▲

The preceding presentation for the cat population can be adapted to any count response on a population consisting of different individuals. For a process

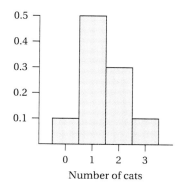

FIGURE 15.1 Probability histogram for the cat population.

that generates a count response, the population is taken to have this same form: a list of possible values and their probabilities. Depending upon how much the researcher knows about the process, the population may be completely known or unknown. The following two examples illustrate these ideas.

EXAMPLE 15.2

Let X denote the number obtained by casting an ordinary six-sided die. The possible values of the random variable X are

$$1, 2, 3, 4, 5, \text{ and } 6,$$

and the probabilities of these values are denoted by

$$P(X = 1), P(X = 2), P(X = 3), P(X = 4), P(X = 5), \text{ and } P(X = 6).$$

If the researcher is willing to assume that the die is balanced, these six probabilities are the same and therefore equal to one-sixth. (Dice can be manufactured that are balanced or almost balanced, as evidenced by the fact that casinos make a profit from the game craps regardless of the player's betting strategy. On the other hand, as the backgammon scene in the movie *Octopussy* demonstrates, it can be foolhardy to assume automatically that all dice are balanced.) Assuming that the die is balanced, the probability distribution of X, or in other words, the population, is given by the following:

x	$P(X = x)$
1	$\frac{1}{6}$
2	$\frac{1}{6}$
3	$\frac{1}{6}$
4	$\frac{1}{6}$
5	$\frac{1}{6}$
6	$\frac{1}{6}$
Total	1

It is easy to draw a probability histogram for this population (details are not given). ▲

EXAMPLE 15.3

Twice a week, Lotto America officials select six numbers from the integers from 1 to 54. Cash prizes (in whole dollars only) are given to persons holding tickets matching 4, 5, or 6 of the selected numbers. Let X denote the prize given for matching exactly four numbers. *If* an analyst is willing to assume that the officials select six numbers at random *and* the players select their numbers at random, the exact sampling distribution of X can be obtained (details not given). Although it may be reasonable to assume the officials select numbers at random, it has been demonstrated in many studies that Lotto players, as a group, select some numbers more than others. Thus, the sampling distribution of X, the population, is assumed to be unknown. ▲

Measurement data are more complicated. Consider a population of one million women, and let the response be a woman's height in inches. The precision of the measuring device is a nuisance. For example, if one researcher, Sally, measures each woman's height to the nearest 1 inch and another researcher, Juanita, measures each woman's height to the nearest 0.1 inch, each will have a different population (probability distribution), even though they are studying the same women! Although Juanita's population is arguably better than Sally's, because its measurements are more precise, it is not the best population; someone else could measure the women even more precisely. The usual practice is to assume that the actual population is given by a smooth curve. The smooth curve can be viewed as the population for the exact values of the response if there were no limitations to the precision of measuring devices—the smooth curve is the underlying truth that Sally and Juanita are approximating. Alternatively, the smooth curve can be viewed as a mathematical convenience; an approximation to the populations of Sally and Juanita, much as the standard normal curve is a smooth curve approximation to the sampling distribution of a number of test statistics and other random variables. In any event, the concept of a population as a smooth curve is needed as a mathematical device to prove many of the results in the remainder of this book. (See the various discussions throughout the remainder of this book concerning the importance of mathematical assumptions.) The smooth curve corresponding to a population is called its **probability density function**, abbreviated pdf, which is read "pea dee eff." The plural is abbreviated pdf's and is read "pea dee effs." Figure 15.2 presents perhaps the simplest pdf; it is called the uniform, or rectangular, pdf on the interval [0, 1].

Just like a probability histogram, a pdf can be used to obtain probabilities by computing areas. For example, for the pdf in Figure 15.2, the probability of obtaining a value between 0.25 and 0.75 is equal to the area under the pdf

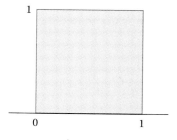

FIGURE 15.2 Uniform (or rectangular) probability density function on the interval [0,1].

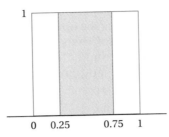

FIGURE 15.3 *Obtaining probabilities by computing areas. The area of the shaded region is the probability of obtaining a value between 0.25 and 0.75 when sampling from the uniform pdf on [0,1].*

between these two values. This area is displayed in Figure 15.3; because the region of interest is a rectangle, its area is easy to compute. The area of a rectangle is its base times its height; in this case the base is $0.75 - 0.25 = 0.50$ and the height is one, yielding an area of $0.50(1) = 0.50$. Thus, the probability of obtaining a value between 0.25 and 0.75 is equal to 0.50.

All pdf's satisfy the following three conditions:

1. The graph of a pdf is always equal to or above the horizontal axis. (Recall that this is a characteristic of probability histograms.)
2. The proportion of the population with response values between any two numbers is equal to the area under the pdf between the two numbers. (There is a similar result for probability histograms.)
3. The total area under a pdf equals 1. (This is also true for a probability histogram; it reflects the fact that the total probability must equal 1.)

Unless the pdf is a simple geometric shape, such as the rectangle in Figure 15.2, the computation of areas can be difficult, requiring methods of calculus. Fortunately, various areas are tabulated for the pdf's that are most important in practice (this will be expanded upon later).

EXAMPLE 15.4 Hybrids

In practice, the distinction between count and measurement data can be blurred. For example, consider the 1993 income for each household in the United States. Since the smallest unit of U.S. currency is the cent, a household's income is determined by counting the number of cents earned during the year. It would be horrendously unwieldy to visualize this population as a table with millions of rows or as a probability histogram with millions of rectangles. Instead, the population is idealized to be (is approximated by) an unknown smooth curve. ▲

15.1.1 Estimation of a Population

It is important to remember that, in practice, a researcher does not know the population (that is, does not know the exact probability histogram or pdf). The researcher will select a random sample from the population and use the resulting

TABLE 15.2 Results of a Random Sample of Size 100 from the Cat Population

Value	Frequency	Relative Frequency
0	9	0.09
1	44	0.44
2	35	0.35
3	12	0.12
Total	100	1.00

data to draw conclusions about the population. The method is illustrated with the cat population introduced earlier.

A researcher studying the cat population would not know that the population is given by the probability histogram in Figure 15.1. Suppose the researcher selects a random sample of size $n = 100$ from the population and obtains the data summarized in Table 15.2. These data can be presented as a density scale histogram, as in Figure 15.4. The density scale histogram can be viewed as an estimate of the population. For the cat population the accuracy of the estimate can be assessed because the population is known. A convenient method of comparison is to draw the estimate and the (true) probability histogram in the same picture, as shown in Figure 15.5. Unfortunately, if the population is not known, as is typical in practice, the accuracy of the estimate cannot be assessed in this way. At this point it is natural for you to suspect that the next topic to be covered is the method for using the estimate, the density scale histogram, combined with some measure of variation to obtain some kind of confidence set for the true population. It would be great for a researcher to be able to say

> I am 95 percent confident that the true population is equal to one of the possible populations in the confidence set.

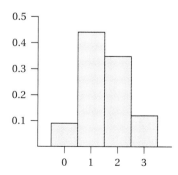

FIGURE 15.4 Density scale histogram of the random sample of size 100 from the cat population.

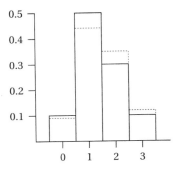

FIGURE 15.5 Comparison of the population (solid lines) and its estimate (dashed lines) for the cat population.

(By analogy, for a dichotomous response, the researcher can be 95 percent confident that the value of p, which determines the population, is somewhere between two numbers computed from the data.) Unfortunately, mathematical statisticians have demonstrated that a confidence set cannot be obtained unless additional assumptions are made about the structure of the population. Therefore statisticians have focused their attention on how to perform inference for two particular numerical characteristics of a population, the mean and median.

The qualitative conclusions on the difficulties with estimating the cat population, a population for a count variable, apply to measurement variables. Namely, a density scale histogram is a reasonable estimate of the pdf, but a confidence set for the pdf cannot be obtained. Thus, inferential attention is focused on the mean or median of the pdf. Therefore, it is useful to present some properties of these two characteristics of the pdf. The ability to compute the mean and median (and the often needed standard deviation, σ) of an arbitrary pdf typically involves calculus and is therefore beyond the scope of this text. The facts below, however, enable the analyst to determine the exact value of the mean and median for some pdf's and the approximate values for others, without performing any computations.

The population median, denoted by ν (the Greek letter nu), is particularly easy to visualize for a pdf. It is the number with the property that the areas to its left and to its right both equal $\frac{1}{2}$. Analogous to the result for probability histograms, the mean of a pdf is equal to its center of gravity. Thus, if a pdf is symmetric about a point, both the population mean and median are equal to that point. There is no simple rule for looking at a pdf and seeing the value of the population standard deviation. It suffices to remember that the more dispersion in the pdf, the larger is the standard deviation. Figure 15.6 illustrates these facts for a number of pdf's.

A feature common to all pdf's is surprising to many people. The feature will be illustrated with a random variable X that has the uniform pdf on the interval [0,1] (see Figure 15.2). Consider the computation of the probability that X equals any particular number, say 0.25. Following the rule for pdf's, this probability equals the area under the uniform curve between the numbers 0.25 and 0.25.

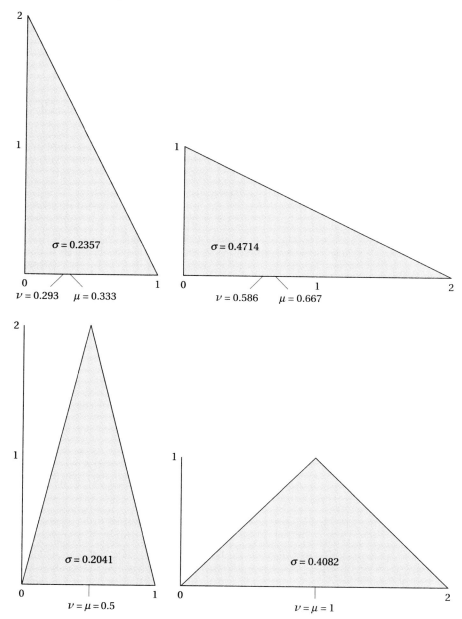

FIGURE 15.6 Four probability density functions with their medians (ν), means (μ), and standard deviations (σ).

In other words, the probability equals the area of the line segment drawn above 0.25 in Figure 15.3. The area of a line segment equals 0; thus,

$$P(X = 0.25) = 0.$$

This argument can be extended; if X is any random variable with a pdf, then

$$P(X = b) = 0 \quad \text{for every number } b. \tag{15.1}$$

Equation (15.1) can be puzzling and unnatural if it is interpreted incorrectly. For example, continuing an example introduced earlier in this chapter, let X denote the height of a woman selected at random from a population of college women and let $b = 65$ inches. Literally, Equation 15.1 states that $P(X = 65) = 0$. In words, it is impossible to select a woman 65 inches tall. This may come as a surprise to any woman who believes she is 65 inches tall! This apparent paradox can be explained as follows. First, if a woman states that she is 65 inches tall, it is implicit that she means 65 inches tall to some unstated level of precision. Suppose the level of precision is that height is measured to the nearest $\frac{1}{2}$ inch. Then the everyday statement, "I am 65 inches tall," actually means, "My height is between 64.75 and 65.25 inches." (Most people don't like to talk this way.) Thus, using the pdf, the probability of selecting a woman who *states* that she is 65 inches tall would be the area under the pdf between the values 64.75 and 65.25. This area is not 0.

15.1.2 The Family of Normal Curves

Undoubtedly the most important pdf's are the members of the family of normal curves. In addition to satisfying the conditions for being a pdf, normal curves have the following properties:

- Every normal curve is bell-shaped and symmetric.
- A normal curve is completely determined by two numbers: its point of symmetry (which equals its mean and its median) and its standard deviation.

The standard normal curve is a member of the family of normal curves; its mean is 0 and its standard deviation is 1. Suppose X is a random variable with its probability distribution given by a normal curve with mean equal to μ and standard deviation equal to σ. This is written as

$$X \sim N(\mu, \sigma).$$

With this notation,

$$Z \sim N(0, 1)$$

signifies that the random variable Z has the standard normal curve for its sampling distribution. Figure 15.7 presents the $N(10, 1)$ and $N(11, 0.5)$ pdf's.

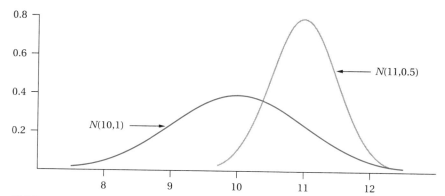

FIGURE 15.7 The N(10, 1) and N(11, 0.5) probability density functions.

 Suppose that $X \sim N(\mu, \sigma)$. It can be shown that the standardized version of X,

$$Z = \frac{X - \mu}{\sigma},$$

has the standard normal curve for its sampling distribution.

The utility of this fact is illustrated in the following example.

EXAMPLE 15.5: IQ Scores

Subjects' raw scores on intelligence tests are converted to an IQ score in such a manner that the population of IQ scores can be well approximated by a normal curve with mean and standard deviation equal to 100 and 15, respectively. Suppose a member of Mensa (an organization of people with high IQ scores) wants to know what proportion of the population has an IQ score above 130. First, the question must be written symbolically; let X denote the IQ score of a randomly selected person. According to the information given above,

$$X \sim N(100, 15).$$

The Mensa member wants to determine

$$P(X > 130).$$

This probability can be obtained by standardizing X:

$$P(X > 130) = P\left(\frac{X - 100}{15} > \frac{130 - 100}{15}\right) = P(Z > 2).$$

Thus, the desired probability is the area under the standard normal curve to the right of two. From Table A.2, this area equals 0.0028. ▲

15.1.3 Two Final Cautions

All inference methods in the current and remaining chapter of this book have been derived by mathematical statisticians. In order to obtain answers, these researchers had to make some assumptions about the population. Often, the assumptions are quite severe. For example, it is common to assume that the population is some member of the family of normal curves. As a practical matter, an analyst must be concerned with the validity of the answers obtained from these methods in those situations when the assumptions are false. Put simply, the question is

> How well does the method perform if the assumptions necessary to derive it are false?

This is a difficult question, and it usually has a complex and incomplete answer. Studies of this question are called studies of **robustness**. A method is called **robust** if it gives valid answers even when its assumptions are false. In the remainder of this book, the presentation of each method will include a discussion of its robustness.

The statistical method(s) selected by an analyst should reflect the scientific goals of the inquiry. Statistical expediency should not determine the goals of the study. For example, if the goal of a study is to determine the pdf, the analyst should use a density scale histogram as an estimate of the population, even though a confidence interval and P-value cannot be obtained. This chapter, the remainder of the book, and much statistical methodology develop a great many results on how to infer the values of one or more population means from data. There are two reasons for this great interest in population means:

- Knowledge of the population mean(s) is often very useful in answering the practical question that motivates a study.
- Especially for data from complicated designs or models, the formulas and interpretations of inference for population means are much simpler than for other population characteristics.

Because of the multitude of statistical methods available, both the expert statistician and the beginning student face a tremendous temptation to study the mean. Resist following this impulse mindlessly! Think about each problem carefully and determine its scientific goals; if the goals concern the mean, fine. But if the mean is not of interest, do not study it. Not only should you not study it, but, as the following example demonstrates, you must not let someone else convince you that the mean is all-important.

EXAMPLE 15.6

The following item appeared in *Newsweek* [1]. It was attributed to "a memo to flight attendants on United Airlines—whose DC-8s and DC-10s all show their date of manufacture on a metal plate in the frame of the forward passenger door":

> To avoid increasing a potentially high level of customer anxiety, please use the following responses when queried by customers. Question: How

old is this aircraft? Answer: I'm unaware of the age of this particular aircraft. However, the average age of United's Aircraft is 13.5 years.

If a person boarding a plane is concerned about its age, the ages of all the other planes in the fleet (which contribute to the mean, median, and so on) are irrelevant. Only the age of the plane in question matters. If a person making a ticket reservation is concerned about the age of the plane to be assigned and also is willing to assume that the plane will be assigned at random from the fleet of the airline's planes, then the person would be interested in the *distribution* of ages of the entire fleet—not the mean or median. Knowing that the mean (or median; it is not clear from the quote) is 13.5 is not particularly useful. Perhaps all of the planes are exactly 13.5 years old. Or perhaps one-half are new and one-half are 27 years old. ▲

15.2 INFERENCE FOR THE POPULATION MEDIAN

Recall that ν denotes the median of the population. The methods of this section assume that the population is a pdf. In particular, if X is the response for a randomly selected subject, the mathematical derivations of this section require that

$$P(X < \nu) = P(X > \nu) = 0.50.$$

In other words, it is assumed that the probability of obtaining a response that is exactly equal to ν is 0:

$$P(X = \nu) = 0.$$

(Although this condition on X is true for some populations for count variables, an analyst cannot be certain it is true unless the population is a pdf.)

The natural choice for the point estimate of ν is the observed value of the sample median, $\tilde{x}$, whose computation was explained in Chapter 12. The derivation of the formula for the confidence interval for the median ν is fairly simple for a sample of size $n = 2$; details are given in the following section.

15.2.1 Confidence Interval for n = 2

Suppose a researcher selects a random sample of size 2 from a population. Let X_1 represent the first response obtained and X_2 the second. Define $X_{(1)}$ (note the parentheses in the subscript) to equal the smaller of the two responses, X_1 and X_2, and $X_{(2)}$ to equal the larger. (If the responses are equal, then $X_{(1)} = X_{(2)}$.) The random variables $X_{(1)}$ and $X_{(2)}$ are called the first and second order statistics, respectively, of the sample. As usual, let lower case x's denote the observed values; in particular, $x_{(1)}$ and $x_{(2)}$ are the observed values of the first and second order statistics.

The confidence interval for ν is

$$[x_{(1)}, x_{(2)}]. \tag{15.2}$$

In words, the first order statistic is the lower bound of the interval and the second order statistic is the upper bound. The confidence level for this interval is obtained below.

Upon reflection, one clearly sees that three possible events can occur,

$$C = \{X_{(1)} \leq \nu \leq X_{(2)}\},$$
$$A = \{\nu < X_{(1)} \leq X_{(2)}\},$$

and

$$B = \{X_{(1)} \leq X_{(2)} < \nu\}.$$

Event C is the event that the confidence interval is correct, that is, it contains the number ν; the event A is the event that the confidence interval is above the median; and the event B is the event that the confidence interval is below the median. Of course, because ν is unknown, a researcher never knows which of these events has occurred. Note that $P(C)$ is the coverage probability for the confidence interval in Formula 15.2 and that

$$P(C) + P(A) + P(B) = 1.$$

The value of $P(A)$ will now be computed. The key step is to realize that the event

$$A = \{\nu < X_{(1)} \leq X_{(2)}\},$$

can be written as

$$A = \{\nu < X_1, \nu < X_2\}.$$

In words, the median is smaller than both *ordered* observations if, and only if, it is smaller than both *unordered* observations. This step is important because the multiplication rule can be used for unordered observations, yielding

$$P(A) = P(\nu < X_1)P(\nu < X_2).$$

By the assumption of this section, each of the probabilities on the right side of this equation equals 0.5. Thus,

$$P(A) = 0.5 \times 0.5 = 0.25.$$

By a similar argument (see Exercise 18),

$$P(B) = 0.25.$$

Since

$$P(C) + P(A) + P(B) = 1,$$

it follows, by substitution, that

$$P(C) + 0.25 + 0.25 = 1,$$

or $P(C) = 0.50$. In words, the confidence level of the interval $[x_{(1)}, x_{(2)}]$ is 50 percent.

15.2.2 Confidence Interval for Arbitrary n

The technique of the previous section can be easily extended to values of n larger than 2, although the details get messy. First, some notation is needed. A random sample of size n is selected from the population. The values of the n responses are denoted by

$$X_1, X_2, \ldots, X_n.$$

Set $X_{(1)}$ (note the parentheses) equal to the smallest of the responses, $X_{(2)}$ equal to the second smallest, and so on, and $X_{(n)}$ equal to the largest. Thus,

$$X_{(1)}, X_{(2)}, \ldots, X_{(n)},$$

are the responses ordered from smallest to largest; they are collectively referred to as the **order statistics.** Further, $X_{(1)}$ is called the first order statistic, $X_{(2)}$ is called the second order statistic, and so on. As usual, lowercase letters represent the numerical values actually obtained in the sample. Thus,

$$x_{(1)}, x_{(2)}, \ldots, x_{(n)}$$

denote the observed values of the order statistics.

For sample sizes n less than or equal to 20, Table A.7 gives the exact confidence level for various confidence intervals. For example, according to Table A.7, if $n = 10$, the confidence interval $[x_{(1)}, x_{(10)}]$ has confidence level equal to 99.8 percent. This is an extremely high confidence level; a researcher may prefer the shorter confidence interval that a lower confidence level would yield. The next two entries in Table A.7 indicate that for $n = 10$ the interval $[x_{(2)}, x_{(9)}]$ has confidence level 97.9 percent and the interval $[x_{(3)}, x_{(8)}]$ has confidence level 89.1 percent.

EXAMPLE 15.7

The above ideas are illustrated with some artificial data. The family of **lognormal** pdf's is an important collection of populations. There are two features of the lognormal family that are important in this text:

- If a random variable X has a lognormal pdf for its sampling distribution, X must be positive.
- All lognormal pdf's are skewed to the right.

A lognormal pdf is determined by the values of two numbers (called parameters). Figure 15.8 is a graph of the lognormal pdf with its first parameter equal to 5 and its second parameter equal to 1. (Technical note: The *order* of the parameters matters; the lognormal pdf with parameters 5 and 1 is different from the lognormal pdf with parameters 1 and 5. The name *lognormal* comes from the following

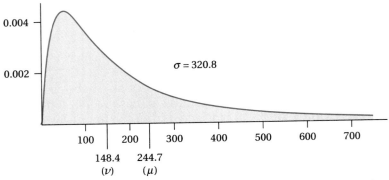

FIGURE 15.8 Example of a probability density function skewed to the right: the lognormal pdf with parameters 5 and 1.

fact: If the sampling distribution of a random variable X is a lognormal pdf, the random variable Y whose value is equal to the logarithm of X—remember that X is always positive, so its logarithm is always defined—has for its sampling distribution a normal pdf.) Note that the lognormal pdf in Figure 15.8 is rather severely skewed to the right. Using calculus one can show that the median of this population is $\nu = 148.4$ and that the mean and standard deviation of this population are $\mu = 244.7$ and $\sigma = 320.8$. A random sample of size $n = 10$ from this population yielded

$$350 \quad 76 \quad 19 \quad 141 \quad 323 \quad 25 \quad 304 \quad 77 \quad 157 \quad 195$$

These numbers give the following 10 observed order statistics:

$x_{(1)}$	$x_{(2)}$	$x_{(3)}$	$x_{(4)}$	$x_{(5)}$	$x_{(6)}$	$x_{(7)}$	$x_{(8)}$	$x_{(9)}$	$x_{(10)}$
19	25	76	77	141	157	195	304	323	350

The 99.8 percent confidence interval for the population median ν is

$$[x_{(1)}, x_{(10)}] = [19, 350].$$

The 97.9 percent confidence interval for ν is

$$[x_{(2)}, x_{(9)}] = [25, 323].$$

The 89.1 percent confidence interval for ν is

$$[x_{(3)}, x_{(8)}] = [76, 304].$$

Note that each of these three intervals contains the actual value of ν, namely 148.4. ▲

For $n > 20$ there is an accurate approximation method for obtaining a confidence interval for the median. Its use will now be described. First, the researcher specifies a target value for the confidence level; typically 80, 90, 95, 98, or 99 percent. This specification yields a value for z from Table A.3.

Next, the researcher obtains the sample values of the order statistics,

$$x_{(1)}, x_{(2)}, \ldots, x_{(n)}.$$

The third step is the computation of

$$k' = \frac{n+1}{2} - \frac{z\sqrt{n}}{2}.$$

Usually, k' is not an integer; round it *down* to the next integer and call the result k (if k' is an integer, then $k = k'$). The confidence interval is

$$[x_{(k)}, x_{(n+1-k)}].$$

In words, the confidence interval extends from a lower limit of the kth order statistic to an upper limit of the $(n+1-k)$th order statistic.

EXAMPLE 15.8

Example 15.7 presented a random sample of size 10 from the population of Figure 15.8. Table 15.3 presents the order statistics of a random sample of size $n = 50$ from the same population. For a 99 percent confidence interval, $z = 2.576$ and

$$k' = \frac{n+1}{2} - \frac{z\sqrt{n}}{2} = \frac{51}{2} - \frac{2.576\sqrt{50}}{2} = 25.5 - 9.11 = 16.39.$$

Thus, $k = 16$ and

$$n + 1 - k = 50 + 1 - 16 = 35.$$

The approximate 99 percent confidence interval for the population median ν is

$$[x_{(16)}, x_{(35)}] = [89, 222],$$

which is considerably narrower than the 99.8 percent confidence interval for $n = 10$, which was shown to be [19, 350]. For a 98 percent confidence interval, $z = 2.326$ and

$$k' = \frac{n+1}{2} - \frac{z\sqrt{n}}{2} = \frac{51}{2} - \frac{2.326\sqrt{50}}{2} = 25.5 - 8.22 = 17.28.$$

TABLE 15.3 Order Statistics for a Random Sample of Size 50 from the Lognormal PDF with Parameters 5 and 1

14	39	40	40	44	57	59	60	63	67
72	73	74	79	88	89	93	98	102	106
109	111	112	119	129	130	135	141	162	165
165	191	206	219	222	230	296	303	306	315
328	375	444	510	545	566	720	733	763	770

Thus, $k = 17$,
$$n + 1 - k = 50 + 1 - 17 = 34,$$
and the approximate 98 percent confidence interval is
$$[x_{(17)}, x_{(34)}] = [93, 219],$$
which is considerably narrower than the 97.9 percent confidence interval for $n = 10$, which was shown to be [25, 323]. Finally, for a 90 percent confidence interval, $z = 1.645$ and
$$k' = \frac{n+1}{2} - \frac{z\sqrt{n}}{2} = \frac{51}{2} - \frac{1.645\sqrt{50}}{2} = 25.5 - 5.82 = 19.68.$$
Thus, $k = 19$,
$$n + 1 - k = 50 + 1 - 19 = 32,$$
and the approximate 90 percent confidence interval is
$$[x_{(19)}, x_{(32)}] = [102, 191],$$
which is considerably narrower than the 89.1 percent confidence interval for $n = 10$, which was shown to be [76, 304]. Note that each of these three confidence intervals includes the value $\nu = 148.4$. ▲

15.2.3 Robustness of the Confidence Interval

If the population is not a pdf, then possibly $P(X = \nu)$ is not 0. (For a count response, a positive proportion of the population may have a response value equal to the median.) If this happens, a consequence is that $P(X < \nu)$ or $P(X > \nu)$ is smaller than $\frac{1}{2}$. The net effect of this is to make the coverage probability of the confidence interval even larger than it would be if the population were a pdf. (If interested, you can verify this for the case when $n = 2$.) For example, for $n = 10$, Table A.7 states that the interval $[x_{(3)}, x_{(8)}]$ has a confidence level of 89.1 percent. If the population is not a pdf, the actual confidence level may be higher than 89.1 percent. This is the sort of discrepancy analysts love! To know that the actual (unknown) confidence level is larger than the level if the assumption is true is certainly preferable to knowing it might be smaller. This is one of the best robustness results in statistics.

EXAMPLE 15.9

Tetris is a video game that can be played on the Nintendo Entertainment System. A player rotates and translates falling shapes comprised of 4 blocks to form complete horizontal rows of 10 blocks. Each completed row disappears from the screen and blocks above drop into the vacated space to allow room for more blocks. A player's score equals the number of rows completed before the

TABLE 15.4 *Tetris Scores from 1990 in Time Order (Left to Right)*

| 94 | 103 | 114 | 100 | 106 | 100 | 111 | 75 | 74 | 95 | 98 | 51 | 94 |
| 70 | 81 | 101 | 110 | 85 | 93 | | 90 | 112 | 90 | 107 | 106 | 73 |

screen overflows. After every 10 completed rows, the speed of the falling shapes increases, making the game more difficult. I became an avid (not skilled, but avid) player and after several weeks concluded that I had stopped improving. It was the winter of 1990 and I wanted to salvage something from my time, so I collected some data. Specifically, I played Tetris 25 times and recorded my scores. A game of Tetris takes about 15 minutes, so the data were collected on five different days, with five games played each day. The data constitute a time series with the complication that the lags are unequal (the lags between days are longer than the lags within days). Table 15.4 presents the 25 scores in the order they were obtained (chronological order). These data are analyzed in this text as a random sample from an unknown population. As discussed in the introduction to this chapter, this assumption is equivalent to the assumptions of stationarity and independence. In other words, it is assumed that my ability did not change during the time of the study (stationarity) and that no score was influenced by other scores (independence). If the series were longer, it would be possible to use the ideas in Section 15.4 to investigate the validity of this assumption.

Table 15.5 presents the 25 Tetris scores ordered from smallest to largest. Figure 15.9 presents a frequency histogram of the 25 scores. The histogram reveals one small outlier, the score of 51. (The dot plot, not shown here, does not reveal any other interesting features of the data set.) The sample mean, median, and standard deviation of the 25 scores are

$$\bar{x} = 93.32, \quad \tilde{x} = 95, \quad \text{and} \quad s = 15.60.$$

TABLE 15.5 *Ordered Tetris Scores from 1990*

| 51 | 70 | 73 | 74 | 75 | 81 | 85 | 90 | 90 | 93 | 94 | 94 | 95 |
| 98 | 100 | 100 | 101 | 103 | 106 | 106 | 107 | 110 | 111 | 112 | 114 | |

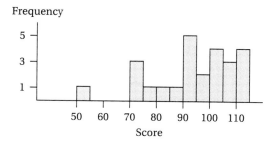

FIGURE 15.9 *Frequency histogram of 25 Tetris scores from 1990.*

The methods of this section can be used to estimate the median, ν, of the unknown population. First, the value of the sample median, 95, is a point estimate of the median of the unknown population. Because the sample size is larger than 20, the approximate confidence interval method can be used. For a 95 percent confidence interval, $z = 1.96$ and

$$k' = \frac{n+1}{2} - \frac{z\sqrt{n}}{2} = \frac{25+1}{2} - \frac{1.96\sqrt{25}}{2} = 13 - 4.9 = 8.1.$$

This is not an integer, so it is rounded down to give $k = 8$. Next, this value of k is substituted into the formula for the confidence interval $[x_{(k)}, x_{(n+1-k)}]$ yielding

$$[x_{(8)}, x_{(25+1-8)}] = [x_{(8)}, x_{(18)}].$$

Finally, a quick examination of Table 15.5 reveals that the observed values of the 8th and 18th order statistics are 90 and 103. Thus, the 95 percent confidence interval for the population median is [90, 103]. Of course, a Tetris score is a count variable, so the robustness result of this section indicates that the actual confidence level may be larger than the nominal level, which is approximately 95 percent. (Technical note: By extending the method that generated Table A.7, it can be shown that the exact confidence level for the interval $[x_{(8)}, x_{(18)}]$ for $n = 25$ and a pdf is 95.6 percent. Thus, the actual (unknown) confidence level for the Tetris study is 95.6 percent or larger.) ▲

15.2.4 Hypothesis Test for the Median

Following the framework first introduced in Chapter 6 for a population model, the first step of the hypothesis test is the specification of the special value of interest and the choice of hypotheses. If there is no special value of interest, a hypothesis test should not be conducted. The null hypothesis states that the population median, ν, equals the special value of interest, which is denoted ν_0. As usual, there are three choices for the alternative: the actual median may exceed, fall short of, or differ from the hypothesized value, referred to, respectively, as the first, second, and third alternatives. Below is a display of the three possibilities:

$$H_0: \nu = \nu_0 \qquad H_0: \nu = \nu_0 \qquad H_0: \nu = \nu_0$$
$$H_1: \nu > \nu_0 \qquad H_1: \nu < \nu_0 \qquad H_1: \nu \neq \nu_0$$

The test of this section should be used only for measurement responses and then only if the researcher has a sufficiently precise measuring device to ensure that each observation is either smaller or larger than the special value of interest. For example, if a researcher wants to test the null hypothesis that the median height of college women is, say, 65 inches, it is imperative that he or she determines for each woman whether she is taller or shorter than 65 inches. The reason for this restriction is made apparent shortly.

Based on the above restriction, each observation in the sample can be assumed to be either larger or smaller, but not equal to, the special value of interest. Label an observation a success if it is larger than the special value of interest, and a failure if it is smaller. With this terminology, the sample

$$X_1, X_2, \ldots, X_n$$

can be viewed as a sequence of successes and failures. In fact, it is a sequence of Bernoulli trials with p equal to the (unknown) proportion of the population that is larger than ν_0. Let Y be the total number of successes in the sample; the random variable Y has a binomial distribution with parameters n and p.

Next, note that if the null hypothesis is correct and the median equals the special value of interest, then $p = 0.50$. Thus, the null hypothesis,

$$H_0: \nu = \nu_0,$$

is equivalent to

$$H_0: p = 0.50.$$

Further, consider the first alternative, $\nu > \nu_0$. For example, for the women's height study mentioned above, suppose the actual median is 65.3 inches, which, of course, is larger than the special value of interest, 65 inches. Clearly, since one-half of the women are taller than 65.3 inches, more than one-half are taller than 65 inches. Thus, the first alternative implies that p is larger than 0.5. Similarly, the second alternative, $\nu < \nu_0$, implies that $p < 0.50$, and the third alternative, $\nu \neq \nu_0$, implies that $p \neq 0.50$.

In summary, the original three choices for the pair of hypotheses,

$$\begin{array}{ccc} H_0: \nu = \nu_0 & H_0: \nu = \nu_0 & H_0: \nu = \nu_0 \\ H_1: \nu > \nu_0 & H_1: \nu < \nu_0 & H_1: \nu \neq \nu_0 \end{array}$$

are mathematically equivalent to the following pairs of hypotheses, respectively:

$$\begin{array}{ccc} H_0: p = 0.50 & H_0: p = 0.50 & H_0: p = 0.50 \\ H_1: p > 0.50 & H_1: p < 0.50 & H_1: p \neq 0.50. \end{array}$$

The test of these hypotheses was presented in Chapter 6.

15.2.5 Robustness of the Test

It is absolutely imperative that each observation in the sample is either a success or a failure; that is, larger or smaller than the special value of interest. (This requirement can be relaxed, but in a way that is not simple and is beyond the scope of this text.) Thus, the test cannot be applied to count data (unless the special value of interest is not a possible value of the response—for example, a fraction—though it is arguably unnatural to select a fraction as the special value of interest for the population median of a count variable). For measurement data the test should be used only if the measuring device is sufficiently precise to enable the researcher to classify each observation a success or failure.

EXERCISES 15.2

1. Use Table A.7 to find the exact confidence level of the following confidence intervals for the population median:

   (a) $[x_{(2)}, x_{(8)}]$ for $n = 9$
   (b) $[x_{(6)}, x_{(14)}]$ for $n = 19$

2. Use Table A.7 to find the exact confidence level of the following confidence intervals for the population median:

   (a) $[x_{(5)}, x_{(13)}]$ for $n = 17$
   (b) $[x_{(3)}, x_{(10)}]$ for $n = 12$

3. Use Table A.7 to find the confidence interval for the population median that comes closest to satisfying the goals below. Give the exact confidence level for the interval you select.

   (a) 95 percent confidence for $n = 14$
   (b) 90 percent confidence for $n = 10$

4. Use Table A.7 to find the confidence interval for the population median that comes closest to satisfying the goals below. Give the exact confidence level for the interval you select.

   (a) 95 percent confidence for $n = 18$
   (b) 90 percent confidence for $n = 16$

5. Refer to the data in Exercise 1 on p. 401. Assume that each data set is a random sample from a population.

   (a) Find the 89.1 percent confidence interval for the population median time it took Brian to run 1 mile while wearing combat boots.
   (b) Repeat part (a) for Brian's data on wearing jungle boots.

6. Refer to the Tetris data presented in Table 15.5. Assuming that the 1990 data is a random sample from a population, construct the 80 percent confidence interval for the population median.

7. Refer to the data in Exercise 3 on p. 402. Assume that each data set is a random sample from a population.

   (a) Find the approximate 95 percent confidence interval for the population median distance Jennifer could hit a ball with an aluminum bat.
   (b) Repeat part (a) for Jennifer's data on the wooden bat.

8. Refer to the data in Exercise 4 on p. 402. Assume that each data set is a random sample from a population.

   (a) Use Table A.7 to find a confidence interval with level equal to approximately 95 percent for the population median of the number of rounds required by Doug to score 301 with his own darts. What is the confidence level given in Table A.7? In view of the discussion of robustness in the text, how does the actual confidence level compare to the value in Table A.7?
   (b) Repeat part (a) for Doug's scores with the bar darts.

9. Refer to the data in Exercise 5 on p. 403. Assume that each data set is a random sample from a population.

   (a) Use Table A.7 to find the 93.8 percent confidence interval for the population median time Kymn required to row the equivalent of 2,000 meters on the ergometer on the small gear, vent closed setting.
   (b) Repeat part (a) for the large gear, vent open setting.

10. Table 15.6 presents the dollar values of 208 prizes given on the television show "The Price Is Right." (These are from the portion of the show in which four contestants bid in an attempt to come closest to the actual price of the prize without going over it.) Assuming that these numbers are a random sample from a population, construct the 95 percent confidence interval for the population median.

11. Refer to the data in Exercise 11 on p. 405. Assume that each data set is a random sample from a population.

    (a) Find the approximate 95 percent confidence interval for the population median score with the first treatment.

TABLE 15.6 Dollar Values of 208 Prizes on "The Price Is Right"

413	413	455	455	459	465	495	495	499	499	500	500	504
510	515	520	529	529	529	530	535	570	570	579	585	589
599	600	650	660	665	665	695	698	699	699	699	700	795
799	799	800	800	813	814	819	819	819	829	848	849	850
855	855	879	880	884	895	899	899	899	900	907	909	932
950	950	975	976	976	979	987	990	998	998	998	999	999
999	999	1,000	1,000	1,000	1,000	1,000	1,009	1,019	1,020	1,020	1,044	1,045
1,048	1,049	1,078	1,080	1,080	1,080	1,099	1,099	1,099	1,100	1,100	1,139	1,150
1,167	1,169	1,169	1,170	1,175	1,191	1,198	1,199	1,200	1,200	1,200	1,200	1,200
1,200	1,200	1,207	1,220	1,222	1,250	1,250	1,253	1,253	1,253	1,277	1,282	1,282
1,285	1,295	1,295	1,299	1,325	1,325	1,345	1,347	1,360	1,395	1,395	1,399	1,399
1,400	1,449	1,449	1,450	1,450	1,450	1,450	1,450	1,450	1,465	1,482	1,485	1,498
1,499	1,499	1,499	1,500	1,500	1,500	1,500	1,500	1,515	1,537	1,539	1,547	1,547
1,547	1,550	1,550	1,550	1,550	1,551	1,556	1,563	1,565	1,568	1,590	1,595	1,595
1,599	1,599	1,599	1,600	1,600	1,618	1,622	1,650	1,650	1,650	1,672	1,695	1,698
1,698	1,699	1,736	1,754	1,759	1,760	1,767	1,780	1,807	1,870	1,895	1,979	1,995

(b) Repeat part (a) for the data on the second treatment.

12. Table 12.11 (p. 428) presents an ordered list of 153 cash prizes for matching four of six numbers in Lotto America. Assuming that these numbers are a random sample from a population, construct the 95 percent confidence interval for the population median.

13. Refer to the golf data presented in Exercise 13 on p. 406.
    (a) Assume that the data on a 3-wood are a random sample from a population. Construct the 95 percent confidence interval for the median of the population.
    (b) Assume that the data on a 3-iron are a random sample from a population. Construct the 95 percent confidence interval for the median of the population.

14. Table 12.12 (p. 429) presents an ordered list of 152 cash prizes for matching five of six numbers in Lotto America. Assuming that these numbers are a random sample from a population, construct the 95 percent confidence interval for the population median.

15. Refer to the data on car acceleration times presented in Exercise 15 on p. 406.
    (a) Assume that the data with the windows open are a random sample from a population. Construct the 95 percent confidence interval for the median of the population.
    (b) Assume that the data with the windows closed are a random sample from a population. Construct the 95 percent confidence interval for the median of the population.

16. *The Baseball Encyclopedia* [2] lists career records for every man who has ever played major league baseball. A random sample of 200 from the population of position players was selected; the number of career games for these players is given in Table 12.6 (p. 394). Use these data to find the 95 percent confidence interval for the median of the population.

17. Refer to the data you collected for Exercise 25 on p. 410. Assume that each data set is a random sample from a population.
    (a) Find the approximate 95 percent confidence interval for the population median score with the first treatment.
    (b) Repeat part (a) for the data on the second treatment.

18. In Section 15.2.1 the event B is defined as
$$B = \{X_{(1)} \leq X_{(2)} < \nu\}.$$
Use the multiplication rule to show that $P(B) = 0.25$.

15.3 INFERENCE FOR THE POPULATION MEAN

Throughout Section 15.3 it is assumed that a random sample of size n is selected from a population that has an unknown mean, denoted μ, and an unknown standard deviation, denoted σ. The values of the n responses are denoted by the random variables

$$X_1, X_2, \ldots, X_n.$$

The random variables $\bar{X}$ and S are the sample mean and sample standard deviation, respectively. The observed values of the response are denoted by

$$x_1, x_2, \ldots, x_n,$$

and the observed values of the sample mean and standard deviation are denoted by $\bar{x}$ and s.

The mean of the sample is a natural point estimate of the mean of the population. In order to construct a confidence interval for, or perform a hypothesis test on, the value of the mean of the population, it is necessary to study the sampling distribution of the point estimate, the sample mean. This is a very difficult problem; not only is the exact sampling distribution often tedious to determine, but it depends on the population, which is assumed to be unknown. Fortunately, there are simple formulas for the expected value and standard error of the sampling distribution of the sample mean.

- The expected value is equal to the mean of the population,
$$E(\bar{X}) = \mu.$$

- The variance is equal to the variance of the population divided by the sample size,
$$\text{Var}(\bar{X}) = \frac{\sigma^2}{n}.$$

- The standard error (standard deviation) is equal to the standard deviation of the population divided by the square root of the sample size,
$$\text{SE}(\bar{X}) = \frac{\sigma}{\sqrt{n}}.$$

The utility of these equations is that

Given the population distribution, although it may be tedious to obtain the exact sampling distribution of $\bar{X}$, it is easy to obtain the expected value, variance, and standard error of the sampling distribution of $\bar{X}$.

It follows from the equations above that the standardized version of $\bar{X}$, denoted Z, is given by

$$Z = \frac{\bar{X} - \mu}{\sigma/\sqrt{n}}.$$

Fractions within fractions make formulas look more formidable than they are and can cause computational errors; thus, they should be eliminated whenever possible, unless the cure is worse than the disease. Rewriting the previous equation gives

$$Z = \frac{\sqrt{n}(\bar{X} - \mu)}{\sigma}. \tag{15.3}$$

In this section interest is focused on μ; not surprisingly, the presence of the unknown σ in the formula for Z is troublesome. Define G to be Z with the unknown population standard deviation, σ, replaced by the sample standard deviation, S:

$$G = \frac{\sqrt{n}(\bar{X} - \mu)}{S} \tag{15.4}$$

One of the most famous results in probability theory, the central limit theorem, is that the sampling distribution of Z given in Equation 15.3 can be well approximated by the standard normal curve if the sample size n is sufficiently large. Slutsky showed that, in addition, the sampling distribution of G given in Equation 15.4 can be well approximated by the standard normal curve if the sample size n is sufficiently large. Slutsky's result can be used to obtain approximate confidence intervals for μ and approximate P-values for hypothesis testing for μ. The main practical difficulty with using Slutsky's result is the following:

> How large must n be in order for the standard normal curve approximation to be considered good? Unfortunately, the answer to this question depends on (typically) unknown characteristics of the population. For any sample size, no matter how large, one can always manufacture a population for which the standard normal curve approximation is very poor for that sample size.

This issue is examined in Section 15.3.3, on robustness.

In 1908 Gosset addressed the problem of determining the *exact* sampling distribution of the random variable G given in Equation 15.4. In order to solve this problem, Gosset had to make an assumption about the population under study. He assumed that the population is given by a normal curve whose mean and standard deviation are unknown.

With his assumption, Gosset was able to compute the exact sampling distribution of G. He discovered that the sampling distribution varied with the sample size; for example, the sampling distribution of G for $n = 5$ is different from the sampling distribution of G for $n = 10$. The distributions derived by Gosset are called the *t*-distributions, or sometimes the Student's *t*-distributions because Gosset published his findings under the name Student. Instead of "the *t*-distribution for a sample of size n," the standard terminology is "the *t*-distribution with $n - 1$ degrees of freedom," written $t(n - 1)$. Remember that the number of degrees of freedom is equal to the sample size minus 1.

The exact mathematical formulas for the *t*-distributions are not important for the goals of this text; the following features are important. Refer to Figure 15.10, which presents two different *t*-distribution curves (or *t*-curves) and the standard normal curve.

1. The total area under a *t*-distribution curve is 1.
2. Every *t*-distribution curve is symmetric about 0. Therefore, for example, the area under the curve to the right of 1.50 equals the area under the curve to the left of -1.50.
3. As the number of degrees of freedom increases, the amount of spread in the *t*-distribution curve decreases. Visually, as the number of degrees of freedom increases, the *t*-curve becomes higher at its peak (located above 0) and lower in its tails. A *t*-curve, however, is never higher at its peak or lower in its tails than the standard normal curve.
4. As the number of degrees of freedom increases, the *t*-distribution curve becomes more difficult to distinguish from the standard normal curve. Thus, if the number of degrees of freedom is greater than 30, the *t*-distribution curve can be well approximated by the standard normal curve. (Note: The threshold of 30 for approximating the *t*-distribution curves by the standard normal curve is popular, but it is not universally used. Some researchers prefer to replace 30 by 60, 120, or a larger number.)

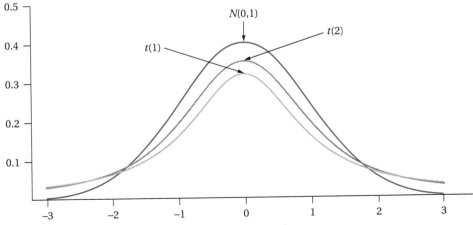

FIGURE 15.10 The t(1), t(2), and standard normal curves.

15.3 Inference for the Population Mean · 547

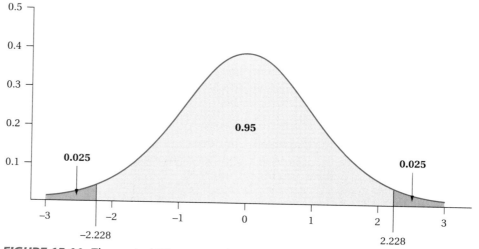

FIGURE 15.11 The central 95 percent of the area under the t-distribution curve with 10 degrees of freedom.

Selected probabilities under the t-distribution curves are given in Table A.6. Table A.6 is not as informative as Table A.2 is for the standard normal curve; it contains only one line of numbers, as opposed to two pages, for each curve. It is specifically designed for constructing confidence intervals and bounding P-values. The next example illustrates how to read Table A.6.

EXAMPLE 15.10

The entry in the row corresponding to 10 degrees of freedom and the column headed by .025 of Table A.6 is 2.228. This means that the area under the $t(10)$ curve to the right of 2.228 equals 0.025. By symmetry, the area under this curve to the left of -2.228 also equals 0.025. Thus, the area under that curve between -2.228 and 2.228 equals 0.95 (see Figure 15.11). ▲

15.3.1 Confidence Interval for μ

Suppose a random sample of size $n = 11$ is selected from a population described by a normal curve. Then

$$G = \frac{\sqrt{n}(\bar{X} - \mu)}{S}$$

is a random variable, and its probabilities are computed from the $t(10)$ curve. For instance, as demonstrated in the preceding example,

$$P(-2.228 \leq \frac{\sqrt{n}(\bar{X} - \mu)}{S} \leq 2.228) = 0.95.$$

The expression inside the probability statement can be rearranged to yield

$$P(\bar{X} - 2.228\frac{S}{\sqrt{n}} \le \mu \le \bar{X} + 2.228\frac{S}{\sqrt{n}}) = 0.95.$$

After the data are collected and the values of the sample mean and sample standard deviation are computed,

$$\bar{x} \pm 2.228 \times \frac{S}{\sqrt{n}}$$

is a 95 percent confidence interval for μ when $n = 11$.

The above argument can be generalized with respect to both the confidence level and the sample size.

EXAMPLE 15.11

With $n = 11$, according to Table A.6, the area under the t-curve with 10 degrees of freedom to the right of 2.764 is 0.01; by symmetry, the area to the left of -2.764 is 0.01. Therefore, the area between -2.764 and 2.764 is equal to 0.98. Thus,

$$P(-2.764 \le \frac{\sqrt{n}(\bar{X} - \mu)}{S} \le 2.764) = 0.98.$$

As argued above, this shows that the 98 percent confidence interval for μ when $n = 11$ is given by

$$\bar{x} \pm 2.764 \times \frac{S}{\sqrt{n}}.$$

Similarly, the 99 percent confidence interval for μ when $n = 11$ is

$$\bar{x} \pm 3.169 \times \frac{S}{\sqrt{n}}.$$

The 90 percent confidence interval for μ when $n = 11$ is

$$\bar{x} \pm 1.812 \times \frac{S}{\sqrt{n}}.$$

Finally, the 80 percent confidence interval for μ when $n = 11$ is

$$\bar{x} \pm 1.372 \times \frac{S}{\sqrt{n}}.$$

The meaning of the labels in the bottom row of Table A.6 is now apparent. ▲

EXAMPLE 15.12

Suppose $n = 21$; the various confidence intervals for μ can be obtained as above by replacing the entry from the row labeled 10 in Table A.6 by the entry in the row labeled 20. For example, an 80 percent confidence interval for μ when $n = 21$ is

$$\bar{x} \pm 1.325 \times \frac{s}{\sqrt{n}}. \blacktriangle$$

The above examples can be generalized as follows:

Assume that a random sample of size n is selected from a normal population. Let $\bar{x}$ and s denote the observed values of sample mean and standard deviation. A confidence interval for the population mean μ is given by

$$\bar{x} \pm t \times \frac{s}{\sqrt{n}}, \qquad (15.5)$$

where t is the entry in the row corresponding to $n - 1$ degrees of freedom and the appropriate column of Table A.6. This interval is called the t-distribution, or t-curve, confidence interval.

Table A.6 gives values of t in Formula 15.5 for 30 or fewer degrees of freedom. In this text, whenever there are more than 30 degrees of freedom, the t-curve is approximated by the standard normal curve and the t in Formula 15.5 is replaced by z from Table A.3. For convenience, the values of z for confidence intervals also appear in the bottom row of Table A.6. (Note: The bottom row of Table A.6 is labeled ∞, which stands for infinity. This notation reflects the fact that the t curve approaches a standard normal curve as n becomes large.)

EXAMPLE 15.13

Refer to the 1990 Tetris study introduced in Example 15.9. Recall that the sample size, sample mean, and sample standard deviation are

$$n = 25, \quad \bar{x} = 93.32, \quad \text{and} \quad s = 15.60.$$

Because a Tetris score is a count variable, the unknown population that represents the process that generated the 1990 scores cannot be a normal curve. In addition, the histogram of the sample data presented in Figure 15.9 (p. 539) lends little support to a supposition that the unknown population can be approximated well by a member of the family of normal curves. Nevertheless, Formula 15.5 will be used to obtain a confidence interval for the mean of the unknown population. The validity of the answer obtained is investigated below in Section 15.3.3.

For convenience, Formula 15.5 is reproduced here:

$$\bar{x} \pm t \times \frac{s}{\sqrt{n}}.$$

The values of $\bar{x}$, s, and n are given above; for a sample of size $n = 25$, the number of degrees of freedom is

$$n - 1 = 25 - 1 = 24.$$

Thus, for a 95 percent confidence interval, from Table A.6, $t = 2.064$. Substituting these values into the formula for the confidence interval gives

$$93.32 \pm 2.064 \times \frac{15.60}{\sqrt{25}} = 93.32 \pm 6.44 = [86.88, 99.76]$$

as the 95 percent confidence interval for the population mean μ. In words, at the 95 percent confidence level I can assert that the process that generated my scores at Tetris in 1990 had a population mean whose value was between 86.88 and 99.76. ▲

A discussion of the robustness of the confidence interval based on the t-distribution is presented after the following material on hypothesis testing.

15.3.2 Hypothesis Testing for μ

Following the framework first introduced in Chapter 6 for a population model, the first step of the hypothesis test is the specification of the special value of interest, denoted μ_0, and the choice of hypotheses. If there is no special value of interest, a hypothesis test should not be conducted. The null hypothesis is that the population mean μ equals the special value of interest. As usual, there are three choices for the alternative; the population mean may exceed, fall short of, or differ from the special value of interest, referred to, respectively, as the first, second, or third alternative. Here is a display of the three possibilities:

$$H_0: \mu = \mu_0 \qquad H_0: \mu = \mu_0 \qquad H_0: \mu = \mu_0$$
$$H_1: \mu > \mu_0 \qquad H_1: \mu < \mu_0 \qquad H_1: \mu \neq \mu_0$$

The second step of the hypothesis test is the specification of the test statistic and the determination of its null sampling distribution. Since the hypotheses involve the mean of the population, it is natural that the test statistic involves the sample mean, $\bar{X}$. Actually, it is more convenient to work with a centered and scaled version of $\bar{X}$, namely G, defined in Equation 15.4 and reproduced here for convenience:

$$G = \frac{\sqrt{n}(\bar{X} - \mu)}{S}.$$

According to the work of Gosset, if the population is a normal curve, the sampling distribution of G is the t-curve with $n - 1$ degrees of freedom. G is not suitable for a test statistic, because its formula contains the unknown μ. If the null hypothesis is true, the unknown μ in the numerator of G can be replaced by the known μ_0. Making this change in G yields the test statistic T,

$$T = \frac{\sqrt{n}(\bar{X} - \mu_0)}{S}, \qquad (15.6)$$

whose null sampling distribution is the t-curve with $n - 1$ degrees of freedom, assuming that the population is given by a normal curve.

Below are the rules of evidence, first stated for $\bar{X}$ and then for the test statistic T. The first set of rules is intuitively reasonable; the second set is mathematically equivalent to the first and is more useful for finding the P-value. For $\bar{X}$ the rules are the following:

- The larger the value of $\bar{X}$, the stronger is the evidence in support of the first alternative, $\mu > \mu_0$.
- The smaller the value of $\bar{X}$, the stronger is the evidence in support of the second alternative, $\mu < \mu_0$.
- The further the value of $\bar{X}$ from μ_0, in either direction, the stronger is the evidence in support of the third alternative, $\mu \neq \mu_0$.

For T the rules are the following:

- The larger the value of T, the stronger is the evidence in support of the first alternative, $\mu > \mu_0$.
- The smaller the value of T, the stronger is the evidence in support of the second alternative, $\mu < \mu_0$.
- The further the value of T from 0, in either direction, the stronger is the evidence in support of the third alternative, $\mu \neq \mu_0$.

Let t denote the observed value of the test statistic T; then

- For the first alternative ($\mu > \mu_0$) the P-value equals the area to the right of t under the t-curve with $n - 1$ degrees of freedom.
- For the second alternative ($\mu < \mu_0$) the P-value equals the area to the left of t under the t-curve with $n - 1$ degrees of freedom.
- For the third alternative ($\mu \neq \mu_0$) the P-value equals the area to the right of $|t|$ plus the area to the left of $-|t|$ under the t-curve with $n - 1$ degrees of freedom.

Fortunately, the t-distribution curve is symmetric about 0. Thus, the above three rules for obtaining the P-value can be rewritten in the following familiar way (familiar, that is, except for the fact that the standard normal curve is replaced by the t-distribution curve as the reference).

- For the first alternative ($\mu > \mu_0$) the P-value equals the area to the right of t under the t-curve with $n - 1$ degrees of freedom.
- For the second alternative ($\mu < \mu_0$) the P-value equals the area to the right of $-t$ under the t-curve with $n - 1$ degrees of freedom.
- For the third alternative ($\mu \neq \mu_0$) the P-value equals twice the area to the right of $|t|$ under the t-curve with $(n - 1)$ degrees of freedom.

Table A.6 can be used to obtain areas under t-distribution curves to the right of selected numbers. Thus, Table A.6 can be used to obtain—or bound—the P-value for this section's problems on hypothesis testing.

Obtaining the P-value from Table A.6 is a bit complicated; the method will be illustrated for each of the three possible alternatives. The method is similar to the technique, introduced in Chapter 10, for finding a P-value for a chi-squared test by using Table A.5.

The First Alternative: $H_1: \mu > \mu_0$ This is the easiest of the three alternatives. The P-value equals the area under the $t(n-1)$ curve to the right of t. The following algorithm indicates how to compute this area.

1. Read across the entries in the row corresponding to $n-1$ degrees of freedom in Table A.6. If t is equal to an entry in the row, the area to the right of t equals the value at the head of that column. Otherwise, one must resort to the invisible column argument introduced in Chapter 10.

2. If t is less than every entry in the row corresponding to $n-1$ degrees of freedom, the area to the right of t is larger than 0.25.

3. If t is greater than every entry in the row corresponding to $n-1$ degrees of freedom, the area to the right of t is smaller than 0.005.

4. If none of the first three cases apply, t lies between two adjacent entries in the row corresponding to $n-1$ degrees of freedom. In this situation, the area to the right of t lies between the headings of these two adjacent entries.

EXAMPLE 15.14

The purpose of this example is to illustrate the above algorithm; thus, no description of a study is given. Moreover, the special value of interest, μ_0, is not given, nor are the values of $\bar{x}$ and s; only the numbers required to obtain the P-value are given, namely, the alternative, the sample size, and the value t of the test statistic.

1. Suppose the alternative is $\mu > \mu_0$, $n = 19$, and $t = 1.33$. The value of t appears in the row corresponding to 18 degrees of freedom in the column headed 0.10, giving

$$P = 0.10.$$

2. Suppose the alternative is $\mu > \mu_0$, $n = 5$, and $t = -0.38$. The value of t is less than every entry in the row corresponding to 4 degrees of freedom in Table A.6. Therefore,

$$P > 0.25.$$

(You may have noticed that since the area to the right of 0 is equal to 0.50 by the symmetry of any t-curve, the area to the right of any negative number must exceed 0.50. As a practical matter, knowing that the P-value exceeds 0.25 usually is as useful as knowing it exceeds 0.50.)

3. Suppose the alternative is $\mu > \mu_0$, $n = 15$, and $t = 3.25$. The value of t is greater than every entry in the row corresponding to 14 degrees of freedom in Table A.6. Therefore,

$$\mathbf{P} < 0.005.$$

4. Suppose the alternative is $\mu > \mu_0$, $n = 9$, and $t = 2.50$. The value of t falls between two adjacent entries in the row corresponding to 8 degrees of freedom, namely, 2.306 and 2.896. These numbers lie in the columns labeled 0.025 and 0.01, respectively. Therefore,

$$0.01 < \mathbf{P} < 0.025. \; \blacktriangle$$

The Second Alternative: $H_1: \mu < \mu_0$ The P-value equals the area under the $t(n-1)$ curve to the right of $-t$. Thus, the above algorithm can be applied to the number $-t$ to yield the P-value.

EXAMPLE 15.15

This example demonstrates how the above algorithm for finding the area under a t-distribution curve to the right of a number can be used to obtain or bound the P-value for the second alternative.

1. Suppose the alternative is $\mu < \mu_0$, $n = 15$, and $t = -2.00$. The P-value is the area under the $t(14)$ curve to the right of $-t = 2.00$. An easy application of the above algorithm gives

$$0.025 < \mathbf{P} < 0.05.$$

2. Suppose the alternative is $\mu < \mu_0$, $n = 20$, and $t = -4.00$. Then

$$\mathbf{P} < 0.005.$$

3. Suppose the alternative is $\mu < \mu_0$, $n = 10$, and $t = -0.60$. Then

$$\mathbf{P} > 0.25. \; \blacktriangle$$

The Third Alternative: $H_1: \mu \neq \mu_0$ The P-value equals twice the area under the $t(n-1)$ curve to the right of $|t|$. Thus, the above algorithm can be applied to the number $|t|$ to yield the P-value. Remember to double the value, or bounds, obtained from the algorithm.

EXAMPLE 15.16

This example demonstrates how the above algorithm for finding the area under a t-distribution curve to the right of a number can be used to obtain or bound the P-value for the third alternative.

1. Suppose the alternative is $\mu \neq \mu_0$, $n = 15$, and $t = -2.00$. Let A denote the area under the $t(14)$ curve to the right of $|t| = 2.00$. Note that the P-value equals $2A$. An application of the algorithm gives

$$0.025 < A < 0.05.$$

Therefore,

$$2(0.025) < 2A < 2(0.05), \quad \text{or}$$

$$0.05 < \mathbf{P} < 0.10.$$

Note that if A can only be bounded, the P-value is bounded by twice the bound(s) on A.

2. Suppose that the alternative is $\mu \neq \mu_0$, $n = 12$, and $t = 2.40$. Let A denote the area under the $t(11)$ curve to the right of $|t| = 2.40$. As above,

$$0.01 < A < 0.025 \quad \text{and} \quad 0.02 < \mathbf{P} < 0.05.$$

3. Suppose that the alternative is $\mu \neq \mu_0$, $n = 12$, and $t = -4.40$. Let A denote the area under the $t(11)$ curve to the right of $|t| = 4.40$. As above,

$$A < 0.005 \quad \text{and} \quad \mathbf{P} < 0.01.$$

4. Suppose that the alternative is $\mu \neq \mu_0$, $n = 22$, and $t = 0.40$. Let A denote the area under the $t(21)$ curve to the right of $|t| = 0.40$. As above,

$$A > .25 \quad \text{and} \quad \mathbf{P} > 0.50. \; \blacktriangle$$

EXAMPLE 15.17

In Example 15.13 a 95 percent confidence interval for the population mean of the 1990 Tetris scores was obtained. This example uses the same data to illustrate the formulas for hypothesis testing given in this section.

Suppose an analyst wants to test the null hypothesis that the population mean equals 100 versus the two-sided (third) alternative. Recall from the earlier example that the sample size, sample mean, and sample standard deviation for the 1990 Tetris data are

$$n = 25, \quad \bar{x} = 93.32, \quad \text{and} \quad s = 15.60.$$

Substituting these values into Formula 15.6 for the test statistic yields

$$t = \frac{\sqrt{n}(\bar{x} - \mu_0)}{s} = \frac{\sqrt{25}(93.32 - 100)}{15.60} = -2.14.$$

For the third alternative, let A equal the area under the $t(24)$ curve to the right of $|t| = 2.14$. Using the above algorithm,

$$0.01 < A < 0.025 \quad \text{and} \quad 0.02 < \mathbf{P} < 0.05. \; \blacktriangle$$

15.3.3 Robustness

It is easier to discuss robustness for confidence intervals; qualitatively similar results can be obtained for hypothesis testing. Continue to assume that the data are the result of a random sample from a population for a numerical variable, but make no further assumptions about the population. In particular, the population *is not* assumed to be a member of the family of normal curves.

Regardless of the assumptions made and what, in fact, is true, an analyst can certainly compute the interval

$$\bar{x} \pm t \times \frac{s}{\sqrt{n}}$$

and *call* it a confidence interval for the population mean. The confidence level of this interval, however, is unknown. Although the confidence level is unknown, it is convenient to define the expression **nominal confidence level.** The nominal confidence level is the value in Table A.6 in the bottom row of the column that was used to find t. In other words, the nominal confidence level would be the (actual) confidence level if the population is a normal curve. The robustness of the t-curve confidence interval,

$$\bar{x} \pm t \times \frac{s}{\sqrt{n}},$$

is assessed by comparing the actual confidence level to the nominal level. The first two results, presented in the next two examples, are extremely discouraging.

EXAMPLE 15.18: Too Little Variation

Suppose a population has the following sampling distribution:

x	$P(X = x)$
0	0.9998
1	0.0001
2	0.0001
Total	1.0000

Suppose a researcher selects a random sample of size $n = 1{,}000$ from this population. Let B be the event that each subject sampled gives a response of 0. By the multiplication rule,

$$P(B) = (0.9998)^{1{,}000} = 0.82.$$

If the event B occurs, both the sample mean and sample standard deviation equal 0, giving a 99 percent confidence interval equal to

$$\bar{x} \pm t \times \frac{s}{\sqrt{n}} = 0 \pm 2.576 \times \frac{0}{\sqrt{1,000}} = [0, 0].$$

It can be shown that the population mean is 0.0003. Thus, if the event B occurs, the t-curve confidence interval is incorrect. In short, the 99 percent confidence interval has an 82 percent or greater chance of being wrong! ▲

EXAMPLE 15.19: Outrageous Outliers

Suppose a population is given by the following sampling distribution:

x	$P(X = x)$
0	0.4999
1	0.4999
3,000	0.0002
Total	1.0000

Two of every 10,000 population members give the outrageous outlier, 3,000, for their response. Suppose a researcher selects a random sample of size $n = 1,000$ from this population. Let B be the event that all responses equal either 0 or 1. By the multiplication rule,

$$P(B) = (0.9998)^{1,000} = 0.82.$$

If the event B occurs, clearly the sample mean is no larger than 1; further, it can be shown that the sample standard deviation is at most 0.5003 (the algebra is beyond the scope of this book). The nominal 99 percent confidence interval for μ has upper bound

$$\bar{x} + 2.576 \times \frac{s}{\sqrt{n}} \leq 1 + 2.576 \times \frac{0.5003}{\sqrt{1,000}} = 1 + 0.041 = 1.041.$$

It can be shown that the population mean μ equals 1.099, which is, of course, larger than 1.041. In words, the nominal 99 percent confidence interval has an 82 percent or greater chance of being incorrect! ▲

The previous two examples indicate that even for a large sample size, $n = 1,000$, the t-curve confidence interval is not robust against populations with too little variation or outrageous outliers. Here are the important features of these examples:

1. Contrary to popular belief, a large sample size does not guarantee that the t-curve confidence interval is robust. The above examples can be easily modified to show that for any sample size, no matter how large, a population can be found with either too little variation or outrageous outliers for which the t-curve confidence interval is not robust.

2. The above examples are artificial; how serious a problem does too little variation or outrageous outliers pose to a researcher?

    - It is difficult to imagine a competent researcher not anticipating that there is little variation in the population of interest. In addition, if little or no variation occurs in the *sample*, there is very strong evidence that too little variation occurs in the *population*. The real moral of Example 15.18 is that a huge amount of data is required to study adequately a population that exhibits little variation.

    - Again, it can be argued that a competent researcher would anticipate the presence of outrageous outliers in the population. Experience, however, has taught me to be cautious. For example, about 400 undergraduates compose the population of students who, in any given year, enroll in Statistics 201 at the University of Wisconsin-Madison. For the variable "annual income" one might assume that there are no outrageous outliers in this population. Presumably, this assumption is usually reasonable, but a few years ago the population included Mr. Al Toon, a highly paid all-pro wide receiver with the New York Jets football team. Mr. Toon was clearly an outrageous outlier for this population.

3. Contrary to popular belief, having *sample data* that can be considered well behaved is no guarantee that the t-curve confidence interval is robust. The difficulty in Example 15.19 is that there are outliers in the *population*. Certainly, the presence of an outrageous outlier in the sample can convince the analyst that there are outrageous outliers in the population, but the absence of outliers in the sample, especially a small sample, is little assurance against outliers in the population. In short, robustness is not solely a mathematical issue. The researcher must use knowledge about the population being studied to assess whether outrageous outliers are possible.

4. One way to deal with the actual or possible presence of outrageous outliers in a population is to switch attention from the population mean to the population median. The validity of inference for the population median, presented in the previous section, is not affected by outliers. Of course, statistical expediency should not determine the goals of a study. In many applications, however, the presence of outrageous outliers will make the mean an undesirable measure of center. In Example 15.19 the population mean, 1.099, is larger than 99.98 percent of the population values;

that is an odd characteristic for a center! On the other hand, consider the population of the dollar value of claims paid by insurance companies in a given year. Insurance companies are interested in the population total, which equals the population size times the population mean. Thus, even though there are likely to be outrageous outliers in the population, the population mean remains the value of interest. (One way insurance companies handle the problem of outrageous outliers is to obtain legal limits to their liability. For example, U.S. law limits the total liability responsibility for a nuclear power plant disaster.)

The remainder of this section examines the robustness of the t-curve confidence interval against deviations from the assumption of a normal pdf other than too little variation or outrageous outliers.

EXAMPLE 15.20

Suppose a researcher selects a random sample of size $n = 10$ from the cat population in Table 15.1 (p. 523). Let $\bar{x}$ and s denote the observed values of the sample mean and sample standard deviation. Using Table A.6 to find $t = 1.383$, the nominal 80 percent confidence interval for the mean of the cat population is

$$\bar{x} \pm 1.383 \times \frac{s}{\sqrt{10}}.$$

The actual confidence level is the probability that

$$\bar{X} - 1.383 \times \frac{S}{\sqrt{10}} \leq \mu \leq \bar{X} + 1.383 \times \frac{S}{\sqrt{10}}. \tag{15.7}$$

The computation of this probability is very difficult and is not attempted here. This probability, however, can be estimated quite easily by a computer simulation experiment, introduced in Chapter 3. Let p denote the actual confidence level for the t-curve confidence interval; in other words, p is the probability that the event in Expression 15.7 occurs. A run of a simulation experiment consists of having a computer select 10 subjects at random from the cat population and compute the nominal 80 percent confidence interval. If the simulated confidence interval includes the population mean, 1.4, it is labeled a success; otherwise it is labeled a failure. It is easy to see that the runs of this simulation experiment satisfy the conditions for Bernoulli trials with probability of success equal to p, the actual confidence level, as defined above. Let m denote the number of runs in a simulation experiment (the symbol n is already being used—it is the sample size for the study). Further, let the random variable Y denote the number of runs that yield successes, and let y denote the observed value of Y. From Chapter 5,

$$Y \sim \text{Bin}(m, p).$$

From Chapter 6, a point estimate of p is given by

$$\hat{p} = \frac{y}{m},$$

and

$$\hat{p} \pm 1.96\sqrt{\frac{\hat{p}\hat{q}}{m}}$$

is a 95 percent confidence interval for p (where, as usual, $\hat{q} = 1 - \hat{p}$).

A simulation experiment with $m = 5,000$ runs was conducted for the current problem, a random sample of size $n = 10$ from the cat population, and yielded 4,006 successes. Therefore, the point estimate of the actual confidence level is

$$\hat{p} = \frac{y}{m} = \frac{4,006}{5,000} = 0.8012.$$

This answer is very encouraging; even though the cat population is not a normal curve, the estimate of the actual confidence level, 0.8012, is very close to the nominal level, 80 percent. In words, the t-curve confidence interval appears to be very robust for this problem. This simulation study can be analyzed further by computing a 95 percent confidence interval for p; the result is

$$\hat{p} \pm 1.96\sqrt{\frac{\hat{p}\hat{q}}{m}} = 0.8012 \pm 1.96\sqrt{\frac{0.8012(0.1988)}{5,000}}$$

$$= 0.8012 \pm 0.0111 = [0.7901, 0.8123].$$

All of the values in this interval are very close to the nominal level, 80 percent.

The simulation study described above was repeated 19 times, for each combination of five confidence levels, 80, 90, 95, 98, and 99 percent, with four different sample sizes, n equal to 10, 20, 30, and 100. The results are presented in Table 15.7. Note that this table presents point estimates of the actual

TABLE 15.7 The 5,000-Run Simulation Point Estimate, $\hat{p}$, of the Actual Confidence Level, p, for a Random Sample of Size n from the Cat Population

Nominal Confidence Level	n			
	10	20	30	100
0.8000	0.8012	0.8038	0.7920	0.8018
0.9000	0.9026	0.9012	0.8938	0.8952
0.9500	0.9490	0.9456	0.9438	0.9494
0.9800	0.9788	0.9784	0.9730	0.9772
0.9900	0.9940	0.9896	0.9850	0.9874

confidence level, but not confidence interval estimates of the level. The results are very encouraging; in every case considered, the point estimate of the actual confidence level is within 1 percentage point of the nominal value. ▲

Statisticians have conducted a large number of simulation studies like the one presented above for the cat population. As discussed earlier, too little variation or outliers can ruin the performance of the t-curve confidence interval regardless of the sample size. In addition, the simulation studies have revealed that, as a general guideline,

> If the sample size is moderately large, about 10 or more, the inference procedures based on the t-distribution perform quite well unless there is severe skewness in the population.

Another way statisticians say this is

> If the sample size is moderately large, about 10 or more, the t-curve confidence interval is **robust** against mildly or moderately skewed populations.

(Recall that the cat population is mildly skewed to the right.)

In practice, of course, no one tells a researcher whether the population is severely skewed. Sometimes the researcher's knowledge of the population creates a strong suspicion of skewness. For example, even without collecting data, it seems natural to suspect that each of the following populations is skewed to the right:

- The age of college students
- The annual income of households in the United States

For these and other such populations, the researcher may decide that inference for the population median is more appropriate than inference for the population mean.

The following example is designed to give some insight into the effects of severe skewness on the t-distribution confidence interval.

EXAMPLE 15.21

I conducted a simulation experiment to investigate the performance of the t-distribution confidence interval when sampling from the lognormal population with parameters 5 and 1, which is graphed in Figure 15.8 (p. 536). Recall that this pdf is extremely skewed to the right. For each of the five popular nominal confidence levels, 80, 90, 95, 98, and 99 percent, and six sample sizes, $n = 10, 20, 30, 100, 150,$ and 200, a simulation experiment of $m = 5,000$ runs was conducted. Each run resulted in either a success or a failure, according to whether the computed confidence interval included the population mean $\mu = 244.7$. The proportion of successes for each combination of nominal confidence level and sample size is presented in Table 15.8. The results of this ta-

TABLE 15.8 The 5,000-Run Point Estimate, $\hat{p}$, of the Actual Confidence Level, p, for a Random Sample of Size n from the Lognormal Population with Parameters 5 and 1

Nominal Confidence Level	n					
	10	20	30	100	150	200
0.8000	0.7042	0.7498	0.7504	0.7674	0.7836	0.7892
0.9000	0.7946	0.8340	0.8418	0.8582	0.8756	0.8848
0.9500	0.8410	0.8764	0.8858	0.9092	0.9254	0.9306
0.9800	0.8846	0.9146	0.9208	0.9450	0.9574	0.9630
0.9900	0.9080	0.9336	0.9392	0.9622	0.9680	0.9728

ble are quite sobering. For $n = 10$, the five estimated confidence levels are 9 to 11 percentage points below the nominal levels. This is a very serious discrepancy. For example, the nominal 95 percent confidence interval, selected presumably because the analyst is willing to accept only a 5 percent chance of making an error, is estimated to have only an 84.1 percent actual confidence level. As the sample size increases, the differences between the nominal and actual levels decrease, but at a disappointingly slow pace. For $n = 20$ the estimated levels are 6 to 8 percentage points below the nominal levels, and for $n = 30$ the estimated levels are 5 to 7 percentage points below the nominal levels. Even a large increase in n, to 100, gives estimated levels that are 3 to 5 percentage points below the nominal levels. For $n = 150$ the estimated levels are 2 to 3 percentage points below the nominal levels. Finally, for $n = 200$, the estimated levels are all within 2 percentage points of the nominal levels; this seems reasonably close, but still pales in comparison to the results obtained for the cat population with a sample as small as size $n = 10$. ▲

EXAMPLE 15.22

In Examples 15.13 and 15.17, earlier in this section, Gosset's results were used to construct a confidence interval for and perform a hypothesis test on the mean of the population for the 1990 Tetris scores. In view of the results of this section, are these earlier inferences valid? An absolute answer to this question cannot be obtained, but it is instructive to examine the three critical issues: does too little variation, outrageous outliers, or severe skewness occur in the population?

The variation in the sample data indicates that the first issue, too little variation in the population, is not a concern for the 1990 Tetris study. Next, a Tetris score can never be smaller than 0, and I am convinced that a score above 200 is a physical impossibility for me. (I would have to be asleep to score 0; 51 is my worst score ever—after the first week of playing, of course—and my best score ever is 145.) Compared with the variation in the sample—from a low of 51 to a high of 114—the values 0 and 200, even if they exist in the population, can hardly be labeled outrageous outliers. Finally, severe skewness is the most difficult to assess. Neither the sample data nor my understanding of the game of Tetris lead me to suspect that the population is severely skewed, but I may be wrong. In summary, I believe that the inference procedures based on the t-curve are reasonably robust for the 1990 Tetris population. ▲

EXERCISES 15.3

1. Find the 95 percent t-distribution confidence interval for each of the following hypothetical results:

   (a) $n = 7$, $\bar{x} = 32.44$, and $s = 7.03$.
   (b) $n = 28$, $\bar{x} = 63.51$, and $s = 12.32$.
   (c) $n = 100$, $\bar{x} = 93.83$, and $s = 33.04$.

2. Find the 95 percent t-distribution confidence interval for each of the following hypothetical results:

   (a) $n = 22$, $\bar{x} = 13.25$, and $s = 2.73$.
   (b) $n = 13$, $\bar{x} = 25.22$, and $s = 10.51$.
   (c) $n = 50$, $\bar{x} = 43.14$, and $s = 16.27$.

3. Use the t-distribution to obtain the P-value for the following hypothetical results:

   (a) $n = 22$ and $t = 2.27$ with the first alternative.
   (b) $n = 22$ and $t = 2.27$ with the third alternative.
   (c) $n = 12$ and $t = -1.93$ with the second alternative.
   (d) $n = 12$ and $t = -1.93$ with the third alternative.
   (e) $n = 16$ and $t = 3.03$ with the first alternative.
   (f) $n = 16$ and $t = 3.44$ with the third alternative.

4. Use the t-distribution to obtain the P-value for the following hypothetical results:

   (a) $n = 14$ and $t = 2.66$ with the first alternative.
   (b) $n = 14$ and $t = 2.66$ with the third alternative.
   (c) $n = 27$ and $t = -1.58$ with the second alternative.
   (d) $n = 27$ and $t = -1.58$ with the third alternative.
   (e) $n = 6$ and $t = 3.78$ with the first alternative.
   (f) $n = 6$ and $t = 3.42$ with the third alternative.

5. Refer to the data in Exercise 1 on p. 401, a CRD with $n_1 = n_2 = 10$. Assume that each data set is a random sample from a population.

   (a) Find the approximate 95 percent confidence interval for the population mean time it took Brian to run 1 mile while wearing combat boots. (Hint: $\bar{x} = 333.0$ seconds, and $s = 8.18$ seconds.)
   (b) Repeat part (a) for Brian's data on wearing jungle boots. (Hint: $\bar{x} = 319.5$ seconds, and $s = 7.93$ seconds.)

6. Refer to the Tetris data presented in Table 15.5 (p. 539). Assuming that the 1990 data make up a random sample from a population, construct an 80 percent confidence interval for the population mean. (Hint: $\bar{x} = 93.32$, $s = 15.60$, and $n = 25$.) Comment on the robustness of this answer.

7. Refer to the data in Exercise 3 on p. 402, a CRD with $n_1 = n_2 = 40$. Assume that each data set is a random sample from a population.

   (a) Find the approximate 95 percent confidence interval for the population mean distance Jennifer could hit a ball with an aluminum bat. (Hint: $\bar{x} = 168.88$ feet, and $s = 45.98$ feet.)
   (b) Repeat part (a) for Jennifer's data on the wooden bat. (Hint: $\bar{x} = 149.05$ feet, and $s = 22.98$ feet.)

8. Refer to the data in Exercise 4 on p. 402, a CRD with $n_1 = n_2 = 20$. Assume that each data set is a random sample from a population.

(a) Find the approximate 95 percent confidence interval for the population mean number of rounds needed by Doug to score 301 with his own darts. (Hint: $\bar{x} = 18.6$ rounds, and $s = 4.00$ rounds.)

(b) Repeat part (a) for Doug's data on bar darts. (Hint: $\bar{x} = 21.2$ rounds, and $s = 5.04$ rounds.)

9. Refer to the data in Exercise 5 on p. 403, a CRD with $n_1 = n_2 = 5$. Assume that each data set is a random sample from a population.

(a) Find the approximate 95 percent confidence interval for the population mean time Kymn required to row the equivalent of 2,000 meters on the ergometer on the small gear, vent closed setting. (Hint: $\bar{x} = 491.4$ seconds, and $s = 1.82$ seconds.)

(b) Repeat part (a) for the large gear, vent open setting. (Hint: $\bar{x} = 484.2$ seconds, and $s = 3.42$ seconds.)

10. Table 15.6 (p. 543) presents the dollar values of 208 prizes given on the television show "The Price Is Right." Assuming that these numbers are a random sample from a population, construct the 95 percent confidence interval for the population mean. (Hint: $\bar{x} = \$1,140.30$, and $s = \$388.40$.) Comment on the robustness of this answer.

11. Refer to the data in Exercise 11 on p. 405, a CRD with $n_1 = n_2 = 40$. Assume that each data set is a random sample from a population.

(a) Find the approximate 95 percent confidence interval for the population mean score on the first treatment. (Hint: $\bar{x} = 35.50$ points, and $s = 24.53$ points.)

(b) Repeat part (a) for the data on the second treatment. (Hint: $\bar{x} = 30.42$ points, and $s = 21.70$ points.)

12. Table 12.11 (p. 428) presents an ordered list of 153 cash prizes for matching four of six numbers in Lotto America. Assuming that these numbers are a random sample from a population, construct the 95 percent confidence interval for the population mean. (Hint: $\bar{x} = \$40.67$, and $s = \$6.50$.) Comment on the robustness of this answer.

13. Refer to the golf data presented in Exercise 13 on p. 406, a CRD with $n_1 = n_2 = 40$.

(a) Assume that the data on a 3-wood are a random sample from a population. Construct the 95 percent confidence interval for the mean of the population. (Hint: $\bar{x} = 106.87$ yards, and $s = 29.87$ yards.)

(b) Assume that the data on a 3-iron are a random sample from a population. Construct the 95 percent confidence interval for the mean of the population. (Hint: $\bar{x} = 98.18$ yards, and $s = 28.33$ yards.)

14. Table 12.12 (p. 429) presents an ordered list of 152 cash prizes for matching five of six numbers in Lotto America. Assuming that these numbers are a random sample from a population, construct a 95 percent confidence interval for the population mean. (Hint: $\bar{x} = \$1,068.50$, and $s = \$276.80$.) Comment on the robustness of this answer.

15. Refer to the data on car acceleration times presented in Exercise 15 on p. 406, a CRD with $n_1 = n_2 = 15$.

(a) Assume that the data with the windows open are a random sample from a population. Construct the 95 percent confidence interval for the mean of the population. (Hint: The sample mean and sample standard deviation equal 7.94 seconds and 0.2913 seconds, respectively.)

(b) Assume that the data with the windows closed are a random sample from a population. Construct the 95 percent confidence interval for the mean of the population. (Hint: The sample mean and sample standard deviation equal 7.10 seconds and 0.3253 seconds, respectively.)

16. Table 12.6 (p. 394) presents the total number of games played for a random sample of 200 major league baseball players. Use these data to find the 95 percent confidence interval for the mean of the population. (Hint: $\bar{x} = 354.0$ games, and $s = 560.4$ games.) Comment on the robustness of this answer.

17. Refer to the data you collected for Exercise 25 on p. 410. Assume that each data set is a random sample from a population.

(a) Find the approximate 95 percent confidence interval for the population mean score with the first treatment.

(b) Repeat part (a) for the data on the second treatment.

TABLE 15.9 Career Winning Proportion of 74 Major League Pitchers

0.000	0.000	0.000	0.000	0.000	0.000	0.000	0.000	0.000	0.000
0.091	0.133	0.200	0.200	0.200	0.206	0.250	0.273	0.278	0.286
0.300	0.300	0.300	0.304	0.314	0.333	0.333	0.350	0.368	0.392
0.407	0.408	0.414	0.429	0.429	0.434	0.436	0.444	0.456	0.463
0.476	0.500	0.500	0.500	0.500	0.500	0.500	0.500	0.500	0.519
0.521	0.521	0.534	0.537	0.538	0.544	0.548	0.548	0.565	0.571
0.571	0.571	0.583	0.583	0.587	0.600	0.600	0.600	0.627	0.643
0.667	0.667	0.690	1.000						

18. A random sample of 74 was selected from the population of major league pitchers who obtained at least one decision (win or loss) during their career. For any such pitcher, his career winning proportion is defined as his number of wins divided by his number of decisions (decisions equals number of wins plus number of losses). Table 15.9 presents the 74 winning proportions in numerical order.

   (a) Construct the 95 percent confidence interval for the population median winning proportion.

   (b) Construct the 95 percent confidence interval for the population mean winning proportion. (Hint: The sample mean and standard deviation are 0.3938 and 0.2140, respectively.)

   (c) A baseball fan looks at your answer to part (b) and comments, "Every game has one winning pitcher and one losing pitcher, so the population mean winning proportion must be 0.500." Comment. (Hint: Can you explain why it is not surprising that 10 men have a winning proportion of 0.000, but only one has a winning proportion of 1.000?)

15.4 INFERENCE FOR TIME SERIES DATA*

This section is optional. In addition, the presentation in this section assumes that you are familiar with the material in the optional Chapter 14.

Chapter 14 illustrated the time series plot, autocorrelation, and smoothing for three examples:

- The biweekly high temperatures in Madison during 1988
- The maximum number of victories by a National League pitcher for the 51 seasons from 1937 to 1987
- The daily high-temperature forecast errors in Madison during August 1988

Recall that there is an extremely strong seasonal pattern in the high-temperature data, some evidence of a pattern in the pitching data, and no visible pattern in the forecast error data.

Time series data would be easier to analyze if all series satisfied the following condition, called the assumption of stationarity and independence, abbreviated as the SI assumption.

The SI Assumption: *The n values in a time series are obtained by selecting a random sample from a population.*

If the SI assumption is true for a given series, the methods of this chapter can be used to analyze the series. In particular, if the observations are selected at random from a fixed population, the order of the data is irrelevant and can be ignored during the analysis. There are two broad categories of ways the SI assumption can be violated. First, the population sampled can change over time. For example, its center, spread, or some other characteristic may change. Second, whether the population changes or not, the values of certain responses are possibly correlated with the values of others. For example, a response may be highly correlated with its immediate neighbors.

The goal of this section is to provide some practical advice on how to decide whether or not to make the SI assumption for a particular time series. The techniques will be illustrated with the three time series studied earlier.

The first step is to think about the process that generates the data and ask

Is the SI assumption plausible for this process?

For example, for the biweekly high-temperature data, is it plausible to assume that high temperatures in Madison in January and July are obtained by selecting values, at random, from the same population? Even without examining the data, I know from 42 years of living in the midwestern United States that it is absurd to make such an assumption. Thus, the SI assumption is discarded for this series; inference must be performed, if at all, with methods specifically designed for time series data.

For the time series of the maximum number of victories, there are many reasons to suspect that the SI assumption is not valid. Some reasons were given in Chapter 14, and you may be able to think of others. None that I can think of, however, is so compelling that I would discard the assumption without, at the very least, looking at the data. Finally, for the forecast error data, I cannot manufacture a convincing conjecture about the process that would suggest the SI assumption is false.

The second step is to examine the data to assess, informally, their evidence against the SI assumption. As discussed in Chapter 14, the time series plot of the National League victories reveals a flurry of high responses during the 1960s and a downward trend since 1972. Of course, these patterns may be due to chance, but I would recommend not making the SI assumption for this series for two reasons:

1. There are good methods for analyzing time series data; they are, unfortunately, beyond the scope of this text.

2. The patterns in the time series plot motivated me to do some further investigation into the data set. Before examining the time series, I had suspected that the population changed subtly over time as a result, primarily, of changes in baseball strategy. For instance, over time relief pitchers have become increasingly important; this has led to a reduction in the number of decisions, both victories and losses, for starting pitchers. In addition, pitchers do not start as many games during the season as they once did. For example, when Sandy Koufax led the National League with 26 and 27 victories in 1965 and 1966, he started 41 and completed 27 games each year. In 1987, Rick Sutcliffe, who led the league with 18 victories, started just 34 games and completed only 6. This extra information not contained in the time series plot lends support to discarding the SI assumption.

In contrast to the pitchers' series, an examination of the forecast error series reveals no obvious patterns. Thus, the SI assumption remains plausible.

The third step consists of examining the data a number of different ways. These ways are introduced below and are illustrated for an expanded forecast error series. Instead of analyzing the errors in August 1988, consider all errors from that summer, from June 21 to September 20. The reason for this change is that these new techniques require many more than 31 observations to have a decent chance of discovering evidence against the SI assumption. The *Wisconsin State Journal* is not printed on July 4 or Labor Day; consequently the series consists of only 89 observations. A time series plot of the 89 observations, not presented here, does not reveal any obvious pattern.

One way to investigate a change in the level of the series (the center of the population) is to compare the first half of the data with the second half. Figure 15.12 presents dot plots of the first 45 observations and the last 44. The dot plots suggest that the first half has a larger center and the second half has more spread, but the differences are not dramatic. Table 15.10 contains selected summary statistics for the two halves. The measures of spread, s, IQR, and R, and center, $\bar{x}$ and $\tilde{x}$, echo the differences noted in the dot plots. In particular, the sample means are 0.133 and -0.386; in the first half of the summer the forecasts were slightly too large, on average, and in the second half they were somewhat too small. A method is presented in Chapter 16 that enables one to decide whether or not this difference in means is statistically significant; it is not.

Of course, there is nothing magical about comparing the first half of the series to the second half. The series can be divided into thirds, fourths, and so on; see the exercises for details. The qualitative conclusions obtained above for comparing the two halves of the series are also reached for the series divided into thirds or fourths. As a general guideline, do not divide the series into pieces consisting of fewer than 20 observations.

In addition to comparing pieces of the series, the analyst should compute the **autocorrelation function.** The autocorrelations in Table 15.11 were obtained with the Minitab statistical software package. The first column in the table lists

15.4 Inference for Time Series Data* • 567

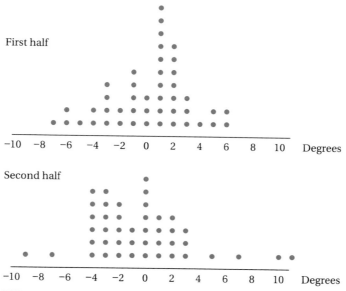

FIGURE 15.12 Dot plots for the first and second halves of the time series of forecast errors during summer 1988.

TABLE 15.10 Summary Statistics for the First and Second Halves of the Time Series of Forecast Errors during Summer 1988

Half	n	Mean	Median	Standard Deviation	IQR	Range
First	45	0.133	1.000	3.123	4.00	13
Second	44	−0.386	−0.500	3.853	4.75	20

TABLE 15.11 The Autocorrelation Function for the Time Series of Forecast Errors during Summer 1988

Lag	Autocorrelation	Lag	Autocorellation	Lag	Autocorrelation
1	0.097	8	−0.075	14	0.036
2	0.028	9	−0.088	15	0.111
3	−0.050	10	0.096	16	−0.022
4	−0.154	11	−0.195	17	−0.011
5	0.094	12	0.032	18	0.183
6	−0.178	13	0.196	19	0.078
7	−0.189				

the lag numbers, 1, 2, and so on. (Recall that lag was defined in Section 14.2.) The second column gives the value of the autocorrelation for the corresponding lag. According to Table 15.11, the lag 1 autocorrelation is 0.097, the lag 2 autocorrelation is 0.028, and so on. There is a simple way to decide whether any of these autocorrelations are real; it involves the (possibly repeated) application of the following fact:

> Let r denote any autocorrelation. If the SI assumption is true, the sampling distribution of
>
> $$z = \sqrt{n} \times r$$
>
> can be well approximated by the standard normal curve.

Consider the autocorrelation of lag 1. Its numerical value for the forecast series is 0.097. By the above result, if the SI assumption is correct, then

$$z = \sqrt{n} \times r = \sqrt{89}(0.097) = 0.92$$

can be viewed as being obtained by selecting a value at random from a population governed by the standard normal curve. Obtaining a value of 0.92 from a standard normal curve is certainly not surprising; thus, the evidence in this correlation against the SI assumption is not compelling. If, on the other hand, the value of

$$z = \sqrt{n} \times r$$

is far from 0 in either direction, larger than +2 or smaller than −2, then considerable doubt is cast upon the SI assumption and the assumption usually should not be made. (Technical note: The above analysis of the autocorrelations can be placed in a formal hypothesis-testing framework; such an exercise is better left to a careful development of time series methods.)

A caution: Minitab computes many autocorrelations (19 for the forecast series). Consider the following analysis strategy:

> Find the autocorrelation that is farthest from 0; for that autocorrelation, r, compute
>
> $$z = \sqrt{n} \times r$$
>
> and compare this value with +2 and −2.

Before we criticize this strategy, it will be applied to the forecast error series. In Table 15.11, the autocorrelation farthest from 0 is $r = 0.196$, which occurs for lag 13 and

$$z = \sqrt{n} \times r = \sqrt{89} \times 0.196 = 1.85.$$

Although this value does not exceed +2, it does come close; does it cast some doubt on the SI assumption? No, because clearly the focusing on a winner fallacy, introduced in Section 6.6.1, is operating. By computing the value of z for the lag 13 autocorrelation because it is farthest from 0, the analyst is implicitly computing 18 other values of z, one for each autocorrelation. About 95 percent of the area under the standard normal curve falls between $+2$ and -2; thus, even if the SI assumption is true, with 19 autocorrelations it would not be surprising to find one or two of the values of

$$z = \sqrt{n} \times r$$

outside this range.

EXERCISES 15.4

1. Refer to the summer 1988 high-temperature forecast errors discussed in the text. Table 15.12 presents the 89 observations in time order (reading left to right)

TABLE 15.12 Eighty-nine High-Temperature Forecast Errors, in Time Order, for Summer 1988

-4	-2	0	-1	1	3	-1	2	-1	1
0	-1	-1	1	2	2	0	5	2	-3
-5	1	-6	6	-2	1	1	6	3	-3
2	1	1	2	-4	2	4	-7	-3	-6
1	5	-3	3	1	2	5	-2	2	-3
0	1	-4	0	-2	-3	-4	-1	-4	2
10	0	1	0	2	0	-3	-2	-4	0
1	-7	1	-3	3	3	-2	-3	-1	-2
-3	-4	3	-4	-9	-1	0	11	7	

Construct three dot plots, one for the first 30 observations, another for the second 30, and one for the last 29. Describe what they reveal.

2. Refer to the previous exercise. Construct four dot plots, one for the first 23 observations, another for the second 22, one for the next 22, and one for the final 22. Describe what they reveal.

3. For the data in Table 15.12, $\bar{x} = -0.124$ and $s = 3.493$. Assume that these data represent a random sample from a process.

   (a) Construct a 95 percent confidence interval for the mean of the population.

   (b) Test the null hypothesis that the mean of the population is 0 versus the third alternative.

15.5 PREDICTIONS*

As discussed in Chapter 6, predictions are more commonly used for process data, but the formulas apply equally well to populations that are collections of distinct individuals. There are two common methods of prediction: the first assumes that the population is represented by a normal curve, and the second assumes only that the population is represented by a pdf. These methods are examined in separate sections below.

Regardless of the assumptions about the population, the data are assumed to consist of a random sample of size n from a population. If the data are from a time series, the analyst should use the methods of the previous section to examine critically the assumption of a random sample, the SI assumption. The random variables

$$X_1, X_2, \ldots, X_n$$

denote the n responses, and

$$x_1, x_2, \ldots, x_n$$

denote their observed values. The analyst wants to predict the value of a future observation from the population. The future observation, denoted X, is assumed to be independent of the n observations in the data set. As in Chapter 6, an analyst may want a point prediction or an interval prediction of the value of X.

15.5.1 Predictions for a Normal Population

Let the random variables $\bar{X}$ and S denote the sample mean and sample standard deviation and let $\bar{x}$ and s denote their observed values.

To build intuition, it is easiest to start with an unrealistically simple case. If both the population mean, μ, and population standard deviation, σ, are known, the point prediction of X is equal to μ, and the interval prediction of X is given by

$$\mu \pm z \times \sigma, \tag{15.8}$$

where z is determined by the desired probability of a correct prediction, and its value is obtained from Table A.3. By assumption, the random variable X has the $N(\mu, \sigma)$ curve for its sampling distribution. Therefore, the directive to use the center of the sampling distribution, μ, as the point prediction seems reasonable. It is easy to prove that the prediction interval given in Formula 15.8 has the proper probability of being correct. To see this, note that the probability the prediction interval is correct equals

$$P(\mu - z \times \sigma \leq X \leq \mu + z \times \sigma).$$

This can be rewritten as

$$P(-z \leq \frac{X - \mu}{\sigma} \leq z).$$

The center expression in this string of inequalities is the standardized version of X. Recall that the standardized version of a normally distributed random variable has for its sampling distribution the standard normal curve. Thus, this last probability equals the area under the standard normal curve between $-z$

and z and can be obtained from Table A.2 or A.3. This completes the proof that the interval given in Formula 15.8 has the stated probability of being correct (including the future observation).

The following example illustrates the above ideas.

EXAMPLE 15.23

As discussed earlier in this chapter, population IQ scores form a normal curve with mean $\mu = 100$ and standard deviation $\sigma = 15$. Let X denote the IQ of a randomly selected population member. By the above result, 100 is the point prediction of X, and the 95 percent prediction interval is

$$100 \pm 1.96 \times 15 = 100 \pm 29.4 = [70.6, 129.4].$$

This prediction interval is too wide to be of much practical use. (Any person interested in IQ scores would view a person with an IQ of 70.6 as importantly different from a person with an IQ of 129.4.) ▲

The second case considered in this section is somewhat more realistic. If the population mean, μ, is unknown, but the population standard deviation, σ, is known, the point prediction of X is equal to $\bar{x}$, and the interval prediction of X is given by

$$\bar{x} \pm z\sigma \sqrt{1 + \frac{1}{n}}, \qquad (15.9)$$

where z is determined by the desired probability of a correct prediction, and its value is obtained from Table A.3.

A comparison of Formulas 15.8 and 15.9 shows that there are two consequences of not knowing the population mean. First, the point prediction must be a computable number, so the unknown μ is replaced by its point estimate, $\bar{x}$. Second, this replacement increases the uncertainty in the prediction and results in the addition of the term $1/n$ under the square root in the formula for the prediction interval. (If this term is dropped, the expression

$$z\sigma \sqrt{1 + \frac{1}{n}}$$

becomes the smaller value,

$$z\sigma \sqrt{1} = z\sigma,$$

the same expression in the earlier formula when μ is known.)

Finally, the most realistic case is considered.

If both the population mean, μ, and the population standard deviation, σ, are unknown, the point prediction of X is equal to $\bar{x}$, and the interval prediction of X is given by

$$\bar{x} \pm ts \sqrt{1 + \frac{1}{n}}, \qquad (15.10)$$

where t is determined by the desired probability of a correct prediction, and its value is obtained from the row corresponding to $n - 1$ degrees of freedom and the appropriate column (as determined by the labels in the bottom row) of Table A.6.

A comparison of Formulas 15.9 and 15.10 shows that the effect of not knowing σ is that it is replaced by its point estimate, s, and that the standard normal curve is replaced by the t-curve as the reference curve.

EXAMPLE 15.24

The computation of Formula 15.10 can be illustrated with the Tetris data introduced in Example 15.9 (p. 538). Recall that

$$n = 25, \quad \bar{x} = 93.32, \quad \text{and} \quad s = 15.60.$$

The point prediction for another, 26th, Tetris score is equal to $\bar{x} = 93.32$, and the 95 percent prediction interval is given by

$$\bar{x} \pm ts\sqrt{1 + \frac{1}{n}} = 93.32 \pm 2.064(15.60)\sqrt{1 + \frac{1}{25}}$$
$$= 93.32 \pm 32.84 = [60.48, 126.16].$$

Since a Tetris score must be an integer, these values can be rounded (other strategies are possible) to yield 93 as the point prediction and [60, 126] as the 95 percent prediction interval for a future 1990 Tetris score. These answers have two frustrating features:

1. The interval is so wide that it is of little value.
2. I did not obtain a 26th Tetris score to compare with the point and interval predictions.

The second feature can be circumvented by using the first 24 scores to predict the 25th score. The sample mean and sample standard deviation of the first 24 Tetris scores (details not given) are

$$\bar{x} = 94.17 \quad \text{and} \quad s = 15.34.$$

Thus, the point estimate, after rounding, of the 25th Tetris score is 94. Substituting into Formula 15.10 (remembering that there are now 23 degrees of freedom) yields

$$94.17 \pm 2.069(15.34)\sqrt{1 + \frac{1}{24}} = 94.17 \pm 32.39 = [61.78, 126.56],$$

which can be rounded to [62, 127]. A reference to Table 15.4 (p. 539) reveals that the actual 25th Tetris score was 73; it is within the prediction interval, but much smaller than the point prediction.

Each probability assigned to a prediction interval in this section is computed on the assumption that the population is a normal curve. If this assumption is false, the stated probabilities can be very different from the actual probabilities. As discussed earlier in Example 15.13 (p. 549), it might not be reasonable to assume that the unknown population for my Tetris scores can be well approximated by a normal curve. Thus, the 95 percent label assigned to the Tetris prediction intervals perhaps should not be interpreted too literally.

If an analyst suspects the validity of the assumption of a normal population, the method of Section 15.5.2 should be used. ▲

15.5.2 Predictions for an Arbitrary Unknown PDF

Let the random variables

$$X_{(1)}, X_{(2)}, \ldots, X_{(n)}$$

denote the order statistics of the random sample and let

$$x_{(1)}, x_{(2)}, \ldots, x_{(n)}$$

denote their observed values. Let the random variable $\tilde{X}$ denote the sample median and let $\tilde{x}$ denote its observed value. The point prediction of X is the observed value of the sample median, $\tilde{x}$.

The idea behind the derivation of the prediction interval can be illustrated quite easily for a sample of size $n = 2$. Consider the three random variables

$$X_1, X_2, \text{ and } X,$$

the observed data and the future observation the analyst is trying to predict. It can be shown that the following six orderings of these variables all have the same probability of occurring, namely, $\frac{1}{6}$.

- $X < X_1 < X_2$.
- $X < X_2 < X_1$.
- $X_1 < X_2 < X$.
- $X_2 < X_1 < X$.
- $X_1 < X < X_2$.
- $X_2 < X < X_1$.

Note that for the last two of these orderings, the future observation, X, falls between the two observed data values. In other words,

$$P(X_{(1)} \le X \le X_{(2)}) = 0.333,$$

making $[x_{(1)}, x_{(2)}]$ a 33.3 percent prediction interval for X.

The above reasoning can be extended to any value of the sample size n. Suppose a population is given by any pdf. For any positive integer j less than or equal to $n/2$, the probability that the prediction interval

$$[x_{(j)}, x_{(n+1-j)}] \tag{15.11}$$

includes the future value X is equal to

$$\frac{n+1-2j}{n+1}. \qquad (15.12)$$

EXAMPLE 15.25

The 1990 Tetris data introduced in Example 15.9 (p. 538) can be used to illustrate the above fact. First, the 25 scores can be used to predict a future, 26th, score. The point prediction of the future score is the sample median, $\tilde{x} = 95$.

Two different prediction intervals are obtained below. Setting $j = 1$ in Formula 15.11 and using the order statistics presented in Table 15.5 (p. 539) yields

$$[x_{(j)}, x_{(n+1-j)}] = [x_{(1)}, x_{(25+1-1)}] = [x_{(1)}, x_{(25)}] = [51, 114]$$

as a prediction interval. By Formula 15.12, this interval has probability equal to

$$\frac{n+1-2j}{n+1} = \frac{25+1-2}{25+1} = \frac{24}{26} = 0.923$$

of containing the future value.

As another example, setting $j = 2$ in the result gives

$$[x_{(j)}, x_{(n+1-j)}] = [x_{(2)}, x_{(25+1-2)}] = [x_{(2)}, x_{(24)}] = [70, 112]$$

as a prediction interval with probability

$$\frac{n+1-2j}{n+1} = \frac{25+1-4}{25+1} = \frac{22}{26} = 0.846$$

of containing the future value.

Alternatively, the first 24 scores can be used to predict the 25th score. The order statistics of the first 24 scores are

51 70 74 75 81 85 90 90 93 94 94 95
94 100 100 101 103 106 106 107 110 111 112 114

Setting $j = 1$ in Result 5.8 gives

$$[x_{(j)}, x_{(n+1-j)}] = [x_{(1)}, x_{(24+1-1)}] = [x_{(1)}, x_{(24)}] = [51, 114]$$

as a prediction interval with probability

$$\frac{n+1-2j}{n+1} = \frac{24+1-2}{24+1} = \frac{23}{25} = 0.92$$

of containing the 25th score. Also, setting $j = 2$ gives

$$[x_{(j)}, x_{(n+1-j)}] = [x_{(2)}, x_{(24+1-2)}] = [x_{(2)}, x_{(23)}] = [70, 112]$$

as a prediction interval with probability

$$\frac{n+1-2j}{n+1} = \frac{24+1-4}{24+1} = \frac{21}{25} = 0.84$$

of containing the 25th score. The actual value of the 25th score is 73; both prediction intervals are correct. ▲

EXERCISES 15.5

1. Table 15.6 (p. 543) presents the dollar values of 208 prizes given on the television show "The Price Is Right." Assume that these values represent a random sample from a process. Let X denote the value of a future prize.

    (a) Assuming the prizes have a normal distribution, find the point prediction and 95 percent prediction interval for the value of X. (Hint: $\bar{x} = \$1{,}140.30$, and $s = \$388.40$.)

    (b) Dropping the assumption that the prizes have a normal distribution, find the point prediction and 95 percent prediction interval for the value of X.

2. Table 12.11 (p. 428) presents an ordered list of 153 cash prizes for matching four of six numbers in Lotto America. Assume that these numbers are a random sample from a population. Let X denote the value of a future prize.

    (a) Assuming that the prizes have a normal distribution, find the point prediction and 95 percent prediction interval for the value of X. (Hint: $\bar{x} = \$40.67$, and $s = \$6.50$.)

    (b) Dropping the assumption that the prizes have a normal distribution, find the point prediction and 95 percent prediction interval for the value of X.

3. One way to get a rough evaluation of a normal population prediction interval is to count how many of the actual observations lie in the prediction interval. For example, the normal population 95 percent prediction interval for a future Tetris score is [60, 126]; an examination of Table 15.5 (p. 539) reveals that 24 out of 25, or 96 percent, of the 25 observations lie in this interval. Use this technique to evaluate your answer to Exercise 1.

4. Table 12.12 (p. 429) presents an ordered list of 152 cash prizes for matching five of six numbers in Lotto America. Assume that these numbers are a random sample from a population. Let X denote the value of a future prize.

    (a) Assuming that the prizes have a normal distribution, find the point prediction and 95 percent prediction interval for the value of X. (Hint: $\bar{x} = \$1{,}068.50$, and $s = \$276.80$.)

    (b) Dropping the assumption that the prizes have a normal distribution, find the point prediction and 95 percent prediction interval for the value of X.

5. Refer to the data in Exercise 1 on p. 401, a CRD with $n_1 = n_2 = 10$. Assume that each data set is a random sample from a population. Assume that Brian obtained one more independent observation from each population and that we want to predict the value of that observation.

    (a) Assuming that the population is represented by a normal curve, find the 95 percent prediction interval for the time it took Brian to run 1 mile while wearing combat boots. (Hint: $\bar{x} = 333.0$ seconds, and $s = 8.18$ seconds.)

    (b) Repeat part (a) dropping the assumption of a normal population.

    (c) Repeat part (a) for combat boots replaced by jungle boots. (Hint: $\bar{x} = 319.5$ seconds, and $s = 7.93$ seconds.)

    (d) Repeat part (c) dropping the assumption of a normal population.

6. Figure 15.12 (p. 567) presents 89 high-temperature forecast errors from summer 1988 in Madison, Wisconsin. Assume that these numbers are a random sample from a population. Let X denote the value of a summer 1989 forecast error. Assume that the same process that generated the 1988 errors is appropriate for 1989.

    (a) Assuming that the errors have a normal distribution, find the point prediction and 95 percent prediction interval for the value of X. (Hint: $\bar{x} = -0.124°$, and $s = 3.493°$.)

    (b) Dropping the assumption that the errors have a normal distribution, find the point prediction and 95 percent prediction interval for the value of X.

7. Refer to the data in Exercise 3 on p. 402, a CRD with $n_1 = n_2 = 40$. Assume that each data set is a random sample from a population. Assume that Jennifer obtained one more independent observation from each population and that we want to predict the value of that observation.

   (a) Assuming that the population is represented by a normal curve, find the 95 percent prediction interval for the distance Jennifer hit the ball with an aluminum bat. (Hint: $\bar{x} = 168.88$ feet, and $s = 45.98$ feet.)

   (b) Repeat part (a) dropping the assumption of a normal population.

   (c) Repeat part (a) for the aluminum bat replaced by a wooden bat. (Hint: $\bar{x} = 149.05$ feet, and $s = 22.98$ feet.)

   (d) Repeat part (c) dropping the assumption of a normal population.

8. Refer to the data in Exercise 4 on p. 402, a CRD with $n_1 = n_2 = 20$. Assume that each data set is a random sample from a population. Assume that Doug obtained one more independent observation from each population and that we want to predict the value of that observation.

   (a) Assuming that the population is represented by a normal curve, find the 95 percent prediction interval for the response Doug obtained with his darts. (Hint: $\bar{x} = 18.6$ rounds, and $s = 4.00$ rounds.)

   (b) Repeat part (a) dropping the assumption of a normal population.

   (c) Repeat part (a) for Doug's darts replaced by bar darts. (Hint: $\bar{x} = 21.2$ rounds, and $s = 5.04$ rounds.)

   (d) Repeat part (c) dropping the assumption of a normal population.

9. Refer to the data in Exercise 5 on p. 403, a CRD with $n_1 = n_2 = 5$. Assume that each data set is a random sample from a population.

   (a) Assuming that the population is represented by a normal curve, use Kymn's first four times on the small gear, vent closed setting to obtain the 95 percent prediction interval for her fifth time. Is the interval correct? (Hint: $\bar{x} = 491.0$ seconds, and $s = 1.83$ seconds.)

   (b) Repeat part (a) for the large gear, vent open setting. Is the interval correct? (Hint: $\bar{x} = 483.75$ seconds, and $s = 3.77$ seconds.)

10. *The Baseball Encyclopedia* [2] lists career records for every man who has ever played major league baseball. A random sample of 200 from the population of position players was selected; the number of career games for these players is given in Table 12.6 (p. 394). The distribution of these data is strongly skewed to the right. Suppose another player will be selected at random from this population. Construct a prediction interval for the number of career games played by this man. Choose the prediction interval that has probability of being correct as close to 95 percent as possible.

11. Refer to the data in Exercise 11 on p. 405, a CRD with $n_1 = n_2 = 40$. Assume that each data set is a random sample from a population. Assume that Christine obtained one more independent observation from each population and that we want to predict the value of that observation.

    (a) Assuming that the population is represented by a normal curve, find the 95 percent prediction interval for the score Christine obtained with the first treatment. (Hint: $\bar{x} = 35.50$ points, and $s = 24.53$ points.)

    (b) Repeat part (a) dropping the assumption of a normal population.

    (c) Repeat part (a) for the second treatment. (Hint: $\bar{x} = 30.42$ points, and $s = 21.70$ points.)

    (d) Repeat part (c) dropping the assumption of a normal population.

REFERENCES

1. "Perspectives: Overheard," *Newsweek*, March 20, 1989, p. 19.

2. J. L. Reichler, ed., *The Baseball Encyclopedia* (New York: Macmillan, 1988).

Numerical Data from Two Sources

16.1 COMPARISON OF MEANS FOR A CRD
16.2 INDEPENDENT RANDOM SAMPLES
16.3 THE RANDOMIZED PAIRS DESIGN
16.4 A POPULATION MODEL FOR PAIRED DATA

Chapter 16

In Chapter 1, the four components of an experimental design were introduced:

1. The selection of the subjects to be included in the study
2. The specification of the treatments to be compared
3. The specification of the response to be obtained from each subject
4. The specification of the method by which subjects are assigned to treatments

The first three chapters of this book considered designs for which

1. No assumptions are made about the method of deciding which subjects are included in the study.
2. Two treatments are compared.
3. The response from each subject is dichotomous.
4. The subjects are assigned to treatments by randomization.

Subsequent chapters extended this simple design in a number of directions, for example, to subjects selected at random from a population, to a multicategory response, to studies that cannot have randomization (observational studies), and to designs that formally incorporate one or more factors.

The first section of this chapter examines the design of the first three chapters for studies in which the response is numerical rather than dichotomous. As in Part I it is called the completely randomized design, abbreviated CRD. Section 16.2 presents and analyzes the two-population problem for independent random samples. (Thus, Section 16.2 extends to a numerical response material in Chapters

580 · Chapter 16 / Numerical Data from Two Sources

7 and 11.) Section 16.3 extends to a numerical response the randomized pairs design of Chapter 4. Section 16.4 presents a population model for the design of Section 16.3.

16.1 COMPARISON OF MEANS FOR A CRD

The methods of this section are introduced with two examples; one with trials for subjects and one with individuals for subjects.

EXAMPLE 16.1

For her class project, M. Cathy Blum applied her new knowledge to her hobby, running. Cathy has two 1-mile routes she runs: one is through a park, and the other is at her local high school. She is interested in determining whether she runs the same on each route or not. To investigate this, she conducted a CRD with six trials, three on each route.

Cathy placed three tickets marked "park" and three marked "high school" in a hat. Each day she selected a ticket at random, without replacement, and ran the route named on the ticket. On each trial Cathy ran the route as fast as she could. The response is the time, in seconds, she required to complete the route. The data are

High school:	530	521	539
Park:	528	520	527

Dot plots of these data are presented in Figure 16.1. Note that Cathy tended to run faster (lower responses) on the park route. ▲

The above example can be generalized as follows. A researcher compares two treatments by obtaining responses from n subjects. The subjects are assigned to treatments by randomization; the number of subjects assigned to the first treatment is denoted by n_1, and the number assigned to the second treatment by n_2. For Cathy's study, let the high school route be the first treatment and let the park route be the second treatment; clearly, $n = 6$, $n_1 = 3$, and $n_2 = 3$.

The null hypothesis is that the two treatments, in Cathy's study the two routes, have the same effect on the response. If the null hypothesis is true, the variation

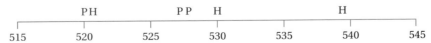

FIGURE 16.1 Dot plots of the times, in seconds, to run 1 mile. H is for the high school route and P is for the park route.

in the observations results from some cause(s) other than the treatments. For instance, in Cathy's study perhaps the variation in the response is due to daily variation in the weather and her energy level. In particular, if the null hypothesis is true, the response time for each trial would have remained unchanged even if the other route had been traversed. For example, one day Cathy required 539 seconds to complete the high school route. If the null hypothesis is true, then if she had run the park route that day her time still would have been 539 seconds.

The three possible alternatives are

1. Treatment 1 tends to yield larger responses than treatment 2. This alternative is denoted by $>$ and is referred to as the first alternative.
2. Treatment 1 tends to yield smaller responses than treatment 2. This alternative is denoted by $<$ and is referred to as the second alternative.
3. Treatment 1 tends either to yield larger responses than treatment 2 or to yield smaller responses than treatment 2. This alternative is denoted by $\neq$ and is referred to as the third alternative.

To this point, the above framework matches the development for a dichotomous response, except for the somewhat different wordings of the alternative hypotheses. For a dichotomous response the test statistic is obvious: simply compare the proportion of successes on the two treatments. For a numerical response, the choice of a test statistic is not clear. Below are two of many possibilities:

- Compare the sample means: the park route has a sample mean of 525 seconds, and the high school route has a sample mean of 530 seconds.
- Compare the sample medians: the park route has a sample median of 527 seconds, and the high school route has a sample median of 530 seconds.

In this text, only a comparison of sample means will be considered. Let $\bar{X}$ and $\bar{Y}$ denote, respectively, the sample means for the first and second treatments and let $\bar{x}$ and $\bar{y}$ denote their observed values. For Cathy's study,

$$\bar{x} = \frac{530 + 521 + 539}{3} = 530$$

and

$$\bar{y} = \frac{528 + 520 + 527}{3} = 525.$$

The test statistic is the difference of the sample means and is denoted by U,

$$U = \bar{X} - \bar{Y}.$$

The observed value of U is denoted by u; for Cathy's study,

$$u = \bar{x} - \bar{y} = 530 - 525 = 5.$$

Three approaches to finding the sampling distribution of the test statistic are presented in this section:

- The exact sampling distribution can be obtained by brute force.
- An approximate sampling distribution can be obtained by a simulation experiment.
- The sampling distribution of a scaled version of U can be approximated by the t-distribution.

The rules of evidence are as follows:

- The larger the value of the test statistic U, the stronger is the evidence in support of the first alternative ($>$).
- The smaller the value of the test statistic U, the stronger is the evidence in support of the second alternative ($<$).
- The farther the value of the test statistic U is from 0, in either direction, the stronger is the evidence in support of the third alternative ($\neq$).

These rules of evidence yield the following formulas for the P-value:

- For the first alternative ($>$),
$$\mathbf{P} = P(U \geq u).$$

- For the second alternative ($<$),
$$\mathbf{P} = P(U \leq u).$$

- For the third alternative ($\neq$),
$$\mathbf{P} = P(U \geq |u|) + P(U \leq -|u|).$$

It can be shown that if $n_1 = n_2$, a common occurrence in a controlled experiment, then the sampling distribution of the test statistic U is symmetric about 0. This fact simplifies the computation of the P-value for the third alternative; the above rule can be replaced by

$$\mathbf{P} = 2P(U \geq |u|) \quad \text{if } n_1 = n_2.$$

The computation of **P** is illustrated below for Cathy's study.

Cathy was interested in finding any difference between the routes and therefore selected the third alternative. As shown earlier, her observed value of the test statistic is 5. It is tedious to obtain the entire sampling distribution by brute force; it is more efficient to compute only the P-value,

$$\mathbf{P} = 2P(U \geq 5).$$

Recall from Chapter 2 that probabilities are induced by the chance mechanism of assigning subjects to treatments. Thus, the P-value is equal to

$$2 \times \frac{\text{The number of assignments that give } \bar{x} - \bar{y} \geq 5}{\text{The total number of assignments}}.$$

First, the total number of assignments is obtained. Number the six trials (days) in Cathy's experiment 1, 2, 3, 4, 5, and 6. The 20 ways to divide these six subjects into two treatment group of equal sizes are listed below. To read this list, note that, for example, 123 means that subjects numbered 1, 2, and 3 are assigned to the first treatment, leaving subjects 4, 5, and 6 for the second treatment.

$$
\begin{array}{cccccccccc}
123 & 124 & 125 & 126 & 134 & 135 & 136 & 145 & 146 & 156 \\
234 & 235 & 236 & 245 & 246 & 256 & 345 & 346 & 356 & 456
\end{array}
$$

Next, the number of assignments that yield

$$\bar{x} - \bar{y} \geq 5$$

must be determined. Note that because the six response values are fixed, any increase in $\bar{x}$ will cause a decrease in $\bar{y}$. Thus, the number of assignments that give $\bar{x} - \bar{y} \geq 5$ is the same as the number of assignments that give $\bar{x} \geq 530$, the value obtained in the actual experiment. It is helpful to combine the two data sets and order the values from smallest to largest:

$$520 \quad 521 \quad 527 \quad 528 \quad 530 \quad 539$$

It is clear from an inspection of this list that four assignments give a value of $\bar{x}$ of 530 or more, namely, the assignments that put the responses

$$(528, 530, 539), (527, 530, 539), (521, 530, 539), \text{ or } (527, 528, 539)$$

on treatment 1. (The sample means for these are 532.3, 532.0, 530.0, and 532.0, respectively.)

Combining these results gives

$$\mathbf{P} = 2 \times \frac{4}{20} = 0.40.$$

There is evidence in her data that Cathy ran faster in the park, but the evidence is not statistically significant for the third alternative.

A second approach is to conduct a simulation experiment to obtain an approximate P-value. A 10,000-run simulation experiment gave 3,986 values of $\bar{x} - \bar{y}$ that were either greater than or equal to 5 or less than or equal to -5. This gives

$$\frac{3{,}986}{10{,}000} = 0.3986$$

as an approximate P-value. This approximation is very close to the exact value, 0.4.

The third, and final, approach involves working with a scaled version of the test statistic U and requires some additional notation. Let S_X and S_Y denote the sample standard deviations of the observations on the first and second treatments, respectively. Define the **pooled sample variance**, S_p^2, by

$$S_p^2 = \frac{(n_1 - 1)S_X^2 + (n_2 - 1)S_Y^2}{n_1 + n_2 - 2}.$$

(Technical note: The pooled sample variance is a **weighted average** of the individual sample variances with the weight assigned to a variance proportional to its degrees of freedom—its sample size less 1.) Note that if $n_1 = n_2$, a common occurrence for a controlled study, the pooled sample variance is simply the arithmetic average of the sample variances. The **pooled sample standard deviation** is defined to equal the positive square root of the pooled sample variance:

$$S_p = \sqrt{S_p^2}.$$

For Cathy's study, it can be shown that the observed values of the sample variances are

$$s_X^2 = 81 \quad \text{and} \quad s_Y^2 = 19.$$

Thus, the observed values of the pooled sample variance and pooled sample standard deviation are

$$s_p^2 = \frac{81 + 19}{2} = 50$$

and

$$s_p = \sqrt{50} = 7.071.$$

Define the variable T by the equation

$$T = \frac{U}{S_p \sqrt{1/n_1 + 1/n_2}}.$$

The observed value of T is

$$t = \frac{u}{s_p \sqrt{1/n_1 + 1/n_2}}.$$

In the special case of a balanced design, the formulas for T and t can be rewritten as

$$T = \frac{\sqrt{n}\, U}{2 S_p} \quad \text{and} \quad t = \frac{\sqrt{n}\, u}{2 s_p},$$

if $n_1 = n_2$, where $n = n_1 + n_2$ is the total number of subjects.

The third approach is built around approximating the sampling distribution of T with the t-distribution with $n - 2$ degrees of freedom.

- For the first alternative ($>$) the approximate P-value is equal to the area under the $t(n - 2)$ curve to the right of t.
- For the second alternative ($<$) the approximate P-value is equal to the area under the $t(n - 2)$ curve to the right of $-t$.
- For the third alternative ($\neq$) the approximate P-value is equal to twice the area under the $t(n - 2)$ curve to the right of $|t|$.

If the number of degrees of freedom, $n - 2$, exceeds 30, the t-curve in the above rules is replaced by the standard normal curve.

For Cathy's study,

$$t = \frac{5}{7.071\sqrt{1/3 + 1/3}} = 0.87.$$

Cathy chose the third alternative, so her approximate P-value is twice the area under the $t(4)$ curve to the right of 0.87. From Table A.6 the area to the right of 0.87 is bounded by 0.10 and 0.25. Thus, the approximate P-value is between twice these bounds, namely 0.20 and 0.50.

The Minitab statistical software package has a command that computes exact areas under t-curves. This command yields 0.2167 as the area under the $t(4)$ curve to the right of 0.87. Thus, twice 0.2167, or 0.4334, is a more precise approximation of the P-value. Note that this approximation is quite close to the exact P-value, 0.40.

Below is another example, to help solidify these new ideas and methods.

EXAMPLE 16.2

For his class project, Dave B. studied the effect of alcohol on athletic performance. The subjects in his study were 10 of his friends. The response was the time, in seconds, required to complete a short obstacle course. The 10 subjects were divided into two groups of equal size by randomization. The five persons assigned to treatment 1 were asked to drink two beers as fast as they could tolerate. Five minutes later they started the obstacle course. The five persons on treatment 2 simply ran the course.

Dave selected the first alternative, that alcohol would increase the time required to complete the course. The results of the experiment were as follows:

	Alcohol (Treatment 1)				
Name	Brad	Sam	Stu	Mike	Mark
Time	50	53	47	51	51

	No Alcohol (Treatment 2)				
Name	Dan	Pat	Alex	Jenny	Dave
Time	51	48	50	49	43

Figure 16.2 presents dot plots of these data. The data provide evidence that the consumption of alcohol diminishes athletic performance. For the first treatment

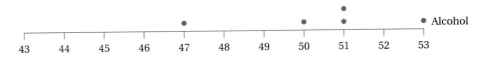

FIGURE 16.2 Dot plots of the times, in seconds, to complete an obstacle course.

group, that with alcohol, the sample mean, variance, and standard deviation are

$$\bar{x} = 50.4, \quad s_X^2 = 4.8, \quad \text{and} \quad s_X = 2.191 \text{ (seconds)};$$

for the second group, that without alcohol, the sample mean, variance, and standard deviation are

$$\bar{y} = 48.2, \quad s_Y^2 = 9.7, \quad \text{and} \quad s_Y = 3.114 \text{ (seconds)}.$$

Note that the data provide evidence in support of Dave's belief, namely, the mean time without alcohol is lower than the mean time with alcohol. The observed value of the test statistic U is

$$u = \bar{x} - \bar{y} = 50.4 - 48.2 = 2.2.$$

There are 252 different possible assignments of subjects to treatments; too many to study by brute force here. It can be shown, however, that 44 of the 252 possible assignments give a value of $\bar{x} - \bar{y}$ that is greater than or equal to the observed difference, 2.2; thus, the exact P-value for the first alternative is

$$P(U \geq 2.2) = \frac{44}{252} = 0.1746.$$

Thus, the evidence in Dave's data is not sufficiently strong to reject the null hypothesis of no treatment effect.

Alternatively, it is quite easy, with access to a computer, to perform a 10,000-run simulation experiment. Of the 10,000 simulated values of $\bar{x} - \bar{y}$, 1,727 were greater than or equal to 2.2; thus, the simulation estimate of the P-value is

$$\frac{1{,}727}{10{,}000} = 0.1727.$$

This is a very good approximation to the exact P-value, 0.1746.

Finally, the t-distribution can be used to obtain an approximate P-value. Because the sample sizes are equal,

$$s_p^2 = \frac{4.8 + 9.7}{2} = 7.25 \quad \text{and} \quad s_p = \sqrt{7.25} = 2.693.$$

The observed value of the test statistic is

$$t = \frac{\sqrt{n}(\bar{x} - \bar{y})}{2s_p} = \frac{\sqrt{10}(2.2)}{2(2.693)} = 1.29.$$

For the first alternative, the approximate P-value is the area under the $t(8)$ curve to the right of 1.29. From Table A.6, this area is between 0.10 and 0.25. With the help of Minitab, this area is found to equal 0.1166. This answer is not a very good approximation of the exact P-value, 0.1746. ▲

16.2 INDEPENDENT RANDOM SAMPLES

Suppose a researcher wants to compare the distributions of a numerical variable in two populations. This goal is very ambitious—too ambitious, in fact, unless one has a very large amount of data. Usually, a researcher will focus on the more manageable problem of comparing the means of the populations.

Section 7.1.1 discussed several different ways independent random samples can be obtained from two populations. These ideas did not depend on the fact that the response was dichotomous. In particular, the same ideas apply to a numerical response. You should review that material before proceeding.

The means of the first and second populations are denoted μ_X and μ_Y, respectively, and are both assumed to be unknown. The respective population standard deviations are denoted σ_X and σ_Y, also both unknown. A random sample of size n_1 is selected from the first population, and an independent random sample of size n_2 is selected from the second population. Let $n = n_1 + n_2$. The sample mean and sample variance of the data from the first population are denoted by $\bar{X}$ and S_X^2. The sample mean and sample variance of the data from the second population are denoted by $\bar{Y}$ and S_Y^2.

Inference procedures, either confidence intervals or hypotheses tests, focus on the difference $\mu_X - \mu_Y$. The natural point estimator of this quantity is $\bar{X} - \bar{Y}$.

It can be shown that the sampling distribution of $\bar{X} - \bar{Y}$ has the following properties:

$$E(\bar{X} - \bar{Y}) = \mu_X - \mu_Y \quad \text{and} \quad \text{Var}(\bar{X} - \bar{Y}) = \frac{\sigma_X^2}{n_1} + \frac{\sigma_Y^2}{n_2}.$$

Thus, the standardized version of $\bar{X} - \bar{Y}$ is

$$W = \frac{(\bar{X} - \bar{Y}) - (\mu_X - \mu_Y)}{\sqrt{(\sigma_X^2/n_1) + (\sigma_Y^2/n_2)}}. \tag{16.1}$$

There are three different possible analyses, corresponding to three different sets of assumptions. The most restrictive assumption yields a familiar result.

16.2.1 Case 1: Normal Populations with Equal Variances

Assume that each population distribution is a normal curve and that

$$\sigma_X^2 = \sigma_Y^2.$$

For notational simplicity, denote this common value of the variances by σ^2.

The first issue to be addressed is the estimation of the common variance, σ^2. It can be shown mathematically that the best estimate of σ^2 is

$$S_p^2 = \frac{(n_1 - 1)S_X^2 + (n_2 - 1)S_Y^2}{n_1 + n_2 - 2},$$

the pooled sample variance defined in the previous section.

Next, replace the unknown variances in the formula for W, Equation 16.1, by the estimate S_p^2, and denote the result by W_1:

$$W_1 = \frac{(\bar{X} - \bar{Y}) - (\mu_X - \mu_Y)}{\sqrt{(S_p^2/n_1) + (S_p^2/n_2)}}.$$

Usually, S_p^2 is factored out of the square root to give

$$W_1 = \frac{(\bar{X} - \bar{Y}) - (\mu_X - \mu_Y)}{S_p \sqrt{(1/n_1) + (1/n_2)}}. \tag{16.2}$$

It can be shown, under the assumptions of this first case, that the sampling distribution of W_1 is a t-curve with $n - 2$ degrees of freedom. Thus, confidence intervals and hypothesis tests can be obtained easily using a modification of the arguments of Chapter 15. More precisely:

With the assumptions of this case, a confidence interval for $\mu_X - \mu_Y$ is given by

$$(\bar{x} - \bar{y}) \pm t s_p \sqrt{\frac{1}{n_1} + \frac{1}{n_2}}, \tag{16.3}$$

where t is equal to the entry in the row corresponding to $n - 2$ degrees of freedom and the appropriate column of Table A.6. If the number of degrees of freedom, $n - 2$, is larger than 30, t can be replaced by z for the standard normal curve from Table A.3.

Hypothesis tests, as usual, take a bit longer to present. The null hypothesis states that the two population means are equal. In symbols this means that $\mu_X = \mu_Y$, which can also be written as $\mu_X - \mu_Y = 0$. The latter representation can be interpreted as saying that the special value of interest for $\mu_X - \mu_Y$ is 0.

First, the researcher must select one of the following three pairs of hypotheses,

$H_0: \mu_X - \mu_Y = 0$ $H_0: \mu_X - \mu_Y = 0$ $H_0: \mu_X - \mu_Y = 0$
$H_1: \mu_X - \mu_Y > 0$ $H_1: \mu_X - \mu_Y < 0$ $H_1: \mu_X - \mu_Y \neq 0$

Alternatively, the hypotheses may be written as, respectively,

$$H_0: \mu_X = \mu_Y \qquad H_0: \mu_X = \mu_Y \qquad H_0: \mu_X = \mu_Y$$
$$H_1: \mu_X > \mu_Y \qquad H_1: \mu_X < \mu_Y \qquad H_1: \mu_X \neq \mu_Y$$

The test statistic is obtained by replacing $\mu_X - \mu_Y$ in W_1 by its special value of interest, 0. The result is

$$T_1 = \frac{\bar{X} - \bar{Y}}{S_p \sqrt{(1/n_1) + (1/n_2)}}. \tag{16.4}$$

It follows that if the null hypothesis is true, the sampling distribution of T_1 is the t-distribution with $n - 2$ degrees of freedom. The rules of evidence are

- The larger the value of T_1, the stronger is the evidence in support of the first alternative, $\mu_X > \mu_Y$.
- The smaller the value of T_1, the stronger is the evidence in support of the second alternative, $\mu_X < \mu_Y$.
- The farther the value of T_1 is from 0, in either direction, the stronger is the evidence in support of the third alternative, $\mu_X \neq \mu_Y$.

Let t_1 denote the observed value of the test statistic T_1.

- For the first alternative ($\mu_X > \mu_Y$) the P-value equals the area under the $t(n - 2)$ curve to the right of t_1.
- For the second alternative ($\mu_X < \mu_Y$) the P-value equals the area under the $t(n - 2)$ curve to the right of $-t_1$.
- For the third alternative ($\mu_X \neq \mu_Y$) the P-value equals twice the area under the $t(n - 2)$ curve to the right of $|t_1|$.

If the number of degrees of freedom, $n - 2$, exceeds 30, the t-curve in the above rules is replaced by the standard normal curve.

These methods are illustrated with the following example.

EXAMPLE 16.3

The twenty-five scores I obtained playing Tetris in 1990 are presented in Table 15.4 (p. 539). After these data were collected, I stopped playing Tetris for about one year until the winter of 1991. After two days of practice, I repeated the protocol from the previous year and obtained 25 more Tetris scores. The two sets of data are compared below. For the descriptive comparison, no further assumptions are required. For an inferential comparison, however, the following terminology and assumptions are adopted.

The first population represents the process that generated the 1991 scores, and the second population represents the process that generated the 1990 scores. The sample data are viewed as independent random samples from the populations. The populations are assumed to be normal curves with equal

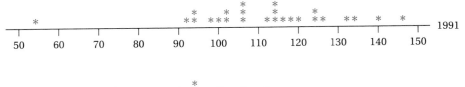

FIGURE 16.3 Dot plots of Tetris scores.

variances. The population means and the common variance are assumed to be unknown.

Figure 16.3 and Table 16.1 present dot plots and selected summary statistics for the two samples. Some conclusions follow easily from an examination of this figure and table.

- The dot plots indicate that the scores improved from 1990 to 1991. There is one small outlier and no large outliers each year. Ignoring the outliers, the spread looks approximately the same each year.
- The summary statistics, like the dot plots, indicate that the center shifted. From 1990 to 1991, the sample mean increased by

$$\bar{x} - \bar{y} = 111.08 - 93.32 = 17.76$$

points and the sample median increased by

$$113 - 95 = 18$$

points.
- The summary statistics indicate some increase in spread from 1990 to 1991. The sample standard deviation grew by 20 percent, from 15.60 to 18.83; the sample interquartile range grew by 2 percent, from 23 to 23.5; and the sample range grew by 44 percent, from 63 to 91. (It may seem strange to label increases of 20 percent and 44 percent "some increase"; as shown in the exercises, much of these increases is attributable to the one outlier in 1991 being more extreme than the one outlier in 1990.)

TABLE 16.1 Summary Statistics for the Tetris Study

Year	n	Mean	Median	Standard Deviation	Minimum	Maximum	Q_1	Q_3	IQR
1991	25	111.08	113.0	18.83	54	145	100.5	124.0	23.5
1990	25	93.32	95.0	15.60	51	114	83.0	106.0	23.0

Formula 16.3 can be used to obtain the 95 percent confidence interval for the increase in the population mean from 1990 to 1991. In order to use this formula, the pooled standard deviation must be computed. Because the sample sizes are equal,

$$s_p^2 = \frac{s_X^2 + s_Y^2}{2} = \frac{(18.83)^2 + (15.60)^2}{2} = 298.96,$$

and

$$s_p = \sqrt{s_p^2} = \sqrt{298.96} = 17.29.$$

In addition, t must be obtained from the t-distribution with

$$n_1 + n_2 - 2 = 25 + 25 - 2 = 48$$

degrees of freedom. Since the number of degrees of freedom exceeds 30, t is replaced by $z = 1.96$ from Table A.3. Substitute the appropriate numbers into Formula 16.3,

$$(\bar{x} - \bar{y}) \pm t s_p \sqrt{\frac{1}{n_1} + \frac{1}{n_2}},$$

to obtain

$$(111.08 - 93.32) \pm 1.96(17.29)\sqrt{\frac{1}{25} + \frac{1}{25}} = 17.76 \pm 9.59 = [8.17, 27.35].$$

In words, at the 95 percent confidence level the mean increase in my score falls somewhere between 8.17 and 27.35 points.

It is also possible to conduct a hypothesis test on the change in means. The observed value of the test statistic

$$T_1 = \frac{\bar{X} - \bar{Y}}{S_p \sqrt{(1/n_1) + (1/n_2)}}$$

is

$$t_1 = \frac{17.76}{17.29 \sqrt{1/25 + 1/25}} = 3.63.$$

Because the number of degrees of freedom, 48, is larger than 30, the P-value can be obtained from the standard normal curve. For the third alternative hypothesis, that the mean changed, the P-value equals twice the area under the standard normal curve to the right of 3.63. From Table A.2,

$$\mathbf{P} < 2(0.0002) = 0.0004.$$

The change in means is highly statistically significant. ▲

16.2.2 Case 2: Normal Populations

As in Section 16.2.1—case 1—each population distribution is assumed to be represented by a normal curve. In case 2, however, no assumption is made about the equality of the population variances.

The unknown population variances in the standardized version of $\bar{X} - \bar{Y}$, Equation 16.1, are replaced by the corresponding sample variances. This replacement gives

$$W_2 = \frac{(\bar{X} - \bar{Y}) - (\mu_X - \mu_Y)}{\sqrt{(S_X^2/n_1) + (S_Y^2/n_2)}}. \tag{16.5}$$

Under the assumptions of this case, it can be shown that the sampling distribution of W_2 can be approximated by a t-curve with ν degrees of freedom. The value of ν is obtained by first computing

$$\nu_1 = \left(\frac{S_X^2}{n_1} + \frac{S_Y^2}{n_2}\right)^2 \quad \text{and} \quad \nu_2 = \frac{S_X^4}{n_1^2(n_1 - 1)} + \frac{S_Y^4}{n_2^2(n_2 - 1)} \tag{16.6}$$

and then computing

$$\nu = \frac{\nu_1}{\nu_2}. \tag{16.7}$$

The above formula for ν usually does not yield an integer; the common practice is to round it off to the nearest integer. (Many persons find the computation of ν to be a formidable task, especially in an exam setting. However, in real life— real life is not an in-class exam—it is difficult to imagine a situation in which a researcher would go to the trouble of conducting a study, yet not take the time to compute the value of ν. Many statistical computing packages, including Minitab, compute the value of ν for the user.)

If the assumptions of this case are true, an approximate confidence interval for $\mu_X - \mu_Y$ is given by

$$(\bar{x} - \bar{y}) \pm t \sqrt{\frac{S_X^2}{n_1} + \frac{S_Y^2}{n_2}}, \tag{16.8}$$

where t is equal to the entry in the row corresponding to ν degrees of freedom and the appropriate column of Table A.6. If ν is greater than 30, t can be replaced by z of the standard normal curve in Table A.3.

For a hypothesis test, the three possible pairs of hypotheses were given in Section 16.2.1. The test statistic is obtained by replacing $\mu_X - \mu_Y$ in W_2 by its special value of interest, 0. This replacement yields

$$T_2 = \frac{\bar{X} - \bar{Y}}{\sqrt{(S_X^2/n_1) + (S_Y^2/n_2)}}. \tag{16.9}$$

16.2 Independent Random Samples

If the null hypothesis is true, the sampling distribution of T_2 can be approximated by the t-distribution with ν degrees of freedom. The rules of evidence are the same as those given for case 1 above, with the obvious change that T_1 is replaced by the current test statistic T_2.

Let t_2 denote the observed value of the test statistic T_2.
- For the first alternative ($\mu_X > \mu_Y$) the approximate P-value equals the area under the $t(\nu)$ curve to the right of t_2.
- For the second alternative ($\mu_X < \mu_Y$) the approximate P-value equals the area under the $t(\nu)$ curve to the right of $-t_2$.
- For the third alternative ($\mu_X \neq \mu_Y$) the approximate P-value equals twice the area under the $t(\nu)$ curve to the right of $|t_2|$.

If the number of degrees of freedom, ν, exceeds 30, the t-curve in the above rules is replaced by the standard normal curve.

The formulas of this section are illustrated in the following example.

EXAMPLE 16.4

The Tetris study described in Example 16.3 is reanalyzed with the less restrictive assumptions of case 2. Recall that summary statistics are given in Table 16.1.

The most difficult step is the computation of the number of degrees of freedom, ν. With the help of Minitab, it is easy to obtain $\nu = 46$. The formula for a confidence interval for $\mu_X - \mu_Y$,

$$(\bar{x} - \bar{y}) \pm t \sqrt{\frac{s_X^2}{n_1} + \frac{s_Y^2}{n_2}},$$

yields, for the 95 percent confidence level,

$$(111.08 - 93.32) \pm 1.96 \sqrt{\frac{(18.83)^2}{25} + \frac{(15.60)^2}{25}} = 17.76 \pm 9.59 = [8.17, 27.35].$$

This is *exactly* the same answer obtained with the case 1 analysis. This is not a coincidence; the reason for this agreement is explored in Section 16.2.4.

For a hypothesis test, the test statistic

$$T_2 = \frac{\bar{X} - \bar{Y}}{\sqrt{(S_X^2/n_1) + (S_Y^2/n_2)}}$$

has observed value

$$t_2 = \frac{111.08 - 93.32}{\sqrt{[(18.83)^2/25] + [(15.60)^2/25]}} = \frac{17.76}{4.8905} = 3.63.$$

Again, this equals *exactly* the value obtained in case 1. With 46 degrees of freedom, the standard normal curve can be used to obtain the P-value,

$$\mathbf{P} < 2(0.0002) = 0.0004.$$

16.2.3 Case 3: Large Sample Approximation

If the researcher is not willing to adopt the assumptions of either case 1 or case 2 above, he or she can still obtain an approximate confidence interval and P-value provided both sample sizes, n_1 and n_2, are greater than 20.

The unknown population variances in the standardized version of $\bar{X} - \bar{Y}$ (Equation 16.1) are replaced by the corresponding sample variances. This gives

$$W_3 = \frac{(\bar{X} - \bar{Y}) - (\mu_X - \mu_Y)}{\sqrt{(S_X^2/n_1) + (S_Y^2/n_2)}}. \quad (16.10)$$

Note that W_3 is identical to W_2 in Equation 16.5. It can be shown that the sampling distribution of W_3 can be well approximated by a standard normal curve.

If the sample sizes, n_1 and n_2, are both greater than 20, an approximate confidence interval for $\mu_X - \mu_Y$ is given by

$$(\bar{x} - \bar{y}) \pm z\sqrt{\frac{s_X^2}{n_1} + \frac{s_Y^2}{n_2}}. \quad (16.11)$$

where z is the appropriate value from Table A.3.

For a hypothesis test, the three possible pairs of hypotheses were given in Section 16.2.1. The test statistic is obtained by replacing $\mu_X - \mu_Y$ by its special value of interest, 0, in W_3. The result is

$$Z = \frac{\bar{X} - \bar{Y}}{\sqrt{(S_X^2/n_1) + (S_Y^2/n_2)}}. \quad (16.12)$$

If the null hypothesis is true, the sampling distribution of Z can be well approximated by the standard normal curve. The rules of evidence are the same as given in case 1 with the obvious change of T_1 to Z.

Let z denote the observed value of the test statistic Z.

- For the first alternative ($\mu_X > \mu_Y$) the approximate P-value equals the area under the standard normal curve to the right of z.
- For the second alternative ($\mu_X < \mu_Y$) the approximate P-value equals the area under the standard normal curve to the right of $-z$.
- For the third alternative ($\mu_X \neq \mu_Y$) the approximate P-value equals twice the area under the standard normal curve to the right of $|z|$.

EXAMPLE 16.5

The methods of this section could be applied to the Tetris study, but they would give exactly the same results as the case 1 and case 2 analyses. Instead, a new example is considered.

The response is taken to be the absolute value of the same-day high-temperature forecast error in Madison, Wisconsin. More precisely, if f represents the high-temperature forecast for the day appearing in the morning's *Wisconsin State Journal* and h represents the actual high, the response is

$$|f - h|.$$

The purpose of this study is to compare the absolute forecast errors in summer and spring, 1988. Before observing any data, I suspected that the absolute errors would tend to be larger in spring because I believed that weather conditions change more quickly and erratically in spring than in summer in Madison. Figure 16.4 and Table 16.2 present density scale histograms and selected summary statistics of the absolute forecast errors for the two seasons. Some conclusions follow easily from an examination of this figure and this table.

1. The histograms indicate that both distributions are skewed to the right with no outliers. The center of the summer absolute errors appears to be to the left (smaller) than the center for spring. This agrees with my pre-data suspicion.

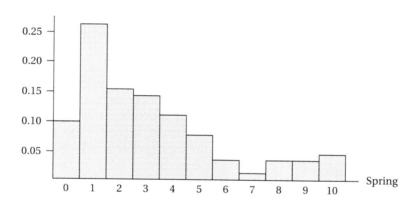
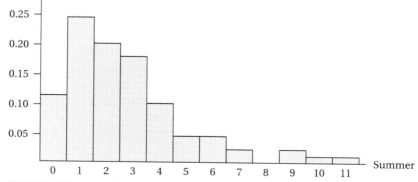

FIGURE 16.4 *Density scale histograms of absolute forecast errors.*

TABLE 16.2 Summary Statistics of the Absolute Forecast Errors

Season	n	Mean	Median	Standard Deviation	Minimum	Maximum	Q_1	Q_3
Spring	91	3.099	2.0	2.688	0	10	1.0	4.0
Summer	89	2.685	2.0	2.299	0	11	1.0	4.0

2. The sample mean for spring is

$$3.099° - 2.685° = 0.414°$$

higher than the sample mean for summer. The sample medians are the same for the two seasons. As demonstrated for many data sets, the value of the median can be a misleading measure of center for an extremely granular distribution.

3. The sample standard deviation for spring is 17 percent larger than for summer, 2.688 compared with 2.299. Because the distributions are not bell-shaped, it is difficult to interpret this difference. The two seasons have identical quartiles, and hence identical interquartile ranges, and nearly equal ranges.

The absolute forecast errors are, of course, time series data. The methods of Chapter 15 applied to these data indicate that the SI assumption seems reasonable for each season (details are not given). Thus, the sample data are assumed to be the result of independent random samples from two populations. Let population 1 denote the spring forecasts and population 2 denote the summer forecasts. The formula for the 90 percent confidence interval for $\mu_X - \mu_Y$,

$$(\bar{x} - \bar{y}) \pm z \sqrt{\frac{s_X^2}{n_1} + \frac{s_Y^2}{n_2}},$$

yields

$$(3.099 - 2.685) \pm 1.645 \sqrt{\frac{(2.688)^2}{91} + \frac{(2.299)^2}{89}} = 0.414 \pm 0.613 = [-0.199, 1.027].$$

In words, at the 90 percent confidence level the mean absolute error in spring could be as much as 1.027 degrees higher than the mean absolute summer error; on the other hand, the mean for summer could be as much as 0.199 degrees higher than the mean for spring.

A hypothesis test also can be conducted. The test statistic

$$Z = \frac{\bar{X} - \bar{Y}}{\sqrt{(S_X^2/n_1) + (S_Y^2/n_2)}}$$

has observed value

$$z = \frac{0.414}{\sqrt{[(2.688)^2/91] + [(2.299)^2/89]}} = 1.11.$$

For the first alternative hypothesis, that the spring mean is larger, the approximate P-value is the area under the standard normal curve to the right of 1.11. From Table A.2, this area equals 0.1335. Thus, the observed difference between spring and summer forecasts is not statistically significant. ▲

16.2.4 Comparison of Methods

The three cases grow, in part, from the need to estimate the variances in the standard error of $\bar{X} - \bar{Y}$,

$$\sqrt{\frac{\sigma_X^2}{n_1} + \frac{\sigma_Y^2}{n_2}}.$$

For cases 2 and 3 the unknown population variances are replaced by the corresponding sample variances. Thus, the only difference in the formulas for the two cases is that case 2 uses the t-distribution with ν degrees of freedom and case 3 uses the standard normal curve as the reference distribution. Thus, if ν exceeds 30, the answers from cases 2 and 3 are identical, given my dictum that the t-distribution with more than 30 degrees of freedom should be approximated by the standard normal curve. The differences between cases 1 and 2 are more interesting, because they are competitors when either or both sample sizes are small.

Case 2 has a big apparent advantage; it does not require that the population variances be equal. It can be shown, however, that the formulas for case 1 perform well if the population variances are approximately equal. Moreover, if normal populations have variances that are strikingly different, there are big differences in the populations that are not summarized by the means. Hence, the strategy of comparing the means as a method of comparing the populations may be hard to justify, except for its expediency. Expediency is a poor justification!

When the sample sizes are small, it is generally recommended that the researcher should plan the study, if possible, so that the sample sizes n_1 and n_2 are equal. There are several reasons for this advice:

1. If the sample sizes are equal, the estimated standard errors of $\bar{X} - \bar{Y}$ are the same for cases 1 and 2. Thus, any difference between the answers obtained from cases 1 and 2 is solely a result of the difference in degrees of freedom between the two cases.
2. If $n_1 = n_2$ and $s_X = s_Y$, then

$$\nu = n_1 + n_2 - 2.$$

In combination with the previous item, this indicates that in this situation cases 1 and 2 give exactly the same answers. Since exact equality of sample standard deviations is rare, this result would appear to be of limited value; however, if the sample standard deviations are similar in value, the number of degrees of freedom computed under the two cases are close, as demonstrated by the Tetris study of this section, for which $n_1 + n_2 - 2 = 48$ and $\nu = 46$.

3. If the sample sizes are equal, the answers obtained assuming case 1 are robust (as discussed in Chapter 15) against a violation of the assumption of equal variances. If the sample sizes are very different, however, answers obtained assuming case 1 are not robust if the assumption of equal variances is not at least approximately true.

Finally, it can be shown that for any choice of sample sizes and any values of the sample variances,

$$\nu \leq n_1 + n_2 - 2.$$

Thus, case 2 never gives more degrees of freedom than case 1. This fact also is illustrated for the Tetris study, which had

$$\nu = 46 \quad \text{and} \quad n_1 + n_2 - 2 = 48.$$

EXERCISES FOR SECTIONS 16.1 AND 16.2

1. Compute s_p for each of the following hypothetical cases:
   (a) $s_X = 5$, $s_Y = 7$, $n_1 = 11$, and $n_2 = 11$.
   (b) $s_X = 5$, $s_Y = 7$, $n_1 = 2$, and $n_2 = 20$.
   (c) $s_X = 5$, $s_Y = 7$, $n_1 = 20$, and $n_2 = 2$.

2. Compute s_p for each of the following hypothetical cases:
   (a) $s_X^2 = 10$, $s_Y^2 = 8$, $n_1 = 11$, and $n_2 = 11$.
   (b) $s_X^2 = 10$, $s_Y^2 = 8$, $n_1 = 2$, and $n_2 = 20$.
   (c) $s_X^2 = 10$, $s_Y^2 = 8$, $n_1 = 20$, and $n_2 = 2$.

3. Compute the observed value of

$$T = \frac{U}{S_p \sqrt{1/n_1 + 1/n_2}}$$

   for each of the following hypothetical cases:
   (a) $s_X = 5$, $s_Y = 7$, $n_1 = 10$, $n_2 = 10$, $\bar{x} = 30$, and $\bar{y} = 25$.
   (b) $s_X = 5$, $s_Y = 7$, $n_1 = 2$, $n_2 = 20$, $\bar{x} = 30$, and $\bar{y} = 25$.
   (c) $s_X = 5$, $s_Y = 7$, $n_1 = 20$, $n_2 = 2$, $\bar{x} = 30$, and $\bar{y} = 25$.

4. Compute the observed value of

$$T = \frac{U}{S_p \sqrt{1/n_1 + 1/n_2}}$$

   for each of the following hypothetical cases:
   (a) $s_X^2 = 10$, $s_Y^2 = 8$, $n_1 = 10$, $n_2 = 10$, $\bar{x} = 50$, and $\bar{y} = 60$.

(b) $s_X^2 = 10$, $s_Y^2 = 8$, $n_1 = 2$, $n_2 = 20$, $\bar{x} = 50$, and $\bar{y} = 60$.

(c) $s_X^2 = 10$, $s_Y^2 = 8$, $n_1 = 20$, $n_2 = 2$, $\bar{x} = 50$, and $\bar{y} = 60$.

5. Refer to the data in Exercise 1 on p. 401, a CRD with $n_1 = n_2 = 10$. For the responses with the combat boots, the sample mean and sample standard deviation equal 333.0 seconds and 8.18 seconds, respectively. For the responses with the jungle boots, these values are 319.5 seconds and 7.93 seconds.

   (a) Test the null hypothesis that the two types of boots have the same effect on the time it took Brian to run 1 mile versus the alternative that they have different effects. Use the t-distribution to obtain an approximate P-value. Comment.

   (b) Assume that the data come from independent random samples from two populations. Construct the 95 percent confidence interval for the difference in population means (combat minus jungle). Comment.

6. Refer to the comparison of my 1990 and 1991 Tetris scores. In this exercise, the data will be reanalyzed after the deletion of the lowest value from each sample. For the remaining data, $n_1 = 24$, $\bar{x} = 113.46$, $s_X = 13.15$, $n_2 = 24$, $\bar{y} = 95.08$, and $s_Y = 14.91$.

   (a) Construct the 95 percent confidence interval for $\mu_X - \mu_Y$. Assume case 3. Repeat assuming case 1. Comment.

   (b) Find the P-value for the third alternative. Assume case 3.

   (c) Compare your answers to this exercise with those obtained in the text for the entire set of 50 observations.

7. Refer to the data in Exercise 3 on p. 402, a CRD with $n_1 = n_2 = 40$. For the responses with the aluminum bat, the sample mean and sample standard deviation equal 168.88 feet and 45.98 feet, respectively. For the responses with the wooden bat, these values are 149.05 feet and 22.98 feet.

   (a) Test the null hypothesis that the two types of bats have the same effect on the distance the ball travels versus the alternative hypothesis that the aluminum bat gives greater distances. Use the t-distribution to obtain an approximate P-value. Comment.

   (b) Assume that the data come from independent random samples from two populations. Construct the 95 percent confidence interval for the difference in population means (aluminum minus wooden). Comment.

8. The following passage is taken from the abstract of a report on a study of Boston-area radiator shops [1]:

   > Exposure to lead occurs during automobile radiator repair when soldered joints are heated, but this relatively common hazard has received little public recognition. We therefore studied lead exposure among automobile radiator mechanics in the Boston area.
   >
   > Twenty-seven shops were surveyed and most were found to be small and poorly ventilated. Seventy-five workers were interviewed and tested for blood lead and free erythrocyte protoporphyrin levels.

   Fifty-six of the 75 workers were radiator mechanics, and the remaining 19 were not.

   (a) Discuss the appropriateness of assuming that these 75 workers represent independent random samples from two populations. Define the populations and discuss what further information you would like to have about how the samples were selected.

   (b) The radiator mechanics had a mean blood lead level of 37.1 μg/dl with a standard deviation of 13.8 μg/dl. The other workers had a mean of 21.6 μg/dl and a standard deviation of 11.2 μg/dl. On the assumption of independent random samples, construct the 95 percent confidence interval for the difference of the population means. Assume case 2. (Hint: $\nu > 30$.)

   (c) On the assumption of independent random samples, test the null hypothesis that there is no difference in means versus the alternative that the radiator mechanics have a larger population mean than the other workers. Assume case 2.

   (d) The radiator mechanics had a mean age of 39.5 years with a standard deviation of 14.5 years, and the other workers had a mean age of 36.2 years with a standard deviation of 12.3 years. Is there a significant difference in the ages of the two groups? Assume case 2. (Hint: $\nu > 30$.) What is the relevance of this analysis?

9. Refer to the golf data presented in Exercise 13 on p. 406, a CRD with $n_1 = n_2 = 40$. For the data on a 3-wood the sample mean and sample standard deviation are 106.87 yards and 29.87 yards, respectively. For the data on the 3-iron, these values are 98.18 yards and 28.33 yards.

   (a) Test the null hypothesis that the two clubs have the same effect on the distance versus the alternative that a ball travels farther when hit with the wood. Use the t-distribution to obtain an approximate P-value. Comment.

   (b) Assume that the data come from independent random samples from two populations. Construct the 95 percent confidence interval for the difference in population means (wood minus iron). Comment.

10. Refer to the dart scores data presented in Exercise 14 on p. 406, a CRD with $n_1 = n_2 = 15$. For the scores from 10 feet the sample mean and sample standard deviation are 201.5 points and 11.2 points, respectively. For the scores from 12 feet these values are 188.0 points and 15.1 points.

    (a) Test the null hypothesis that the two distances have the same effect on the score versus the alternative that the shorter distance gives higher scores. Use the t-distribution to obtain an approximate P-value. Comment.

    (b) Assume that the data come from independent random samples from two populations. Construct the 95 percent confidence interval for the difference in population means (shorter minus longer distance). Comment.

11. Refer to the data on car acceleration times presented in Exercise 15 on p. 406, a CRD with $n_1 = n_2 = 15$. For the times with the windows open, the sample mean and sample standard deviation equal 7.94 seconds and 0.2913 seconds, respectively. For the times with the windows closed, these values are 7.10 seconds and 0.3253 seconds.

    (a) Test the null hypothesis that the two treatments have the same effect on the time to reach 65 miles per hour versus the alternative that having the windows closed gives shorter times. Use the t-distribution to obtain an approximate P-value. Comment.

    (b) Assume that the data come from independent random samples from two populations. Construct the 95 percent confidence interval for the difference in population means (open minus closed). Comment.

12. The following is taken from the abstract of a Dutch medical study [2]:

    > Previous reports have indicated that coffee consumption may increase serum cholesterol levels. We studied the effects of coffee prepared by two common brewing methods (filtering and boiling) on serum lipid levels in a twelve-week randomized trial.... After a three-week run-in period during which they all consumed filtered coffee, the participants were randomly assigned to one of three groups receiving four to six cups of boiled coffee a day, four to six cups of filtered coffee a day, or no coffee, for a period of nine weeks.

    For the purpose of this exercise, the no-coffee group will be disregarded (the researchers found no statistically significant differences between the filtered-coffee group and the no-coffee group). The subjects were selected from a group of 596 participants in another study. These 596 persons were believed to be a random sample from a population. Each of the 596 people who were aged 18 or older and who habitually drank coffee were invited to participate in the present study. One hundred seven subjects volunteered to be in this study, and 101 completed the entire 12-week program. Before the study all subjects drank filtered coffee.

    (a) Discuss the appropriateness of the assumption that the 101 subjects were a random sample from the population of habitual coffee drinkers.

    (b) This can be viewed as a "before and after" study within each treatment group. For example, for the 33 subjects who drank boiled coffee, the mean increase (from the end of the run-in to the end of the study) in total cholesterol was 0.52 millimoles per liter with a standard deviation of 0.733 millimoles per liter. On the assumption of a random sample, construct the 95 percent confidence interval for the mean increase in total cholesterol.

(c) Refer to the previous question. The 34 subjects who drank filtered coffee had a mean increase of 0.04 millimoles per liter with a standard deviation of 0.744 millimoles per liter. On the assumption of a random sample, construct the 95 percent confidence interval for the mean increase in total cholesterol.

(d) This study can be viewed as comparing two treatments, with the treatments being boiled and filtered coffee and the response being the change in total cholesterol. Test the null hypothesis that the treatments have identical effects versus the third alternative hypothesis. Is the assumption of a random sample needed for the computation of the approximate P-value?

(e) Refer to the previous question. On the assumption of a random sample, construct the 95 percent confidence interval for the difference between the two mean increases in total cholesterol. Assume case 1.

13. The following passage is taken from the abstract of the report on a study of women with osteoporosis [3]:

> Progressive bone loss in osteoporosis results from bone resorption in excess of bone formation. We conducted a double-blind study in 40 women with postmenopausal osteoporosis of therapy with etidronate, a diphosphonate compound that reduces bone resorption by inhibiting osteoclastic activity.
>
> The patients were randomly assigned in equal numbers to receive oral etidronate (400 mg per day) or placebo for two weeks, followed by a 13-week period in which no drugs were given. This sequence was repeated ten times for a total of 150 weeks. Daily oral supplementation with calcium and vitamin D was given throughout the study to both groups.
>
> For each subject, her bone mineral content at the end of the study was compared with the content at baseline to obtain the percentage change. For the etidronate group, the 20 women had a mean change of 5.3 percent with a standard deviation of 7.05 percent. For the placebo group, the mean change was -2.7 percent with a standard deviation of 9.83 percent.

(a) Compare etidronate to the placebo with the appropriate one-sided hypothesis test.

(b) On the assumption that the 40 women were selected at random from a population of women with postmenopausal osteoporosis, compute the 95 percent confidence interval for the difference in means between etidronate and the placebo. Assume case 3.

14. Refer to the coffee study of Exercise 12, in which the total cholesterol level was the response. Repeat the analyses of Exercise 12 with the response equal to LDL (low-density lipoprotein) cholesterol using the following summary statistics. The increase in LDL cholesterol for the boiled-coffee group had a mean of 0.36 and a standard deviation of 1.084. The increase in LDL cholesterol for the filtered coffee group had a mean of -0.03 and a standard deviation of 0.714.

15. Refer to the data you collected for Exercise 25 on p. 410.

(a) Use the t-distribution to find an approximate P-value for testing the null hypothesis that the treatments are equal versus the third alternative.

(b) Assume that the data are independent random samples from two populations. Construct the 95 percent confidence interval for the difference of the means.

(c) Write a few sentences that summarize what you have learned.

16. Do major league pitchers or position players have longer careers? A random sample of 100 was selected from the population of men who pitched in the major leagues. An independent random sample of 100 was selected from the population of men who were position players in the major leagues. For each man the number of years he appeared in at least one game in the major leagues was recorded. For the position players, the sample mean and standard deviation are 4.75 and 4.833, respectively. For the pitchers, these values are 3.66 and 3.903. Find the 95 percent confidence interval for the population mean of the position players minus the population mean of the pitchers. Assume case 3.

16.3 THE RANDOMIZED PAIRS DESIGN

The randomized pairs design was introduced in Chapter 4 for a dichotomous response. This section extends this design to the case of a numerical response. No assumptions are made about how the subjects are selected for inclusion in the study. The next section of this chapter considers the consequences of assuming that subjects are selected at random from a population. The ideas are introduced with an example.

EXAMPLE 16.6

For her class project, Kathy Poquette studied the effect of the weight of the ball on her young son's performance at bowling. Readers interested in bowling will likely find a number of features of her design to criticize; some of the weaknesses are discussed below.

Kathy's subjects were frames, and the response was the total number of pins her son Jesse knocked down with two rolls. The first treatment was a 10-pound ball, and the second was a 12-pound ball. Kathy decided against a CRD because she was concerned with the possibility of a strong time trend: either improvement of performance caused by practice or deterioration caused by fatigue or boredom. The first two frames were combined to form the first pair of trials, the third and fourth frames formed the second pair of trials, and so on for a total of 26 frames and 13 pairs of frames. Within each pair, Kathy randomly assigned one frame to the first treatment (10-pound ball) and the other frame to the second treatment (12-pound ball). The result of her randomization was

> For pairs 1, 2, 5, 7, 9, 11, 12, and 13, the first frame in the pair was assigned to the lighter ball; for the remaining pairs, the first frame was assigned to the heavier ball. ▲

It is convenient to introduce some notation. Let n denote the number of pairs in the study. For Kathy's study, $n = 13$. Let

$$X_1, X_2, \ldots, X_n,$$

denote the responses obtained with the first treatment for pairs $1, 2, \ldots, n$, respectively. Similarly, let

$$Y_1, Y_2, \ldots, Y_n,$$

denote the responses obtained with the second treatment for pairs $1, 2, \ldots, n$, respectively. It is helpful to display these values in an array, as given in the first two rows of Table 16.3. Finally, for each pair, subtract the response obtained on the second treatment from the response obtained on the first treatment. Denote these n differences by

$$D_1, D_2, \ldots, D_n,$$

TABLE 16.3 Structure of Paired Data

	Pair Number			
Treatment	1	2	...	n
1 (X)	X_1	X_2	...	X_n
2 (Y)	Y_1	Y_2	...	Y_n
Difference ($D = X - Y$)	D_1	D_2	...	D_n

as given in the third row of Table 16.3. Note that these differences are obtained by subtracting, entry by entry, the second row from the first row. As usual, observed values of all random variables are denoted by lowercase letters. The data of Kathy's study are in Table 16.4.

The null hypothesis is that there is no difference between treatments. The three possible alternative hypotheses are

- Treatment 1 tends to yield larger responses than treatment 2. This alternative is denoted by $>$ and is referred to as the first alternative.
- Treatment 1 tends to yield smaller responses than treatment 2. This alternative is denoted by $<$ and is referred to as the second alternative.

TABLE 16.4 Data for Kathy's Bowling Study

	Pair Number*						
Treatment†	1	2	3	4	5	6	7
1 (x)	10	10	0	6	0	10	0
2 (y)	1	1	10	9	0	1	7
$d = x - y$	9	9	-10	-3	0	9	-7

	Pair Number*					
Treatment†	8	9	10	11	12	13
1 (x)	0	7	10	10	8	3
2 (y)	8	8	6	9	8	10
$d = x - y$	-8	-1	4	1	0	-7

*The response is the number of pins knocked down in one frame.
†Treatment 1 is bowling with a light ball, and treatment 2 is bowling with a heavy ball.

- Treatment 1 tends either to yield larger responses than treatment 2 or smaller responses than treatment 2. This alternative is denoted by $\neq$ and is referred to as the third alternative.

Note that these alternatives state, respectively, that

- The values of D tend to be positive.
- The values of D tend to be negative.
- The values of D either tend to be positive or tend to be negative.

For the bowling study, Kathy chose the first alternative because she wanted to show that Jesse bowled better with the lighter ball (the first treatment).

As for the CRD, there are a number of possibilities for the test statistic. In this chapter, the test statistic is taken to be the sample mean of the observed differences, here denoted $\bar{D}$. Note that

$$\bar{D} = \bar{X} - \bar{Y};$$

in words, the mean of the values of $D = X - Y$ can be obtained by subtracting the mean of the Y values from the mean of the X values. Thus, the test statistic is the same as the one studied for a CRD. Because the test statistic for the current problem is the same as that for the CRD, it is not surprising to find they have the same rules of evidence. The rules of evidence are

- The larger the value of the test statistic, $\bar{D}$, the stronger is the evidence in support of the first alternative ($>$).
- The smaller the value of the test statistic, $\bar{D}$, the stronger is the evidence in support of the second alternative ($<$).
- The farther the value of the test statistic $\bar{D}$ is from 0, in either direction, the stronger is the evidence in support of the third alternative ($\neq$).

These rules of evidence yield the following formulas for the P-value. As usual, $\bar{d}$ denotes the observed value of the test statistic $\bar{D}$.

- For the first alternative ($>$),

$$\mathbf{P} = P(\bar{D} \geq \bar{d}).$$

- For the second alternative ($<$),

$$\mathbf{P} = P(\bar{D} \leq \bar{d}).$$

- For the third alternative ($\neq$),

$$\mathbf{P} = 2P(\bar{D} \geq |\bar{d}|),$$

because is can be shown that the null sampling distribution of $\bar{D}$ is symmetric about 0.

First, the method of finding the exact sampling distribution of $\bar{D}$ by brute force will be discussed. Consider the first pair in the bowling study. For the first pair, Jesse used the lighter ball in the first frame and scored 10 and used the heavier ball in the second frame and scored 1. If the null hypothesis is true, this difference between frames was not influenced by which ball was used, but simply reflects inherent variability in the process of Jesse's bowling. Thus, if the null hypothesis is true, the observed value of D for Jesse's first pair was equally likely to be $+9$ or -9; it turned out to be $+9$ because the lighter ball had the good fortune of being assigned to the first frame in the pair; if the assignment had been reversed, D would have equaled -9. Similarly, the third pair yielded the values $x = 0$, $y = 10$, and $d = -10$. According to the null hypothesis, the heavier ball had the good fortune of being assigned to the frame in which Jesse performed better.

In order to obtain the sampling distribution of $\bar{D}$, the reasoning of the previous paragraph is applied repeatedly to each pair. It is too tedious to obtain the sampling distribution for Kathy's entire study of 13 pairs; for illustration it is obtained for only the first three pairs.

Each of the first three differences has two possible values, which can be combined in eight ways. The triple (D_1, D_2, D_3) can take on the values

$$(9, 9, 10),\ (9, 9, -10),\ (9, -9, 10),\ (9, -9, -10),$$
$$(-9, 9, 10),\ (-9, 9, -10),\ (-9, -9, 10),\ \text{and}\ (-9, -9, -10).$$

These eight combinations yield the following values for $\bar{D}$:

$$9.33,\ 2.67,\ 3.33,\ -3.33,\ 3.33,\ -3.33,\ -2.67,\ \text{and}\ -9.33,\ \text{respectively}.$$

If the null hypothesis is true, each of these eight combinations is equally likely to occur. Thus, the sampling distribution of $\bar{D}$ is

$\bar{d}$	-9.33	-3.33	-2.67	2.67	3.33	9.33	Total
$P(\bar{D} = \bar{d})$	0.125	0.250	0.125	0.125	0.250	0.125	1.000

(Note that this sampling distribution is symmetric about 0, as claimed earlier.) The actual outcome of the experiment was $(9, 9, -10)$, which yields $\bar{d} = 2.67$. Thus, the exact P-value for the first alternative is

$$\mathbf{P} = P(\bar{D} \geq 2.67) = 0.125 + 0.250 + 0.125 = 0.5.$$

For the entire bowling study, consisting of 13 pairs, it can be shown that

$$\bar{d} = \frac{\sum d}{n} = \frac{-4}{13} = -0.308.$$

The evidence in the data is opposite what Kathy was expecting to find. In the study, her son bowled, on average, better with the heavier ball. For the first alternative,

$$\mathbf{P} = P(\bar{D} \geq -0.308).$$

It is too tedious to obtain this number by brute force. Note, however, that because the sampling distribution of $\bar{D}$ is symmetric around 0, the P-value must exceed 0.5.

A simulation experiment of 100 runs yielded 63 values of $\bar{d}$ greater than or equal to $-\frac{4}{13}$; thus, the simulation estimate of the P-value is

$$\frac{63}{100} = 0.63.$$

Let S_D denote the sample standard deviation of the differences,

$$D_1, D_2, \ldots, D_n.$$

The third approach to obtaining the P-value is to approximate the null sampling distribution of

$$T = \frac{\sqrt{n}\bar{D}}{S_D} \tag{16.13}$$

by the t-distribution with $n - 1$ degrees of freedom:

Let t denote the observed value of T.

- For the first alternative ($>$) the approximate P-value is equal to the area under the $t(n-1)$ curve to the right of t.
- For the second alternative ($<$) the approximate P-value is equal to the area under the $t(n-1)$ curve to the right of $-t$.
- For the third alternative ($\neq$) the approximate P-value is equal to twice the area under the $t(n-1)$ curve to the right of $|t|$.

If the number of degrees of freedom, $n - 1$, exceeds 30, the t-curve in the above rules is replaced by the standard normal curve.

For Kathy's bowling study, it can be shown that the observed value of the sample standard deviation is $s_D = 6.651$. Therefore, the observed value of T in Equation 16.13 is

$$t = \frac{\sqrt{13}(-0.308)}{6.651} = -0.17.$$

For the first alternative, the approximate P-value is the area under the $t(12)$ curve to the right of -0.17. With the help of Minitab, this area is found to be 0.5661.

This section ends with some comments on the Bowling study. The choice of response is questionable. Knocking down all 10 pins with one ball, a strike, is better than knocking them down with two balls, a spare. The above response does not distinguish between these outcomes. In addition, there is a scoring bonus in bowling for a mark (a strike or a spare). Thus, in an actual bowling game a mark and an eight in two frames is preferred to two nines; for the above response, however, these possibilities are treated the same. Finally, Kathy's goal was to help her son score better at bowling. To that end, it is unreasonable to have him switch balls every frame. It would have been better to define a subject to be a game and have Jesse bowl two games each day, one with each ball, for a number of days. (Kathy was aware of these weaknesses; her project was satisfactory given her severe time limitations.)

16.4 A POPULATION MODEL FOR PAIRED DATA

Consider the following situation. There exists a population of subjects and two competing treatments. For example, the population could be all persons with a particular medical condition, and there could be two treatments for the condition. One can imagine placing every member of the population of subjects on the first treatment and obtaining a numerical response from each subject. This collection of responses constitutes the first population. Similarly, the second treatment generates a second population of numerical responses. The purpose of the study is to compare these two populations; in other words, it is to see how the response depends on the treatment. The above scenario is familiar. It appeared in Chapter 7 for a dichotomous response in a study of chronic Crohn's disease.

As before, the researcher begins by selecting a random sample of n subjects to be studied from the population of subjects. In the approach of Chapter 7, the sample of n subjects was divided by randomization into two groups. The subjects in one group were assigned to the first treatment, and the subjects in the other group were assigned to the second treatment. This section considers those studies in which each subject can be given both treatments. This technique is not applicable to the study of Crohn's disease (Why?). If the medical condition is headaches or acne, for example, it is plausible to give each subject both treatments.

It is important to randomize in these studies. The type of randomization varies from study to study. A common practice is to randomize the order of the treatments. For example, in a study of headache remedies, the protocol might call for the first headache to be assigned to one treatment and, after a suitable passage of time, for the next headache to be assigned to the other treatment. In such a study the order of the treatments should be randomized separately for each subject. In an acne study, the person might receive both treatments simultaneously, perhaps one treatment on each side of the face; the assignment of sides of the face to treatments should be made by randomization.

From each subject two numbers are obtained: the response to the first treatment, denoted X, and the response to the second treatment, denoted Y. The methods of Chapter 15 can be applied to the sample values of X to infer characteristics of the first population and to the sample values of Y to infer characteristics of the second population. The population techniques introduced earlier in this chapter, however, may *not* be applied to compare the two populations because the random samples are *not* independent. Instead, the method described below must be used.

Because each subject can be given both treatments, there is a third population of interest: the population of differences

$$D = X - Y.$$

Following the development of Chapter 15, assume that the population of differences is given by a normal curve with unknown mean, μ_D, and unknown standard deviation, σ_D. It can be shown that

$$\mu_D = \mu_X - \mu_Y;$$

in words, the population mean of the differences $D = X - Y$ equals the first population mean minus the second population mean.

The data are a random sample from the population of differences and are denoted

$$D_1, D_2, \ldots, D_n.$$

The sample mean and sample standard deviation of these differences are denoted by $\bar{D}$ and S_D, respectively. As usual, actual observed values are denoted by lowercase letters. Thus,

$$d_1, d_2, \ldots, d_n,$$

are the actual differences observed, and $\bar{d}$ and s_D are the observed values of the sample mean and sample standard deviation of the differences.

It was argued in Chapter 15 (see Formula 15.5, on p. 549) that

$$\bar{x} \pm t \times \frac{s}{\sqrt{n}}$$

gives a confidence interval for a population mean μ. This formula can be applied to the problem of this section; the result is that

$$\bar{d} \pm t \times \frac{s_D}{\sqrt{n}} \tag{16.14}$$

is a confidence interval for μ_D. The number t is taken from the row corresponding to $n - 1$ degrees of freedom and the appropriate column of Table A.6.

The comments on the robustness of this interval given in Chapter 15 apply to the current problem. In particular, if $n \geq 10$, this interval is robust against a mildly or moderately skewed population of differences.

In addition, it was argued in Chapter 15 (see Formula 15.6, p. 550) that

$$T = \frac{\sqrt{n}(\bar{X} - \mu_0)}{S}$$

is the test statistic for the null hypothesis that μ equals the special value of interest, μ_0. For the problem of the current section, since a researcher is interested in the null hypothesis, that $\mu_X = \mu_Y$, the special value of interest for μ_D is 0. Thus, the test statistic for the current problem is

$$T = \frac{\sqrt{n}\bar{D}}{S_D}, \qquad (16.15)$$

which has observed value

$$t = \frac{\sqrt{n}\bar{d}}{s_D}. \qquad (16.16)$$

The P-value is obtained by the rule given in Chapter 15, which is reproduced below for convenience.

- For the first alternative ($\mu_D > 0$ or $\mu_X > \mu_Y$) the approximate P-value is equal to the area under the $t(n-1)$ curve to the right of t.
- For the second alternative ($\mu_D < 0$ or $\mu_X < \mu_Y$) the approximate P-value is equal to the area under the $t(n-1)$ curve to the right of $-t$.
- For the third alternative ($\mu_D \neq 0$ or $\mu_X \neq \mu_Y$) the approximate P-value is equal to twice the area under the $t(n-1)$ curve to the right of $|t|$.

If the number of degrees of freedom, $n-1$, exceeds 30, then the t-curve in the above rules is replaced by the standard normal curve.

Observational Studies Consider the following observational study version of the above. Two numerical responses can be obtained for each member of a population of subjects. A random sample of n subjects is selected from the population, yielding n pairs of numbers:

$$(x_1, y_1), (x_2, y_2), \ldots, (x_n, y_n).$$

This problem has already been studied extensively. One approach is to perform a regression analysis using the methods of Chapter 13. Another is to analyze the differences

$$d_1 = x_1 - y_1, \; d_2 = x_2 - y_2, \ldots, d_n = x_n - y_n,$$

using the methods presented in this section.

Correlation Coefficient Formula 16.14 for a confidence interval for μ_D and Equation 16.15 for the test statistic for testing $\mu_D = 0$ both require the values of $\bar{d}$ and s_D. Occasionally, a researcher does not have these values but does know $\bar{x}$, s_X, $\bar{y}$, and s_Y. If, in addition, the researcher knows the value of the sample correlation coefficient, r, for X and Y, then Formula 16.14 and Equation 16.15 can be used. First, the sample mean of the differences, $\bar{d}$, equals the difference of the sample means:

$$\bar{d} = \bar{x} - \bar{y}.$$

In addition, it can be shown that the sample variance of the differences satisfies the equation

$$s_D^2 = s_X^2 + s_Y^2 - 2rs_X s_Y.$$

After this equation is used to obtain the sample variance, s_D can be obtained by taking the square root.

EXERCISES FOR SECTIONS 16.3 AND 16.4

1. The video game Tetris has been discussed in a number of examples and exercises. The player has the option of seeing the next shape to fall or not (the default value is to see it). Seeing the next shape helps the player plan ahead and should result in a better score. To investigate this, I conducted an RPD with 20 trials; the results are given in Table 16.5. (Hint: For the first treatment, the sample mean is 116.5 and the sample standard deviation is 11.56; for the second treatment, these values are 90.0 and 7.77; for the differences, these values are 26.5 and 11.87.)

    (a) Construct dot plots of the scores with treatment 1, the scores with treatment 2, and the differences. Describe what they reveal.

    (b) A statistician looks at the data and observes that all 10 differences are positive; she states

    > If the treatments are identical, the probability of such an extreme result is $(0.5)^{10} = 0.001$.

    Comment.

    (c) Use the t-distribution to obtain an approximate P-value for the first alternative.

2. Refer to the data presented in Exercise 2 on p. 401, an RPD with $n = 40$. It can be shown that $\bar{d} = 11.50$ seconds and $s_D = 8.82$ seconds.

    (a) Find the approximate P-value for the first alternative. Comment.

    (b) Construct the 95 percent confidence interval for μ_D. Comment. What assumptions are needed for this inference?

3. Refer to Exercise 1.

    (a) Assume that the 10 differences form a random sample from a process. Construct the 95 percent confidence interval for the population mean of the differences.

    (b) Assume that the 10 scores on treatment 1 form a random sample from a process. Construct the 95 percent confidence interval for the population mean of the scores with treatment 1.

TABLE 16.5 Results of an RPD to Study the Importance of the Preview Feature in Tetris

Treatment	Pair									
	1	2	3	4	5	6	7	8	9	10
1 (Preview)	106	112	118	102	112	110	130	110	127	138
2 (No preview)	84	93	86	86	94	88	108	91	79	91
(1) − (2)	22	19	32	16	18	22	22	19	48	47

(c) Assume that the 10 scores on treatment 2 form a random sample from a process. Construct the 95 percent confidence interval for the population mean of the scores with treatment 2.

4. Refer to the data presented in Exercise 18 on p. 407, an RPD with $n = 7$. It can be shown that $\bar{d} = 9.00$ pins and $s_D = 25.68$ pins. (The seven differences are the score with the heavy ball subtracted from the score with the light ball.)

   (a) Find the approximate P-value for the third alternative. Comment.

   (b) Construct the 95 percent confidence interval for μ_D. Comment. What assumptions are needed for this inference?

5. Refer to Exercises 1 and 3.

   (a) Assume (not a valid assumption) that the data had been collected as independent random samples from the two processes: with and without a preview (that is, the population model for a CRD). Construct the 95 percent confidence interval for the difference between means (assume case 1).

   (b) Compare your answer to part (a) to the actual answer obtained in Exercise 3. Do you think a CRD would have been a better design for this study? (Important note: Even if the CRD gives a better answer—a narrower confidence interval—it is not valid to assume that the design was a CRD.)

6. Dan Lauffer performed an RPD to learn more about his swimming ability. A pull-buoy is a Styrofoam device placed between a swimmer's legs to inhibit kicking. In each trial Dan swam one length of a 50-meter pool using the crawl (freestyle) stroke. In the first treatment, Dan swam with a pull-buoy, and in the second treatment, Dan swam without a pull-buoy. For each trial, Dan would swim as fast as possible and the response is the time, in seconds, he needed to swim the length of the pool. Dan's raw data are presented in Table 16.6.

   (a) Construct dot plots of the times with the pull-buoy, the times without the pull-buoy, and the differences. Comment.

TABLE 16.6 Time Required for Dan Lauffer to Swim 50 Meters

Treatment	Pair				
	1	2	3	4	5
With pull-buoy	32.62	32.98	31.92	33.28	33.01
Without pull-buoy	28.10	28.21	29.04	28.16	27.82
With − without	4.52	4.77	2.88	5.12	5.19

   (b) Test the null hypothesis that the treatments have the same effect on the response versus the alternative that the pull-buoy yields slower times. Find the exact P-value. (Hint: There are 32 possible assignments of subjects to treatments.) Use the t-distribution to find an approximate P-value. (Hint: $\bar{d} = 4.50$ seconds, and $s_D = 0.943$ seconds.)

   (c) Assume that the differences are a random sample from a population. Construct the 95 percent confidence interval for the mean of the population.

   (d) Suppose the data were obtained from a CRD instead of an RPD. Further, assume that the data can be viewed as independent random samples. Construct the 95 percent confidence interval for the population mean time with a pull-buoy minus the population mean time without a pull-buoy, assuming case 1. (Hint: The sample means and standard deviations of times are 32.762 seconds and 0.526 seconds, respectively, with the pull-buoy and 28.266 seconds and 0.458 seconds, respectively, without the pull-buoy.) Compare this supposed answer with the previous, appropriate, answer. Do you think the choice of an RPD was beneficial?

7. Refer to the data presented in Exercise 19 on p. 408. It can be shown that $\bar{d} = 20.4$ pins and $s_D = 8.17$ pins. (The five differences were obtained by subtracting the score with two hands from the score with one hand.)

   (a) Find the approximate P-value for the first alternative. Comment.

   (b) Construct the 95 percent confidence interval for μ_D. Comment. What assumptions are needed for this inference?

8. Refer to the data presented in Exercise 20 on p. 408. It can be shown that $\bar{d} = 17.9$ seconds and $s_D = 8.75$ seconds. (The 10 differences were obtained by subtracting the time to climb the hill when not smoking from the time when smoking.)

   (a) Find the approximate P-value for the first alternative. Comment.

   (b) Construct the 95 percent confidence interval for μ_D. Comment. What assumptions are needed for this inference?

9. Wrigley Field, the home of the Chicago Cubs, is widely believed to be a great hitter's park, especially for power hitters. This exercise will attempt to investigate this issue empirically. In 1987 there were 204 home runs hit in Wrigley Field, compared with a mean of 147.3 at the 11 other National League ball parks [4]. This, however, is not a good way to compare stadiums. The owners of the Cubs built their team to take advantage of their home park. For example, they signed Andre Dawson as a free agent before the 1987 season, and Dawson led the league that season with 49 home runs. A better method is to compare the number of home runs hit by both teams in Cubs home games with the number hit by both teams in Cubs road games. Explain why this proposed method is an improvement.

10. Refer to the previous exercise. Table 16.7 presents the number of home runs hit by both teams in all Cubs home and away games during the 20 seasons from 1967 to 1987 (the season shortened by a strike, 1981, is omitted from this data set) [4].

    (a) Construct dot plots of the number of home runs hit in home games, the number hit in away games, and the difference. What do these plots reveal?

    (b) Assume that the 20 differences are a random sample from a process. Construct the 95 percent confidence interval for the mean of the process. (Hint: The sample mean and sample standard deviation of the differences are equal to 46.2 home runs and 24.42 home runs, respectively.)

11. Refer to the previous two exercises. It is widely believed that it is very difficult to hit a home run at the Astrodome, in Houston. Table 16.8 presents the number of home runs hit by both teams in all Astros away and home games during the 20 seasons from 1967 to 1987 (the season shortened by a strike, 1981, is omitted from this data set) [4].

    (a) Construct dot plots of the number of home runs hit in home games, the number hit in away games, and the difference. What do these plots reveal?

TABLE 16.7 Number of Home Runs, Both Teams, in Cubs Games, 1967–1987

Location	\multicolumn{9}{c}{Season}								
	1967	1968	1969	1970	1971	1972	1973	1974	1975	1976
Home	160	166	148	201	144	146	138	139	125	155
Away	110	102	112	121	116	99	107	93	100	73
Home − Away	50	64	36	80	28	47	31	46	25	82
Location	\multicolumn{9}{c}{Season}								
	1977	1978	1979	1980	1982	1983	1984	1985	1986	1987
Home	151	117	151	116	115	140	156	202	168	204
Away	88	80	111	100	112	117	79	104	130	164
Home − Away	63	37	40	16	3	23	77	98	38	40

TABLE 16.8 *Number of Home Runs, Both Teams, in Astros Games, 1967–1987*

Location	Season									
	1967	1968	1969	1970	1971	1972	1973	1974	1975	1976
Away	150	82	125	145	101	134	134	101	107	91
Home	63	52	90	115	45	114	111	93	83	57
Away–home	87	30	35	30	56	20	23	8	24	34

Location	Season									
	1977	1978	1979	1980	1982	1983	1984	1985	1986	1987
Away	151	97	93	96	104	137	123	145	136	166
Home	73	59	46	48	57	54	47	95	105	97
Away–home	78	38	47	48	47	83	76	50	31	69

(b) Assume that the 20 differences are a random sample from a process. Construct the 95 percent confidence interval for the mean of the process. (Hint: The sample mean and sample standard deviation of the differences are equal to 45.7 home runs and 22.74 home runs, respectively.)

12. Refer to the previous three exercises. Figure 16.5 is a time series plot of similar data for the St. Louis Cardinals for the years from 1956 to 1975.

    (a) Comment on the pattern in the time series.

    (b) On May 12, 1966, the Cardinals moved into a new stadium for their home games. In view of this information, repeat part (a).

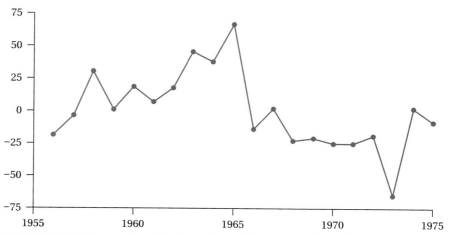

FIGURE 16.5 *Time series plot of number of home runs hit by both teams in home games minus the number of home runs hit by both teams in away games for the St. Louis Cardinals, 1956–1975.*

TABLE 16.9 Number of Wins Minus Number of Losses for 74 Major League Pitchers

−37	−28	−27	−26	−22	−16	−13	−12	−11	−10
−10	−9	−8	−6	−5	−5	−5	−5	−5	−4
−4	−4	−3	−3	−3	−3	−2	−2	−2	−2
−2	−1	−1	−1	−1	−1	−1	−1	−1	−1
−1	0	0	0	0	0	0	0	0	1
1	1	1	1	1	2	2	2	3	4
4	5	5	10	12	16	16	21	21	38
46	58	59	81						

(c) For the last 10 years in the old stadium, the sample mean and standard deviation of the data are 20.4 home runs and 25.73 home runs, respectively. For the first 10 years in the new stadium, these values are −19.8 home runs and 19.15 home runs. Viewing these data as independent random samples, assume case 1 to test the null hypothesis that the means are equal versus the third alternative. Comment.

13. A random sample of 74 was selected from the population of men who pitched in the major leagues and obtained at least one decision (a win or a loss). For each man his number of wins minus his number of losses was obtained. The 74 sorted values are presented in Table 16.9.

    (a) Since each game has one winning pitcher and one losing pitcher, the mean of the population must be 0. Ignore this fact and compute the 95 percent confidence interval for the population mean difference. (Hint: The sample mean and standard deviation of the differences are 1.45 and 18.10, respectively.) Is your confidence interval correct?

    (b) Construct the 95 percent confidence interval for the population median of the differences.

14. Refer to the data you collected for Exercise 26 on p. 410.

    (a) Use the t-distribution to find an approximate P-value for testing the null hypothesis that the treatments are equal versus the third alternative.

    (b) Assume that the differences are a random sample from a population. Construct the 95 percent confidence interval for the population mean of the differences.

    (c) Write a few sentences that summarize what you have learned.

REFERENCES

1. R. Goldman et al., "Lead Poisoning in Automobile Radiator Mechanics," *New England Journal of Medicine*, Vol. 317, July 23, 1987, pp. 214–218.
2. A. Bak and D. Grobbee, "The Effect on Serum Cholesterol Levels of Coffee Brewed by Filtering or Boiling," *New England Journal of Medicine*, Vol. 321, November 23, 1989, pp. 1432–1437.
3. R. Prince et al., "Prevention of Postmenopausal Osteoporosis," *New England Journal of Medicine*, Vol. 325, October 24, 1991, pp. 1189–1195.
4. J. L. Reichler, ed., *The Baseball Encyclopedia* (New York: Macmillan, 1988).

SOLUTIONS TO ODD-NUMBERED EXERCISES

Section 1.1

1. Various correct answers are possible.
3. Various correct answers are possible.
5. Various correct answers are possible.
7. 1, 2, 6, 9, 10, and 11.
9. The ten easiest-to-catch rats are assigned to the first treatment. This could distort the results of the study.
11. If students are able to choose their professor, then the choice made might be related to factors that influence the score on the final exam.

Section 1.2

1. Below are the two bar charts.

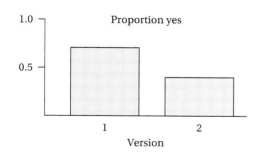

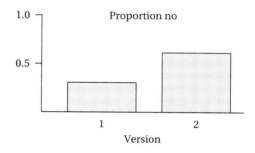

3. The two tables are

Version	S	F	T	S	F	T
B	20	5	25	0.80	0.20	1.00
G	15	10	25	0.60	0.40	1.00
Total	35	15	50			

5. The two tables are

Version	S	F	T	S	F	T
65	4	1	5	0.80	0.20	1.00
35	1	4	5	0.20	0.80	1.00
Total	5	5	10			

615

7. The two tables are

Version	S	F	T	S	F	T
1	17	8	25	0.68	0.32	1.00
2	8	17	25	0.32	0.68	1.00
Total	25	25	50			

9. The two tables are

Version	S	F	T	S	F	T
Friend	18	7	25	0.72	0.28	1.00
You	8	17	25	0.32	0.68	1.00
Total	26	24	50			

11. The two tables are

Version	S	F	T	S	F	T
1	3	12	15	0.20	0.80	1.00
2	13	2	15	0.87	0.13	1.00
Total	16	14	30			

13. The two tables are

Treatment	S	F	T	S	F	T
Tables	14	8	22	0.64	0.36	1.00
No tables	8	14	22	0.36	0.64	1.00
Total	22	22	44			

15. The two tables are

Treatment	S	F	T	S	F	T
Aspirin	30	4	34	0.88	0.12	1.00
Placebo	20	11	31	0.65	0.35	1.00
Total	50	15	65			

17. The two tables are

Treatment	S	F	T	S	F	T
Low	41	6	47	0.87	0.13	1.00
Standard	29	18	47	0.62	0.38	1.00
Total	70	24	94			

Section 1.3

1. Various correct answers are possible.
3. No. The subjects are not assigned to treatments by randomization.
5. The two tables are

Hand	S	F	T	S	F	T
Right	12	13	25	0.48	0.52	1.00
Left	10	15	25	0.40	0.60	1.00
Total	22	28	50			

7. The two tables are

Treatment	S	F	T	S	F	T
Left	20	5	25	0.80	0.20	1.00
Front	8	17	25	0.32	0.68	1.00
Total	28	22	50			

9. The two tables are

Treatment	S	F	T	S	F	T
Open	12	13	25	0.48	0.52	1.00
Closed	4	21	25	0.16	0.84	1.00
Total	16	34	50			

11. The two tables are

Treatment	S	F	T	S	F	T
Short	6	19	25	0.24	0.76	1.00
Long	1	24	25	0.04	0.96	1.00
Total	7	43	50			

13. The two tables are

Treatment	S	F	T	S	F	T
Women's	17	8	25	0.68	0.32	1.00
Men's	21	4	25	0.84	0.16	1.00
Total	38	12	50			

15. The two tables are

Treatment	S	F	T	S	F	T
Right	20	5	25	0.80	0.20	1.00
Left	8	17	25	0.32	0.68	1.00
Total	28	22	50			

17. The two tables are

Treatment	S	F	T	S	F	T
Left	22	3	25	0.88	0.12	1.00
Right	20	5	25	0.80	0.20	1.00
Total	42	8	50			

19. The two tables are

Treatment	S	F	T	S	F	T
Feather	36	14	50	0.72	0.28	1.00
Plastic	48	2	50	0.96	0.04	1.00
Total	84	16	100			

Sections 2.1 and 2.2

1. Skeptic: Thirty-three subjects were destined to show improvement regardless of the treatment they received. It was only by chance that 22 of these subjects received cyclosporine. Advocate: If there were no difference between treatments, then it is extremely unlikely that such strong evidence in favor of cyclosporine would have been obtained.
3. $n_A = 0$.
5. $n_C = 0$.
7. The first alternative ($>$).
9. The third alternative ($\neq$).
11. The third alternative ($\neq$).
13. The second alternative ($<$).
15. The third alternative ($\neq$).
17. I would select the third alternative ($\neq$) because I would be interested in any difference. Others may reasonably disagree. For example, if one felt it was inconceivable that aspirin would be detrimental, then the first alternative ($>$) should be selected.
19. She said she believed her son to be ambidextrous; hence, presumably either difference would be of interest to her.
21. The first alternative ($>$).

23. The first alternative (>). I consider it inconceivable that he could shoot better with his eyes closed.
25. The first alternative (>). I consider it inconceivable that she could shoot better from 40 feet than from 30 feet.
27. The third alternative ($\neq$).
29. The first alternative (>).
31. The third alternative ($\neq$).
33. The archer wanted to decide which type of arrow to use in the rain; either difference would have been of interest.

Section 2.3

1. (a) The twenty assignments are

Assignment	Version 1	Version 2
1	A, B, C	D, E, F
2	A, B, D	C, E, F
3	A, B, E	C, D, F
4	A, C, D	B, E, F
5	A, C, E	B, D, F
6	A, D, E	B, C, F
7	B, C, D	A, E, F
8	B, C, E	A, D, F
9	B, D, E	A, C, F
10	C, D, E	A, B, F
11	A, B, F	C, D, E
12	A, C, F	B, D, E
13	A, D, F	B, C, E
14	A, E, F	B, C, D
15	B, C, F	A, D, E
16	B, D, F	A, C, E
17	B, E, F	A, C, D
18	C, D, F	A, B, E
19	C, E, F	A, B, D
20	D, E, F	A, B, C

(b) The twenty assignments are equally likely; thus, the probability of each one is 0.05.

(c) This event consists of assignments 1, 2, 3, 10, 11, 18, 19, and 20; thus, its probability is 0.4.

(d) This event is the complement of the previous event; thus, its probability is $1 - 0.4 = 0.6$.

3. $P(X = -1) = 0.05$, $P(X = -1/3) = 0.45$, $P(X = 1/3) = 0.45$, and $P(X = 1) = 0.05$.

5. $P(X = -2/3) = 0.2$, $P(X = 0) = 0.6$, and $P(X = 2/3) = 0.2$.

Section 2.4

1. (a) $A = \{1, 3, 5, 7, 9\}$, $B = \{9, 10\}$, $AB = \{9\}$, and $(A \text{ or } B) = \{1, 3, 5, 7, 9, 10\}$.
   (b) $P(A) = 0.5$, $P(B) = 0.2$, $P(AB) = 0.1$, and $P(A \text{ or } B) = 0.6$.
   (c) No, because A and B are not disjoint.

3. (a) 0.4; 0.2.
   (b) $0.4 + 0.3 + 0.2 = 0.9$; $0.1 + 0.4 + 0.3 = 0.8$.
   (c) $1 - 0.1 = 0.9$; $1 - 0.2 = 0.8$.

5. Various correct answers are possible.

Section 2.5

1. 0.50; 0.30; 0.50.
3. -0.30; -0.50; -0.50.
5. (a) 0.0782; 0.3888.
   (b) 0.0232; 0.8019.
   (c) 0.0016; 0.7776.
7. Not statistically significant, not statistically significant; statistically significant, not statistically significant; highly statistically significant, not statistically significant.
9. (a) 0.0712; 0.0116.
   (b) 0.1753; 0.0369.
   (c) 0.0010; 0.0485.
11. Not statistically significant, statistically significant; not statistically significant, statistically significant; highly statistically significant, statistically significant.

Section 2.6

1. $\mathbf{P} = 0.0099$; $\mathbf{P} = 0.0198$.
3. $\mathbf{P} = 0.1935$; $\mathbf{P} = 0.3870$.
5. $R = 5$, $C = 5$, and $O = 4$. $\mathbf{P} = 0.1032$.

Section 3.1

1. Below is the probability histogram. The $P(X = 3)$ is the area of the rectangle above 3. Its base is 1 and its height is 0.3; hence, its area is $1(0.3) = 0.3$. This area agrees with the tabulated value.

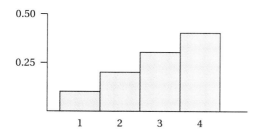

3. Below is the probability histogram. The $P(X = 4)$ is the area of the rectangle above 4. Its base is 1 and its height is 0.2; hence, its area is $1(0.2) = 0.2$. This area agrees with the tabulated value.

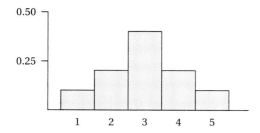

Section 3.2

1. The simulation approximation to the P-value is the sum of the relative frequencies of 0.6 and 1.0: $0.096 + 0.005 = 0.101$. This value is quite close to the exact value, 0.1032.

Section 3.3

1. $\sigma = 0.1309$.
3. $\sigma = 0.1054$.
5. $\sigma = 0.1418$.
7. (a) The equation for Z is $Z = (X - 6)/2$.
   (b) The possible values of Z are $-2, -1, 0,$ and 1.

(c) Below are the two probability histograms. The probability histogram for X is centered at 6, but the one for Z is centered at 0. The probability histogram for Z has much less spread than does the one for X.

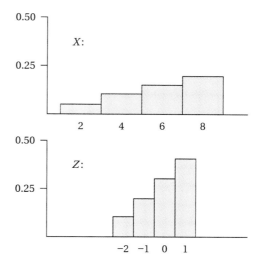

Section 3.4

1. 0.3707, 0.9887, 0.7324, 0.1112, 0.0009, 0.5319, < 0.0002.

3. 0.6293, 0.0113, 0.2676, 0.8888, 0.9991, 0.4681, > 0.9998.

5. 0.0812, 0.6310, 0.4972, 0.4772.

7. $z = 1.79$, giving $2(0.0367) = 0.0734$ for the approximate P-value. There is evidence that using tables in a history text enhances the argument, but the evidence is just short of being statistically significant.

9. $z = 2.25$, giving $2(0.0122) = 0.0244$ for the approximate P-value. A fairly simple treatment, aspirin, gave a success on somewhat more than 7 out of 8 women. Note, however, that a placebo gave a success to somewhat more than 5 out of 8 women. The P-value suggests that this difference does reflect a real therapeutic effect from aspirin.

11. $z = -2.82$, giving 0.0024 as the approximate P-value for the second alternative. The standard dose of lithium is highly statistically significantly superior to the low dose at preventing relapses.

13. $z = 0.76$, giving $2(0.2236) = 0.4472$ for the approximate P-value. There is evidence that Pearl is somewhat better at a canter depart on the left lead, but the evidence is weak.

15. $z = -3.26$, giving $2(0.0006) = 0.0012$ for the approximate P-value. There is overwhelming evidence that the archer shot better with the plastic-vane arrows.

Section 3.5

1. The simulation approximation is better for all studies.

Section 3.6

1. $\mu = 3.0, \sigma^2 = 1.0,$ and $\sigma = 1.0$.
3. $\mu = 3.0, \sigma^2 = 1.20,$ and $\sigma = 1.10$.

Section 4.1

1. The tables of row proportions for blocks 1 and 2, respectively, are

Treatment	S	F	Total
1	0.55	0.45	1.00
2	0.45	0.55	1.00

Treatment	S	F	Total
1	0.75	0.25	1.00
2	0.65	0.35	1.00

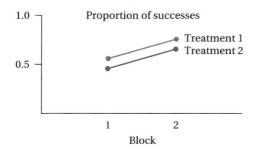

The collapsed table and its row proportions are

Treatment	S	F	Total
1	130	70	200
2	110	90	200

Treatment	S	F	Total
1	0.65	0.35	1.00
2	0.55	0.45	1.00

3. Let V denote version, and T denote total. For the first student,

V	S	F	T	S	F	T
1	11	15	26	0.42	0.58	1.00
2	8	18	26	0.31	0.69	1.00

For the second student,

V	S	F	T	S	F	T
1	5	3	8	0.62	0.38	1.00
2	2	6	8	0.25	0.75	1.00

For the third and fourth students,

V	S	F	T	S	F	T
1	3	2	5	0.6	0.4	1.0
2	1	4	5	0.2	0.8	1.0

For the fifth student,

V	S	F	T	S	F	T
1	3	2	5	0.6	0.4	1.0
2	2	3	5	0.4	0.6	1.0

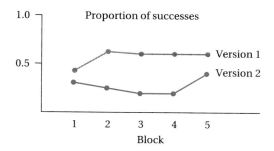

The collapsed table and its row proportions are

V	S	F	T	S	F	T
1	25	24	49	0.51	0.49	1.00
2	14	35	49	0.29	0.71	1.00

Section 4.2

1. (a) For the first block, $z = 1.41$ and $\mathbf{P} = 0.0793$. For the second block, $z = 1.54$ and $\mathbf{P} = 0.0618$.
   (b) For the MH test, $z = 2.08$ and $\mathbf{P} = 0.0188$.
   (c) In each block, there is evidence that the first treatment is better, but the evidence is not statistically significant. The MH test combines the consistent evidence from the two blocks to show that overall the first treatment is statistically significantly better than the second treatment.

3. (a) The approximate P-value for the first student is 0.1949; the exact P-value for the second student is 0.1573; for the other three students each P-value is greater than or equal to 0.2500.
   (b) For the MH test, $z = 2.22$ and $\mathbf{P} = 0.0132$.
   (c) See the comment to Exercise 1, above.

5. The tables of observed counts for each block:

	CD4+ Is					
	Below 200			200–499		
Treatment	S	F	T	S	F	T
1,500	48	3	51	395	11	406
500	48	7	55	394	4	398
Placebo	44	12	56	351	21	372
Total	140	22	162	1,140	36	1,176

7. The tables of row proportions for each block:

	CD4+ Is					
	Below 200			200–499		
Treatment	S	F	T	S	F	T
1,500	0.94	0.06	1	0.97	0.03	1
500	0.87	0.13	1	0.99	0.01	1
Placebo	0.79	0.21	1	0.94	0.06	1

9. For each treatment, subjects with a higher CD4+ count fared better. For either block, the placebo was inferior to either level of zidovudine. For the patients with the higher cell count, the lower dose performed better than the higher dose, but for patients with a lower cell count, the higher dose performed better.

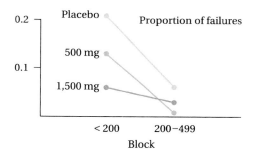

11. Let the high dose be treatment 1. For the low-count block, the test statistic is $z = 1.20$; for the high-count block it is $z = -1.78$. For the MH test, $T = 443$, $E(T) = 444.61$, and $\text{Var}(T) = 5.966$, yielding $z = -0.66$ and
$$P = 2(0.2546) = 0.5092.$$
None of the tests is statistically significant; because the order of effectiveness is different in the two blocks, the MH test, which looks for a consistent pattern, finds little of interest in the data.

13. Let the low dose be treatment 1. For the low-count block, the test statistic is $z = 1.21$; for the high-count block it is $z = 3.63$. For the MH test, $T = 442$, $E(T) = 430.67$, and $\text{Var}(T) = 10.020$, yielding $z = 3.58$ and $\mathbf{P} = 0.0002$. The difference in sample proportions in the low count block is too small to be statistically significant for the small sample sizes. In the high-count block, the difference in sample

proportions is much larger and there is much more data; the result is a highly significant result. The MH test is also highly statistically significant.

15. Let the high dose be treatment 1. For the low-count block, the test statistic is $z = 2.30$; for the high-count block it is $z = 2.06$. For the MH test, $T = 443$, $E(T) = 433.15$, and $\text{Var}(T) = 10.914$, yielding $z = 2.98$ and $P = 0.0014$. In each block the high dose is statistically significantly better than the placebo. Overall, using the MH test, the high dose is highly statistically significantly better than the placebo.

17. Let the combined doses be treatment 1. For the low-count block, the test statistic is $z = 2.12$ which gives $P = 0.0170$; for the high-count block it is $z = 3.50$, which gives $P = 0.0002$. For the MH test, $T = 885$, $E(T) = 870.99$, and $\text{Var}(T) = 11.881$, yielding $z = 4.06$ and $P < 0.0002$. In each block AZT is better than the placebo; the difference is statistically significant in the low-count block and highly statistically significant in the high-count block. The MH test combines the evidence across blocks to yield a very large value of z.

19. Letting failure denote death, the collapsed table and its row proportions are

	Counts			Row Proportion		
Treatment	S	F	T	S	F	T
AZT	144	1	145	0.99	0.01	1
Placebo	118	19	137	0.86	0.14	1
Total	262	20	282			

The test statistic for Fisher's test is $z = 4.30$, yielding $P < 0.0002$.

21. Letting failure denote the development of an opportunistic infection, the collapsed table and its row proportions are

	Counts			Row Proportion		
Treatment	S	F	T	S	F	T
AZT	121	24	145	0.83	0.17	1
Placebo	92	45	137	0.67	0.33	1
Total	213	69	282			

The test statistic for Fisher's test is $z = 3.18$, yielding $P = 0.0007$.

Section 4.3

1. He chose an RPD to avoid the practice, fatigue, and familiarity effects.

	Floor		
Bed	S	F	Total
S	3	2	5
F	0	0	0
Total	3	2	5

1.00; 0.60; 0.250.

3. She chose an RPD to avoid both the practice and familiarity effects. 0.96; 0.44; $z = 3.36$, yielding $P = 0.0004$.

5.

	E		
A	S	F	Total
S	11	10	21
F	4	0	4
Total	15	10	25

0.84; 0.60; the exact P-value is 0.090 for the first alternative.

7.

	Slap		
Wrist	S	F	Total
S	8	8	16
F	5	4	9
Total	13	12	25

0.64; 0.52; the exact P-value is 0.291 for the first alternative.

9.

	Calm		
Yelling	S	F	Total
S	5	32	37
F	11	2	13
Total	16	34	50

0.74; 0.32; $z = 3.20$, yielding **P** $= 0.0007$.

Section 5.2

1. (a) $p = 0.7$; (b) $q = 0.3$; (c) $pq = 0.7(0.3) = 0.21$; (d) $1 - P(X_1 = 0, X_2 = 1) = 1 - qp = 1 - 0.21 = 0.79$.
3. (a) 0.0441; (b) 0.0441; (c) 0.0882.
5. 0.30.
7. (a) 0.1225; (b) 0.4550; (c) 0.4225.

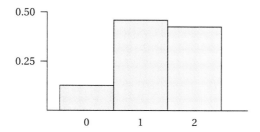

9. (a) 0.2355; (b) 0.3280.
11. (a) 0.2765; (b) 0.3602; (c) 0.2753.
13. (a) 0.1328. (b) 0.0974. (c) 0.7878. (d) 0.8852. (e) 0.3450. (f) 0.2122.
15. (a) $(X - 200)/10$; (b) $(X - 60)/6.481$; (c) $(X - 3200)/25.30$

Section 5.3

1. 0.2122.
3. Let X denote the number correct. $P(X \geq 1) = 1 - P(X = 0) = 1 - (0.75)^5 = 1 - 0.2373 = 0.7627$.

5. The table relating the current trial to the previous trial is

	Current		
Previous	S	F	T
S	2	2	4
F	3	2	5
Total	5	4	9

7. First, there are two possibilities—female and male. Second, there is no reason to believe that the probability of either gender will change over the short time of the study. Third, excepting identical twins, the gender of one baby has no effect on the gender of the next. Identical twins are sufficiently rare that I am not concerned about this slight violation of the third assumption.

9. (a) Muffin had 27 successes in the first half of the study and 30 successes in the second half. This is an increase of only 6.25 percentage points. It does not cast serious doubt on the second assumption of Bernoulli trials.
   (b) The tables of observed counts and row proportions are

Prev.	S	F	T	S	F	T
S	34	22	56	0.61	0.39	1
F	22	17	39	0.56	0.44	1
Total	56	39	95			

There is little evidence that the outcome of one trial influences the outcome of the next trial.

11. Margaret's proportion of successes was almost constant for the two halves of the study. In addition, the following table provides little evidence against the third assumption of Bernoulli trials.

Prev.	S	F	T	S	F	T
S	21	24	45	0.47	0.53	1
F	24	30	54	0.44	0.56	1
Total	45	54	99			

13. The golf pro had a slightly higher success rate during the second half of the study, but his improvement was not enough to make the second assumption suspect. The following table indicates that there is evidence the pro shot better after a miss. (Perhaps it is easier to adjust an error than repeat a success.) (Note: The techniques of Chapter 7 indicate that this pattern is not convincing evidence against the third assumption of Bernoulli trials.)

Prev.	S	F	T	S	F	T
S	40	26	66	0.61	0.39	1
F	25	8	33	0.76	0.24	1
Total	65	34	99			

Section 6.2

1. (a) $0.730 \pm 0.032 = [0.698, 0.762]$.
   (b) $0.450 \pm 0.035 = [0.415, 0.485]$.
3. $0.620 \pm 0.029 = [0.591, 0.649]$.
5. $0.595 \pm 0.133 = [0.462, 0.728]$.
7. $0.617 \pm 0.139 = [0.478, 0.756]$. I am 95 percent confident that the proportion that would not suffer a relapse during the length of time of the study is somewhere between 0.478 and 0.756.
9. (a) $0.420 \pm 0.162 = [0.258, 0.582]$; (b) $0.400 \pm 0.161 = [0.239, 0.561]$.
11. $0.959 \pm 0.019 = [0.940, 0.978]$.
13. (c) $0.650 \pm 0.073 = [0.577, 0.723]$.

Section 6.3

1. The confidence levels are 95 percent, 99 percent, 90 percent, 80 percent, and 98 percent, respectively.
3. Daniel had $\hat{p} = 0.40$; Victor had $\hat{p} = 0.20$.

Section 6.4

1. (a) Oscar. (b) Victor. (c) Victor.
3. $z = 1.25$; the approximate P-value for the first alternative is 0.1056 and for the third alternative, 0.2112.
5. 0.172; 0.011; 0.623.
7. 0.289; 0.008; 0.727.
9. $z = 6.62$ and the approximate P-value is smaller than 0.0002.
11. $z = 1.84$ and the approximate P-value is 0.0658.
13. $z = 5.33$ and the approximate P-value is smaller than 0.0004.

Section 6.5

1. The prediction interval is [4,902, 5,098]. I predict that the *proportion* of heads that will be obtained is between 0.4902 and 0.5098; thus, the proportion should be very close to 0.5. The actual number of heads, however, may differ from 5,000 by as much 98 and still be in the prediction interval.
3. (a) 100; [88, 112]. (b) 104; [84, 124].
5. (a) 27; (b) [19, 35]; (c) The point prediction was low; the prediction interval was correct.

Section 6.6

1. $1 - (0.98)^5 = 0.0961$.
3. (a) 0.50; (b) 0.5987.
4. If the friend tossed the coin only 10 times, then—assuming the coin is fair—the answer is $(0.5)^{10} = 0.0010$. If the friend tossed the coin more than 10 times, the probability is larger.
5. (a) The paper assumed that the probability of being born on October 31 is 1/365 and that the multiplication rule can be used.
   (b) First, the answer should be increased to
   $$(1/365)^3$$
   because a match on any day—not just Halloween—would be considered noteworthy. Next, the parents may have tried to have a baby on their birthday; if so, the probability for the child should change from 1/365 to a considerably larger number, perhaps 1/10. Finally, this is a clear example of "focusing on the winner."

Chapter 7

1. (a) $0.020 \pm$
   $$1.96 \sqrt{\frac{(0.42)(0.58)}{5{,}000} + \frac{(0.4)(0.6)}{5{,}000}} =$$
   $0.020 \pm 0.019 = [0.001, 0.039]$.
   (b) $z = 2.03$ and $\mathbf{P} = 0.0424$.

3. (a) An observational study on finite populations.
   (b) The 95 percent confidence interval for the increase in approval is $0.07 \pm 0.05 = [0.02, 0.12]$.
   (c) $z = 2.77$. The approximate P-value for the third alternative is 0.0056.
   (d) The confidence interval indicates that the proportion who approve increased by between 2 and 12 percentage points. The hypothesis test indicates that the increase is highly statistically significant.

5. (a) Controlled on a sequence of trials.
   (b) 0.080 ± 0.274 or $[-0.194, 0.354]$.

7. $z = 6.35$. The evidence in the data against the third assumption of Bernoulli trials is incredibly strong.

9. (a) $z = 0.22$ and the approximate P-value for the third alternative is 0.8258.
   (b) 0.022 ± 0.197 or $[-0.175, 0.219]$.
   (c) This table provides virtually no evidence against the third assumption of Bernoulli trials. The point estimate suggests that the success probability after a success is slightly larger than the success probability after a failure. The confidence interval shows that the difference between these success probabilities is not estimated very precisely.

11. 0.058 ± 0.034 or $[0.024, 0.092]$. The confidence interval indicates that there was a real decrease in awareness. This is strange but might be explained by the following. The recent law mentioned in each survey actually was passed in early 1982. Perhaps by 1984 it was no longer viewed as a recent law.

13. 0.014 ± 0.043 or $[-0.029, 0.057]$. There does not appear to be any important change in knowledge.

15. -0.050 ± 0.043 or $[-0.093, -0.007]$. The confidence interval indicates that the increase in knowledge was negative. The proportion of the population that knew the correct answer declined by an amount estimated to be between 0.7 and 9.3 percentage points.

Section 8.2

1. (a) Correct positive (CP): A woman has cervical cancer and obtains a positive test result. CN: A woman does not have cervical cancer and obtains a negative test result. FP: A woman does not have cervical cancer but obtains a positive test result. FN: A woman has cervical cancer but obtains a negative test result. A positive result will lead to further treatment; thus, the error of a false positive should be rectified. A false negative will result in a woman not receiving treatment until the disease has progressed further. A false negative is clearly a much more serious error.
   (b) CP: A woman has breast cancer and the test is positive. CN: A woman does not have breast cancer and the test is negative. FP: A woman does not have breast cancer but the test is positive. FN: A woman has breast cancer but the test is negative. Both types of errors are serious, but a false negative is a more serious error than a false positive.

3. (a) The table of population counts:

	B	B^c	Total
A	360	140	500
A^c	240	260	500
Total	600	400	1,000

   (b) The table of population proportions:

	B	B^c	Total
A	0.36	0.14	0.50
A^c	0.24	0.26	0.50
Total	0.60	0.40	1.00

   (c) The table of conditional probabilities of the Bs given the As.

	B	B^c	Total
A	0.72	0.28	1.00
A^c	0.48	0.52	1.00
Unconditional	0.60	0.40	1.00

(d) The table of conditional probabilities of the As given the Bs.

	B	B^c	Unconditional
A	0.60	0.35	0.50
A^c	0.40	0.65	0.50
Total	1.00	1.00	1.00

(e) No, they are dependent.

5. (a) The table of population counts:

	B	B^c	Total
A	200	300	500
A^c	200	300	500
Total	400	600	1,000

(b) The table of population proportions:

	B	B^c	Total
A	0.20	0.30	0.50
A^c	0.20	0.30	0.50
Total	0.40	0.60	1.00

(c) The table of conditional probabilities of the Bs given the As.

	B	B^c	Total
A	0.40	0.60	1.00
A^c	0.40	0.60	1.00
Unconditional	0.40	0.60	1.00

(d) The table of conditional probabilities of the As given the Bs.

	B	B^c	Unconditional
A	0.50	0.50	0.50
A^c	0.50	0.50	0.50
Total	1.00	1.00	1.00

(e) Yes.

7. The table of probabilities:

	B	B^c	Total
A	0.12	0.18	0.30
A^c	0.28	0.42	0.70
Total	0.40	0.60	1.00

9. (a) The table of population proportions:

	B	B^c	Total
A	0.16	0.04	0.20
A^c	0.48	0.32	0.80
Total	0.64	0.36	1.00

(b) The table of conditional probabilities of the As given the Bs.

	B	B^c	Unconditional
A	0.25	0.11	0.20
A^c	0.75	0.89	0.80
Total	1.00	1.00	1.00

11. (a) The table of population proportions:

	B	B^c	Total
A	0.0095	0.0005	0.01
A^c	0.0099	0.9801	0.99
Total	0.0194	0.9806	1.00

(b) The table of conditional probabilities of the Bs given the As.

	B	B^c	Total
A	0.95	0.05	1.00
A^c	0.01	0.99	1.00
Unconditional	0.02	0.98	1.00

(c) The table of conditional probabilities of the As given the Bs.

	B	B^c	Unconditional
A	0.49	0.0005	0.01
A^c	0.51	0.9995	0.99
Total	1.00	1.0000	1.00

(d) $p_{A^c B} = 0.0099$; $p_{B|A^c} = 0.01$; $p_{A^c|B} = 0.51$. If the entire population were tested, just under 1 percent of the tests would yield a false positive; given that the condition is absent, there is a 1 percent chance the test would incorrectly report it is present; given the test is positive, there is a 51 percent chance that the person does not have the condition. This difference demonstrates the importance of knowing which false positive rate is being discussed.

(e) $p_{AB^c} = 0.0005$; $p_{B^c|A} = 0.05$; $p_{A|B^c} = 0.0005$. If the entire population were tested, 1 out of 2,000 tests would yield a false negative. For the subpopulation that has the condition, there is a 5 percent chance the test would incorrectly report it is absent; given the test is negative, there is 1 chance in 2,000 that the person has the condition.

13. (a) The proportion of the population infected with HIV is 1 in 10,000. The screening test never makes a false negative error. If all of the uninfected persons are tested, 5 out of 100,000 will obtain a false positive.

(b) The table of population proportions:

	B	B^c	Total
A	0.00010	0.00000	0.0001
A^c	0.00005	0.99985	0.9999
Total	0.00015	0.99985	1.0000

(c) The table of conditional probabilities of the Bs given the As.

	B	B^c	T
A	1.00000	0.00000	1
A^c	0.00005	0.99995	1
Unconditional	0.00015	0.99985	1

(d) The table of conditional probabilities of the As given the Bs.

	B	B^c	Unconditional
A	0.67	0.00	0.0001
A^c	0.33	1.00	0.9999
Total	1.00	1.00	1.0000

(e) $p_{A^c B} = 0.00005$; $p_{B|A^c} = 0.00005$; $p_{A^c|B} = 0.33$. Refer to the comment in the answer to 11(d), above.

(f) They are all 0. If false negatives are impossible, then the three rates are all 0.

Sections 8.3 and 8.4

1. (a) The 2 × 2 table:

	B	B^c	Total
A	264	876	1,140
A^c	214	986	1,200
Total	478	1,862	2,340

   (b) The table of relative frequencies:

	B	B^c	Total
A	0.113	0.374	0.487
A^c	0.091	0.421	0.513
Total	0.204	0.796	1.000

   (c) The point estimates of the conditional probabilities of drinking status given gender:

	B	B^c	Total
A	0.232	0.768	1.000
A^c	0.178	0.822	1.000
Unconditional	0.204	0.796	1.000

   (d) The point estimates of the conditional probabilities of gender given drinking status:

	B	B^c	Unconditional
A	0.552	0.470	0.487
A^c	0.448	0.530	0.513
Total	1.000	1.000	1.000

   (e) $z = 3.19$ and $P = 0.0014$.

   (f) The 95 percent confidence interval is

   $$(0.232 - 0.178) \pm 1.96 \sqrt{\frac{264(876)}{(1{,}140)^3} + \frac{214(986)}{(1{,}200)^3}} =$$

   $$0.054 \pm 1.96(0.1668) = 0.054 \pm 0.033 =$$

   $$[0.021, 0.087].$$

3. (a)

	Most Important	Not Most Important	Total
Female	0.224	0.263	0.487
Male	0.220	0.293	0.513
Total	0.444	0.556	1.000

   (b)

	Most Important	Not Most Important	Total
Female	0.460	0.540	1.000
Male	0.429	0.571	1.000

   In the sample, a higher proportion of females than males viewed mandatory jail terms to be a most important deterrent. The difference is only 3.1 percentage points.

   (c) $z = 1.41$; thus, $P = 0.1586$.

   (d) 0.031 ± 0.042 or $[-0.011, 0.073]$.

5. (a) I view this as a random sample from the process that caused people to seek "care for a bicycle-related injury at one of the hospitals" in Seattle for the time period of the study. The process is assumed to satisfy the assumptions of bivariate Bernoulli trials as discussed in the text. I would be hesitant to view the data as a random sample from the United States or from Seattle in an extended time period, but these issues are debatable.

(b) The table of conditional probabilities of the Bs (injury type) given the As (helmet use).

	B	B^c	Total
A	0.142	0.858	1.000
A^c	0.398	0.602	1.000
Unconditional	0.352	0.648	1.000

Subjects who were not wearing a helmet had a 25.6-percentage-point-higher rate of head injuries than those who were wearing a helmet.

(c) $z = -5.32$ and the approximate P-value is smaller than 0.0002.

(d) $-0.256 \pm 0.075 = [-0.331, -0.181]$. At the 95 percent confidence level, subjects who were wearing a helmet had a lower probability of sustaining a head injury than those who were not wearing a helmet. The difference in probabilities is estimated to be between 18.1 and 33.1 percentage points.

(e) The biggest weakness is that this is an observational study. Perhaps people who do not wear helmets take more risks on their bikes. Thus, making them wear helmets would not necessarily reduce the probability of a head injury. It also is possible that people do not wear helmets because they are trying to be extra careful. Making them wear helmets—assuming they remain extra careful—might reduce head injuries even more than the current study suggests.

(f) It seems unreasonable to assume that people who arrive at a hospital are a random sample of all bike riders.

7. (a)

Helmet Worn?	Brain Injury?		Total
	Yes	No	
Yes	0.235	0.765	1.000
No	0.436	0.564	1.000
Unconditional	0.421	0.579	1.000

Among subjects who suffered a head injury, those who were wearing a helmet had a 20.1-percentage-point-lower rate of brain injuries than those who were not wearing a helmet.

(b) $z = -1.61$, giving $\mathbf{P} = 0.0537$.

Section 8.5

1. $\hat{p} = 1553/2396 = 0.648$. The 99 percent confidence interval is

$$0.648 \pm 2.576\sqrt{\frac{0.648(0.352)}{2396}} = 0.648 \pm 0.025 = [0.623, 0.673].$$

3. The point estimate is 0.438; the 99 percent confidence interval is 0.438 ± 0.027 or $[0.411, 0.465]$.

5. The point estimate is 0.464; the 99 percent confidence interval is 0.464 ± 0.027 or $[0.437, 0.491]$.

7. $z = 14.77$; the approximate P-value is smaller than 0.0004. The 99 percent confidence interval is $0.219 \pm 0.036 = [0.183, 0.255]$.

9. $z = 13.98$; the approximate P-value is smaller than 0.0004. The 99 percent confidence interval is $0.179 \pm 0.032 = [0.147, 0.211]$.

11. $z = 9.73$; the approximate P-value is smaller than 0.0004. The 99 percent confidence interval is $0.128 \pm 0.033 = [0.095, 0.161]$.

13. $z = -0.56$; the approximate P-value is 0.5754. The 99 percent confidence interval is $-0.008 \pm 0.039 = [-0.047, 0.031]$.

15. $z = -2.15$; the approximate P-value is 0.0316. The 99 percent confidence interval is $-0.025 \pm 0.031 = [-0.056, 0.006]$.

Chapter 9

1. (a) The table of row proportions for the collapsed table is as follows:

Gender	S	F	Total
Female	0.56	0.44	1.00
Male	0.66	0.34	1.00

(b) The table of row proportions for the sample from the first subpopulation:

Gender	S	F	Total
Female	0.66	0.34	1.00
Male	0.70	0.30	1.00

The table of row proportions for the sample from the second subpopulation:

Gender	S	F	Total
Female	0.58	0.42	1.00
Male	0.68	0.32	1.00

The table of row proportions for the sample from the third subpopulation:

Gender	S	F	Total
Female	0.55	0.45	1.00
Male	0.61	0.39	1.00

The table of row proportions for the sample from the fourth subpopulation:

Gender	S	F	Total
Female	0.53	0.47	1.00
Male	0.66	0.34	1.00

(c) In the collapsed table, a higher proportion of men than women know the correct answer. In each subgroup, this relationship is preserved, with the difference ranging from a low of 4 percentage points in subgroup 1 to a high of 13 percentage points in subgroup 4. Among females, as drinking increases, knowledge increases; among males, the pattern is not as simple. Except for subgroup 3, as drinking frequency increases, knowledge increases slightly, but subgroup 3 males have a noticeably low level of knowledge.

Overall, not much insight is gained by examining subpopulations; the collapsed analysis seems reasonable.

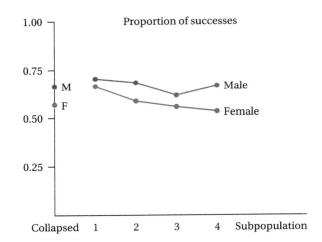

3. The weights for the females are

Category	Frequency	Relative Frequency
1	100	0.088
2	225	0.199
3	543	0.480
4	264	0.233
Total	1,132	1.000

The weights for the males are

Category	Frequency	Relative Frequency
1	229	0.191
2	374	0.312
3	381	0.318
4	213	0.178
Total	1,197	0.999

If the women in the sample drank as much as the men in the sample, then the estimate of their proportion of successes would be

$$0.191(0.66) + 0.312(0.58) + 0.318(0.55) + 0.178(0.53) = 0.576,$$

compared to 56 percent in the collapsed table—not much difference.

If the men in the sample drank as little as the women in the sample, then the estimate of their proportion of successes would be

$$0.088(0.70) + 0.199(0.68) + 0.480(0.61) + 0.233(0.66) = 0.644,$$

compared to 66 percent in the collapsed table—not much difference.

5. $z = -4.82$ giving $P < 0.0004$.

7. Reading left to right and top to bottom, the added entries are 144.61, 32.313, 297, 54.822, -1.95, 140, 154.97, and -2.79. The P-values for the four subgroups, respectively, are 0.4902, 0.0164, 0.0512, and 0.0052. The difference between men and women is highly statistically significant in subgroup 4, statistically significant in subgroup 2, nearly statistically significant in subgroup 3, and not statistically significant in subgroup 1.

9. $T = 634$, $E(T) = 679.73$, and $\text{Var}(T) = 130.91$; this yields $z = -4.00$, which gives $P < 0.0004$. I like the MH analysis best; the z value is not as extreme as in the collapsed table because knowledge does depend, in part, on drinking frequency. The individual subgroup analyses are interesting too; they indicate that among the heaviest drinkers there might not be a gender difference.

11. The weights for female abusers are

Component Table	Frequency	Relative Frequency
1	50	0.303
2	36	0.218
3	27	0.164
4	20	0.121
5,6	1	0.006
7	16	0.097
8	15	0.091
Total	165	1.000

The weights for male abusers are

Component Table	Frequency	Relative Frequency
1	85	0.323
2	10	0.038
3	63	0.240
4	37	0.141
5,6	5	0.019
7	44	0.167
8	19	0.072
Total	263	1.000

13. Recall that, in the collapsed table, 70.3 percent of the victims of female abusers accepted services. If this figure is adjusted by using the male weights, the proportion accepting services becomes

$$0.323(0.72) + 0.038(0.81) + 0.240(0.67) +$$
$$0.141(0.65) + 0.019(0.00) + 0.167(0.62) +$$
$$0.072(0.67) = 0.668.$$

Thus, the 10.2-percentage-point "gender difference" in the collapsed table is reduced to 6.7 percentage points.

Similarly, in the collapsed table 60.1 percent of the victims of male abusers accepted services. If this figure is adjusted by using the female weights, the proportion accepting services becomes

$$0.303(0.55) + 0.218(0.70) + 0.164(0.67) +$$
$$0.121(0.73) + 0.006(0.40) + 0.097(0.57) +$$
$$0.091(0.42) = 0.613.$$

Thus, the 10.2-percentage-point "gender difference" in the collapsed table is reduced to 9.0 percentage points.

Section 10.1

1. The perceived chance increased substantially from 1982 to 1986.

Response	Frequency	Relative Frequency
Very low	46	0.027
Low	338	0.201
About 50 : 50	752	0.448
High	378	0.225
Very high	166	0.099
Total	2,585	1.000

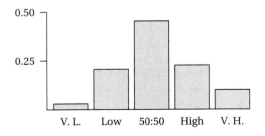

3. (a) Seven.
   (b) Unordered.

(c)

Response	Frequency	Relative Frequency
Milk	21	0.175
Liver	20	0.167
Vegetable	20	0.167
Bacon	19	0.158
Cheese	17	0.142
Poultry	12	0.100
Meat	11	0.092
Total	120	1.001

(d)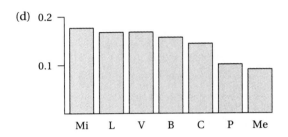

5. (a)

Response	Frequency	Relative Frequency
F	285	0.152
E	281	0.150
B	278	0.149
D	272	0.145
C	257	0.137
G	254	0.136
A	245	0.131
Total	1,872	1.000

(b)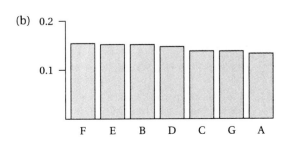

Section 10.2

1. For round and yellow:
   $$0.567 \pm 0.041 = [0.526, 0.608].$$
   For round and green:
   $$0.194 \pm 0.033 = [0.161, 0.227].$$
   For wrinkled and yellow:
   $$0.182 \pm 0.032 = [0.150, 0.214].$$
   For wrinkled and green:
   $$0.058 \pm 0.019 = [0.039, 0.077].$$
3. 0.95.
5. The area is between 0.10 and 0.50.
7. The area is between 0.01 and 0.025.
9. The area is smaller than 0.01.
11. The area is 0.0080. Using Table A.5, the area is smaller than 0.01.

Section 10.3

1. The new analysis gives $\chi^2 = 4.43$ with two degrees of freedom. This gives $0.10 < P < 0.50$. This new analysis does not bother me if I believe the scientist; if, however, the scientist is being directed by the data, this is a variation of the focusing on a winner fallacy.
3. The value of the test statistic is $\chi^2 = 5.767$. The approximate P-value—the area to the right of 5.767 under the chi-squared curve six degrees of freedom—is between 0.10 and 0.50. The data are not statistically significant, but the sample size is small.
5. The value of the test statistic is $\chi^2 = 4.805$. The approximate P-value—the area to the right of 4.805 under the chi-squared curve six degrees of freedom—is between 0.50 and 0.90. The data are not statistically significant, but the sample size is small.
7. The value of the test statistic is $\chi^2 = 3.913$. The approximate P-value—the area to the right of 3.913 under the chi-squared curve six degrees of freedom—is between 0.50 and 0.90. The data are not statistically significant, but the sample size is small (although larger than in Exercises 3 and 5).

Chapter 11

1. I justify using the chi-squared test by assuming that the process of assigning biscuits to boxes satisfies the assumption of multinomial trials. But I do not assume that each box has the same process. (For example, perhaps "Bacon" is more likely with the second process than with the first.) The null hypothesis states that the probabilities of the seven flavors for the first box are identical to the corresponding probabilities for the second box. The alternative states that there is some difference between the processes. Note that, unlike Chapter 10, the hypotheses do not deal with the issue of the individual flavors being equally likely or not. The number of degrees of freedom equals six; from Table A.5 of the Appendix, the approximate P-value is between 0.10 and 0.50. The data are not statistically significant. There is insufficient evidence to reject the belief that the processes are identical.
3. Below are the bar charts for the new response for the four years studied.

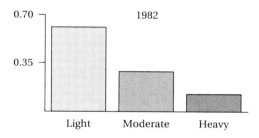

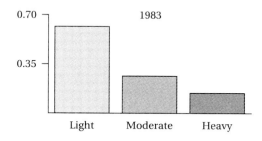

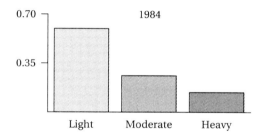

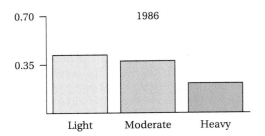

Taken at face value, these figures indicate that there was very little change in the distribution of drinking frequency during the years 1982–84, followed by a huge increase in drinking frequency in 1986. I do not believe that happened. Rather, I suspect that a good proportion of respondents do not read the category labels carefully and instead think that "on a scale of 1 to 4 I am about a 2 on drinking frequency." My conjecture is compatible with the huge increase in drinking in 1986, when two categories corresponded to "heavy," compared to one category in earlier years.

5. (a) Any IV drug use increases dramatically the proportion of HIV antibody–positive subjects; the higher the use, the higher the proportion of HIV antibody–positive subjects.

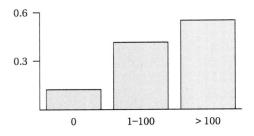

(b)

Number of Injections	HIV-Antibody Status		Total
	Positive	Negative	
0	9	66	75
1–100	98	144	242
>100	71	61	132
Total	178	271	449

(c) 52.3.

(d) 6.69.

(e) Probably not. Persons in a methadone treatment program may be importantly different from those who are not.

(f) $P < 0.01$. I reject the notion that the three populations are identical. Yes.

(g) The values of z are -4.55, -5.92, and -2.47 for the comparison of populations 1 and 2; 1 and 3; and 2 and 3; respectively. Thus, the (one-sided) P-values are all very small, and I conclude that the three populations are all different.

7. (a) The longer the duration of IV drug use, the larger the proportion of HIV-positive individuals.

(b)

Duration	HIV-Antibody Status		Total
	Positive	Negative	
0 mo.	9	66	75
1–50 mo.	81	137	218
>50 mo.	88	70	158
Total	178	273	451

(c) 95.6.

(d) 6.86.

(e) $P < 0.01$. I reject the notion that the three populations are identical. Yes.

(f) The values of z are -4.07, -6.31, and -3.56 for the comparison of populations 1 and 2; 1 and 3; and 2 and 3; respectively. Thus, the P-values are all very small, and I conclude that the three populations are all different.

9. $0.05 < P < 0.10$. No.

11. $P < 0.01$. Yes.

13. For the test of homogeneity, $P < 0.01$, so it is clear that the two distributions are different. Bar charts indicate that the women view the problem as more serious.

15. For the test of homogeneity, **P** < 0.01, so it is clear that the two distributions are different. The bar charts indicate that the responses were more extreme in 1983.

17. For the test of homogeneity, **P** < 0.01, so I reject the notion that the four distributions are identical. The sample proportion correct increases with drinking frequency *except* that group 2 has a slightly lower correct rate than group 1. The pairwise comparisons can be summarized as follows (one-sided alternative): 1 and 2 are not statistically significantly different; 3 and 4 are not statistically significantly different; all other pairwise comparisons give a P-value below 0.01 except the comparison of 1 and 3, which gives 0.0197. In view of the large number of tests, a conservative analyst would conclude that the observed difference between groups 1 and 3 might not be real.

Sections 12.2 and 12.3

1. (a) The dot plots:

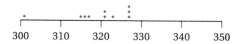

Aluminum		Wood
	8	6
	9	
5	10	
05	11	479
035	12	12377
059	13	2479
125789	14	12479
01569	15	123345679
0377	16	2346778
2447	17	167
269	18	19
2	19	
029	20	0
4	21	
	22	
1	23	
1	24	
	25	
0	26	
	27	
⋮	⋮	⋮
0	35	

(b) Because the distribution is skewed to the right with one very large outlier. (Jennifer said this was the farthest she had ever hit a ball!) Yes, because $\bar{x} = 159.5$. After deleting the largest value, $\bar{x} = 164.23$; it is still larger the median (which becomes 159 after the deletion).

(c) $\bar{x} = 149.05$ and $\tilde{x} = 152.5$.

(d) In agreement with conventional wisdom, Jennifer hit a ball farther with aluminum than wood. The distances with the wooden bat are fairly symmetric with one particularly bad value. By contrast, the aluminum bat's values are skewed to the right.

5. (b) For the first treatment, $\bar{x} = 491.4$ and $\tilde{x} = 492$; and for the second treatment, $\bar{x} = 484.2$ and $\tilde{x} = 485$.

(b) For combat boots, $\bar{x} = 333.0$ and $\tilde{x} = 331.5$; for jungle boots, $\bar{x} = 319.5$ and $\tilde{x} = 321.0$ seconds.

(c) Brian was considerably faster wearing the jungle boots. The times with combat boots are roughly symmetric with no outliers. The times with the jungle boots are skewed to the left with the smallest value as an outlier.

3. (a) The stem and leaf plots are at the top of the next column. Note that seven empty stems have been deleted to save space—be careful not to be misled by this shortcut!

(c) All of Kymn's times on the second treatment were faster than all of her times on the first treatment. For each treatment, the values of the mean and median are similar, but there is too little data to comment on its shape. Her times were slightly more consistent on the first treatment.

(d) She appears to be aerobically very fit.

7. (a) The stem and leaf plots are given below.

No Looking		No Restrictions	
11	18	11	
12	002789	12	
13	2579	13	1255
14	0027	14	26
15	00	15	04479
16	04	16	1134556
17		17	07

(b) The stem and leaf plots are given below.

No Looking		No Restrictions	
11	1	11	
11	8	11	
12	002	12	
12	789	12	
13	2	13	12
13	579	13	55
14	002	14	2
14	7	14	6
15	00	15	044
15		15	79
16	04	16	1134
16		16	556
17		17	0
17		17	7

(c) With the restriction of not looking, $\bar{x} = 135.55$ and $\tilde{x} = 136$; with no restriction, $\bar{x} = 154.35$ and $\tilde{x} = 158$.

(d) The no looking values are skewed to the right and the others are skewed to the left. There are no clear outliers. Amy typed faster without restrictions.

9. (a) The stem and leaf plots are

No Looking		No Restrictions	
7	000	7	
8	067789	8	
9	799	9	
10	8	10	5
11	002225	11	6
12	0	12	2
13		13	0245699
14		14	1247
15		15	356
16		16	001

(b) The stem and leaf plots are

No Looking		No Restrictions	
7	000	7	
7		7	
8	0	8	
8	67789	8	
9		9	
9	799	9	
10		10	
10	8	10	5
11	00222	11	
11	5	11	6
12	0	12	2
12		12	
13		13	024
13		13	5699
14		14	124
14		14	7
15		15	3
15		15	56
16		16	001

(c) With the restriction of not looking, $\bar{x} = 96.05$ and $\tilde{x} = 98$; with no restriction, $\bar{x} = 140.35$ and $\tilde{x} = 140$.

(d) Both distributions are unfamiliar (and look fairly weird), but each has similar values for their mean and median. Amy is much better at typing with no restrictions. Comparing this exercise with Exercise 10 shows that the no looking restriction slows Amy down and leads to many more errors.

11. (b) The stem and leaf plots for the first and second treatments, respectively, are

One at a Time		Three at a Time	
0	22679	0	0003367
1	12889	1	03556777
2	112455679	2	1345778
3	0134688	3	125667
4	0015	4	239
5	13678	5	3569
6		6	0158
7	000	7	
8	1	8	5
9		9	
10		10	
11		11	
12	4	12	

(c) Yes. Both plots are skewed to the right and the plot for aiming has one large-valued outlier. (Incidentally, the sample means are 35.5 and 30.4.)

(d) In addition to the comments in part (c), we have learned that Christine scored somewhat better when she aimed.

13. (a) Below is a density scale histogram for the 3-wood.

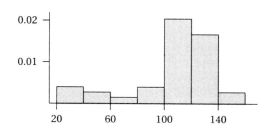

(b) Below is a density scale histogram for the 3-iron.

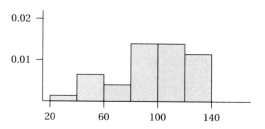

(c) The 3-wood has the larger mean. The medians are 112 for the 3-wood and 99.5 for the 3-iron.

(d) Both distributions are skewed to the left. Very bad shots (less than 40 yards) were more common with the 3-wood. The reason for the additional bad shots with the 3-wood is that its mean is only 8.7 yards larger whereas its median is 12.5 yards larger.

15. (a) Below is the dot plot of the times with the windows open.

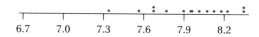

(b) Below is the dot plot of the times with the windows closed.

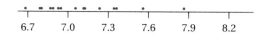

(c) The sample median for the windows open is 7.96 seconds and for the windows closed, 7.05 seconds. The larger mean is for the windows open condition.

(d) With the windows open, the times are skewed to the left, but with the windows closed the times are skewed to the right. Not surprisingly, the times are faster with the windows closed.

17. (b) For the prior knowledge condition, $\bar{x} = 5.5$ and $\tilde{x} = 5.5$; for the no prior knowledge condition, $\bar{x} = 2.6$ and $\tilde{x} = 2$.

(c) The scores are much higher when the subjects have prior knowledge of what to look for.

19. (c) Using one hand, $\bar{x} = 112.6$ and $\tilde{x} = 110$; Using two hands, $\bar{x} = 92.2$ and $\tilde{x} = 94$; and for the differences, $\bar{x} = 20.4$ and $\tilde{x} = 17$.

(d) Alisa was clearly much better when she used one hand. For each treatment the means and medians have similar values, but there is too little data to say much about the shapes, except that the scores with two hands were more consistent than the scores with one hand.

21. (c) With the tennis balls, $\bar{x} = 6.1$ and $\tilde{x} = 6$; with the apples, $\bar{x} = 5.2$ and $\tilde{x} = 5$; and for the differences, $\bar{x} = 0.9$ and $\tilde{x} = 1$.

(d) The dot plot of the times with the balls has a large peak at 6 seconds, and has a long right tail. The times with the apples have a broad peak from 1 to 5 seconds and are strongly skewed to the right. The differences are approximately symmetric. Martha was better at juggling tennis balls.

23. (a) The sample median equals 22 seconds. The distribution has three tall peaks with deep valleys between them. (It was probably difficult to measure accurately to the nearest one-hundredth of a second.)

(b) Below is the stem and leaf plot for the pedal depressed.

8	1 denotes 810
8	1334459
9	123455689
10	112344578
11	000112334489
12	000

(c) At the top of the next column is another stem and leaf plot for the pedal depressed.

(d) The second plot shows that the times are skewed to the left with three substantial peaks.

8	1 denotes 810
8	13344
8	59
9	1234
9	55689
10	112344
10	578
11	0001123344
11	89
12	000

27. The first histogram hides the gaps in the data; the second histogram is superior.

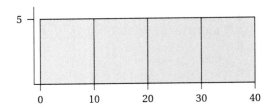

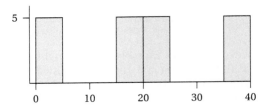

29. In the top picture, the heights of the middle six rectangles do not vary dramatically. There are no gaps and no outliers; the peaks and valleys look like they could be due to chance. The bottom picture reveals much more detail than the top picture. The rectangle centered at $1,000 is huge (over 10 percent of the prices fall in the interval from $975 to $1,025), but the two rectangles on either side are very short. In general, peaks tend to occur at rectangles centered at a round number—at 500, 700, 900, 1,000, 1,100, 1,300, and 1,500. This picture reveals two gaps—at the class intervals centered at 750 and 1,950 dollars. There are some values to the right of the latter gap, but I would not label them outliers.

31. (b) The stem and leaf plot is misleading because it suggests a valley when in fact the distribution is perfectly flat.

Section 12.4

1. (a) For the combat boots, $R = 26$ and IQR $= 8$; and for the jungle boots, $R = 26$ and IQR $= 11$.
   (b) The standard deviations suggest the combat boots have slightly more spread; the IQRs suggest the combat boots have substantially less spread; and the ranges find no difference between treatments. The standard deviation and range are unduly influenced by the one small outlier on jungle boots. The IQR is the best measure of spread for these data.

3. (a) For the aluminum bat, $R = 245$ and IQR $= 46$; and for the wooden bat, $R = 114$ and IQR $= 32$.
   (b) All three measures of spread accurately reflect the greater spread in the aluminum bat data. The standard deviation and especially the range of the aluminum bat data are greatly influenced by the one large-valued outlier.
   (c) For the aluminum bat, $R = 155$ and IQR $= 45$; and for the wooden bat, $R = 86$ and IQR $= 32$. The deletions have a big impact on the ranges and little or none on the IQRs.

5. (a) For the small gear, $R = 4$ and IQR $= 3.5$; and for the large gear, $R = 9$ and IQR $= 6$.
   (b) All three measures of spread accurately reflect the greater variation on the large gear.

7. (a) With no looking allowed, $R = 53$ and IQR $= 20$; and with no restrictions, $R = 46$ and IQR $= 20.5$.
   (b) I think the "no restrictions" data have less spread; thus, I feel that the ranges and standard deviations do a better job than the IQRs at measuring spread for these data sets.

9. (a) With no looking allowed, $R = 50$ and IQR $= 24.5$; and with no restrictions, $R = 56$ and IQR $= 21$.
   (b) The standard deviations and IQRs suggest that the no restriction condition has less spread. The comparison of ranges reaches the opposite conclusion because it is influenced by the long left tail of the no restrictions distribution.

11. (a) For throwing darts one at a time, $R = 122$ and IQR $= 28$; and for throwing darts three at a time, $R = 85$ and IQR $= 31$.
    (b) The measures of spread tell different stories. The range and standard deviation of the one at a time condition are influenced greatly by the one large-valued outlier.

13. (a) With the 3-wood, $R = 125$ and IQR $= 27$; and with the 3-iron, $R = 112$ and IQR $= 34$.
    (b) These are difficult data sets to compare visually; it seems that the distances with the 3 iron are somewhat less dispersed. I prefer the standard deviations because they best reflect the minor difference in spreads between the distributions.

15. (a) With the windows open, $R = 1.00$ and IQR $= 0.50$; and with the windows closed, $R = 1.19$ and IQR $= 0.48$.
    (b) The windows closed data set has one large outlier, which contributes to its larger values of the range and standard deviation. Ignoring the outlier, the window closed data appear to be less dispersed; thus, the IQR is the preferred measure of spread.

19. (b) The sample range is 10 for each data set. Thus, the range does not distinguish between these data sets.
    (c) The IQR is 10 for A and B and 0 for C. The IQR shows that C has the least spread, but it is ineffective in distinguishing between A and B.
    (d) Yes.

21. (a) $77/152 = 0.507$.
    (b) The approximation is
    $$\Phi(0.48) - \Phi(-0.61) = 0.4135;$$
    the approximation is poor, because the distribution is more peaked in the middle than a bell-shaped curve.

23. (a) $R = 496.7$.
    (b) IQR $= 40.7$.
    (c) $R = 303.0$ and IQR $= 39.75$. The deletion of a single observation decreases the range by 39 percent, but decreases the IQR by only 2 percent.
    (d) The larger value is for the entire data set.

Section 12.5

1. The dot plot is skewed to the right. Based on the shape, I expect the mean to be larger. The sample mean and median both equal 3; I was misled by the shape because of the granularity.

3. (a) 4. (b) 4.5. (c) Knowing the medians of the individual data sets is *not* sufficient information to obtain the median of the combined data set.

Section 13.1

1. 3; −2; 13.

3. (a, b)

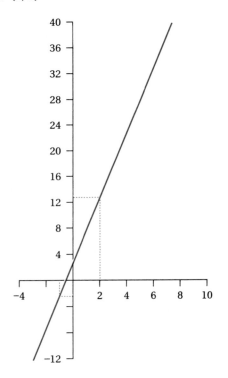

Section 13.2

1. (b) The distribution is strongly skewed to the left with one or two small outliers.

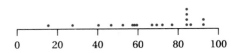

(c) Like the dot plot, the stem and leaf plot reveals that the distribution is strongly skewed to the left with one or two small outliers. The stem and leaf plot also contains a second peak in the class interval 50–60.

1	5
2	7
3	
4	06
5	2789
6	79
7	27
8	44446
9	223

(d) Except for one small outlier, the distribution of the stars for the public is approximately symmetric around 3.45. The stars for the critics have much lower values and are much more dispersed than the stars for the public.

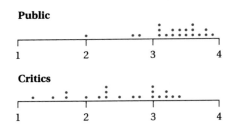

(e) There is one isolated case, *The Bonfire of the Vanities*. There is a very weak relationship between the critics' and the viewers' ratings, although it does appear to be direct. $r = 0.36$.

3. (a) The cases with two or three incidents could be labeled isolated. The relationship between X and Y is indirect, roughly linear, and fairly strong.

(b) There is no data for temperature below 53 degrees, so it is difficult to make a prediction. For the temperatures for which there is data, decreasing temperatures are bad (every launch below 65 degrees had at least one incident). Given the simple physics of how rings operate, it is hard to imagine they would work better at lower temperatures. Thus, it seems foolish to launch at 31 degrees. (If I had to go up, I would want the temperature to be above 65.)

(c) The relationship could be viewed as curved or linear, but weak. I don't find this restricted data to be convincing.

5. (a) The lightest spider is an isolated case.

   (b) $r = 0.055$.

   (c) The PH&Ws are about the same weight as the small hunters, but the former have much lower heart rates. The PH&Ws are smaller than the other three types and have lower heart rates than the LHs and WWs. The tarantulas have somewhat lower heart rates than the PH&Ws. The PH&Ws have the weakest linear relationship between size and heart rate for the five types of spiders.

7. (a) The plot reveals a strong direct relationship with no isolated cases.

   (b) $r = 0.86$.

9. (b) Yes; the cases (59,76), (60,72), (62,75), and (67,77) are isolated.

   (c) The counts are 19, 10, 11, and 3 in quadrants 1, 2, 3, and 4, respectively. Thirty observations give evidence of a direct relationship, compared to only 13 that give evidence of an indirect relationship.

   (d) $r = 0.41$.

11. If all cases lie on a horizontal straight line, then all cases have the same value for y. Thus, $s_Y = 0$, which violates our restriction. (This is not "cheating"; if there is no variation in the values of Y, then it is impossible to study how Y depends on X.)

Section 13.3

1. (a) The ratings by the public have a larger center and are less dispersed than the ratings by the critics.

   (b) i. $\hat{y} = 3.32 + 0.247(x - 2.52)$.

   ii. $\hat{y} = 3.44$ for *Ghost*; $\hat{y} = 3.49$ for *Havana*.

3. (a) $\hat{y} = 393.33 + 18.371(x - 75.11)$. $\hat{y} = 393.33 + 18.371(74 - 75.11) = 372.9$. The actual fish activity for this case is about 23 units below the predicted value.

   (b) $\hat{y} = 75.11 + 0.04026(x - 393.33)$. $\hat{y} = 75.11 + 0.04026(350 - 393.33) = 73.4$. The actual water temperature for this case is about 0.6 degrees above the predicted value.

5. (a) $\hat{y} = 69.78 + 0.6088(x - 64)$. $\hat{y} = 69.78 + 0.6088(66 - 64) = 71.0$. The predicted value is equal to the actual value!

   (b) $\hat{y} = 64 + 0.2775(x - 69.78)$. $\hat{y} = 64 + 0.2775(71 - 69.78) = 64.3$. The predicted value is 1.7 inches below the actual value. This is a bit surprising to many people; even though "Mom" predicts "Son" perfectly, "Son" does not predict "Mom" perfectly.

7. The following table is useful for comparing the two lines.

x	y	$\hat{y}$	$(y - \hat{y})^2$	$\ddot{y}$	$(y - \ddot{y})^2$
1	3	2	1	1	4
1	1	2	1	1	0
2	2	3	1	3	1
2	3	3	0	3	0
3	4	4	0	5	1
Total			3		6

The line with the smaller sum of squared errors, $\hat{y}$, is better.

9. Even with no intervention by the "wizard," most of the children with the lowest scores would improve because of the regression effect.

Section 13.4

1. The two plots with $r = 0$ because X and e are uncorrelated for any data set.

3. (a) The two smallest values of the residuals might be classified as outliers. Except for the possible outliers, the scatterplot of the residuals versus X looks okay.

   (b) *Bonfire of the Vanities*; *Havana*.

   (c) Dropping 10 percent of the data results in a 15 percent increase in the correlation coefficient. This change is not very great.

5. (a) $r = +\sqrt{0.311} = 0.558$.

   (b) $\hat{y} = 117 + 12.3(3.6) = 161$.

   (c) $e = 153 - 161 = -8$.

7. (a) $r = +\sqrt{0.917} = 0.958$.
   (b) For Philadelphia, $x = -1.3$; thus, $\hat{y} = 41 + 2.50(-1.3) = 38$. For Chicago, $x = 10.4$; thus, $\hat{y} = 41 + 2.50(10.4) = 67$.
   (c) For Philadelphia, $e = 35 - 38 = -3$. For Chicago, $e = 67 - 67 = 0$.

9. For each year $\bar{y}$ must equal 41 because each team played 82 games. In addition, by the definition of X it follows that $\bar{x}$ must equal 0 each year. Thus, each year the prediction equation must be equal to 41 plus "something" times x. It is noteworthy that the slope remained almost constant over the two seasons. In words, an increase of one point in the mean difference per game translates into about 2.5 additional victories.

Chapter 14

1. (a) The first and last three years of his career, his home run numbers were (by his standards) low. During the remainder of his career, his output was fairly consistent.
   (b) The smoothed series tells a similar story.
   (c) I prefer the original series because changing values (smoothing) does not make it significantly easier to interpret the series.
   (d) 0.405.
   (e) By the last two years of his career, Aaron's abilities had deteriorated badly. During his prime, there seems to have been no autocorrelation of lag one.

3. (a) There is a great deal of up-and-down behavior to the series and some evidence that scores decreased during the second half of the study.

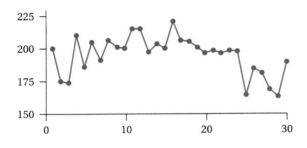

(b) The series is relatively smooth and it indicates that the deviations decreased in the second half of the study.

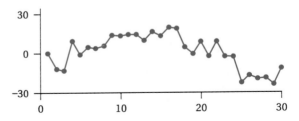

(c) As mentioned above, the second series is smoother. The second is better for detecting a time trend because it adjusts for the fairly large difference in the metasubject's abilities at the two distances.

Section 15.2

1. (a) 96.1 percent;
   (b) 93.6 percent.

3. (a) $[x_{(4)}, x_{(11)}]$; 94.3 percent.
   (b) $[x_{(3)}, x_{(8)}]$; 89.1 percent.

5. (a) Because $n < 20$, the exact confidence level method can be used. From Table A.7, the interval $[x_{(3)}, x_{(8)}]$ has confidence level equal to 89.1 percent. For these data the interval equals $[329, 337]$.
   (b) The 89.1 percent confidence interval equals $[316, 327]$.

7. (a) $k' = 14.3$; $k = 14$; the confidence interval is $[x_{(14)}, x_{(27)}]$, which equals $[148, 174]$.
   (b) $k' = 14.3$; $k = 14$; the confidence interval is $[x_{(14)}, x_{(27)}]$, which equals $[141, 159]$.

9. (a) $[489, 493]$.
   (b) $[479, 488]$.

11. (a) $k' = 14.3$; $k = 14$; the confidence interval is $[x_{(14)}, x_{(27)}]$, which equals $[24, 40]$.
    (b) $k' = 14.3$; $k = 14$; the confidence interval is $[x_{(14)}, x_{(27)}]$, which equals $[17, 36]$.

13. (a) $k' = 14.3$; $k = 14$; the confidence interval is $[x_{(14)}, x_{(27)}]$, which equals $[107, 122]$.
    (b) $k' = 14.3$; $k = 14$; the confidence interval is $[x_{(14)}, x_{(27)}]$, which equals $[92, 108]$.

15. (a) Because $n < 20$, the exact confidence level method can be used. From Table A.7, the interval $[x_{(4)}, x_{(12)}]$ has confidence level equal to 96.5 percent. For these data the interval equals $[7.67, 8.17]$.
    (b) The 96.5 percent confidence interval equals $[6.86, 7.34]$.

Section 15.3

1. (a) $32.44 \pm 6.50 = [25.94, 38.94]$.
   (b) $63.51 \pm 4.78 = [58.73, 68.29]$.
   (c) $93.83 \pm 6.48 = [87.35, 100.31]$.

3. (a) $0.01 < P < 0.025$.
   (b) $0.02 < P < 0.05$.
   (c) $0.025 < P < 0.05$.
   (d) $0.05 < P < 0.10$.
   (e) $P < 0.005$.
   (f) $P < 0.01$.

5. (a) $333.0 \pm 2.262(8.18)/\sqrt{10} = 333.0 \pm 5.9 = [327.1, 338.9]$.
   (b) $319.5 \pm 2.262(7.93)/\sqrt{10} = 319.5 \pm 5.7 = [313.8, 325.2]$.

7. (a) $168.88 \pm 1.96(45.98)/\sqrt{40} = 168.88 \pm 14.25 = [154.63, 183.13]$.
   (b) $149.05 \pm 1.96(22.98)/\sqrt{40} = 149.05 \pm 7.12 = [141.93, 156.17]$.

9. (a) $491.4 \pm 2.776(1.82)/\sqrt{5} = 491.4 \pm 2.3 = [489.1, 493.7]$.
   (b) $484.2 \pm 2.776(3.42)/\sqrt{5} = 484.2 \pm 4.2 = [480.0, 488.4]$.

11. (a) $35.50 \pm 1.96(24.53)/\sqrt{40} = 35.50 \pm 7.60 = [27.90, 43.10]$.
    (b) $30.42 \pm 1.96(21.70)/\sqrt{40} = 30.42 \pm 6.72 = [23.70, 37.14]$.

13. (a) $106.87 \pm 1.96(29.87)/\sqrt{40} = 106.87 \pm 9.26 = [97.61, 116.13]$.
    (b) $98.18 \pm 1.96(28.33)/\sqrt{40} = 98.18 \pm 8.78 = [89.40, 106.96]$.

15. (a) $7.94 \pm 2.145(0.2913)/\sqrt{15} = 7.94 \pm 0.16 = [7.78, 8.10]$.
    (b) $7.10 \pm 2.145(0.3253)/\sqrt{15} = 7.10 \pm 0.18 = [6.92, 7.28]$.

Section 15.4

1. There are some minor differences between the dot plots (for example, the third has somewhat more spread), but nothing striking.

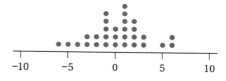

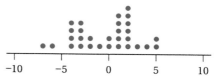

3. (a) $-0.124 \pm 1.96\dfrac{3.493}{\sqrt{89}} = -0.124 \pm 0.726 = [-0.850, 0.602]$.

   (b) $t = \dfrac{\sqrt{89}(-0.124 - 0)}{3.493} = -0.33$.

   Because the degrees of freedom, 88, is larger than 30, the approximate P-value can be obtained from the standard normal curve; for the third alternative it is 0.7414.

Section 15.5

1. (a) The sample mean equals 1,140.3, but since prices are always reported as a whole dollar amount, the point prediction is $1,140. The 95 percent prediction interval is

$$1,140.3 \pm 1.96(388.4)\sqrt{1 + 1/208} =$$
$$1,140.3 \pm 763.1 = [377, 1{,}903].$$

   (b) The point prediction is the sample median, 1158.5, rounded to either $1158 or $1159. By trial and error, $j = 5$ gives 95.2 percent confidence and $j = 6$ gives 94.3 percent; thus, I will use $j = 5$. The prediction interval is

$$[x_{(5)}, x_{(204)}] = [459, 1{,}807].$$

3. The prediction interval includes 206 of the 208 data points, or 99.0 percent. The prediction interval is too wide because the assumption of a normal population is not supported by the data.

5. (a) The 95 percent prediction interval is

$$333.0 \pm 2.262(8.18)\sqrt{1 + 1/10} =$$
$$333.0 \pm 19.4 = [313.6, 352.4].$$

   (b) The widest interval is $[x_{(1)}, x_{(10)}] = [321, 347]$, but the probability this interval is correct equals only 81.8 percent.

   (c) The 95 percent prediction interval is

$$319.5 \pm 2.262(7.93)\sqrt{1 + 1/10} =$$
$$319.5 \pm 18.8 = [300.7, 338.3].$$

   (d) The widest interval is $[x_{(1)}, x_{(10)}] = [301, 327]$, but the probability this interval is correct equals only 81.8 percent.

7. (a) The 95 percent prediction interval is

$$168.88 \pm 1.96(45.98)\sqrt{1 + 1/40} =$$
$$168.88 \pm 91.24 = [77.64, 260.12].$$

   (b) The widest interval is $[x_{(1)}, x_{(40)}] = [105, 350]$, and the probability this interval is correct equals 95.1 percent.

   (c) The 95 percent prediction interval is

$$149.05 \pm 1.96(22.98)\sqrt{1 + 1/40} =$$
$$149.05 \pm 45.60 = [103.45, 194.65].$$

   (d) The widest interval is $[x_{(1)}, x_{(40)}] = [86, 200]$, and the probability this interval is correct equals 95.1 percent.

9. (a) The 95 percent prediction interval is

$$491.0 \pm 3.182(1.83)\sqrt{1 + 1/4} =$$
$$491.0 \pm 6.5 = [484.5, 497.5].$$

   The actual fifth value was 493; thus, this interval is correct.

   (b) The 95 percent prediction interval is

$$483.75 \pm 3.182(3.77)\sqrt{1 + 1/4} =$$
$$483.75 \pm 13.41 = [470.34, 497.16].$$

   The actual fifth value was 486; thus, this interval is correct.

11. (a) The 95 percent prediction interval is

$$35.50 \pm 1.96(24.53)\sqrt{1 + 1/40} =$$
$$35.50 \pm 48.68 = [0, 84].$$

   (Remember that negative scores are not possible. The outlier at 124 has a big impact on this analysis.)

   (b) The widest interval is $[x_{(1)}, x_{(40)}] = [2, 124]$, and the probability this interval is correct equals 95.1 percent.

   (c) The 95 percent prediction interval is

$$30.42 \pm 1.96(21.70)\sqrt{1 + 1/40} =$$
$$30.42 \pm 43.06 = [0, 73].$$

   (d) The widest interval is $[x_{(1)}, x_{(40)}] = [0, 85]$, and the probability this interval is correct equals 95.1 percent.

Sections 16.1 and 16.2

1. (a) 6.083; (b) 6.914; (c) 5.119.

3. (a) 1.838; (b) 0.975; (c) 1.317.

5. (a) $s_p = 8.06$;

$$t = \frac{\sqrt{20}(13.5)}{2(8.06)} = 3.75.$$

The approximate P-value is less than 0.01. Brian is faster when wearing jungle boots and the difference is highly statistically significant.

(b) $13.5 \pm 2.101(8.06)\sqrt{2/10} = 13.5 \pm 7.6 = [5.9, 21.1]$. At the 95 percent confidence level, one can conclude that Brian's mean time in jungle boots is between 5.9 and 21.1 seconds lower than his mean time in combat boots.

7. Note that all three cases give the same answer.

(a) $s_p = 36.35$;

$$t = \frac{\sqrt{80}(19.83)}{2(36.35)} = 2.44.$$

The approximate P-value is 0.0073. The ball travels further when hit with an aluminum bat and the difference is highly statistically significant.

(b) $19.83 \pm 1.96(36.35)\sqrt{2/40} = 19.83 \pm 15.93 = [3.90, 35.76]$. At the 95 percent confidence level, one can conclude that the mean distance for Jennifer with the aluminum bat is between 2.98 and 35.76 feet greater than her mean distance with the wooden bat.

9. Note that all three cases give the same answer.

(a) $s_p = 29.11$;

$$t = \frac{\sqrt{80}(8.69)}{2(29.11)} = 1.34.$$

The approximate P-value is 0.0901. The observed difference between the wood and the iron is not statistically significant.

(b) $8.69 \pm 1.96(29.11)\sqrt{2/40} = 8.69 \pm 12.76 = [-4.07, 21.45]$. At the 95 percent confidence level, one can conclude that the mean for the wood is as much as 21.45 yards larger than the mean for the iron or the mean for the iron is as much as 4.07 yards larger than the mean for the wood.

11. (a) $s_p = 0.3088$;

$$t = \frac{0.84}{0.1128} = 7.45.$$

Using the t-distribution curve with 28 degrees of freedom indicates that the approximate P-value is smaller than 0.005. The mean time is significantly smaller with the windows closed.

(b) $0.84 \pm 2.048(0.3088)\sqrt{1/15 + 1/15} = 0.84 \pm 0.23 = [0.61, 1.07]$. The population mean time with the windows open is, at the 95 percent level, between 0.61 and 1.07 seconds less than the population mean time with the windows closed.

13. (a)

$$s_p^2 = \frac{(7.05)^2 + (9.83)^2}{2} = 73.17.$$

$$s_p = \sqrt{73.17} = 8.55.$$

The test statistic equals

$$\frac{5.3 + 2.7}{8.55\sqrt{1/20 + 1/20}} = 2.96.$$

The approximate P-value is 0.0015.

(b) $(5.3 + 2.7) \pm 1.96(8.55)\sqrt{1/20 + 1/20} = 8.0 \pm 5.3 = [2.7, 13.3]$.

Sections 16.3 and 16.4

1. (a) The scores on the first treatment are skewed to the right. The scores on the second treatment have one large outlier. The differences have two gaps and two or three large outliers.

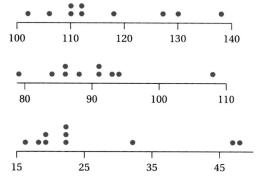

(b) She is correct. The exact one-sided P-value is 0.001.

(c) $$t = \frac{\sqrt{10}(26.5)}{11.87} = 7.06.$$

The approximate P-value is smaller than 0.005.

3. (a) $26.5 \pm (2.262)(11.87)/\sqrt{10} = 26.5 \pm 8.5 = [18.0, 35.0]$.

(b) $116.5 \pm (2.262)(11.56)/\sqrt{10} = 116.5 \pm 8.3 = [108.2, 124.8]$.

(c) $90.0 \pm (2.262)(7.77)/\sqrt{10} = 90.0 \pm 5.6 = [84.4, 95.6]$.

5. (a) $s_p^2 = ((11.56)^2 + (7.77)^2)/2 = 97.00$; $s_p = \sqrt{97.00} = 9.85$. The 95 percent confidence interval is $(116.5 - 90.0) \pm 2.101(9.85)\sqrt{1/10 + 1/10} = 26.5 \pm 9.3 = [17.2, 35.8]$.

(b) The confidence interval for this exercise is wider. No. The RPD appears to be a better design.

7. (a) $t = (\sqrt{5})(20.4)/8.17 = 5.58$. The approximate P-value is smaller than 0.005. The observed difference between using two hands and one hand is highly statistically significant.

(b) $20.4 \pm (2.776)(8.17)/\sqrt{5} = 20.4 \pm 10.1 = [10.3, 30.5]$. The population mean score using one hand minus the population mean score using two hands is between 10.3 and 30.5 pins. This analysis assumes that the differences in the data set are the result of taking a random sample from the population that represents the process that generates the differences. This process is assumed to be stationary (it does not change), and the trials are assumed to be independent.

9. With the suggested analysis, each team in the league will appear the same number of times in the games at Wrigley Field and in the games at the other stadiums. This will help guard against other factors that influence the number of home runs. For example, if the Cubs had many home run hitters, then it would have many home run hitters in its home and road games.

11. (a) Each plot shows a tremendous amount of variation in the number of home runs. The plots are very flat with large gaps in the middle of the distributions.

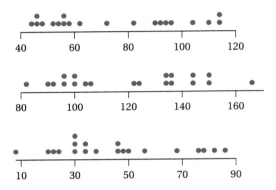

(b) $45.7 \pm (2.093)(22.74)/\sqrt{20} = 45.7 \pm 10.6 = [35.1, 56.3]$.

13. (a) $1.45 \pm (1.96)(18.10)/\sqrt{74} = 1.45 \pm 4.12 = [-2.67, 5.57]$. The confidence interval is correct because it includes 0.

(b) $k' = 29.1$ and $k = 29$. The 95 percent confidence interval is $[x_{(29)}, x_{(46)}] = [-2, 0]$.

APPENDIX

A.1 EXACT P-VALUE FOR FISHER'S TEST

Table A.1 enables you to compute the exact P-value for Fisher's test provided the design is balanced ($n_1 = n_2$) and the common value of n_1 and n_2 is 14 or smaller. In order to use Table A.1, you must determine the values of R, C, and O for your data, using the following algorithm:

1. R is equal to the common value of n_1 and n_2.

2. C is equal to the smaller of m_1 and m_2.

3. For the first alternative ($>$),

   (a) If $m_1 \leq m_2$, then $O = a$; otherwise, $O = d$.

   (b) Locate the entry in the "Prob." column of Table A.1 for the above values of R, C, and O. The entry equals the P-value. If the above values of R, C, and O cannot be located in the table, the P-value is greater than or equal to 0.2500.

   For the second alternative ($<$),

   (a) If $m_1 \leq m_2$, then $O = c$; otherwise, $O = b$.

   (b) Locate the entry in the "Prob." column of Table A.1 for the above values of R, C, and O. The entry equals the P-value. If the above values of R, C, and O cannot be located in the table, the P-value is greater than or equal to 0.2500.

   For the third alternative ($\neq$),

   (a) If $m_1 \leq m_2$, then O equals the larger of a and c; otherwise, O equals the larger of b and d.

   (b) Locate the entry in the "Prob." column of Table A.1 for the above values of R, C, and O. Two times the entry equals the P-value. If the above values of R, C, and O can not be located in the table, the P-value is greater than or equal to 0.5000.

TABLE A.1 Exact P-value for Fisher's Test

R	C	O	Prob.	R	C	O	Prob.	R	C	O	Prob.
2	2	2	0.1667	8	6	6	0.0035	10	9	9	0.0001
				8	6	5	0.0594	10	9	8	0.0027
3	2	2	0.2000	8	7	7	0.0007	10	9	7	0.0349
3	3	3	0.0500	8	7	6	0.0203	10	9	6	0.1849
				8	7	5	0.1573	10	10	10	0.0000
4	2	2	0.2143	8	8	8	0.0001	10	10	9	0.0005
4	3	3	0.0714	8	8	7	0.0051	10	10	8	0.0115
4	4	4	0.0143	8	8	6	0.0660	10	10	7	0.0894
4	4	3	0.2429								
				9	2	2	0.2353	11	2	2	0.2381
5	2	2	0.2222	9	3	3	0.1029	11	3	3	0.1071
5	3	3	0.0833	9	4	4	0.0412	11	4	4	0.0451
5	4	4	0.0238	9	5	5	0.0147	11	5	5	0.0175
5	5	5	0.0040	9	5	4	0.1470	11	5	4	0.1554
5	5	4	0.1032	9	6	6	0.0045	11	6	6	0.0062
				9	6	5	0.0656	11	6	5	0.0743
6	2	2	0.2273	9	7	7	0.0011	11	7	7	0.0019
6	3	3	0.0909	9	7	6	0.0249	11	7	6	0.0317
6	4	4	0.0303	9	7	5	0.1674	11	7	5	0.1807
6	5	5	0.0076	9	8	8	0.0002	11	8	8	0.0005
6	5	4	0.1212	9	8	7	0.0076	11	8	7	0.0119
6	6	6	0.0011	9	8	6	0.0767	11	8	6	0.0913
6	6	5	0.0400	9	9	9	0.0000	11	9	9	0.0001
				9	9	8	0.0017	11	9	8	0.0038
7	2	2	0.2308	9	9	7	0.0283	11	9	7	0.0402
7	3	3	0.0962	9	9	6	0.1735	11	9	6	0.1935
7	4	4	0.0350					11	10	10	0.0000
7	5	5	0.0105	10	2	2	0.2368	11	10	9	0.0010
7	5	4	0.1329	10	3	3	0.1053	11	10	8	0.0150
7	6	6	0.0023	10	4	4	0.0433	11	10	7	0.0992
7	6	5	0.0513	10	5	5	0.0163	11	11	11	0.0000
7	7	7	0.0003	10	5	4	0.1517	11	11	10	0.0002
7	7	6	0.0146	10	6	6	0.0054	11	11	9	0.0045
7	7	5	0.1431	10	6	5	0.0704	11	11	8	0.0431
				10	7	7	0.0015	11	11	7	0.1974
8	2	2	0.2333	10	7	6	0.0286				
8	3	3	0.1000	10	7	5	0.1749	12	2	2	0.2391
8	4	4	0.0385	10	8	8	0.0004	12	3	3	0.1087
8	5	5	0.0128	10	8	7	0.0099	12	4	4	0.0466
8	5	4	0.1410	10	8	6	0.0849	12	5	5	0.0186

TABLE A.1 (continued) Exact P-value for Fisher's Test

R	C	O	Prob.	R	C	O	Prob.	R	C	O	Prob.
12	5	4	0.1584	13	7	5	0.1891	14	7	7	0.0029
12	6	6	0.0069	13	8	8	0.0008	14	7	6	0.0384
12	6	5	0.0775	13	8	7	0.0151	14	7	5	0.1923
12	7	7	0.0023	13	8	6	0.1008	14	8	8	0.0010
12	7	6	0.0343	13	9	9	0.0002	14	8	7	0.0165
12	7	5	0.1854	13	9	8	0.0056	14	8	6	0.1043
12	8	8	0.0007	13	9	7	0.0484	14	9	9	0.0003
12	8	7	0.0136	13	9	6	0.2055	14	9	8	0.0064
12	8	6	0.0965	13	10	10	0.0001	14	9	7	0.0516
12	9	9	0.0002	13	10	9	0.0018	14	9	6	0.2099
12	9	8	0.0047	13	10	8	0.0207	14	10	10	0.0001
12	9	7	0.0447	13	10	7	0.1131	14	10	9	0.0022
12	9	6	0.2002	13	11	11	0.0000	14	10	8	0.0230
12	10	10	0.0000	13	11	10	0.0005	14	10	7	0.1182
12	10	9	0.0014	13	11	9	0.0077	14	11	11	0.0000
12	10	8	0.0180	13	11	8	0.0554	14	11	10	0.0007
12	10	7	0.1069	13	11	7	0.2142	14	11	9	0.0092
12	11	11	0.0000	13	12	12	0.0000	14	11	8	0.0601
12	11	10	0.0003	13	12	11	0.0001	14	11	7	0.2200
12	11	9	0.0061	13	12	10	0.0024	14	12	12	0.0000
12	11	8	0.0894	13	12	9	0.0236	14	12	11	0.0002
12	12	12	0.0000	13	12	8	0.1189	14	12	10	0.0032
12	12	11	0.0001	13	13	13	0.0000	14	12	9	0.0271
12	12	10	0.0017	13	13	12	0.0000	14	12	8	0.1259
12	12	9	0.0196	13	13	11	0.0006	14	13	13	0.0000
12	12	8	0.1102	13	13	10	0.0085	14	13	12	0.0000
				13	13	9	0.0576	14	13	11	0.0009
13	2	2	0.2400	13	13	8	0.2169	14	13	10	0.0107
13	3	3	0.1100					14	13	9	0.0642
13	4	4	0.0478	14	2	2	0.2407	14	13	8	0.2247
13	5	5	0.0196	14	3	3	0.1111	14	14	14	0.0000
13	5	4	0.1609	14	4	4	0.0489	14	14	13	0.0000
13	6	6	0.0075	14	5	5	0.0204	14	14	12	0.0002
13	6	5	0.0801	14	5	4	0.1630	14	14	11	0.0035
13	7	7	0.0026	14	6	6	0.0080	14	14	10	0.0285
13	7	6	0.0365	14	6	5	0.0824	14	14	9	0.1284

A.2 AREA UNDER THE STANDARD NORMAL CURVE TO THE RIGHT OF z

TABLE A.2 Area under the Standard Normal Curve to the Right of z

z	0.00	0.01	0.02	0.03	0.04	0.05	0.06	0.07	0.08	0.09
−3.5	.9998	.9998	.9998	.9998	.9998	.9998	.9998	.9998	.9998	.9998
−3.4	.9997	.9997	.9997	.9997	.9997	.9997	.9997	.9997	.9997	.9998
−3.3	.9995	.9995	.9995	.9996	.9996	.9996	.9996	.9995	.9995	.9995
−3.2	.9993	.9993	.9994	.9994	.9994	.9994	.9994	.9992	.9993	.9993
−3.1	.9990	.9991	.9991	.9991	.9992	.9992	.9992	.9992	.9993	.9993
−3.0	.9987	.9987	.9987	.9988	.9988	.9889	.9989	.9989	.9990	.9990
−2.9	.9981	.9982	.9982	.9983	.9984	.9984	.9985	.9985	.9986	.9986
−2.8	.9974	.9975	.9976	.9977	.9977	.9978	.9979	.9979	.9980	.9981
−2.7	.9965	.9966	.9967	.9968	.9969	.9970	.9971	.9972	.9973	.9974
−2.6	.9953	.9955	.9956	.9957	.9959	.9960	.9961	.9962	.9963	.9964
−2.5	.9938	.9940	.9941	.9943	.9945	.9946	.9948	.9949	.9951	.9952
−2.4	.9918	.9920	.9922	.9925	.9927	.9929	.9931	.9932	.9934	.9936
−2.3	.9893	.9896	.9898	.9901	.9904	.9906	.9909	.9911	.9913	.9916
−2.2	.9861	.9864	.9868	.9871	.9875	.9878	.9881	.9884	.9887	.9890
−2.1	.9821	.9826	.9830	.9834	.9838	.9842	.9846	.9850	.9854	.9857
−2.0	.9772	.9778	.9783	.9788	.9793	.9798	.9803	.9808	.9812	.9817
−1.9	.9713	.9719	.9726	.9732	.9738	.9744	.9750	.9756	.9761	.9767
−1.8	.9641	.9649	.9656	.9664	.9671	.9678	.9686	.9693	.9699	.9706
−1.7	.9554	.9564	.9573	.9582	.9591	.9599	.9608	.9616	.9625	.9633
−1.6	.9452	.9463	.9474	.9484	.9495	.9505	.9515	.9525	.9535	.9545
−1.5	.9332	.9345	.9357	.9370	.9382	.9394	.9406	.9418	.9429	.9441
−1.4	.9192	.9207	.9222	.9236	.9251	.9265	.9279	.9292	.9306	.9319
−1.3	.9032	.9049	.9066	.9082	.9099	.9115	.9131	.9147	.9162	.9177
−1.2	.8849	.8869	.8888	.8907	.8925	.8944	.8962	.8980	.8997	.9015
−1.1	.8643	.8665	.8686	.8708	.8729	.8749	.8770	.8790	.8810	.8830
−1.0	.8413	.8438	.8461	.8485	.8508	.8531	.8554	.8577	.8599	.8621
−0.9	.8159	.8186	.8212	.8238	.8264	.8289	.8315	.8340	.8365	.8389
−0.8	.7881	.7910	.7939	.7967	.7995	.8023	.8051	.8078	.8106	.8133
−0.7	.7580	.7611	.7642	.7673	.7703	.7734	.7764	.7794	.7823	.7852
−0.6	.7257	.7291	.7324	.7357	.7389	.7422	.7454	.7486	.7517	.7549
−0.5	.6915	.6950	.6985	.7019	.7054	.7088	.7123	.7157	.7190	.7224
−0.4	.6554	.6591	.6628	.6664	.6700	.6736	.6772	.6808	.6844	.6879
−0.3	.6179	.6217	.6255	.6293	.6331	.6368	.6406	.6443	.6480	.6517
−0.2	.5793	.5832	.5871	.5910	.5948	.5987	.6026	.6064	.6103	.6141
−0.1	.5398	.5438	.5478	.5517	.5557	.5596	.5636	.5675	.5714	.5753
−0.0	.5000	.5040	.5080	.5120	.5160	.5199	.5239	.5279	.5319	.5359

TABLE A.2 (continued) Area under the Standard Normal Curve to the Right of z

z	0.00	0.01	0.02	0.03	0.04	0.05	0.06	0.07	0.08	0.09
0.0	.5000	.4960	.4920	.4880	.4840	.4801	.4761	.4721	.4681	.4641
0.1	.4602	.4562	.4522	.4483	.4443	.4404	.4364	.4325	.4286	.4247
0.2	.4207	.4168	.4129	.4090	.4052	.4013	.3974	.3936	.3897	.3859
0.3	.3821	.3783	.3745	.3707	.3669	.3632	.3594	.3557	.3520	.3483
0.4	.3446	.3409	.3372	.3336	.3300	.3264	.3228	.3192	.3156	.3121
0.5	.3085	.3050	.3015	.2981	.2946	.2912	.2877	.2843	.2810	.2776
0.6	.2743	.2709	.2676	.2643	.2611	.2578	.2546	.2514	.2483	.2451
0.7	.2420	.2389	.2358	.2327	.2297	.2266	.2236	.2206	.2177	.2148
0.8	.2119	.2090	.2061	.2033	.2005	.1977	.1949	.1922	.1894	.1867
0.9	.1841	.1814	.1788	.1762	.1736	.1711	.1685	.1660	.1635	.1611
1.0	.1587	.1562	.1539	.1515	.1492	.1469	.1446	.1423	.1401	.1379
1.1	.1357	.1335	.1314	.1292	.1271	.1251	.1230	.1210	.1190	.1170
1.2	.1151	.1131	.1112	.1093	.1075	.1056	.1038	.1020	.1003	.0985
1.3	.0968	.0951	.0934	.0918	.0901	.0885	.0869	.0853	.0838	.0823
1.4	.0808	.0793	.0778	.0764	.0749	.0735	.0721	.0708	.0694	.0681
1.5	.0668	.0655	.0643	.0630	.0618	.0606	.0594	.0582	.0571	.0559
1.6	.0548	.0537	.0526	.0516	.0505	.0495	.0485	.0475	.0465	.0455
1.7	.0446	.0436	.0427	.0418	.0409	.0401	.0392	.0384	.0375	.0367
1.8	.0359	.0351	.0344	.0336	.0329	.0322	.0314	.0307	.0301	.0294
1.9	.0287	.0281	.0274	.0268	.0262	.0256	.0250	.0244	.0239	.0233
2.0	.0228	.0222	.0217	.0212	.0207	.0202	.0197	.0192	.0188	.0183
2.1	.0179	.0174	.0170	.0166	.0162	.0158	.0154	.0150	.0146	.0143
2.2	.0139	.0136	.0132	.0129	.0125	.0122	.0119	.0116	.0113	.0110
2.3	.0107	.0104	.0102	.0099	.0096	.0094	.0091	.0089	.0087	.0084
2.4	.0082	.0080	.0078	.0075	.0073	.0071	.0069	.0068	.0066	.0064
2.5	.0062	.0060	.0059	.0057	.0055	.0054	.0052	.0051	.0049	.0048
2.6	.0047	.0045	.0044	.0043	.0041	.0040	.0039	.0038	.0037	.0036
2.7	.0035	.0034	.0033	.0032	.0031	.0030	.0029	.0028	.0027	.0026
2.8	.0026	.0025	.0024	.0023	.0023	.0022	.0021	.0021	.0020	.0019
2.9	.0019	.0018	.0018	.0017	.0016	.0016	.0015	.0015	.0014	.0014
3.0	.0013	.0013	.0013	.0012	.0012	.0011	.0011	.0011	.0010	.0010
3.1	.0010	.0009	.0009	.0009	.0008	.0008	.0008	.0008	.0007	.0007
3.2	.0007	.0007	.0006	.0006	.0006	.0006	.0006	.0005	.0005	.0005
3.3	.0005	.0005	.0005	.0004	.0004	.0004	.0004	.0004	.0004	.0003
3.4	.0003	.0003	.0003	.0003	.0003	.0003	.0003	.0003	.0003	.0002
3.5	.0002	.0002	.0002	.0002	.0002	.0002	.0002	.0002	.0002	.0002

A.3 CENTRAL AREA UNDER THE STANDARD NORMAL CURVE

TABLE A.3 *Selected Central Areas under the Standard Normal Curve*

Central Area (Confidence Level) (Prediction Level)	z
80%	1.282
90%	1.645
95%	1.960
98%	2.326
99%	2.576

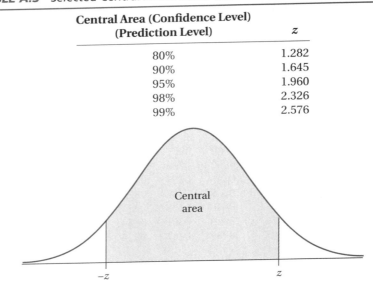

- The area under the standard normal curve between -1.282 and $+1.282$ is 0.80.
- The area under the standard normal curve between -1.645 and $+1.645$ is 0.90.
- The area under the standard normal curve between -1.960 and $+1.960$ is 0.95.
- The area under the standard normal curve between -2.326 and $+2.326$ is 0.98.
- The area under the standard normal curve between -2.576 and $+2.576$ is 0.99.

A.4 EXACT P-VALUE FOR McNEMAR'S TEST

Table A.4 yields the exact P-value for McNemar's test, for the test of $H_0: p = 0.5$ of Chapter 6, and for the test of the value of the median of Chapter 15. To use Table A.4, you must first determine R and O for your data.

- For McNemar's test, $R = b + c$ and $O = b$.
- For the tests of Chapters 6 and 15, R equals the sample size and O equals the number of successes.

TABLE A.4 Exact P-values for McNemar's Test

R	O	Alternative >	<	≠	R	O	Alternative >	<	≠	R	O	Alternative >	<	≠
1	0	1.000	0.500	1.000	8	2	0.965	0.145	0.289	12	0	1.000	0.000	0.000
	1	0.500	1.000	1.000		3	0.855	0.363	0.727		1	1.000	0.003	0.006
2	0	1.000	0.250	0.500		4	0.637	0.637	1.000		2	0.997	0.019	0.039
	1	0.750	0.750	1.000		5	0.363	0.855	0.727		3	0.981	0.073	0.146
	2	0.250	1.000	0.500		6	0.145	0.965	0.289		4	0.927	0.194	0.388
3	0	1.000	0.125	0.250		7	0.035	0.996	0.070		5	0.806	0.387	0.774
	1	0.875	0.500	1.000		8	0.004	1.000	0.008		6	0.613	0.613	1.000
	2	0.500	0.875	1.000							7	0.387	0.806	0.774
	3	0.125	1.000	0.250	9	0	1.000	0.002	0.004		8	0.194	0.927	0.388
4	0	1.000	0.062	0.125		1	0.998	0.020	0.039		9	0.073	0.981	0.146
	1	0.938	0.312	0.625		2	0.980	0.090	0.180		10	0.019	0.997	0.039
	2	0.688	0.688	1.000		3	0.910	0.254	0.508		11	0.003	1.000	0.006
	3	0.312	0.938	0.625		4	0.746	0.500	1.000		12	0.000	1.000	0.000
	4	0.062	1.000	0.125		5	0.500	0.746	1.000					
5	0	1.000	0.031	0.062		6	0.254	0.910	0.508	13	0	1.000	0.000	0.000
	1	0.969	0.188	0.375		7	0.090	0.980	0.180		1	1.000	0.002	0.003
	2	0.812	0.500	1.000		8	0.020	0.998	0.039		2	0.998	0.011	0.022
	3	0.500	0.812	1.000		9	0.002	1.000	0.004		3	0.989	0.046	0.092
	4	0.188	0.969	0.375							4	0.954	0.133	0.267
	5	0.031	1.000	0.062	10	0	1.000	0.001	0.002		5	0.867	0.291	0.581
6	0	1.000	0.016	0.031		1	0.999	0.011	0.021		6	0.709	0.500	1.000
	1	0.984	0.109	0.219		2	0.989	0.055	0.109		7	0.500	0.709	1.000
	2	0.891	0.344	0.688		3	0.945	0.172	0.344		8	0.291	0.867	0.581
	3	0.656	0.656	1.000		4	0.828	0.377	0.754		9	0.133	0.954	0.267
	4	0.344	0.891	0.688		5	0.623	0.623	1.000		10	0.046	0.989	0.092
	5	0.109	0.984	0.219		6	0.377	0.828	0.754		11	0.011	0.998	0.022
	6	0.016	1.000	0.031		7	0.172	0.945	0.344		12	0.002	1.000	0.003
						8	0.055	0.989	0.109		13	0.000	1.000	0.000
7	0	1.000	0.008	0.016		9	0.011	0.999	0.021					
	1	0.992	0.062	0.125		10	0.001	1.000	0.002	14	0	1.000	0.000	0.000
	2	0.938	0.227	0.453							1	1.000	0.001	0.002
	3	0.773	0.500	1.000	11	0	1.000	0.000	0.001		2	0.999	0.006	0.013
	4	0.500	0.773	1.000		1	1.000	0.006	0.012		3	0.994	0.029	0.057
	5	0.227	0.938	0.453		2	0.994	0.033	0.065		4	0.971	0.090	0.180
	6	0.062	0.992	0.125		3	0.967	0.113	0.227		5	0.910	0.212	0.424
	7	0.008	1.000	0.016		4	0.887	0.274	0.549		6	0.788	0.395	0.791
						5	0.726	0.500	1.000		7	0.605	0.605	1.000
8	0	1.000	0.004	0.008		6	0.500	0.726	1.000		8	0.395	0.788	0.791
	1	0.996	0.035	0.070		7	0.274	0.887	0.549		9	0.212	0.910	0.424
						8	0.113	0.967	0.227		10	0.090	0.971	0.180
						9	0.033	0.994	0.065		11	0.029	0.994	0.057
						10	0.006	1.000	0.012		12	0.006	0.999	0.013
						11	0.000	1.000	0.001		13	0.001	1.000	0.002
											14	0.000	1.000	0.000

A.5 PERCENTAGE POINTS OF χ^2 DISTRIBUTIONS

TABLE A.5 Percentage Points of χ^2 Distributions

df	\multicolumn{9}{c}{Area to the Right}
	0.99	0.975	0.95	0.90	0.50	0.10	0.05	0.025	0.01
1	.0002	.001	.004	.02	.45	2.71	3.84	5.02	6.63
2	.02	.05	.10	.21	1.39	4.61	5.99	7.38	9.21
3	.11	.22	.35	.58	2.37	6.25	7.81	9.35	11.34
4	.30	.48	.71	1.06	3.36	7.78	9.49	11.14	13.28
5	.55	.83	1.15	1.61	4.35	9.24	11.07	12.83	15.09
6	.87	1.24	1.64	2.20	5.35	10.64	12.59	14.45	16.81
7	1.24	1.69	2.17	2.83	6.35	12.02	14.07	16.01	18.48
8	1.65	2.18	2.73	3.49	7.34	13.36	15.51	17.53	20.09
9	2.09	2.70	3.33	4.17	8.34	14.68	16.92	19.02	21.67
10	2.56	3.25	3.94	4.87	9.34	15.99	18.31	20.48	23.21
11	3.05	3.82	4.57	5.58	10.34	17.28	19.68	21.92	24.72
12	3.57	4.40	5.23	6.30	11.34	18.55	21.03	23.34	26.22
13	4.11	5.01	5.89	7.04	12.34	19.81	22.36	24.74	27.69
14	4.66	5.63	6.57	7.79	13.34	21.06	23.68	26.12	29.14
15	5.23	6.26	7.26	8.55	14.34	22.31	25.00	27.49	30.58
16	5.81	6.91	7.96	9.31	15.34	23.54	26.30	28.85	32.00
17	6.41	7.56	8.67	10.09	16.34	24.77	27.59	30.19	33.41
18	7.01	8.23	9.39	10.86	17.34	25.99	28.87	31.53	34.81
19	7.63	8.91	10.12	11.65	18.34	27.20	30.14	32.85	36.19
20	8.26	9.59	10.85	12.44	19.34	28.41	31.41	34.17	37.57
21	8.90	10.28	11.59	13.24	20.34	29.62	32.67	35.48	38.93
22	9.54	10.98	12.34	14.04	21.34	30.81	33.92	36.78	40.29
23	10.20	11.69	13.09	14.85	22.34	32.01	35.17	38.08	41.64
24	10.86	12.40	13.85	15.66	23.34	33.20	36.42	39.36	42.98
25	11.52	13.12	14.61	16.47	24.34	34.38	37.65	40.65	44.31
26	12.20	13.84	15.38	17.29	25.34	35.56	38.89	41.92	45.64
27	12.88	14.57	16.15	18.11	26.34	36.74	40.11	43.19	46.96
28	13.56	15.31	16.93	18.94	27.34	37.92	41.34	44.46	48.28
29	14.26	16.05	17.71	19.77	28.34	39.09	42.56	45.72	49.59
30	14.95	16.79	18.49	20.60	29.34	40.26	43.77	46.98	50.89

A.6 PERCENTAGE POINTS OF t-DISTRIBUTIONS

TABLE A.6 Percentage Points of t-distributions

df	Area to the Right					
	0.25	0.10	0.05	0.025	0.01	0.005
1	1.000	3.078	6.314	12.706	31.821	63.657
2	.816	1.886	2.920	4.303	6.965	9.925
3	.765	1.638	2.353	3.182	4.541	5.841
4	.741	1.533	2.132	2.776	3.747	4.604
5	.727	1.476	2.015	2.571	3.365	4.032
6	.718	1.440	1.943	2.447	3.143	3.707
7	.711	1.415	1.895	2.365	2.998	3.499
8	.706	1.397	1.860	2.306	2.896	3.355
9	.703	1.383	1.833	2.262	2.821	3.250
10	.700	1.372	1.812	2.228	2.764	3.169
11	.697	1.363	1.796	2.201	2.718	3.106
12	.695	1.356	1.782	2.179	2.681	3.055
13	.694	1.350	1.771	2.160	2.650	3.012
14	.692	1.345	1.761	2.145	2.624	2.977
15	.691	1.341	1.753	2.131	2.602	2.947
16	.690	1.337	1.746	2.120	2.583	2.921
17	.689	1.333	1.740	2.110	2.567	2.898
18	.688	1.330	1.734	2.101	2.552	2.878
19	.688	1.328	1.729	2.093	2.539	2.861
20	.687	1.325	1.725	2.086	2.528	2.845
21	.686	1.323	1.721	2.080	2.518	2.831
22	.686	1.321	1.717	2.074	2.508	2.819
23	.685	1.319	1.714	2.069	2.500	2.807
24	.685	1.318	1.711	2.064	2.492	2.797
25	.684	1.316	1.708	2.060	2.485	2.787
26	.684	1.315	1.706	2.056	2.479	2.779
27	.684	1.314	1.703	2.052	2.473	2.771
28	.683	1.313	1.701	2.048	2.467	2.763
29	.683	1.311	1.699	2.045	2.462	2.756
30	.683	1.310	1.697	2.042	2.457	2.750
40	.681	1.303	1.684	2.021	2.423	2.704
60	.679	1.296	1.671	2.000	2.390	2.660
120	.677	1.289	1.658	1.980	2.358	2.617
∞	.674	1.282	1.645	1.960	2.326	2.576
Confidence level		80%	90%	95%	98%	99%

A.7 CONFIDENCE INTERVALS FOR THE POPULATION MEDIAN

TABLE A.7 Confidence Intervals for the Population Median

n	Confidence Interval	Confidence Level	n	Confidence Interval	Confidence Level
2	$[x_{(1)}, x_{(2)}]$	50.0%	14	$[x_{(2)}, x_{(13)}]$	99.8%
			14	$[x_{(3)}, x_{(12)}]$	98.7%
3	$[x_{(1)}, x_{(3)}]$	75.0%	14	$[x_{(4)}, x_{(11)}]$	94.3%
			14	$[x_{(5)}, x_{(10)}]$	82.0%
4	$[x_{(1)}, x_{(4)}]$	87.5%			
			15	$[x_{(3)}, x_{(13)}]$	99.3%
5	$[x_{(1)}, x_{(5)}]$	93.8%	15	$[x_{(4)}, x_{(12)}]$	96.5%
			15	$[x_{(5)}, x_{(11)}]$	88.2%
6	$[x_{(1)}, x_{(6)}]$	96.9%			
6	$[x_{(2)}, x_{(5)}]$	78.1%	16	$[x_{(3)}, x_{(14)}]$	99.6%
			16	$[x_{(4)}, x_{(13)}]$	97.9%
7	$[x_{(1)}, x_{(7)}]$	98.4%	16	$[x_{(5)}, x_{(12)}]$	92.3%
7	$[x_{(2)}, x_{(6)}]$	87.5%	16	$[x_{(6)}, x_{(11)}]$	79.0%
8	$[x_{(1)}, x_{(8)}]$	99.2%	17	$[x_{(3)}, x_{(15)}]$	99.8%
8	$[x_{(2)}, x_{(7)}]$	93.0%	17	$[x_{(4)}, x_{(14)}]$	98.7%
			17	$[x_{(5)}, x_{(13)}]$	95.1%
9	$[x_{(1)}, x_{(9)}]$	99.6%	17	$[x_{(6)}, x_{(12)}]$	85.7%
9	$[x_{(2)}, x_{(8)}]$	96.1%			
9	$[x_{(3)}, x_{(7)}]$	82.0%	18	$[x_{(4)}, x_{(15)}]$	99.2%
			18	$[x_{(5)}, x_{(14)}]$	96.9%
10	$[x_{(1)}, x_{(10)}]$	99.8%	18	$[x_{(6)}, x_{(13)}]$	90.4%
10	$[x_{(2)}, x_{(9)}]$	97.9%			
10	$[x_{(3)}, x_{(8)}]$	89.1%	19	$[x_{(4)}, x_{(16)}]$	99.6%
			19	$[x_{(5)}, x_{(15)}]$	98.1%
11	$[x_{(1)}, x_{(11)}]$	99.9%	19	$[x_{(6)}, x_{(14)}]$	93.6%
11	$[x_{(2)}, x_{(10)}]$	98.8%	19	$[x_{(7)}, x_{(13)}]$	83.3%
11	$[x_{(3)}, x_{(9)}]$	93.5%			
			20	$[x_{(4)}, x_{(17)}]$	99.7%
12	$[x_{(2)}, x_{(11)}]$	99.4%	20	$[x_{(5)}, x_{(16)}]$	98.8%
12	$[x_{(3)}, x_{(10)}]$	96.1%	20	$[x_{(6)}, x_{(15)}]$	95.9%
12	$[x_{(4)}, x_{(9)}]$	85.4%	20	$[x_{(7)}, x_{(14)}]$	88.5%
13	$[x_{(2)}, x_{(12)}]$	99.7%			
13	$[x_{(3)}, x_{(11)}]$	97.8%			
13	$[x_{(4)}, x_{(10)}]$	90.8%			

INDEX

A

Addition rule, 61–62
  and multiplication rule, 156
  and sampling distribution, 62
  and test statistic, 62
  Venn diagram, 61
Advocate's Argument, 40
AIDS-IP study
  collapsed table and row proportions, 121
  Fisher's test to analyze RBD, 123–25
  MH test to calculate P-value, 127
  as randomized block design (RBD), 116–19
Alternative hypotheses, 42–46
  choosing, 42–46, 240
  choosing for two dichotomous responses, 274–75
  defined for chi-squared goodness of fit test, 335
  defined for two population proportions, 240
Arithmetic average, 292
  defined, 108–9, 291
  weighted average, 108–9, 292
Autocorrelation, 507–10
  function for a time series, 566–69
  lags, 509

B

Balanced design, 7
Bar chart, 14–15
Batting averages study, 442–47
  correlation coefficient calculated, 454
  data, 442–44
  frequency histograms, 444
  notation for, 444
  players' names, 442–44
  regression line, 470–72
  residual, 481–82
  scatterplot, 446, 471
  summary statistics, 445
Best prediction line. *See* Regression line
Bell-shaped distribution, 422–25
Bell-shaped dot plot, 382
Bernoulli trials, 167–76
  assumptions for, 167–70, 289, 324
  bivariate, 272–73
  defined, 324
  false assumptions, 174–76
  independent trials, 168–70
  observational study, 245
Binomial (sampling or probability) distribution, 157–64
  defined, 157–58
  notation, 159
  probability histograms, 162–64
Bird, Larry. *See* Free throw study
Birthday coincidences (as example of random sampling), 180–82
Blind, 19–20
Blocks, 115
Bonferroni's result, 224–26
Box plot, 416–18
Brute force, 148, 155

C

Canada geese study, 342–45
Catcher study, 134–36
Cell, defined, 18
Census, 146
Central limit theorem, 545
Chance mechanism, 51–53
Chi-squared curves, 328–33
Chi-squared goodness of fit test, 334–47
  computation of E, 335–37
  explained as hypothesis test, 334–42
Chi-squared test of homogeneity, 351–67
Chronic Crohn's Disease study
  alternative hypotheses, choice of, 44–46
  confidence interval, 237
  data, 234
  follow-up, 22
  medical study, 20–21, 116
  point estimate, 236–37
  practical importance, 248
  probability histogram, 89–91, 95–96
  P-value, 71, 101, 106, 241
  random sample assumption, 251–52
  table of row proportions, 20
Class intervals, 386–87, 390
Class study, 51–55
Coefficient of determination, 477–78
  time series, 508
Collapsed tables, 121–22, 138–39, 289–300
Colloquium study
  confidence interval, 188
  CRD, 6
  experimental design, 3–4
  Fisher's test, 39–40
  point estimate of P-value, 71, 187
  probability histogram, 87, 89–92, 95
  P-value, 106
  randomization, 6–10

Colloquium study *(continued)*
  rule of evidence, 65–66
  sampling methods, 6–10, 145–46
  simulation experiment, 85–87
  table of row proportions, 14
  test statistic, 50
  2 × 2 contingency table, 14
Complement rule, 62
Completely randomized design (CRD)
  defined, 6
  means compared, 580–87
    alternatives, 581
    null hypothesis, 580–81
    P-value
      approximation by simulation experiment, 583–84
      formulas, 583–84
    test statistic formula, 581–82
  RPD compared, 129
Conditional probabilities, 259–67, 269–74
Confidence intervals
  analyses, number of, 220–26
  comparing conditional probabilities, 276–78
  confidence level, choice of, 202
  dependent and independent analyses, 224–226
  derivation of, 189–91
  general guideline, 278
  half-width, 198–202
  interpretation of, 188–89, 195–98, 220–26, 284
  mean difference, 547–50, 608
  median, 533–40
  multicategory response, 325–28
  one proportion, 187–92
  terminology, 201–2
  two proportions, 235–38, 281–84
  width, 198–202
  z, 191–92
Contingency table
  format 1, 17–18
  format 2, 137
  format 3, 272
  2 × 2 defined, 13
Continuity correction, 107, 422
Control group, 18–19
Controlled study, 229–31
Correct negatives and positives. *See* Screening tests
Correlation coefficient, 442, 450–56
Count response, 522–25

CRD. *See* Completely randomized design
Cricket study, 465–67, 469–73, 476

D

Data, defined, 3
Dating Study
  confidence interval, 237
  hypothesis testing, 241
  independent random samples (assumption of), 251
  observational study on finite populations, 230
  point estimate of $p_1 - p_2$, 236
  practical importance, 248
  results, 233, 245–46
  two population study, 245–46
Degrees of freedom (df)
  goodness of fit test, 346–47
  test of homogeneity, 356
Density scale histogram
  area, 387, 389
  defined, 387
  estimate of PDF, 526–28
  use of, 389, 395–97
Dependent analyses, 220–21, 223–24
Dependent events, 265
Descriptive statistics, defined, 3
Deviation, 111, 418–20
Dichotomous box, 152
Dichotomous responses, 6–7, 257–84
  adjusting for a factor, 294
  conditional probability, 259–67
  screening tests, 259–64
Direct relationships, 447, 452, 454
Disjoint events, 61
Distribution, 380–81
Dot plot
  characteristics, 381–85
  compared to histogram, 390
  defined, 381
Double blind study, 20–21

E

$E(X)$, 164, 190. *See also* Mean
  goodness of fit test, 336–37
  test of homogeneity, 354–55
Elder abuse study
  collapsed tables, 307
  component tables, 306, 308
  hypothesis test, 304–9
  interaction graph, 308

Empirical rule, 422–25
Empty event, 60–61
Equally likely case, 52, 168
Error
  estimated standard error, 201–2
  false negative and false positive. *See* Screening tests
Error bounds, 106, 190, 201
ESP testing
  hypothesis testing, 203–11
  hypothesis testing for independent analyses, 221–23
Event
  defined, 51
  empty or impossible, 60–61
  notation, 51
Expected value, defined, 90
Experimental design, 3–5, 145

F

Factorial, 159–60
Factors
  adjusting for, 289–309
  determination of, 8–9
  influence on response, 8–9
False negative and false positive. *See* Screening tests
Familiarity effect, 134
Finite population
  defined, 145–46, 322
  practical considerations of random sampling, 178–80
  standard terminology, 146–47
  study of, 145, 205
Fisher, Sir Ronald A., 40, 365. *See also* Fisher's test
Fisher's test, 39–72. *See also* Advocate's Argument; Alternative hypotheses; Null hypothesis; P-value; Rule of evidence; Sampling distribution; Skeptic's Argument; Test statistic
  alternative choices, 126
  blocks, 138–39
  chance mechanisms, 55–56
  Mantel-Haenszel test, 123–26
  schematic representation, 56
  step 1, 42–47
  step 2, 49–57
  step 3, 65–67
  step 4, 68–72

Focusing-on-a-winner fallacy, 222
Follow-up, 22
Fourth alternative, 47
Free throw study
   collapsing tables, 289–94
   data tables, 290
   standardized rate, 299
   three strategies, 304
Frequency histogram, 387, 389

G

Generalizations in studies, 145–46
General guidelines
   chi-squared curve approximation, 338, 356
   with or without replacement, 150
   SNC approximation, 106–7, 240
General's dilemma, 5–6
Gosset, 545–46
Gould, Stephen Jay, 496–98
Granular distribution, 381, 430

H

Half-width (h)
   confidence interval values, 198–202
   formula, 199
   and p, 199–201
   and sample size, 200
Height study, dot plots, 382, 383
High temperature forecast study
   autocorrelation of lag 1, 510
   Bernoulli trials, 172–74, 231–32
   confidence interval, 188, 238
   correlation coefficient, 503
   data, 235
   dot plot, 504
   observational study, 231–32
   point estimate, 186, 236, 238
   point predictions, 218
   practical importance, 248
   P-value approximation, 242
   scatterplot, 510
   smoothing, 511–14
   standard deviation, 420–25
   time series plots, 503–4
Histogram
   characteristics, 389–92
   choice of, 389
   defined, 385
   density scale. *See* Density scale histogram
   obtaining, 386–92
   probability. *See* Probability histogram
   relative frequency. *See* Relative frequency
Hypothesis testing
   analyses
      dependent, 223–24
      independent, 221–23
   Fisher's test, 39–72
   mean, 550–54
   numerical response RPD, 603–607
   special value of interest p_0, 203–13
   steps of, 42
   two dichotomous responses, 274–75
   two proportions
      adjusting for a factor, 300–309
      Bernoulli trials, 239–45
      finite, 239–45

I

Identically distributed, 153
Inconceivable, 44
Independence of trials, 242–43
Independent analyses, 220, 221–23
Independent events, 264
Independent random samples (assumption of), 233, 251–52, 587–98
Independent random variables, 157, 264–65
Indirect relationships, 447, 449–50, 454
Inferential statistics, defined, 3
Infidelity study, 11
   collapsed table, 122
   Mantel-Haenszel test, 129–30
   probability histograms, 79–80, 88–91
   probability rules, 62–63
   P-values, 69, 71, 106, 129–30
   randomized block design, 119–22
   table of row proportions, 16, 122
   test statistic, 91–93
   2 × 2 contingency table, 16
Infinite population, 167, 322
Interaction graph, 119, 295–96
Interquartile range, 415–16, 418
Invisible column solution for Table A.5, 329–33
Isolated cases, 446–47, 485–89

L

Lag, 509
Least squares prediction line, 470. *See also* Regression line

Linear function, 439–40
Literary Digest poll, 147–48, 252
Lognormal PDFs, 535–38, 561

M

Mantel-Haenszel test (MH test)
   adjusting for a factor, 300
   alternatives, 126
   blocks combined, 124
   Fisher's test and, 124–28
   hypotheses, 126–27
   inappropriate usage, 138–39
   McNemar's test, 140, 212–13
   misleading conclusions, 129–30
   notation, 124–27
   null hypothesis, 42, 126
   P-value, 129–30
   RBD analyzed, 123–24
   rule of evidence, 128–29
   table calculations, 127
   test statistic, 127
   variance, 125
McNemar's test, 140, 212–13
Mean, 90–92, 399–401
   binomial distributions, 164
   formula, 108–9
   pdf, 528–29
   probability histogram, 90, 110
   weighted averages, 108–9
Median, 533–41
   hypothesis test for, 540–41
Medical studies, 18–20, 116
Mendel, Gregor Johann, and pea study, 325–26, 334–39
Metasubject, 30–31
MH test. *See* Mantel-Haenszel test
Multicategory data, 294, 315–21
Multinomial trials, 324–25
Multiplication rule, 155–57, 168, 170, 264
   for conditional probabilities, 267

N

Normal curves, 530. *See also* Standard normal curve
Normal populations, 587–93
Null hypothesis
   assuming true, 53
   matched pairs of subpopulations, 300
   notation, 42
   power, statistical, 248–51

Normal hypothesis *(continued)*
  rejection of, 71, 249–51
  test of homogeneity, 353–54
  two dichotomous responses, 274–75
  two populations, 239–40
Numerical variable, 379–80

O

O
  goodness of fit test, 335–39
  test of homogeneity, 354–55
Observational studies, 229–31
  adjusting for a factor, 294–300
  not CRD, 12, 32
Observed counts, 335
Ordered multicategory response, 318–21
Outliers
  causes of, 482–83
  defined, 385–86
  effects of, 425

P

Paired data, 602–610
Pairwise comparisons, 358–59, 363–65
Pearson, Karl, 365
Percentage points, defined, 14
Pitcher victory study, 505–7, 509, 512–13
Placebo, 19
Placebo effect, 19
Point estimate, 186–87, 189, 235–38, 283–84, 326
Point estimator, 189
Point prediction, 215–19
Pooled sample standard deviation, 584, 591
Pooled sample variance, 583–84, 588
Population
  defined, 145, 521–22
  estimation of, 526–30
  finite, defined, 145–46
  infinite, defined, 146
Population box, 146–49, 151–52
Population mean. *See* Mean
Population median. *See* Median
Population models
  finite population, two dichotomous variables, 257
  multiple populations, 229–52, 359–61

one population, two responses, 362–65
  paired data, 607–610
Practical importance, 247–48
Prediction interval, 215–19, 570–74
Prediction line, 465–70. *See also* Regression line
Prednisone-Duchenne's Muscular Dystrophy study (P-DMD study), 352–58
Principle of least squares, 468–69
Prisoner study, 11–12, 16–17, 71–72, 106
Probability
  chance mechanism, 51
  computations of
    random sample with replacement, 152
    two dichotomous responses, 258
    two random variables, 152–60
    estimated, 269–74
  Fisher's test, 50–51
  interpretation of, 259–61
  notation, 55, 323–24
  rules, 60–63
Probability density function (pdf), 525–26
Probability histogram, 79–82, 88–93
P-value
  approximation of
    chi-squared curve, 338, 340
    consequences of, 107
    t-distribution, 551–54, 582, 584–87, 606, 609
    simulation, 86, 105–7, 606
    standard normal curve, 101–3, 105–7, 123–24, 210–11, 240–41, 275–76, 303
  defined, 43, 49, 68–69
  exact solution, 75–76, 105–7, 209
  Fisher's test, 65–69
  interpretation of, 70–72, 130
  inverse rule, as an, 70
  Mantel-Haenszel test, 127–30
  McNemar's test, 140
  notation, 68, 70–72
  rounding, 71
  sampling distribution, 93–94
  statistical significance, 70, 72

Q

q, point estimate of, 186

R

Random assignment. *See* Randomization
Randomization
  medical studies, 18–20
  method, 6, 7–11
  purpose, 8–9, 18–20, 31
Randomized block design, 115–40
  block comparisons, 119–20
  collapsed table, 121–22
  Fisher's test on blocks, 123–24
  interaction graphs, 117, 119
  medical studies, 115–17
Randomized pairs design (RPD), 134–40
  CRD, 134
  defined, 134
  notation, 136, 138
  numerical response, 602–607
  outcomes, 136
Random sampling
  with replacement, 148–50
  without replacement, 148–50
Random variable, defined, 56
Range, 414–15
RBD. *See* Randomized block design
Regression effect, 473–75
Regression line, 470–73
Relative frequency
  distribution, 316–18
  histogram, 386–91
  vs. probability, 59–60
  table, 386
Residual, 481–84
Residual plot, 484–85
Response
  categories, 352–53
  defined, 5–6
  dichotomous, 6, 315–16
  multicategory, 7, 315–16
  numerical, 7, 315–16
  prediction of, 442
Robustness, 532, 538–40, 555–62
Rolling Stones study
  density scale histogram, 388
  distribution, 386, 389
  dot plot, 386
  frequency histogram, 387, 390
  outliers, 385–86
  relative frequency histogram, 388
Row proportions, table of, 14
Rule of evidence
  alternatives (1, 2, 3), 65–67, 128–29
  chi-squared goodness of fit test, 338

Rule of evidence *(continued)*
  Fisher's test, 65–67
  special value of interest p_0, 205–7
Rules of probability, 60–63
Running median of three smoother, 511
Runs, 83–87

S

Sample, 146, 147–48
  interquartile range, 415–18
  mean, 399–401, 544–45
  median, 400–401
  proportion of collapsed table, 292–93
  range, 414–15
  size, 14, 149–50
  space, 51
  standard deviation, 418–25, 545
Sample size, defined, 14
Sample space, defined, 51
Sampling
  replacement, with and without, 148–50
  studies, four types of, 232–33
Sampling distribution
  Bernoulli trials, 167–70
  chi-squared goodness of fit test, 335, 337
  determination of, methods, 74–76
  formula, 159–60
  of sample mean, 544–45
  of test statistic, 54–57, 75, 79–112
Scatterplot
  analysis, 445–50
  correlation coefficient, 445
  counting method, 452
  defined, 445
  residuals, 481–85
Screening tests, 259–64
Sensitivity (of a screening test), 264
Sequence of trials
  bivariate Bernoulli trials, assumptions of, 272–73
  RPD, 134
  subjects, 28
  time trend, 31
SI assumption, 565
Simpson's paradox, 289–94
Simulation experiment, 83–87, 150–51, 250–51
Skeptic's Argument, 39–40
Skewed to right or left, 388

Slutsky, 191, 239, 545
Smoothing, 511–14
Special value of interest p_0, 204, 211–13
Specificity of a screening test, 264
Spider study, 448–50
Spread, 90–91
SSE (sum of squared errors), 468, 476–77, 484
SSTO, 444, 476–77
Standard deviation, 90–93, 110–12, 164, 189–90
Standard error, 190
Standardization of a random variable, 90–93
Standardized rates, 297–300
Standard normal curve
  binomial distribution, approximation of, 189–91
  characteristics of, 96–100, 530
  general guideline for approximation, 191, 275
  probability, approximation of, 163–65
  probability histogram, approximation of, 163–65
  P-value approximation, 129
Statistical power, 248–51
Statistical significance, defined, 70
Stem and leaf plot, 392–93
Strategies (analysis), 303–4
  first, 300–301
  second, 301
  third, 301–3. *See also* Mantel-Haenszel test
Student (Gosset), 545–46
Student height study, 380–83
Studies, 5, 229–32
Subjects
  defined, 5
  dichotomous variables, 257–58
  types of, 40–41, 51, 212–13
  numerical variables, 439–98
Subpopulations, 300
Superpopulation, 230–31, 243
Survey, 146–51, 179
Symmetry, probability histograms, 81, 88, 90, 162–65

T

Table (appendix), notation and use of
  A.1, 75–76, 240
  A.2, 97–101

A.3, 236
A.4, 140, 212–13, 283
A.5, 329–33
A.6, 547–54
A.7, 535
Table of observed frequencies, 270
Table of population counts, 258
Table of population proportions (probabilities), 258–59
Table of probabilities, 267
Table of row proportions, 14, 23
Table of relative frequencies, 270–71
t-curve, 546–62
Test of homogeneity, 353–61
Test statistic
  approximation by chi-square curve, 355–56
  defined, 50, 204–5
  hypothesis testing, 50–57, 65–67
  interpretation of value, 50
  notation, 50, 124
  P-value, 68–69, 209–11
  sampling distribution, 55–56
  special value of interest p_0, 204
  test of homogeneity, 355–56
Three-point basket study
  alternative hypotheses, choice of, 43–44
  Bernoulli trials, 231
  CRD, 28–29
  $p_1 - p_2$, 236, 238
  practical importance, 247–48
  P-value, 71, 106, 241
  results, 234
  statistical power, 248–51
Tied subjects, 381
Time order, 501
Time series
  defined, 501
  dot plot, 504, 505
  inference for, 564–69
  SI assumption, 565–66
Time series plot
  analysis, 502–5, 507
  correlation coefficient, 503–5
Time trend, 31, 134
Treatments
  defined, 5
  subject types, 40–41
  test of homogeneity, 354
Trials. *See* Sequence of trials
Two points in time, 243–45

Two population means
  case 1 (normal populations with equal variances), 588–91
  case 2 (normal populations), 592–93
  case 3 (large sample approximation), 594–97
  compared, 597–98

U

Unconditional probability, 264
Unordered multicategory response, 318–19

V

Variables, 146, 257–58, 442
Variance, Var (X), 90, 110

W

Weighted average, 108–9, 291, 298–99, 584
Weights, 108–9, 291
"What happens next" paradigm, 44
Whiskers, 416
Wisconsin Driver Survey
  collapsed tables, 295
  confidence intervals, 188, 196, 326–28
  data, 315–9, 359–65
  dependent analyses, 220–21
  finite population model with one multicategory response, 317, 323
  half-width, 198–200
  hypothesis testing for two dichotomous responses, 275–76
  point estimates, 186, 270–72, 326–27
  random sample assumption, 251
  relative frequency distribution, 270, 298, 316–18
  standardized rates, 298–300
  survey example, 145–46
  test of homogeneity, 359–65
Wisconsin state lottery, 319–21, 345–46
World series games, 384–85, 390–91

Z

z
  formula, 102–3, 240
  Mantel-Haenszel test, 123–24
  P-value computation, 211
  standardized variable, 92–94
  standardized version of test statistic, 212–13

Percentage Points of χ^2 Distributions

df	\multicolumn{8}{c}{Area to the Right}
	0.99	0.975	0.95	0.90	0.50	0.10	0.05	0.025	0.01
1	.0002	.001	.004	.02	.45	2.71	3.84	5.02	6.63
2	.02	.05	.10	.21	1.39	4.61	5.99	7.38	9.21
3	.11	.22	.35	.58	2.37	6.25	7.81	9.35	11.34
4	.30	.48	.71	1.06	3.36	7.78	9.49	11.14	13.28
5	.55	.83	1.15	1.61	4.35	9.24	11.07	12.83	15.09
6	.87	1.24	1.64	2.20	5.35	10.64	12.59	14.45	16.81
7	1.24	1.69	2.17	2.83	6.35	12.02	14.07	16.01	18.48
8	1.65	2.18	2.73	3.49	7.34	13.36	15.51	17.53	20.09
9	2.09	2.70	3.33	4.17	8.34	14.68	16.92	19.02	21.67
10	2.56	3.25	3.94	4.87	9.34	15.99	18.31	20.48	23.21
11	3.05	3.82	4.57	5.58	10.34	17.28	19.68	21.92	24.72
12	3.57	4.40	5.23	6.30	11.34	18.55	21.03	23.34	26.22
13	4.11	5.01	5.89	7.04	12.34	19.81	22.36	24.74	27.69
14	4.66	5.63	6.57	7.79	13.34	21.06	23.68	26.12	29.14
15	5.23	6.26	7.26	8.55	14.34	22.31	25.00	27.49	30.58
16	5.81	6.91	7.96	9.31	15.34	23.54	26.30	28.85	32.00
17	6.41	7.56	8.67	10.09| 16.34	24.77	27.59	30.19	33.41
18	7.01	8.23	9.39	10.86| 17.34	25.99	28.87	31.53	34.81
19	7.63	8.91	10.12	11.65	18.34	27.20	30.14	32.85	36.19
20	8.26	9.59	10.85	12.44	19.34	28.41	31.41	34.17	37.57
21	8.90	10.28	11.59	13.24	20.34	29.62	32.67	35.48	38.93
22	9.54	10.98	12.34	14.04	21.34	30.81	33.92	36.78	40.29
23	10.20	11.69	13.09	14.85	22.34	32.01	35.17	38.08	41.64
24	10.86	12.40	13.85	15.66	23.34	33.20	36.42	39.36	42.98
25	11.52	13.12	14.61	16.47	24.34	34.38	37.65	40.65	44.31
26	12.20	13.84	15.38	17.29	25.34	35.56	38.89	41.92	45.64
27	12.88	14.57	16.15	18.11	26.34	36.74	40.11	43.19	46.96
28	13.56	15.31	16.93	18.94	27.34	37.92	41.34	44.46	48.28
29	14.26	16.05	17.71	19.77	28.34	39.09	42.56	45.72	49.59
30	14.95	16.79	18.49	20.60	29.34	40.26	43.77	46.98	50.89